RENEWALS 458-4574

DATE DUE			
GAYLORD			PRINTED IN U.S.A.

HANDBOOK OF STATISTICS
VOLUME 8

Handbook of
Statistics

VOLUME 8

General Editor

C. R. Rao

NORTH-HOLLAND
AMSTERDAM · LONDON · NEW YORK · TOKYO

Statistical Methods in
Biological and Medical Sciences

Edited by

C. R. Rao

Department of Mathematics and Statistics
University of Pittsburgh, Pittsburgh, PA, USA

R. Chakraborty

Center for Demographic and Population Genetics
University of Texas Graduate School of Biomedical Sciences
Houston, TX, USA

1991

NORTH-HOLLAND
AMSTERDAM · LONDON · NEW YORK · TOKYO

ELSEVIER SCIENCE PUBLISHERS B.V.
Sara Burgerhartstraat 25
P.O. Box 211, 1000 AE Amsterdam, The Netherlands

Distributors for the United States and Canada:

ELSEVIER SCIENCE PUBLISHING COMPANY INC.
655 Avenue of the Americas
New York, N.Y. 10010, USA

Library of Congress Cataloging-in-Publication Data

Statistical methods in biological and medical sciences / edited by
   C.R. Rao, R. Chakraborty.
      p.   cm. —— (Handbook of statistics; v. 8)
   Includes bibliographical references.
   Includes index.
   ISBN 0-444-88095-X: Df 1.290.00
   1. Medical statistics.   2. Biometry.   I. Rao, C. Radhakrishna
(Calyampudi Radhakrishna), 1920-   .   II. Chakraborty, Ranakut,
1946-   .   III. Series.
   [DNLM: 1. Statistical —— methods.   WA 950 S7965]
RA409.S684   1991
610'.72 —— dc20
DNLM/DLC
for Library of Congress   90—7671
                                                                      CIP

ISBN: 0-444-88095-X

Printed in The Netherlands

Preface

In the tradition of its predecessors, the main purpose of this volume of the *Handbook of Statistics* is the dissemination of statistical methodologies in the area of biological and medical sciences. The chapters were written by specialists with considerable experience in the application of statistical techniques for investigating biological issues. A few remarks may be worth mentioning to set the background for some of the chapters.

Biological and medical sciences have been served with statistical rigor from the very beginning of the development of statistics. Fisher, Pearson, Neyman and others developed the principles of estimation and hypotheses testing mostly considering specific biological questions. The subject of population genetics initially gained mathematical rigor principally from the work of Fisher, Haldane and Wright, and subsequently Kimura, Nei, Ewens, Karlin and others showed that the stochastic considerations of biological problems give tremendous insight in the study of evolution. Epidemiological studies of risk factors of common diseases also offer a great opportunity to study statistical properties of categorical data. Anthropologists find the theory of discriminant analysis inevitable in studying the biological affinity of populations. Modern medical advances in clinical trials and surgical procedures enabled interesting applications of survival analysis for examining the efficacies of new drugs or medical procedures. It might be mentioned that even the complexity of the process of carcinogenesis was also first examined from statistical analysis of epidemiological data on cancer risks.

The chapters of this volume, divided into four parts, illustrate some recent work on many of these issues. In the first, five chapters discuss some genetic and epidemiological problems. Rice et al. reviewed statistical methods relevant for studying inheritance of qualitative traits. The genetic basis of such traits is often confounded by age, gender, and environmental covariates, and hence their effects must be adjusted to tease out the genetic component of variation. These authors show that the concepts of logistic regression and the LISREL model can be exploited in such studies. Ewens discusses an important problem of observational data, where by necessity the investigator must consider only a portion of the sampling universe. Although the history of ascertainment bias correction is quite old in the statistical literature, and it led to an important class of distributions, called Weighted Distributions, Ewens shows that assignment of arbitrary ascertainment probabilities and non-independence of the probands can easily be incorporated in complex segregation analysis of qualitative as well as quantitative traits. D. C. Rao's chapter deals with the important method of Path Analysis, which is strictly a causal structural analysis of dependence between variables. Originally developed by Wright, this methodology had been 're-discovered' and popularized by Rao, Morton, and others during the mid-seventies. The most significant genetic project of today is the *Human Genome Project*, whose ultimate goal is to understand the genetic organization of the human genome. An important component of this project is to understand how different genes are linked (arranged) on chromosomes within a genome. Linkage analysis first started with experimental organisms where progenies could be generated through directed breeding experiments, whereby the proximity of genes could be studied by direct observations on recombinant or non-recombinant progenies. The gene organization in humans cannot be studied through such principles for the lack of data on outcomes of directed matings. As a result, the statistical tools of estimation and hypothesis testing play a significant role in human linkage analysis. Lathrop and Lalouel discuss such methods in the fourth chapter. In the last chapter of Part I, Breslow discusses epidemiological study designs and provides a synopsis of the current trend of epidemiological research. His chapter also illustrates the application of categorical data analysis in one of the landmark epidemiological surveys, the Framingham survey.

The next six chapters in Part II cover applications of statistics in anthropology and evolutionary biology. Balakrishnan and Ambagaspitiya consider the classical problem of identification and discrimination with multivariate observations. The application section of this chapter also illustrates the utility of such methods for dichotomous traits. Next Chakraborty and Danker-Hopfe discuss the classic problem of population genetics, the analysis of population structure. These authors provide a comprehensive review of the existing methods of estimation of fixation indices, and conclude that in spite of the philosophical differences of formulation of such estimation procedures, they all yield virtually identical estimates. These authors show that a decomposition of diversity indices is a special form of Categorical Analysis of Variance (CATANOVA). Thompson considers another important problem, the determination of biological relationships between individuals from genetic data. The chapter by the co-editors discuss the measurement of genetic variation for evolutionary studies, stressing that the different measures of genetic variation within and between populations can be formulated under a unified framework. The question of estimation of genetic variation is also addressed and large sample properties of various measures are described. Saitou's chapter on phylogenetic tree reconstruction provides another important application of statistics in evolutionary studies. The last contribution of this part deals with some statistical models for sex ratio evolution in which Lessard reviews the relationships between male and female fitnesses and sex ratio considering genetic models of evolution of sex ratio.

The next two chapters in Part III relate to the area of cancer biology. Moolgavkar presents the stochastic models of carcinogenesis that provide a synthesis of the biology of cell growth, cancer epidemiology, and the stochastic nature of normall cell differentiation. The application section of this chapter is a good mixture of experimental and observational epidemiology of cancer risks. Gart's chapter considers a specialized problem of the application of score methodology for confidence interval estimation of one-hit curves that also allows the comparison of one-hit curves across samples or populations.

Part IV contains three chapters that deal with the subject of medical statistics. In a case study of nephropathy, Kardaun's first chapter considers the point estimation of survival functions. The technique of incorporation of missing value in the estimation of survival functions brings a distinct novelty in this work. In the next chapter, Kardaun considers the confidence interval estimation of survival function using a decision theoretic approach. The concept of effective lifetime distributions can be of substantial prognostic value in medical statistics. In the last chapter, Bock and Toutenburg discuss the important subject of sample size determination in clinical research, examining the sampling adequacies and limitations in medical statistical studies.

In summary, the wide variety of applications of statistical methodologies is conspicuously depicted through the sixteen chapters. Most of the chapters discuss the statistical principles in conjunction with specific applications. As a result biological as well as statistical readers should benefit from this exposition, and obtain the current state of knowledge in these areas. Many authors emphasized the open problems that might lead to further work on these subjects.

We are thankful to all authors for their co-operation and patience during the preparation of the volume. Our special thanks are due to Professors M. Nei, M. Boehnke, K. M. Weiss, B. S. Weir, E. Boerwinkle, C. F. Sing and Mr. L. Jin for their constructive suggestions during the review of the chapters. Miss S. A. Barton's untiring efforts to keep the voluminous files in order must be acknowledged without which the work would have proceeded much slower. We thank our publisher. Elsevier Scientific Publishers Inc. for their kind cooperation. In essence, this is truly a collaborative project, and we are thankful to all whose names are associated with this volume.

C. R. Rao
R. Chakraborty

Table of Contents

PART IV. MEDICAL STATISTICS

Contributors

R. S. Ambagaspitiya, *Department of Mathematics and Statistics, McMaster University, Hamilton, Ontario, Canada L8S 4K1* (Ch. 6)

N. Balakrishnan, *Department of Mathematics and Statistics, McMaster University, Hamilton, Ontario, Canada L8S 4K1* (Ch. 6)

J. Bock, *Department of Clinical Research, Biometrics, c/o F. Hoffman-La Roche Ltd, CH-4002 Basel, Switzerland* (Ch. 16)

N. Breslow, *Department of Biostatistics, SC-32, University of Washington, Seattle, WA 98195, USA* (Ch. 5)

R. Chakraborty, *Center for Demographic and Population Genetics, University of Texas Graduate School of Biomedical Sciences, P.O. Box 20334, Houston, TX 77225, USA* (Ch. 7, Ch. 9)

H. Danker-Hopfe, *Department of Human Biology and Physical Anthropology, Faculty of Biology, University of Bremen, D-2800 Bremen 33, Germany* (Ch. 7)

W. J. Ewens, *Department of Mathematics, Monash University, Clayton, Victoria 3168, Australia* (Ch. 2)

J. J. Gart, *Mathematical Statistics and Applied Mathematics Section, Biostatistics Branch, National Cancer Institute, Bethesda, MD 20892, USA* (Ch. 13)

O. J. W. F. Kardaun, *Max-Planck Institute for Plasmaphysics, Boltzmannstrasse 2, D-8046 Garching bei München, Germany* (Ch. 14, 15)

J. M. Lalouel, *Howard Hughes Medical Institute, University of Utah, Salt Lake City, UT 84132, USA* (Ch. 4)

G. M. Lathrop, *Centre d'Etude du Polymorphisme Humain, 27 rue Juliette Dodu, 75010 Paris, France* (Ch. 4)

S. Lessard, *Department of Mathematics and Statistics, University of Montreal, C.P. 6128, Succursale A, Montreal, Quebec HC3 3J7, Canada* (Ch. 11)

S. O. Moldin, *Department of Psychiatry, Washington University School of Medicine, 216 South Kingshighway, St. Louis, MO 63110, USA* (Ch. 1)

S. H. Moolgavkar, *Division of Public Health Sciences, Fred Hutchinson Cancer Research Center, 1124 Columbia Street, Seattle, WA 98104, USA* (Ch. 12)

R. Neuman, *Department of Psychiatry, Washington University School of Medicine, 216 South Kingshighway, St. Louis, MO 63110, USA* (Ch. 1)

C. R. Rao, *Center for Multivariate Analysis, Department of Statistics, 123 Ponds Laboratory, The Pennsylvania State University, University Park, PA 16802, USA* (Ch. 9)

D. C. Rao, *Division of Biostatistics and Departments of Psychiatry and Genetics,*

Washington University School of Medicine, 660 South Euclid Avenue, Box 8067, St. Louis, MO 63110, USA (Ch. 3)

J. Rice, *Department of Psychiatry and Division of Biostatistics, Washington University School of Medicine, 216 South Kingshighway, St. Louis, MO 63110, USA* (Ch. 1)

N. Saitou, *Department of Anthropology, Faculty of Science, The University of Tokyo, Hongo, Bunkyo-ku, Tokyo 113, Japan* (Ch. 10)

E. A. Thompson, *Department of Statistics, GN22, University of Washington, Seattle, WA 98195, USA* (Ch. 8)

H. Toutenburg, *University of Regensburg, 8400 Regensburg, Germany* (Ch. 16)

C. R. Rao and R. Chakraborty, eds., *Handbook of Statistics, Vol. 8*
© Elsevier Science Publishers B.V. (1991) 1–27

Methods for the Inheritance of Qualitative Traits

John Rice, Rosalind Neuman and Steven O. Moldin

1. Introduction

The field of genetic epidemiology focuses on the inheritance of common disease in man. Inheritance is used in its broad sense to include both genetic and environmental (cultural) transmission. Cultural transmission results from parents imparting attitudes or social climate. In addition, attention is given to shared environmental effects for, say, siblings who are reared contemporaneously, and who have increased similarity due to non-transmissible factors.

Recognizing that environments cannot be randomized (as in animal work), that family members share environmental as well as genetic sources of resemblance, and that a major goal in the analysis of common diseases is to quantify and understand environmental similarity, it is necessary to formulate complex models that include both genetic and nongenetic types of variation and that allow for ancillary biological and environmental covariates. This modeling has as it roots the traditions of quantitative genetics and multifactorial inheritance that follow from the classical work of Fisher (1918), rather than from the traditions of epidemiology. However, methods in the latter discipline are becoming more important in understanding the time–place clustering of disease within families.

In what follows, we emphasize the modeling of the inheritance for a qualitative trait using one particular model in genetic epidemiology based on an underlying liability distribution with thresholds. Accordingly, we do not attempt to describe the myriad of models and approaches utilized in genetic epidemiology, but rather attempt to highlight only a few. In addition, we present two newer approaches based on the logistic model and on structural equations.

This research was supported in part by USPHS grants MH-37685, MH-31302, MH-43028, MH-25430 and NIMH Research Training Grant MH-14677 (RN and SOM).

2. Liability/threshold concepts and population distributions

2.1. Multifactorial model

We assume that the observed trait (called the phenotype) is qualitative, and refer to individuals as affected or unaffected. Often, affected individuals may be sub-classified as mild or severe, or subclassified according to clinical picture. It is assumed that there is a single continuous variable X, termed the liability to develop the disorder, on which all relevant genetic and environmental factors act additively.

In the basic formulation of the multifactorial model (Falconer, 1965, 1967; Reich et al., 1972; Curnow and Smith, 1975) it is assumed that X is normally distributed with mean 0 and variance 1, $X \sim N(0, 1)$, and that there is a threshold T as in Figure 1(a) with individuals with liability scores above T being affected. Alternatively, there may be multiple thresholds (Reich et al., 1972) as in Figure 1(b) with individuals with scores between threshold values representing milder phenotypic classes of affection. In the standard multifactorial model, the liability X is determined by both genetic and environmental factors.

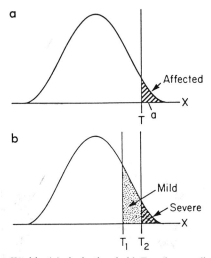

Fig. 1. Liability distribution X with: (a) single threshold T and mean liability of affecteds denoted by a; and (b) two thresholds T_1 and T_2 used to model severity.

With one threshold, the lifetime morbid risk K_p (also referred to as the 'popula-tion prevalence' or simply 'prevalence' in the genetic literature) is given by

$$K_p = \int_T^\infty \phi(x)\,dx, \tag{2.1}$$

where ϕ is the density of the standard normal random variable. The mean a and variance σ^2 of the liability of affected individuals are given by

$$a = \phi(T)/K_p \qquad (2.2)$$

and

$$\sigma^2 = 1 - a(a - T). \qquad (2.3)$$

The above concept of an abrupt threshold may be replaced by that of a risk function (Edwards, 1969; Curnow and Smith, 1975). Let $R(x)$ denote the probability that an individual with liability score x is affected. The prevalence in the population is then given by

$$K_p = \int_{-\infty}^{\infty} R(x)\,\phi(x)\,dx. \qquad (2.4)$$

With an abrupt threshold, the probability of being affected, $R_1(x)$, given ones liability value is a step function with $R_1(x) = 0$ if $x \leqslant T$ and $R_1(x) = 1$ if $x > T$. Then the prevalence K_p of affected individuals in the population is

$$K_p = \int_{T}^{\infty} \phi(x)\,dx \int_{-\infty}^{\infty} \phi(x)\,R_1(x)\,dx. \qquad (2.5)$$

Edwards (1969) proposed a risk function of the form:

$$R_2(x) = a\,e^{bx}, \quad a, b > 0. \qquad (2.6)$$

With this function, the distribution of liability for affected individuals is itself normal (in contrast to the abrupt threshold model), and simplifies subsequent calculations. However, $R_2(x)$ exceeds 1 for large values of x, so that $R_2(x)$ is not truly a risk function, but with judicious choices for a and b, the probability of having x greater than $-(\log a)/b$ may be small enough to obviate this difficulty.

Curnow and Smith (1975) suggest a cumulative normal risk function of the form

$$R_3(x) = \Phi\left\{\frac{(x - T)}{\sigma}\right\}, \qquad (2.7)$$

where Φ is the distribution function of the standard normal. Let $Z \sim N(0, \sigma^2)$ be independent from X and consider $Y = X + Z$. Then the probability that $Y > T$ equals the probability that $Z > T - X$, which is $\Phi\{(x - T)/\sigma\}$. That is, the abrupt threshold model for the variable Y is mathematically equivalent to the cumulative normal risk model for the variable X.

2.2. *Single major locus model*

A key assumption in the multifactorial model is that the joint distribution in liability of family members is multivariate normal. Since X is unobserved it suffices that there is a transformation which achieves this.

One way that the normality assumption can fail is that there is a gene of major effect. That is, suppose there were a locus with two alleles, A and a, where locus refers to the position of a gene on a chromosome and allele refers to one of the alternative genes at that locus. Since humans have two copies of each (autosomal) chromosome, there are three possible genotypes: AA, Aa, aa. If the liability is the sum of effects due to this major gene g and a normally distributed residual ε, $X = g + \varepsilon$, then the distribution in the population will be as depicted in Figure 2.

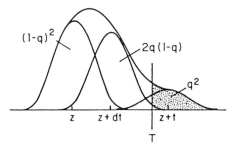

Fig. 2. Liability distributions resulting from a single major locus in Hardy–Weinberg equilibrium. Note the incomplete penetrance of the genotypes and the skewed composite distribution.

Let q denote the frequency of the allele a in the population. Under the assumption of random mating, the probabilities for the three genotypes, AA, Aa, aa are $(1 - q)^2$, $2(1 - q)q$, and q^2, respectively. We assume that the distribution within a genotype is normal with means z, $z + dt$, and $z + t$, respectively, and a common variance E (where E is the variance of ε). We follow the notation of Morton and MacLean (1974) and Lalouel and Morton (1981), with the displacement between the homozygotes denoted by t and dominance measured by d. Note that the composite distribution is a mixture (or commingling) of three distributions and will in general be skewed. For a quantitative measure, the mean and variance of X will be parameters of the model. For qualitative data, we take X to have mean 0 and variance 1, so that

$$z = -q^2 t - 2q(1 - q)td \tag{2.8}$$

and

$$E = 1 - q^2(z + t)^2 - 2q(1 - q)(z + td)^2 - (1 - q)^2 z^2 . \tag{2.9}$$

The penetrances of the fenotypes, f_1, f_2, f_3, are defined as the probability that individuals of genotypes AA, Aa, aa, respectively, are affected. Accordingly, the model can also be described in terms of the four parameters q, f_1, f_2, f_3. If ε is uncorrelated between family members, this model is referred to as the single major locus (SML) model (Reich et al., 1972; Elston, 1979).

3. Familial resemblance

3.1. Multifactorial model

In the multifactorial model, familial resemblance is quantified by the correlations in liability between family members. As noted below, these correlations may be parameterized in terms of additive genetic effects, etc.

For a family of size s, it is assumed that the joint distribution in liability is an s-variate normal with correlations matrix ρ. The probability of observing the family with specified phenotypes requires evaluation of integrals

$$\int_{I_1} \cdots \int_{I_s} \phi(y_1, \ldots, y_s; \rho)\, dy_1 \cdots dy_s, \tag{3.1}$$

where ϕ denotes the appropriate s-variate normal density function and each I_i is of the form $(-\infty, T)$ or $(T, +\infty)$ depending on whether the i-th individual is unaffected or affected, respectively.

Unfortunately, except in special cases, a practical algorithm to evaluate integrals of the multivariate normal is not available. We discuss two situations commonly seen in the genetics literature.

3.1.1. The bivariate case

Let (Y_1, Y_2) have a bivariate normal distribution with each Y_1 having mean 0 and variance 1, and with the correlation between them equal to ρ_{12}, and let T_1 and T_2 be the threshold values for Y_1 and Y_2, respectively. Denoting $K_{p1} = \Pr(Y_1 \geqslant T_1)$ by P_1, we recall the following standard results:

$$a_1 = E[Y_1 | Y_1 \geqslant T_1] = \phi(T_1)/P_1, \tag{3.2}$$

$$\sigma_1^2 = \mathrm{Var}[Y_1 | Y_1 \geqslant T_1] = 1 - a_1(a_1 - T_1), \tag{3.3}$$

$$a_{2|1} = E[Y_2 | Y_1 \geqslant T_1] = \rho_{12} a_1, \tag{3.4}$$

$$\sigma_{2|1}^2 = \mathrm{Var}[Y_2 | Y_1 \geqslant T_1] = 1 - \sigma_{12}^2 a_1(a_1 - T_1). \tag{3.5}$$

Here, we will think of Y_1 and Y_2 as liability distribution for a pair of relatives. In the standard polygenic model, the correlations between full siblings would be $\frac{1}{2}h^2$, etc. (Falconer, 1981).

Thus, the mean and variance of relatives of 'affected' individuals are $\rho_{12} a_1$ and $1 - \rho_{12}^2 a_1(a_1 - T_1)$, respectively. The distribution of such relatives is not normal unless $\rho_{12} = 0$, but may be approximately so under certain conditions, so that an approximation $P_2 = \Pr(Y_1 \geqslant T_1, Y_2 \geqslant T_2)$ is given by $P_2 = P_1 \Phi_c(Z_2)$, where the standardized threshold Z_2 is given by $Z_2 = (T_2 - a_{2|1})/\sigma_{2|1}$, and $\Phi_c = 1 - \Phi$. This approximation is discussed by Pearson (1903), Reich et al. (1972), Mendell and Elston (1974), Smith and Mendell (1974) and Rice et al. (1979), and has been found to be quite good, especially for small ρ_{12}.

Given current technology, computer programs to evaluate the bivariate normal integral are readily available, so that no approximation is necessary. However, the above approach has been widely utilized and discussed in the genetics literature. Moreover, many data summaries are given in terms of K_p, the population prevalence, and K_R, the risk to relatives of an affected individual, for various degrees of genetic relationship. Using the above bivariate approach, the correlation in liability is then calculated in a closed algebraic form using the above approximation.

Letting $T_{2|1} = (T_2 - a_{2|1})/\sigma_{2|1}$, the estimate is given by

$$\hat{\rho}_{12} = \frac{T_1 - T_{2|1}\sqrt{1 - (T_1^2 - T_{2|1}^2)(1 - T_1/a_1)}}{a_1 + T_{2|1}^2(a_1 - T_1)}. \tag{3.6}$$

The values of T_1, $T_{2|1}$ and a_1, may be computed (or looked up in the tables of Falconer, 1965), and ρ_{12} estimated using the above formula.

3.1.2. Multivariate case
We assume that the joint distribution in liability for a family of s individuals is an s-variate normal. Let $\phi(x_1, \ldots, x_s; \rho)$ denote the density function of the standard s-variate normal with correlation matrix ρ. The probability P_F that all s individuals are affected is then

$$P_F = \int_{-\infty}^{\infty} \cdots \int_{-\infty}^{\infty} \phi(x_1, \ldots, x_s; \rho) R_i(x_1) \cdots R_i(x_s) \, dx_1 \cdots dx_s, \tag{3.7}$$

where $i = 1, 2, 3$ corresponds to one of the risk functions described above. Probabilities of a family containing unaffected individuals may be computed by replacing $R_i(x)$ by $1 - R_i(x)$ for those unaffected. The functions R_i may be made to depend on the individual's sex, age or severity class (Curnow and Smith, 1975).

Estimation of the correlations between relatives (or a set of genetic parameters which parameterize ρ) requires evaluation of integrals of the form given in equation (3.7). In the case of an abrupt threshold (R_1), these are integrals of the multivariate normal density over rectangular regions. For the model of Curnow and Smith (R_3), the integrals may be expressed in terms of an abrupt threshold model with an appropriate correlation matrix ρ^*. If the correlations between x-values of relatives are due to additive genetic effects at many loci, then the correlation matrix depends only on the degree of relationship between individuals (i.e., $\rho^* = 1$ for monozygotic twins, $\rho^* = \frac{1}{2}$ for first degree relatives, etc.). In this situation only the μ and σ in R_3 need to be estimated. When the correlation matrix is of a special form, the integral of equation (3.1) can be reduced to one of lower dimension (Lange, Westlake and Spence, 1976) or in some cases to a one-dimensional integral (Curnow and Dunnett, 1962; Curnow, 1972).

Indeed, Curnow and Dunnett (1962) have noted that when all $\rho_{ij} = \rho$, and all $T_i = T$, then the multiple integral of equation (3.1) reduces to the single integral

$$P_F = \int_{-\infty}^{\infty} \phi(t) \, \Phi_c^s [(T + t\sqrt{\rho})/\sqrt{1 - \rho}] \, dt . \tag{3.8}$$

However, general evaluation of integrals with an arbitrary correlation matrix remains a difficulty in higher dimensions due to the computer time required. This is compounded since evaluation is often in the context of an iterative maximum likelihood search for parameter estimation.

3.2. The generalized single major locus model

In the generalized single major locus (SML) model, relevant genetic variation is due to the presence of two alleles at a single locus (Reich et al., 1972; Elston and Campbell, 1971; Elston, 1979). As noted earlier, the penetrance of a genotype (the proportion of individuals of a given genotype who are affected) is denoted f_1, f_2, f_3 for genotypes AA, Aa, and aa, respectively. When these f's differ from 0 or 1, we have the case of incomplete penetrance. Letting q denote the frequency in the population of the allele a, the SML model is defined by four parameters: q, f_1, f_2, and f_3. However, not all four parameters can be estimated from data consisting of proband-relative pairs (James, 1971), although this can be overcome by having a separate estimate of the population prevalence, by analysis of pedigrees, or by having multiple thresholds. Risch (1990) extends the James treatment to consider the risk ratio K_R/K_p in single locus and multilocus models.

The SML model can be applied to large pedigrees (Elston and Stewart, 1971) and, moreover, information on chromosomal markers may also be used in analysis to establish linkage of a marker to that major locus (Ott, 1985).

For a family of size s, the likelihood of a pedigree may be expressed as

$$L = \sum_{g_1} \cdots \sum_{g_s} \Pr(x_1, \ldots, x_s | g_1, \ldots, g_s) \, \Pr(g_1, \ldots, g_s) , \tag{3.9}$$

where x_i and g_i denote the phenotype and genotype of the i-th member. Under the assumption of random mating and no inbreeding, this reduces to

$$L = \sum_{g_1} \cdots \sum_{g_s} \prod \Pr(x_i | g_i) \, \Pr(g_i | \cdot) , \tag{3.10}$$

where $\cdot$ represents either (a) the parental genotypes if the parents are included, or (b) the empty set if the person marries in (i.e. does not have parents in the set $\{1, \ldots, s\}$).

Elston and Stewart (1971) noted that if individuals are ordered so that parents precede their children, then the likelihood may be written as a telescoping sum

$$L = \sum \Pr(x_1|g_1) \Pr(g_1|\cdot) \cdots \sum \Pr(x_s|g_s) \Pr(g_s|\cdot). \tag{3.11}$$

This algorithm is referred to as the Elston and Stewart algorithm and is the basic approach in current linkage analysis problems. By starting at the 'bottom' of a pedigree, the inner summations may be progressively computed and saved in an efficient manner.

Elston and Stewart (1971) introduced general transmission probabilities of transmitting an A allele for individual of genotype AA, Aa, and aa. For a single major locus, these τ parameters would, of course, be $1, \frac{1}{2}, 0$, respectively, so that the Mendelian hypothesis may be tested against the broad class of alternatives represented by these general τ values. The program PAP (Hasstedt and Cartwright, 1979) permits analysis of extended pedigrees allowing for complex ascertainment and more than one locus which determine the phenotype.

3.3. The mixed model

The mixed model allows for a single major locus with a multifactorial background. As in the SML, we have $X = g + \varepsilon$, but in this model we assume that the joint distribution of ε among family members, conditioned on their genotypes, is multivariate normal. In the model of Morton and McLean (1974), the background was polygenic (although, they did allow for a common environment of rearing for siblings). Whereas, their approach was for nuclear families, Ott (1979) described pedigree analysis for a polygenic background, but this approach is limited to small pedigrees (about size 10) by numerical considerations.

Lalouel and Morton (1981) introduced the concept of a pointer, a person external to a nuclear family who leads to its ascertainment, which enables analysis of pedigrees by breaking them into their constituent nuclear parts. The non-Mendelian transmission parameters of Elston and Stewart (1971) have been incorporated into the mixed model (Lalouel et al., 1983) and a computer program POINTER is available which implements these procedures.

In summary, there have been three basic models used in genetic epidemiology to model familial resemblance. The first is the multifactorial model in which the liability to manifest the disorder is the cumulative effect of many risk factors of small effect. The second is the single major locus model in which there is a gene of major effect (in the sense that there is a wide separation in the mean liability of homozygous AA and aa individuals), and familial resemblance is accounted for by Mendelian transmission. However, there is environmental variation within a genotype that accounts for the reduced penetrance (i.e., not everyone with the same genotype as the same phenotype) of that locus. Thirdly, there is the mixed model in which there is a major single locus with a heritable background. Other approaches that involve two locus models, or that involve genetic linkage between a major locus for the trait and marker loci, will not be discussed here.

4. Bivariate multifactorial transmission models

Given the much greater information content of quantitative as compared to qualitative information in the analysis of family data (MacLean et al., 1975; Elston
et al., 1975), there are limitations when using dichotomous phenotypes (affected
or unaffected) exclusively to study the complex inheritance of familial diseases
that do not follow classical Mendelian patterns of inheritance.

Adjunct consideration of both qualitative (affection status) and quantitative (a
correlated liability indicator) information to define a bivariate phenotype (Lalouel
et al., 1985; Borecki et al., 1989) can increase considerably the power of segra-
gation analysis for modeling familial resemblance (Moldin et al., 1990). The
greater information content afforded through bivariate segregation analysis is
especially desirable for the genetic analysis of illnesses where unaffected individuals can not be graded according to liability. We consider two models for bivariate
segregation analysis, as implemented by the computer program POINTER (Morton
and MacLean, 1974; Lalouel et al., 1983; Morton et al., 1983) and YPOINT
(Lalouel et al., 1985).

4.1. The POINTER model

The bivariate model implemented in the computer program POINTER assumes that
a measurable quantitative trait x' is correlated with liability to affection y' (where
the symbol $'$ denotes an unstandardized variable). Affected individuals are those
whose liability is greater than a threshold value. For the general mixed model, the
trait x' results from an autosomal diallelic locus g; a multifactorial (genetic and/or
cultural) transmissible background c'; and an uncorrelated residual environmental
contribution e', which may be correlated within a sibship. We assume that the
major locus g has two alleles A and a (the gene frequencies of 'a' and 'A' are q
and $1 - q$, respectively), with relative proportions of the three genotypes AA, Aa,
and aa consistent with Hardy–Weinberg equilibrium and equal to $(1 - q)^2$,
$2q(1 - q)$, and q^2, respectively. A convenient parameterization of the bivariate
mixed model is given by: V, the variance of x'; u, the mean of x'; q, the gene
frequency of the susceptibility allele a; t, the displacement between the two homozygote means; d, the dominance parameter ($d = 0$, recessive; $d = 1$, dominant); H,
the proportion of the variance attributable to multifactorial (polygenic or cultural)
effects; and w, random environmental effects.

Consider the case of multifactorial transmission only of liability and the correlated quantitative trait under the POINTER model, such that

$$x' = c' + e' .$$

The relationship between x' and y' is such that

$$y' = x' + w' \quad (y' = c' + e' + w'),$$

with variances related by

$$\mathrm{Var}(y') = \mathrm{Var}(x') + \mathrm{Var}(w').$$

The within-person correlation between liability to affection and the quantitative trait (Morton and MacLean, 1974) is

$$r_w = \sqrt{\mathrm{Var}(x')/(\mathrm{Var}(x') + \mathrm{Var}(w'))}.$$

We can standardize the above equations such that

$$x = h_{11x}c + \varepsilon_x e, \quad x \sim N(0, 1),$$

and

$$y = r_w h_x c + r_w \varepsilon_x e + \sigma w, \quad y \sim N(0, 1).$$

A path diagram is presented in figure 3, for two parents (p) of form $\{i = 1, 2\}$ and one offspring (o). The parent–offspring correlation on the quantitative trait, with $H_x = h_x^2$, is

$$r_{x_{p_i} x_o} = \tfrac{1}{2}h_x^2, \tag{4.1}$$

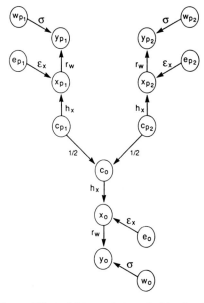

Fig. 3. Path diagram depicting multifactorial transmission of a bivariate (discrete-continuous) phenotype under the POINTER model. Subscripts p_1, p_2, and o denote parent 1, parent 2, and offspring, respectively; x, y, c, and w denote the quantitative trait, liability to affection, transmissible multifactorial component, and random environmental effects, respectively.

the correlation between parental liability and the offspring quantitative trait is

$$r_{y_{\text{P}}, x_{\text{o}}} = \tfrac{1}{2} h_x^2 r_{\text{w}} ,$$

(4.2)

and the parent–offspring correlation in liability ($\tfrac{1}{2} h_y^2$), with $H_y = h_y^2$, is

$$r_{y_{\text{P}}, y_{\text{o}}} = \tfrac{1}{2} h_x^2 r_{\text{w}}^2 .$$

(4.3)

This model makes the following important assumptions:

(i) the metric trait serves to define the disease (such as blood glucose levels in diabetes), and measures liability directly. It follows that knowledge of x is more informative than knowledge of y, with the underlying genotype being more highly correlated with the quantitative trait than with a qualitative determination of affection status;

(ii) the mode of transmission of both liability and the quantitative trait is constrained to be the same; and

(iii) since it follows from equation (4.3) that

$$h_y^2 = h_x^2 r_{\text{w}}^2 ,$$

(4.4)

the heritability in liability will always be less than the heritability of the quantitative trait when $r_{\text{w}} < 1.0$.

4.2. The YPOINT model

The other bivariate segregation model we consider is an extension of the POINTER model and is implemented in the computer program YPOINT. We assume x' is a measurable quantitative trait correlated with underlying liability y, where affected individuals are those whose liability is greater than a threshold value. The model assumes that x' and y result from an autosomal diallelic locus g, multifactorial component c', and random environmental factors e'. The model postulates that the same three factors—g, c', and e'—exert specific effects on x' and y. The component w', with variance W, represents a residual specific to liability. Different displacements, dominance effects, and heritabilities are considered for x' and y via common genetic factors. With respect to the x' trait, the model is parameterized in terms of: E_x, the residual variance; u, the mean; dominance, d; displacement, t; q, the gene frequency of the a allele; and $H_x = h_x^2$, the multifactorial heritability. Correspondingly for y, four additional parameters are defined: d_y, t_y, $H_y = \alpha^2 C_x / V_y$ (with $V_y = 1$), and residual variance, $E_y = \beta^2 E_x$.

Consider the use of multifactorial transmission only of liability and the correlated quantitative trait, under the YPOINT model, such that

$$x' = c' + e'$$

(4.5)

and

$$y = \alpha c' + \beta e' + w' ,$$

(4.6)

so that the variances (V) of x' and y are

$$V_x = C_x + E_x \tag{4.7}$$

and

$$V_y = \alpha^2 C_x + \beta^2 E_x + W = 1 . \tag{4.8}$$

When we assume that each term is normally distributed with mean zero and unit variance, the resulting standardized equations from the path diagram in Figure 4 for two parents (p) of form $\{i = 1, 2\}$ and one offspring (o) are

$$x = h_x c + \varepsilon e , \quad x \sim N(0, 1) \tag{4.9}$$

and

$$y = h_y c + \varepsilon_y e + \sigma w , \quad y \sim N(0, 1) . \tag{4.10}$$

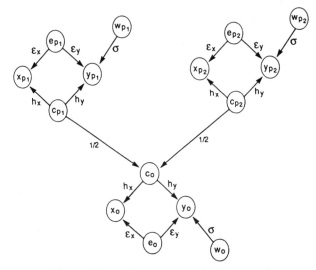

Fig. 4. Path diagram depicting multifactorial transmission of a bivariate (discrete-continuous) phenotype under the YPOINT model. Subscripts p_1, p_2, and o denote parent 1, parent 2, and offspring, respectively; x, y, c, e, and w denote the quantitative trait, liability to affection, transmissible multifactorial component, random environmental effects, and residual effects specific to liability, respectively.

Given that $H_x = h_x^2$ and $H_y = h_y^2$, with $h_y^2 = \alpha^2 C_x / V_y$ and $h_x^2 = C_x / V_x$, we see that $\alpha = h_y / h_x \sqrt{V_x}$ and $h_y = \alpha h_x \sqrt{V_x}$. In terms of the YPOINT parameters, since $E_y = \beta^2 E_x$ (Morton et al., 1983, p. 97) and $\varepsilon_x = \sqrt{E_x} / \sqrt{V_x}$ from the path diagram, then $\varepsilon_y = \beta \varepsilon_x \sqrt{V_x} = \sqrt{E_y}$. Following our earlier notation, the parent–offspring correlations on the quantitative trait ($H_x = h_x^2$) and liability ($H_y = h_y^2$) given by

$$r_{x_{p_i} x_o} = \tfrac{1}{2} h_x^2 \tag{4.11}$$

and

$$r_{y_{P_i} y_o} = \tfrac{1}{2} h_y^2 . \tag{4.12}$$

The correlation between parental liability and the offspring quantitative trait is

$$r_{y_{P_i} x_o} = \tfrac{1}{2} h_y h_x , \tag{4.13}$$

The within–person correlation between liability and the quantitative trait is

$$r_w = r_{y_{P_i} x_{P_i}} = r_{y_o x_o} = h_x h_y + \varepsilon_x \varepsilon_y . \tag{4.14}$$

By substitution with YPOINT parameters for the β terms, we see that

$$r_w = h_x h_y + (\sqrt{E_x} \sqrt{E_y}) / \sqrt{V_x} . \tag{4.15}$$

Given that $\alpha = h_y/(h_x \sqrt{V_x})$, we see that

$$h_y = \alpha h_x \sqrt{V_x} . \tag{4.16}$$

The YPOINT model makes the important assumption that both the quantitative trait and liability share common polygenic factors such that if $h_x^2 = 0$, then $h_y^2 = 0$. Likewise, under the general YPOINT model, the parameters are constrained so that if $h_y^2 \neq 0$ then $h_x^2 \neq 0$.

5. Logistic regressive approach

In the above models, the disease phenotype corresponds to deviant values on an underlying liability distribution. In this section we consider a more recent model formulated in terms of a logistic model.

Consider a dichotomous outcome variable Y, with $Y = 1$ corresponding to 'affected' and $Y = 0$ corresponding to 'unaffected', and let X be a variable related to the probability p of being affected. A natural transformation of p is the logit,

$$\theta = \text{logit}(p) = \log\left(\frac{p}{1 - p}\right) = \log\left(\frac{\Pr(Y = 1)}{\Pr(Y = 0)}\right), \tag{5.1}$$

where $\Pr(\ )$ denotes the probability of the event in parentheses.

Note that p is bounded between 0 and 1, whereas $\text{logit}(p)$ has a range from $-\infty$ to $+\infty$. In the logistic model, we assume that logit of p has a linear regression on X, i.e.,

$$\text{logit}(p) = \alpha + \beta X . \tag{5.2}$$

Exponentiating both sides of the above equation, we have

$$p/(1 - p) = e^{\alpha + \beta X} . \tag{5.3}$$

Note that if X is dichotomous and takes on the values 0 and 1, then the odds for being affected are multiplied by e^β for $X = 1$ as compared to $X = 0$ so that e^β may be interpreted as a (partial) odds ratio. Finally, solving for p above, we have

$$p = e^{\alpha + \beta X}/(1 + e^{\alpha + \beta X}). \tag{5.4}$$

In general, for a vector $X = (X_1, \ldots, X_n)$ of covariates, we have

$$\text{logit}(p) = \alpha + \beta_1 X_1 + \beta_2 X_2 + \cdots + \beta_n X_n. \tag{5.5}$$

Equation (5.5) is the logistic model as applied in epidemiologic settings. Now consider a set of n relatives with binary outcomes $Y = (Y_1, \ldots, Y_n)$ and $X = (X_1, \ldots, X_n)$, where each X_i is an explanatory variable associated with Y_i. We can write $\text{Pr}(Y)$ as (Bonney, 1986, 1987):

$$\text{Pr}(Y|X) = \text{Pr}(Y_1|X)\,\text{Pr}(Y_2|Y_1, X) \cdots \text{Pr}(Y_n|Y_1, \ldots, Y_{n-1}, X). \tag{5.6}$$

For simplicity, we shall assume

$$\text{Pr}(Y_i|Y_1, \ldots, Y_{i-1}, X) = \text{Pr}(Y_i|Y_1, \ldots, Y_{i-1}, X_i). \tag{5.7}$$

That is, conditional on the Y values, only the covariate for individual i is needed in predicting the phenotype of individual i.

In the regressive logistic model, it is assumed that each term in the right-hand side of (5.6) is given by a logistic equation. For example, in the class A model proposed by Bonney, the logit θ_i of an individual depends only on the phenotype of his parents and his covariate values, i.e.,

$$\text{Pr}(Y_i|Y_1, \ldots, Y_{i-1}, X_i) = \text{Pr}(Y_i|Y_{\text{F}}, Y_{\text{M}}, X_i), \tag{5.8}$$

where the father and mother are in the set $Y_1, \ldots, Y_{i-1}$, i.e.,

$$\theta_i = \alpha + \gamma_{\text{F}} Y_{\text{F}} + \gamma_{\text{M}} Y_{\text{M}} + \beta X_i. \tag{5.9}$$

A major locus can be introduced by

$$P(Y|X) = \sum P(G)P(Y|G, X), \tag{5.10}$$

where the summation is over all possible genotypes. The formulation of $P(G)$ may be given in terms of the τ_g parameters of Elston and Stewart (1971) and is implemented in the computer program SAGE (Elston et al., 1986). A general discussion of the logistic model and its applications in genetic analyses is given by Bonney (1986).

In applications of equation (5.6) it is necessary to allow for missing values for two reasons. The terms in equation (5.6) have different numbers of predictors, and coefficients are related across these equations. Second, for example, in a class A model (equation 5.9), a parental phenotype may be unavailable. One way to allow for this is to define variables Z (Bonney, 1986) by

$$Z = \begin{cases} 1 & \text{if } Y = 1, \\ 0 & \text{if unknown}, \\ -1 & \text{if } Y = 0. \end{cases} \qquad (5.11)$$

In this case, $e^{ZB} = 1$ if $Z = 0$, so that the odds of affection in the predicted relative [see equation (5.3)] do not change if the status of the predictive relative is unknown. However, this approach is problematic (Rice and Neuman, unpublished manuscript) in that the coefficient for one parent (with the other parent missing) cannot be identical to that when both parents are present. That is, the partial odds ratios are different depending on the number of variables, and this cannot be obviated by a simple recoding.

Using (5.11), equation (5.9) would become

$$\theta_i = \alpha + \gamma_F Z_F + \gamma_M Z_M + \beta X_i. \qquad (5.12)$$

However, we consider the case with no missing values and random mating. For a family consisting of a child (person 3) and his parents (persons 1 and 2), the three logits are given by

$$\theta_1 = \alpha_1 + \beta X_1,$$
$$\theta_2 = \alpha_1 + \beta X_2, \qquad (5.13)$$
$$\theta_3 = \alpha_2 + \gamma_1 Y_1 + \gamma_2 Y_2 + \beta X_3,$$

or in matrix form

$$\begin{bmatrix} \theta_1 \\ \theta_2 \\ \theta_3 \end{bmatrix} = \begin{bmatrix} 1 & 0 & 0 & 0 & X_1 \\ 1 & 0 & 0 & 0 & X_2 \\ 0 & 1 & Y_1 & Y_2 & X_3 \end{bmatrix} \begin{bmatrix} \alpha_1 \\ \alpha_2 \\ \gamma_1 \\ \gamma_2 \\ \beta \end{bmatrix}. \qquad (5.14)$$

In general, for a sibship with size $n - 2$ siblings, we have $\theta = A\lambda$, where A is a matrix of the form

$$\begin{bmatrix} 1 & 0 & 0 & 0 & X_1 \\ 1 & 0 & 0 & 0 & X_2 \\ & & \cdots & & \\ 1 & 1 & Y_1 & Y_2 & X_n \end{bmatrix} \qquad (5.15)$$

Here α_1 is $\log[K_p/(1 - K_p)]$ and α_2 is $\log[K_{RUU}/(1 - K_{RUU})]$ where K_{RUU} is the risk to children of two unaffected parents.

Note that this model is formally equivalent to the standard logistic model with n independent observations with the columns of A representing explanatory variables. That is, the standard computer programs for fitting the logistic model may be used for the regressive model. However, if a latent major locus is also considered, then a specialized program such as SAGE must be used.

6. The LISREL model

Structural equation models have long been used to describe relationships among variables and to test putative cause and effect models involving both measured (observable) and latent (unobservable) variables. These relationships are analyzed by decomposing the variances and covariances of the measured variables into terms involving a set of unknown parameters which are used to model the effects of the latent variables. This general method of analysis is often referred to as covariance structure analysis (Jöreskog, 1978). Methods of analyzing covariance structures involve minimizing the difference between sample covariances and the covariances predicted by the model. This contrasts with, for example, linear regression, which involves minimizing functions of observed and predicted values of regression coefficients. LISREL is a computer program which was developed to estimate the unknown parameters in a set of structural equations which were developed by Jöreskog (1973). The program has become so widely used that the structural equation model introduced by Jöreskog has become known as the LISREL model.

We begin with a description of the LISREL model and then briefly discuss the identification and estimation of the parameters of the model. We shall also discuss methods that are used to assess the fit of the model. The utility of this approach in genetic modeling will be illustrated with a simple application.

6.1. The mathematical model

In its most general form, the LISREL model consists of two parts: the measurement model and the structural equation model. The measurement model, containing two factor analytic models, relates the observed variables to a set of latent variables. The structural equation model specifies the causal relationships among the latent variables.

Assume all observed and unobserved variables are measured as deviations from their means. Let $y' = (y_1, \ldots, y_p)$ and $x' = (x_1, \ldots, x_q)$ be a set of observed vectors, and let $\eta' = (\eta_1, \ldots, \eta_m)$ and $\xi' = (\xi_1, \ldots, \xi_n)$ be a set of latent vectors.

The measurement model relates x and y to ξ and η, respectively, by the following sets of equations:

$$x = \Lambda_x \xi + \varepsilon, \tag{6.1}$$

$$y = \Lambda_y \eta + \delta, \tag{6.2}$$

where $\Lambda_x(q \times n)$ and $\Lambda_y(p \times m)$ are matrices of regression coefficients of y on η and x on ξ; $\varepsilon' = (\varepsilon_1, \ldots, \varepsilon_q)$ and $\delta' = (\delta_1, \ldots, \delta_p)$ are vectors of error terms.

The structural equation model assumes that η and ξ are related by the equations:

$$\eta = B\eta + \Gamma\xi + \zeta, \tag{6.3}$$

where B is an $(m \times m)$ matrix of coefficients relating the endogenous (dependent) variables, η, to one another, and Γ is an $(m \times n)$ matrix of coefficients relating the exogenous (independent) variables, ξ, to the endogenous variables. Here $\zeta' = (\zeta_1, \zeta_2, \ldots, \zeta_m)$ is a random vector of errors.

The following assumptions are made concerning the above relationships:

– ε, δ and ζ are all mutually uncorrelated, that is $\text{Cov}(\varepsilon_i, \delta_j) = 0$, $\text{Cov}(\varepsilon_i, \zeta_k) = 0$ and $\text{Cov}(\delta_j, \zeta_k) = 0$, for every $i = 1, \ldots, q, j = 1, \ldots, p$,

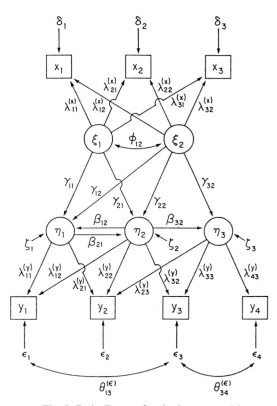

Fig. 5. Path diagram for the LISREL model.

and $k = 1, \ldots, m$. However, ε, δ, and ζ may be correlated within themselves.

- ε is uncorrelated with η, δ is uncorrelated with ξ, and ζ is uncorrelated with ξ. $\qquad$ (6.4)
- B has zeros in the diagonal, hence no η_k has an effect on itself.
- $I - B$ is nonsingular; that is, there is no redundancy among these equations.

Four covariance matrices can be defined. Let $\Phi(n \times n)$ be the covariance matrix of ξ, and let $\Psi(m \times m)$, $\theta_\varepsilon(q \times q)$ and $\theta_\delta(p \times p)$ be the covariance matrices of the error vectors, ζ, ε, and δ, respectively. Since all variables are measured as deviations from their means, $\Phi = E(\xi\xi')$, $\Psi = E(\zeta\zeta')$, $\theta_\varepsilon = E(\varepsilon\varepsilon')$ and $\theta_\delta = E(\delta\delta')$.

The full model used by LISREL can be illustrated by the path diagram in Figure 5. The path diagram, developed by Wright (1921, 1934), is a useful tool for communicating the essential ideas of the model. In the path diagram, observed variables are enclosed in squares and unobserved variables are represented by circles. Arrows indicate which variables are to be considered a function of other variables; two-headed arrows indicate correlations between variables.

By using a path diagram to specify the model, we can easily write down the corresponding structural and measurement equations in matrix form

$$
\begin{bmatrix} \eta_1 \\ \eta_2 \\ \eta_3 \end{bmatrix} = \begin{bmatrix} 0 & \beta_{12} & 0 \\ \beta_{21} & 0 & 0 \\ 0 & \beta_{32} & 0 \end{bmatrix} \begin{bmatrix} \eta_1 \\ \eta_2 \\ \eta_3 \end{bmatrix} + \begin{bmatrix} \gamma_{11} & \gamma_{12} \\ \gamma_{21} & \gamma_{22} \\ 0 & \gamma_{32} \end{bmatrix} \begin{bmatrix} \xi_1 \\ \xi_2 \end{bmatrix} + \begin{bmatrix} \zeta_1 \\ \zeta_2 \\ \zeta_3 \end{bmatrix},
$$

$$
\begin{bmatrix} y_1 \\ y_2 \\ y_3 \\ y_4 \end{bmatrix} = \begin{bmatrix} \lambda_{11} & \lambda_{12} & 0 \\ \lambda_{21} & \lambda_{22} & \lambda_{23} \\ 0 & \lambda_{32} & \lambda_{33} \\ 0 & 0 & \lambda_{43} \end{bmatrix} \begin{bmatrix} \eta_1 \\ \eta_2 \\ \eta_3 \end{bmatrix} + \begin{bmatrix} \varepsilon_1 \\ \varepsilon_2 \\ \varepsilon_3 \\ \varepsilon_4 \end{bmatrix},
$$

$$
\begin{bmatrix} x_1 \\ x_2 \\ x_3 \end{bmatrix} = \begin{bmatrix} \lambda_{11} & \lambda_{12} \\ \lambda_{21} & \lambda_{22} \\ \lambda_{31} & \lambda_{32} \end{bmatrix} \begin{bmatrix} \xi_1 \\ \xi_2 \end{bmatrix} + \begin{bmatrix} \delta_1 \\ \delta_2 \\ \delta_3 \end{bmatrix}.
$$

The four covariance matrices are the two diagonal matrices

$$
\Theta = \text{diag}(\theta_{11}^{(\delta)}, \theta_{22}^{(\delta)}, \theta_{33}^{(\delta)}),
$$

$$
\Psi = \text{diag}(\psi_{11}, \psi_{22}, \psi_{33}),
$$

$$\Phi = \begin{bmatrix} \phi_{11} & \phi_{12} \\ \phi_{21} & \phi_{22} \end{bmatrix}, \qquad \Theta_{\varepsilon} = \begin{bmatrix} \theta_{11} & 0 & \theta_{13} & 0 \\ 0 & \theta_{22} & 0 & 0 \\ \theta_{31} & 0 & \theta_{33} & \theta_{34} \\ 0 & 0 & \theta_{43} & \theta_{44} \end{bmatrix}.$$

Let Σ be the population covariance matrix for the observed variables x and y:

$$\Sigma = E\left[\begin{pmatrix} y \\ x \end{pmatrix} \begin{pmatrix} y' & x' \end{pmatrix} \right] = E\begin{bmatrix} yy' & yx' \\ xy' & xx' \end{bmatrix}, \tag{6.5}$$

where $\begin{pmatrix} y \\ x \end{pmatrix}$ is the $((p+q) \times 1)$ vector formed by combining the elements of y and x into a single vector.

By substituting equations (6.1)–(6.3) into (6.5) and using the prior assumptions (6.4), we may write Σ as $\Sigma(\theta)$ where

$$\Sigma(\theta) = \begin{bmatrix} \Lambda_y(I-B)^{-1}(\Gamma\Phi\Gamma' + \Psi)(I-B)^{-1'}\Lambda_y' + \Theta_{\varepsilon} & \Lambda_y(I-B)^{-1}\Gamma\Phi\Lambda_x' \\ \Lambda_x\Phi\Gamma'(I-B)^{-1'}\Lambda_y' & \Lambda_x\Phi\Lambda_x' + \Theta_{\delta} \end{bmatrix},$$

and θ contains parameters of the model to be estimated, that is the elements contained in the eight parameter matrices, B, Γ, Λ_x, Λ_y, Φ, Ψ, Θ_{ε}, and Θ_{δ}. If the model is correct and the population parameters are known, each element of Σ should equal its corresponding element in $\Sigma(\theta)$.

Let S be a sample estimate of Σ, i.e. S is the observed covariance matrix. The objective of LISREL is to estimate those values of the parameters that minimize the 'difference' between S and the predicted covariance matrix $\Sigma(\hat{\theta})$ where the difference between the two matrices is determined by the method of estimation as discussed below. Table 1 contains a summary of the parameter matrices used by LISREL.

Table 1
Definitions of the matrices used by LISREL

Matrix (name)	Matrix element definitions	
Λ_x (Lambda-X)	$\lambda_{ij}^{(x)}$	Path coefficient from ξ_j to x_i
Λ_y (Lambda-Y)	$\lambda_{ij}^{(y)}$	Path coefficient from η_j to y_i
B (Beta)	β_{ij}	Path coefficient from η_j to η_i
Γ (Gamma)	γ_{ij}	Path coefficient from ξ_j to η_i
Φ (Phi)	ϕ_{ij}	Covariance between ξ_i and ξ_j
Θ_{δ} (Theta-Delta)	$\theta_{ij}^{(\delta)}$	Covariance between δ_i and δ_j
Θ_{ε} (Theta-Epsilon)	$\theta_{ij}^{(\varepsilon)}$	Covariance between ε_i and ε_j
Ψ (PSI)	ψ_{ij}	Covariance between ζ_i and ζ_j

Up to this point we have tacitly assumed that the latent and observed variables are continuous. This assumption is not necessary. LISREL is able to handle qualitative or ordinal as well as quantitative data. However, qualitative data should not be analyzed using a covariance or a Pearson correlation matrix. (LISREL is able to analyze correlation matrices as well as covariance matrices.) When the data contain variables that are continuous as well as variables that are ordinal, the matrix analyzed should consist of the Pearson correlation for two continuous variables, the polychoric correlation (Kendall and Stuart, 1979) for two variables that are ordinal, and a polyserial correlation when one variable is continuous and the other is ordinal.

6.2. Identification and estimation of the model

Prior to estimating the parameters of the model, it is necessary to be sure that the model is identified. Identification refers to whether the parameters of the model can be uniquely determined. Identification depends upon the choice of the model and the specification of which parameters are to be fixed to specific values, which parameters are unknown but constrained to be equal to other parameters, and which are free. It may be the case that some, but not all, of the parameters are identified. If, however, all parameters are identified, the whole model is identified; otherwise, the model is non-identified.

In general it is not possible to give necessary and sufficient conditions for identification in the full LISREL model. If possible the identification for each particular problem must be solved. This may involve examining the equation

$$\sigma_{ij} = f_{ij}(\theta), \quad i \leqslant j,$$

where σ_{ij} is the (i, j)-th element of Σ expressed as a function of the vector of unknown parameters θ, and showing that each element of θ can be solved for in terms of the population variances and covariances σ_{ij}.

If t is the number of free parameters in the models, then since there are $\frac{1}{2}(p + q)(p + q + 1)$ equations in t unknowns, a necessary, but certainly not sufficient, condition for identification of all parameters is that

$$t \leqslant \tfrac{1}{2}(p + q)(p + q + 1).$$

For a more complete discussion of the identification problem see Jöreskog and Sörbom (1988).

Once the identification problem is resolved, estimation of the free parameters can proceed. We begin with the observed sample covariance matrix S and estimate the parameters of $\Sigma(\hat{\theta})$ so that $\Sigma(\hat{\theta})$ fits S as closely as possible. There are several numerical methods used to estimate the parameters; the 'classical' methods are unweighted least squares (ULS), generalized least squares (GLS), and maximum likelihood (ML). Weighted least squares (WLS) should be used to estimate the parameters when polychoric and polyserial correlations are analyzed. These methods are designed to minimize the difference between S and the predicted covariance matrix $\Sigma(\hat{\theta})$. The function used to minimize these differences is called a fitting function. The particular form of the fitting function is dependent upon the method used to obtain the estimates.

When ULS is the method used to estimate the parameters, the fitting function is

$$F_{ULS} = \tfrac{1}{2}\mathrm{tr}\left[(S - \Sigma)^2\right].$$

The fitting function for GLS is

$$F_{GLS} = \tfrac{1}{2}\mathrm{tr}\left[(I - S^{-1}\Sigma)^2\right].$$

The ML estimator minimizes

$$F_{ML} = \log|\Sigma| + \mathrm{tr}(S\Sigma^{-1}) - \log|S| - (p + q),$$

where, for a square matrix A, $|A|$ is the determinant of A and $\mathrm{tr}(A)$ is the trace of A, i.e., the sum of the diagonal elements of A. The fitting function for WLS is

$$F_{WLS} = [s - \sigma]' W^{-1}[s - \sigma],$$

where s is a vector of the elements from the lower half of S (including the diagonal elements) containing the polychoric, polyserial, and Pearson correlation coefficients, and σ a vector containing the corresponding elements of Σ. W is an estimator of the asymptotic covariance matrix of S (see Jöreskog and Sörbom, 1988, for details). Notice that all of the fitting functions are an intuitively reasonable way of determining how close S and $\hat{\Sigma}$ are. In each case, if $S = \hat{\Sigma}$, then the fitting function is zero.

The fitting functions all have desirable properties. The ULS fitting function is a consistent estimator without making any assumptions about the distribution of the observed variables x and y. If we assume that the observed variables come from a multinormal distribution then both the GLS and ML estimators are also consistent, asymptotically unbiased, and efficient. In general, closed form solutions do not exist for parameters which minimize the fitting functions. Instead, iterative numerical methods are used to find acceptable solutions.

6.3. *Evaluating the model*

In evaluating the LISREL model we wish to assess the significance of each individual parameter and examine the overall fit of the model by considering the closeness of $\Sigma(\hat{\theta})$ and S. These two assessments are independent in the following sense: $\Sigma(\hat{\theta})$ may be determined to be an excellent fit to S, but the individual parameter estimates may be unacceptable or visa versa.

The first step in assessing the results is to examine the values obtained for the parameters for unreasonable values. Estimated variances should be positive and correlations should be between -1 and 1. Improper solutions can arise from many causes such as sampling variability, inaccurate data collection, or inappropriate form of data analysis. For large sample sizes, the standard errors of each parameter's estimated value may be used to determine the precision of the estimate. If $\hat{\alpha}$ is the estimator of some parameter α, and $\hat{\sigma}$ is the standard error, then $z = \hat{\alpha}/\hat{\sigma}$ is approximately normal with mean 0 and variance 1. Hence z may be used to test whether α is statistically different from 0.

The overall fit of the model may be critiqued by use of a χ^2 measure. The measure used is $(N-1)F_{min}$, where F_{min} is the minimum value of the fitting function and N is the number of observations in the sample. This statistic provides a test of the null hypothesis that the model is correct, against an alternative hypothesis, H_1, that Σ is any covariance matrix. The degrees of freedom associated with this test is:

$$df = \text{number of independent parameters under } H_1$$
$$- \text{ number of independent parameters under } H_0.$$

Since, under H_1, there is one parameter for each element of the symmetric matrix Σ,

$$df = \tfrac{1}{2}(p+q)(p+q+1) - t,$$

where t is the number of independent parameters under H_0.

The assumptions necessary to regard the χ^2 measure as a likelihood ratio test are often violated. However, according to Jöreskog and Sörbom (1988), it is reasonable to regard this measure as a goodness of fit test by considering large χ^2 values as evidence for a poor overall fit and small χ^2 values as an indication of a good fit.

Another use of the χ^2 measure is in comparing nested models. Model 2 is said to be nested in a more general model, model 1, if every free parameter in model 2 is a free parameter in model 1. The goodness of fit of these two models can be compared by taking the difference in their χ^2 values. For large samples, this difference is distributed as a χ^2 with degrees of freedom equal to the difference in degrees of freedom of the two models. If this difference is significant we may conclude that the more general model provides an improvement in the fit of the model. If the difference is not significant, we would conclude that the additional parameters do not add meaning to the model.

Several other measures of goodness of fit which supplement the criteria mentioned are discussed in Jöreskog and Sörbom (1988) and are implemented in LISREL 7.

6.4. Genetic analysis of twin data using the LISREL model

In this section we illustrate how the LISREL model can be used for genetic analysis of twin data. Suppose we have a random sample of monozygotic (MZ) and dizygotic (DZ) twin pairs. Each individual is either affected or unaffected for a particular disease. Let L_1 and L_2 be the underlying, latent variable called liability to illness for twin 1 and twin 2, respectively; an individual is affected if and only if his total liability is greater than a certain threshold. We restrict our discussion to the following simple model which expresses the relationship between L_i, additive genetic effects (A_i), environmental effects shared by twin pairs (C_i), and environmental effects unique to each member of a twin pair (E_i), $i = 1, 2$,

$$L_1 = A_1 + C_1 + E_1 \quad \text{for twin 1},$$

$$L_2 = A_2 + C_2 + E_2 \quad \text{for twin 2}.$$

Without loss of generality, we assume $\text{Var}(L_1) = \text{Var}(L_2) = 1$. The variances of A_i, C_i, and E_i ($i = 1, 2$) are denoted h^2, c^2, and e^2, respectively. The correlation between A_1 and A_2 is 1 for MZ twins and $\frac{1}{2}$ for DZ pairs (see Falconer, 1981, for details). The model assumes no genotype–environment interaction nor any effects due to sex. We also assume

$$\text{Cov}(E_1, E_2) = \text{Cov}(E_i, A_j) = \text{Cov}(E_i, C_j) = 0 \quad \text{for } i, j = 1, 2.$$

The effect of the common environmental factor on the liability is the same for both twins, i.e., $C_1 = C_2$. Figure 6 illustrates this simple genetic model for MZ twins and DZ twins using a path diagram.

In terms of LISREL notation, A_i, C_i, and E_i ($i = 1, 2$) are the six latent exogenous variables denoted ξ_1 through ξ_6; L_1 and L_2 are latent dependent

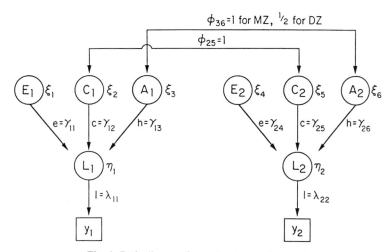

Fig. 6. Path diagram for a simple genetic model.

(endogenous) variables, η_1 and η_2 in LISREL. The y_1 and y_2 represent the observed phenotype, affected or unaffected.

The parameters to be estimated from this model are the contribution of an individual's genotype and environmental exposure, both shared and unique, to liability; that is, we wish to estimate h, c, and e. (Notice, that these parameters are constrained to be equal for MZ and DZ pairs as well as for each individual within a twin pair.) The proportion of the total variance in liability due to the genetic effects, h^2, is called the heritability of the liability. By examining the path diagram, we can easily write down the measurement and structural equations,

$$y_1 = \eta_1 + 0, \qquad y_2 = \eta_2 + 0,$$

$$\eta_1 = e\xi_1 + c\xi_2 + h\xi_3 + 0\xi_4 + 0\xi_5 + 0\xi_6 + 0,$$

$$\eta_2 = 0\xi_1 + 0\xi_2 + 0\xi_3 + e\xi_4 + c\xi_5 + h\xi_6 + 0.$$

The LISREL parameter matrices will be as follows:

$$\Phi = \begin{bmatrix} 1 & 0 & 0 & 0 & 0 & 0 \\ 0 & 1 & 0 & 0 & 1 & 0 \\ 0 & 0 & 1 & 0 & 0 & r \\ 0 & 0 & 0 & 1 & 0 & 0 \\ 0 & 1 & 0 & 0 & 1 & 0 \\ 0 & 0 & r & 0 & 0 & 1 \end{bmatrix},$$

where $r = 1$ for MZ pairs and $r = \frac{1}{2}$ for DZ pairs,

$$\Gamma = \begin{bmatrix} e & c & h & 0 & 0 & 0 \\ 0 & 0 & 0 & e & c & h \end{bmatrix}, \qquad \Lambda_y = \begin{bmatrix} 1 & 0 \\ 0 & 1 \end{bmatrix},$$

Λ_x, B, Φ, Θ_δ, Θ_ε, and Ψ are all zero.

LISREL allows data from more than one group to be analyzed simultaneously. (In this model the MZ and DZ pairs constitute two groups.) Therefore, the estimates of the parameters may be constrained to be equal for each group. The model is easily seen to be identified since the correlation between y_1 and y_2 is $h^2 + c^2$ for MZ pairs and $\frac{1}{2}h^2 + c^2$ for DZ pairs. The LISREL manual (Jöreskog and Sörbom, 1988) gives full details on setting up LISREL runs.

6.5. Summary

The above example should give the reader an indication of the utility of structural equation models for analyzing genetic data. More complex models involving twin and family data can also be handled using the LISREL model (Neale and Martin, 1989; Boomsma et al., 1989). However, it is important to keep in mind that the use of covariance structures in modeling the inheritance in liability to qualitative illnesses is still evolving. There are limitations in the ability of covariance structures to handle certain types of problems. For example, these models would not be appropriate if the sample of twins used in the above example were not a random sample. If the sample had been ascertained because, for example, at least one of the twins were affected, other methods of analysis should be used to estimate the unknown parameters. Even with limitations in the use of these models, covariance structural analysis provides the researcher with another powerful statistical tool for analyzing the transmission of illnesses within families.

7. Discussion

Family studies offer an important strategy for studying common diseases for which family resemblance is high, but for which biological or environmental correlates are not well understood. However, these diseases do not follow classical Mendelian modes of inheritance. Such patterns of inheritance are unlikely to be found when clinical manifestations are heterogeneous or when environmental factors contribute to the expression of an illness. Accordingly, complex statistical models are needed that take into account genetic and environmental variables.

Moreover, these models are necessarily multivariate. For a family of size s, with n measurements per person, we have an ns-variate structure. We have considered only the case where the disease is polychotomous, although there may be continuous covariates measured on each individual. The approach based on an underlying liability with thresholds requires integration of the multivariate normal density functions. For a polygenic model, this integral may be equated to one of lower dimensions, but in general this poses a practical obstacle. The approach to this problem we have described in detail is that implemented in the programs POINTER and YPOINT. Pedigrees are broken up into their constituent nuclear families and treated as independent. This, in fact, should be viewed as an approximation to the correct likelihood. This approach does allow for ascertainment corrections (the 'pointer' is an affected person external to the nuclear family responsible for its ascertainment) and different levels of affection.

The logistic regressive approach avoids the necessity of evaluating the multivariate normal integral. For the class A model, which includes only parental affection statuses, it should be noted that the mixed model is not included as a special case. In the mixed model, offspring are conditionally independent when conditioned on the major locus and the polygenic component, so that conditioning on parental phenotypes (rather than the transmissible component of the phenotype) yields an incompatibility between the two models. Moreover, as noted above the presence of unknown parents imposes a parametric constraint within the logistic regressive model. A major advantage is the ability to consider simultaneous covariates.

The LISREL model itself is quite flexible and has been widely used in twin analyses. There are three limitations that must be considered when applied in genetic epidemiology. First, there is no way to incorporate a single major locus, so that only multifactorial resemblance may be considered. Secondly, data must be grouped by family structure, with parameters related across groups (see our example). This requires separate estimates of the covariance matrix within each group, so that there must be sufficient sample sizes to achieve this. Accordingly, such an approach would not be feasible with families of moderate size. Finally, LISREL requires estimates of the population covariance matrices, so that there is no obvious way to handle families ascertained through the proband method.

In summary, all current methods have their limitations and require some type of model-based assumptions. It is important that the user of these methods recognize these assumptions in the context of his or her application and interpret inferences made within that context.

References

Bonney, G.E. (1986). Regressive logistic models for familial disease and other binary traits. *Biometrics* **42**, 611–625.

Bonney, G.E. (1987). Logistic regression for dependent binary observations. *Biometrics* **43**, 951–973.

Boomsma, D.I., van den Bree, M. B. M., Orlebeke, J. F. and Molenaar, P. C. M. (1989). Resemblances of parents and twins on sports participation and heart rate. *Behav. Genet.* **19**, 123–141.

Borecki, I. B., Rao, D. C., Yaouanq, J. and Lalouel, J. M. (1989). Segregation of genetic hemochromatosis indexed by latent capacity of transferrin. *Am. J. Genet.* **45**, 465–470.

Curnow, R. N. (1972). The multifactorial model for the inheritance of liability to disease and its implications for relatives at risk. *Biometrics* **28**, 931–946.

Curnow, R. N. and Dunnett, C. W. (1962). The numerical evaluation of certain multivariate normal integrals. *Ann. Math. Stat.* **00**, 571–79.

Curnow, R. N. and Smith, C. (1975). Multifactorial models for familial disease. *J. R. Stat. Soc. A* **138**, 131–169.

Edwards, J. H. (1969). Familial predispositions in man. *Br. Med. Bull.* **25**, 58–64.

Elston, R. C. (1979). Major locus analysis for quantitative traits. *Am. J. Hum. Genet.* **31**, 655–661.

Elston, R. and Campbell, M. A. (1971). Schizophrenia: evidence for the major gene hypothesis. *Behav. Genet.* **1**, 3–10.

Elston, R.C. and Stewart, J. (1971). A general model for genetic analysis of pedigree data. *Hum. Hered.* **21**, 523–542.

Elston, R. C., Namboodiri, K. K., Glueck, C. J., Fallat, R., Tsang, R. and Leuba, V. (1975). Study of the genetic transmission of hypercholesterolemia and hypertriglyceridemia in a 195 member kindred. *Ann. Hum. Genet.* **39**, 67–87.

Elston, R.C., Bailey-Wilson, J.E., Bonney, G.E., Keats, B.J. and Wilson, A.F. (1986). S.A.G.E. — A package of computer programs to perform Statistical Analysis for Genetic Epidemiology. Presented at the *7th International Congress of Human Genetics*. Berlin Sept. 22–26.

Falconer, D. S. (1965). The inheritance of liability to certain diseases, estimated from the incidence among relatives. *Ann. Hum. Genet.* **29**, 51–76.

Falconer, D. S. (1967). The inheritance of liability to diseases with variable age of onset, with particular reference to diabetes mellitus. *Ann. Hum. Genet.* **31**, 1–20.

Falconer, D. S. (1981). *Introduction to Quantitative Genetics*, 2nd ed. Longman, London.

Fisher, R.A. (1918). The correlation between relatives on the supposition of Mendelian inheritance. *Trans. Roy. Soc. Edinb.* **52**, 399–433.

Hasstedt, S. and Cartwright, P. (1979). PAP Pedigree Analysis Package, Technical Report No. 13, Department of Medical Biophysics and Computing, University of Utah Medical Center.

James, J. W. (1971). Frequency in relatives for an all-or-none trait. *Ann. Hum. Genet.* **35**, 47–49.

Jöreskog, K. G. (1973). A general method for estimating a linear structural equation system. In: A. S. Goldberger and O. D. Duncan, eds., *Structural equation models in the social sciences*, Academic Press, New York, 85–112.

Jöreskog, K. G. (1978). Structural Analysis of covariance and correlation matrices. *Psychometrika* **43**, 443–477.

Jöreskog, K. G. and Sörbom, D. (1988). *LISREL 7 a guide to the program and applications.* SPSS, Inc.

Kendall, M. and Stuart, A. (1979). *The advanced theory of statistics,* Vol. 2, Inference and relationship. Hafner, New York.

Lalouel, J. M. and Morton, N. E. (1981). Complex segregation analysis with pointers. *Hum. Hered.* **31**, 312–321.

Lalouel, J. M., Le Mignon, L., Simon, M., Fauchet, R., Bourel, M., Rao, D. C. and Morton, N. E. (1985). Genetic analysis of idiopathic hemochromatosis using both qualitative (disease status) and quantitative (serum iron) information. *Am. J. Hum. Genet.* **37**, 700–718.

Lalouel, J. M., Rao, D. C., Morton, N. E. and Elston, R. C. (1983). A unified model for complex segregation analysis. *Am. J. Hum. Genet.* **35**, 816–826.

Lange, K., Westlake, J. and Spence, M. A. (1976). Extensions to pedigree analysis II. Recurrence risk calculations under the polygenic threshold model. *Hum. Hered.* **26**, 337–348.

MacLean, C. J., Morton, N. E. and Lew, R. (1975). Analysis of family resemblance. IV. Operational characteristics of segregation analysis. *Am. J. Hum. Genet.* **27**, 365–384.

Mendell, N. R. and Elston, R. C. (1974). Multifactorial qualitative traits: genetic analysis and prediction of recurrence risks. *Biometrics* **30**, 41–57.

Moldin, S.O., Rice, J.P., Van Eerdewegh, P., Gottesman, I.I. and Erlenmeyer-Kimling, L. (1990). Estimation of disease risk under bivariate models of multifactional inheritance. *Genetic Epidemiology* **7**, 371–386.

Morton, N. E. and MacLean, C. J. (1974). Analysis of family resemblance. III. Complex segregation of quantitative traits. *Am. J. Hum. Genet.* **26**, 489–503.

Morton, N. E., Rao, D. C. and Lalouel, J. M. (1983). *Methods in Genetic Epidemiology.* Karger, New York.

Neale, M. C. and Martin, N. G. (1989). The effects of age, sex and genotype on self-report drunkenness following a challenge dose of alcohol. *Behav. Genet.* **19**, 63–78.

Ott, J. (1979). Maximum likelihood estimation by counting methods under polygenic and mixed models in human pedigrees. *Am. J. Hum. Genet.* **31**, 161–179.

Ott, J. (1985). *Analysis of human genetic linkage,* The Johns Hopkins University Press, Baltimore.

Pearson, K. (1903). On the influence of natural selection on the variability and correlation of organs. *Philos. Trans. R. Soc. London Ser. A* **200**, 1–66.

Reich, T., James, J. W. and Morris, C. A. (1972). The use of multiple thresholds in determining the model of transmission of semicontinuous traits. *Ann. Hum. Genet.* **36**, 163–184.

Rice, J., Reich, T., Cloninger, C. R. and Wette, R. (1979). An approximation to the multivariate normal integral: its application to multifactorial qualitative traits. *Biometrics* **35**, 451–459.

Risch, N. (1990). Linkage strategies for genetically complex traits. I. Multilocus models. *Am. J. Hum. Genet.* **46**, 222–228.

Smith, C. and Mendell, N. R. (1974). Recurrence risks from family history and metric traits. *Ann. Hum. Genet.* **37**, 275–286.

Wright, S. (1921). Correlation and causation. *J. Agric. Res. Washington D.C.* **20**, 557–585.

Wright, S. (1934). The method of path coefficients. *Ann. Math. Biostat.* **5**, 161–215.

C. R. Rao and R. Chakraborty, eds., *Handbook of Statistics, Vol. 8*
© Elsevier Science Publishers B.V. (1991) 29–61

2

Ascertainment Biases and Their Resolution in Biological Surveys

W. J. Ewens

1. Introduction

Fortunately, genetic diseases are rare. Their very rarity, however, causes statistical difficulties in the investigation of their genetic basis. The reason for this is that, in general, simple random sampling is not a practicable approach to obtaining samples with a sufficiently large number of affected individuals. For example, if a disease affects one person in ten thousand and if a sample of two hundred affected individuals is needed for a reasonably precise analysis, then a simple random sample of about two million is needed to obtain this number, usually far beyond the realms of practicability.

The approach used instead by many geneticists is that of 'ascertainment' sampling. Here the entire family of some affected individual is examined: we say that the family has been 'ascertained' through this affected individual. In some cases a well-defined pedigree, rather than merely a family, is ascertained. Assuming that the disease has in fact a genetic basis, other individuals in the family or pedigree are more likely to be affected than individuals taken at random, and thus ascertainment sampling of families or pedigrees can be an efficient method of arriving at a sufficiently large sample of affected individuals. As additional benefits, use of families can often provide genetic information (in particular about linkage) not easily obtained (and possibly not even obtainable) from randomly chosen individuals, sampling individuals within families is often quick and convenient, and the possible allocation of genetic and environmental effects can often be attempted.

It is clear that simple random sampling formulae should not be used for parameter estimation, or hypothesis testing, when the data arise from families ascertained through affected family members. In particular, simple random sampling formulae would grossly overestimate the population frequency of the disease allele (or alleles). But the ascertainment procedure raises problems well beyond this, since the 'sociology' of the ascertainment procedure must be taken into account in the estimation process. This 'sociology' can be encapsulated in the following question: "Is a family with three affected children three times as likely to be sampled, i.e. ascertained, as one with only one affected child?". This question relates not to the genetics of the disease but to the nature of the sampling process, and unfortunately the likelihood from which genetic parameters are estimated depends on the nature of the sampling process, so that any incorrect assumption about this process will lead to biased estimates of genetic parameters. This is the 'ascertainment problem'.

One can imagine cases where the probability that a family is ascertained is proportional to the number of affected children, for example if children are initially sampled at a certain age class in schools: such a case is called 'single ascertainment'. On the other hand, if families are ascertained by searching through registries, then the probability of ascertainment of a family could well be independent of the number of affected children: this is called 'complete ascertainment'. Nor are these the only two possibilities: one can easily imagine cases where the probability of ascertainment of a family increases other than linearly with the number of affected children in it. This problem was recognized very early on by Weinberg (1912a,b, 1928), who made original and outstanding contributions to the identification and resolution of this problem.

The affected individual through whom a family is ascertained is called a 'proband'. While this is a useful concept, several problems associated with it should be mentioned. First, there are many practical cases (for an excellent summary, see Greenberg, 1986) in which no easily identified family member can reasonably be described as the 'proband'. Second, the definition assumes (and this assumption is embodied in several computer packages) that affected individuals act independently in becoming probands: Elandt-Johnson (1971), for example, defines a proband as 'an affected individual who has been detected independently of other members of the family and through whom the family can be ascertained'. It is hard to imagine young children in the care of their parents as acting independently so far as proband behavior is concerned, again leading to some difficulty with the proband concept. Thirdly, it is generally assumed (in particular in computer packages) that there is some common probability (denoted here by π) that an affected individual will become a proband. Thus the possibility of an age-order effect in proband status, or of variation from one family to another in proband probability, is ignored. Morton (1959) identified many of these problems and showed how several of them could be overcome. One of our aims, later, is to do away with the proband concept and hence also of the assumptions with which it is associated.

For simplicity, we assume throughout that, as is often the case in practice, families are ascertained through the children (sibs). Thus, by 'family size' we mean the number of children. Any ascertained family will therefore contain at least one affected child; since however the parents are normally also observed, their affectedness status and also the genetic information found from them are often used in the estimation procedure.

We always carry out our estimations by maximum likelihood, because of the large-sample properties that this method enjoys. However several general difficulties should be noted. The likelihoods used are often complex, being usually found from conditional probabilities, and this has two consequences. First, explicit expressions for MLE's, and their variances, often cannot be found, the estimates being given implicitly as the solutions of a set of simultaneous equations which must be solved numerically. Second, the asymptotic ML theory is sometimes inaccurate even in quite large samples: approximate normality of estimates is often not reached, the true variances of estimates are not given accurately by the information matrix approach, and the estimators themselves are not necessarily approximately unbiased. This should be kept in mind when considering the asymptotic bias and variance calculations derived below. (Fortunately, problems do not arise in our examples with null hypothesis parameter values being at boundary points, and the fact that non-independent or non-identically distributed data arise does not cause difficulties.) Despite these reservations, we often discuss the asymptotic biasedness (or otherwise) of parameter estimates because this provides useful comparisons of the bias properties of different approaches to estimation.

Three points concerning the likelihood should be noted. First, multiplicative constants (often combinatorial coefficients) are usually irrelevant to the estimation process and are often systematically referred to below as 'const'. Second, it is clear that while ascertained families can be thought of as coming from some population, it is often difficult to define what this population is or even to know its size, since the great majority of families will not be ascertained and the geographical area defining the ascertainment sampling process may well be poorly defined. It has, nevertheless, been vigorously argued by Thompson and Cannings (1979) that 'non-observed' families should be thought of as part of the data forming the likelihood. Bailey (1951a) also advocates and uses this approach, and the form of likelihood to which it leads is discussed below. Essentially all other authors use simpler conditional likelihood involving observed data. Ewens (1982) (but see Thompson and Edwards, 1984) shows that the two procedures lead to effectively identical estimates, as is noted in the example below: in practice one then would prefer to use the simpler conditional approach. Finally, the family size distribution (FSD) if known should in principle be used in the likelihood function. In practice this is rarely done, for four reasons. The first is that only small increases in the precision of estimates arise in using the FSD, in particular if this distribution contains parameters that must be estimated simultaneously with genetic parameters. The second is that any error in specifying the FSD leads to biased estimates of genetic parameters (Ewens et al., 1986). The third is that assuming a quite general form $(f_0, f_1, f_2, \ldots)$ for the FSD, where f_i is a free parameter (subject only to $\sum f_i = 1$) leads to the same estimates of genetic parameters as when the FSD is ignored (Hodge, 1985), and the last is that ignoring the FSD generally leads to far simpler estimation equations for genetic parameters than when the FSD is used.

We now turn to a very simple (and classical) example illustrating several of the points made above, as well as introducing the mathematical theory for ascertainment sampling.

2. Theory and an example

For a sample of n families, arrived at by simple random sampling, the likelihood from which genetic parameters are estimated is

$$L = \prod_{i=1}^{n} \Pr(D_i), \tag{2.1}$$

where by D_i we mean the data in family i. For a sample of n families obtained through ascertainment sampling, the appropriate likelihood is

$$L = \prod_{i=1}^{n} \Pr(D_i|A_i) \tag{2.2}$$

$$= \prod_{i=1}^{n} \Pr(D_iA_i)/\Pr(A_i), \tag{2.3}$$

where A_i is the event that family i is ascertained. The ascertainment problem is that the two probabilities in (2.3) depend not only on the genetics, but also on the nature of the sampling scheme, so that any misspecification of the latter will lead to biased estimates of genetic parameters.

We use a simple example considered often in the literature (Fisher, 1934; Bailey, 1951a; Elandt-Johnson, 1971) to illustrate the biases that can arise through misspecification of the ascertainment procedure. Define a family to be 'at risk' if the two parents can produce affected children, and suppose that in any at risk family the children are independently affected, each with probability p. (This would arise approximately if the disease is caused by a rare recessive allele: here essentially all at risk families have both parents heterozygous for the normal and the disease allele, and if the disease is fully penetrant, $p = \frac{1}{4}$.) Suppose that in family i there are s_i children: then the datum D_i is the number of children r_i who are affected.

The value of the probability $\Pr(D_i|A_i)$ depends on the nature of the ascertainment scheme. Under complete ascertainment, the probability of ascertainment is independent of the number of affected sibs, (assuming this number exceeds zero), and thus

$$\Pr(D_i|A_i) = \text{const } p^{r_i}(1-p)^{s_i-r_i}/[1-(1-p)^{s_i}]. \tag{2.4}$$

On the other hand, under single ascertainment, the probability of ascertainment of a family is proportional to the number of affected children, so

$$\Pr(D_iA_i) = \text{const } p^{r_i}(1-p)^{s_i-r_i}r_i,$$

$$\Pr(A_i) = \text{const } \sum_{r_i} r_i \binom{s_i}{r_i} p^{r_i}(1-p)^{s_i-r_i}$$

$$= \text{const } p,$$

and thus

$$\Pr(D_i|A_i) = \text{const}\, p^{r_i-1}(1-p)^{s_i-r_i}. \tag{2.5}$$

As a third possibility, suppose that the probability of ascertainment of a family is proportional to the square of the number of affected children. Then

$$\Pr(A_i) = \text{const} \sum_{r_i} r_i^2 \binom{s_i}{r_i} p^{r_i}(1-p)^{s_i-r_i}$$

$$= \text{const}\,[(s_i p)^2 + s_i p(1-p)],$$

leading to

$$\Pr(D_i|A_i) = \text{const}\, p^{r_i-1}(1-p)^{s_i-r_i}/[(s_i p + 1 - p)]. \tag{2.6}$$

We describe this situation as one of 'quadratic ascertainment', and consider later a situation in which it could arise.

The likelihood of the sample of n families is found from (2.4), (2.5), (2.6), or whatever else might be the appropriate formula (since many possibilities other than complete, single or quadratic ascertainment exist). If complete ascertainment is appropriate, the MLE equation for $\hat{p}$ is, from (2.4),

$$R/\hat{p} = \sum_i s_i/[1 - (1-\hat{p})^{s_i}], \tag{2.7}$$

where R is the total number of affected children in the sample. This equation must be solved numerically: note that it is also the 'method of moments' estimation equation. If single ascertainment is appropriate, we get explicitly from (2.5)

$$\hat{p} = (R-n)/(S-n), \tag{2.8}$$

where S is the total number of children in the n families. The MLE equation deriving from the quadratic case (2.6) is complex and we will not write it down.

We make several comments concerning (2.7) and (2.8). First, the solution of (2.7) is usually very close to an explicit approximating expression first given by Li and Mantel (1968), namely

$$\hat{p} = (R-T)/(S-T), \tag{2.9}$$

where T is the number of families in the sample with exactly one affected child. Since in a family with s_i children,

$$E(r_i) = s_i p/[1 - (1-p)^{s_i}], \tag{2.10}$$

and the probability that there is exactly one affected child is

$$s_i p(1-p)^{s_i-1}/[1 - (1-p)^{s_i}], \tag{2.11}$$

it follows after some algebra that (2.9) is an asymptotically unbiased estimator of p. The second comment is that (2.8) can be thought of as arising by ignoring one affected child in each family and then using simple random sampling formulae: this is the first hint of a property arising often under single ascertainment and we return to it several times later. Third, it is clear that biased estimation will arise if (2.7) is used when (2.8) is appropriate, or vice versa, and that (2.7) and (2.8) both give biased estimates if some form of ascertainment other than complete or single is appropriate.

To check the sizes of the potential biases, we take the simple case where $s_i = 2$ for all i, i.e. all families have two children. Then under complete ascertainment,

$$E(R) = 2np/(2p - p^2),\tag{2.12}$$

$$E(T) = 2np(1 - p)/(2p - p^2),\tag{2.13}$$

so that the Li–Mantel estimator (2.9) is, as noted above, an asymptotically unbiased estimator of p. However the mean value of the 'single ascertainment' estimator (2.8) is $p/(2 - p)$, so under complete ascertainment, this is clearly a biased estimator of p. Conversely, under single ascertainment,

$$E(R) = n(1 + p),\tag{2.14}$$

$$E(T) = n(1 - p),\tag{2.15}$$

and so (2.8) is an unbiased estimator of p while in this case (2.9) is biased, having asymptotic mean $2p/(1 + p)$. Finally, when quadratic ascertainment is the case,

$$E(R) = n(1 + 3p)/(1 + p),$$

$$E(T) = n(1 - p)/(1 + p),$$

and so (2.8) and (2.9) are both biased, being asymptotically unbiased estimators not of p, but of $2p/(1 + p)$ and $4p/(1 + 3p)$, respectively.

A model which allows complete and single ascertainment as particular cases is the 'multiple ascertainment' model, described explicitly by Fisher (1934). Under this model it is assumed that each affected individual independently becomes a proband (defined above) with fixed (but unknown) probability π. Given that there are r affected children in a family, the probability that there are q probands is, under these assumptions,

$$\binom{r}{q} \pi^q (1 - \pi)^{r-q},\tag{2.16}$$

and the probability that the family is ascertained is $1 - (1 - \pi)^r$. It follows that the unconditional probability that a family with s children is ascertained is

$$\sum_r \binom{s}{r} p^r (1 - p)^{s-r} [1 - (1 - \pi)^r] = 1 - (1 - \pi p)^s.\tag{2.17}$$

For this model, there are two frequently used likelihood contributions from any given family. The first ignores the number of probands (other than noting that this number must exceed zero), and is

$$\text{const}\,[1 - (1 - \pi)^r]\,p^r(1 - p)^{s-r}/[1 - (1 - \pi p)^s]\,. \tag{2.18}$$

The second explicitly uses the number q of probands, and is

$$\text{const}\,\pi^q(1 - \pi)^{r-q}p^r(1 - p)^{s-r}/[1 - (1 - \pi p)^s]\,. \tag{2.19}$$

It is interesting to compare (2.18) and (2.19). Clearly (2.19) is the correct likelihood, since it uses all the data (proband and affectedness status of each child) in the family. Possibly (2.18) is used on occasion because of an uncertainty in practice as to which sibs in each family are probands. However, use of (2.18) rather than (2.19) results in estimators with substantially increased standard errors. For example when the sample consists of n families, each with $s = 5$ sibs, and if $p = 0.3$, $\pi = 0.4$, the asymptotic variance of $\hat{p}$ using (2.19) is $0.0603/n$ and using (2.18) is $3.8896/n$, an increase by a factor of 64. For this reason we focus for the moment on (2.19) and consider the ML equations deriving from it, restricting for simplicity to the case where all families sampled are of the same size s. If Q is the total number of probands, R the total number of affected children and n the number of families in the sample, the likelihood is

$$\text{const}\,\pi^Q(1 - \pi)^{R-Q}p^R(1 - p)^{ns-R}[1 - (1 - \pi p)^s]^{-n}\,, \tag{2.20}$$

and this leads to the ML equations

$$R/\hat{p} - (ns - R)/(1 - \hat{p}) - ns\hat{\pi}(1 - \hat{\pi}\hat{p})^{s-1}/[1 - (1 - \hat{\pi}\hat{p})^s] = 0\,, \tag{2.21}$$

$$Q/\hat{\pi} - (R - Q)/(1 - \hat{\pi}) - ns\hat{p}(1 - \hat{\pi}\hat{p})^{s-1}/[1 - (1 - \hat{\pi}\hat{p})^s] = 0\,, \tag{2.22}$$

for $\hat{p}$ and $\hat{\pi}$. These equations must be solved numerically. In the numerical example introduced by Fisher (1934) and considered also by Bailey (1951a), $s = 5$, $n = 340$, $Q = 432$, $R = 623$, and with these values the solution of (2.21) and (2.22) is $\hat{p} = 0.2526$, $\hat{\pi} = 0.4753$. A further round of differentation of the log likelihood leads to a 2×2 information matrix and to variance estimates 1.670×10^{-4} for $\hat{p}$ and 9.634×10^{-4} for $\hat{\pi}$.

There are two interesting comments one can make about these estimates. The first concerns the main theme of this section, namely the biases that can arise through an incorrect assumption about the ascertainment process. This question was already taken up by Fisher (1934), using the numerical example above. Fisher noted that if complete ascertainment is assumed, (his unfortunately named 'proband method'), so that p is estimated from (2.7) rather than (2.21) and (2.22), the estimate of p is $\hat{p} = 0.3086$, significantly in excess of the value found from (2.21) and (2.22). (Fisher did not use (2.21) and (2.22) but a closely approximating method, yielding $\hat{p} = 0.2488$.) Thus if the 'multiple ascertainment' model is correct, significant bias in the estimation of p arises by assuming complete ascertainment (or, equivalently, by assuming $\pi = 1$ within the multiple ascertainment model). We consider later many further examples of biases of this nature.

The second comment concerns 'unobserved' families. As mentioned in Section 1, Bailey (1951a) advocates and uses an approach different from the above. He introduces the further unknown quantity N, the number of families in the population being surveyed, noting that there are then $N - 340$ unobserved families and that the probability that a family is not observed is $(1 - \pi p)^5$. This leads him to use a likelihood

$$\text{const} \binom{N}{340} (1 - \pi p)^{5(N - 340)} \pi^{432} (1 - \pi)^{191} p^{632} (1 - p)^{1077} \qquad (2.23)$$

with now three unknown parameters, N, p and π. The ML equations become

$$623/\hat{p} - 1077/(1 - \hat{p}) - 5(\hat{N} - 340)\hat{\pi}/(1 - \hat{\pi}\hat{p}) = 0, \qquad (2.24)$$

$$432/\hat{\pi} - 191/(1 - \hat{\pi}) - 5(\hat{N} - 340)\hat{p}/(1 - \hat{\pi}\hat{p}) = 0, \qquad (2.25)$$

$$5\log(1 - \hat{\pi}\hat{p}) + \psi(\hat{N} + 1) - \psi(\hat{N} - 339) = 0, \qquad (2.26)$$

where $\psi(x)$ is the digamma function $d\{\log\Gamma(x)\}/dx$. Bailey uses the close approximation $\psi(x + 1) = \log x$ to solve these equations, but does not appear to notice that using this approximation, (2.26) becomes

$$(1 - \hat{\pi}\hat{p})^5 = (\hat{N} - 340)/\hat{N}, \quad \hat{N} = 340/\{1 - (1 - \hat{\pi}\hat{p})^5\}, \qquad (2.27)$$

which when substituted into (2.24) and (2.25) yield (2.21) and (2.22) (with the appropriate numerical values of R, Q, n and s substituted). It is therefore not surprising that Bailey's numerical solutions for $\hat{p}$ and $\hat{\pi}$ are identical to the degree of accuracy considered, to those above deriving directly from (2.21) and (2.22). This supports the view given in the Introduction that there is no value in using the complicated likelihood (2.23). Further, one more round of differentiation of the log likelihood leads to a 3×3 information matrix which gives variance estimates for $\hat{p}$ and $\hat{\pi}$ identical to those found above, again as we expect. This example is sufficient to lead us to use simple likelihoods such as (2.20) rather than more complex likelihoods such as (2.23) in our future calculations.

We close this section with some further comments on the 'multiple ascertainment' model implied by (2.16). This covers both the single and the complete ascertainment cases by allowing $\pi \to 0$, $\pi = 1$ respectively, and under these limiting operations both (2.18) and (2.19) converge to the single and complete ascertainment likelihood contributions (2.4) and (2.5) (note that under complete ascertainment $r = q$, so we ignore the term $(1 - \pi)^{r-q}$ in (2.19)). However there are many situations which are not covered by the 'multiple ascertainment' model. For example, quadratic ascertainment (leading, for example, to (2.6)) lies outside the complete-to-single range and thus cannot be described by this model for any choice of π. Attempts to bring quadratic ascertainment under the 'multiple ascertainment' umbrella by allowing negative π run into difficulties because use of (2.19) then leads to potentially negative probabilities while use of (2.18), while not leading to this problem, suffers from the high variance estimates described earlier.

Because of these and other problems, we now turn to an approach which does not make the restrictive assumptions implicit in this ascertainment model and which, furthermore, does not make use of the potentially troublesome proband concept.

3. Arbitrary ascertainment probabilities

The numerator and denominator in the likelihood contribution (2.18) both contain terms deriving from the ascertainment assumptions implicit in (2.16). Suppose now we are unwilling to make such an explicit assumption about ascertainment probabilities, and indeed have no idea what the probability of ascertainment is for a family with s sibs, r of whom are affected. Suppose that we simply allocated some unknown probability $\alpha(r, s)$ for this probability. Then the likelihood contribution (2.18) would be replaced by

$$\text{const } \alpha(r, s) p^r (1 - p)^{s-r} \bigg/ \left[\sum_{j=1}^{s} \binom{s}{j} p^j (1 - p)^{s-j} \alpha(j, s) \right]. \tag{3.1}$$

Since this contribution depends on the $\alpha(r, s)$ values only through their ratios, we impose the further condition that for each s, $\sum_j \alpha(j, s) = 1$.

Even with this condition, unique parameter estimates do not arise from (3.1). To illustrate this, suppose that in a sample of n families, each having two children, n_1 have one affected child and n_2 have two affected. Writing $\alpha(1, 2) = 1 - \alpha$, $\alpha(2, 2) = \alpha$, the likelihood is

$$\text{const } (1 - \alpha)^{n_1} \alpha^{n_2} p^{n_1 + 2n_2} (1 - p)^{n_1} / [2p(1 - p)(1 - \alpha) + p^2 \alpha]^n \tag{3.2}$$

$$= \text{const} (1 - q)^{n_1} q^{n_2}, \tag{3.3}$$

where

$$q = p\alpha / [2(1 - p)(1 - \alpha) + p\alpha].$$

The expression (3.3) is maximized when $q = n_2/n$, but the form of the function q shows that there are infinitely many (p, α) combinations satisfying this equation. This merely reflects the common-sense observation that a high value of n_2/n can be explained by a high value of the population frequency p or a high ascertainment probability of families with two affected.

The equation $q = n_2/n$ may be rewritten as

$$p^2 \alpha = (n_2/n)[2p(1 - p)(- \alpha) + p^2 \alpha] .$$

If all families are of size s exceeding two, the corresponding equation is

$$\binom{s}{r} p^r(1 - p)^{s-r} \alpha(r, s) = \frac{n_r}{n} \left[\sum_i \binom{s}{i} p^i(1 - p)^{s-i} \alpha(i, s) \right], \qquad (3.4)$$

and again it is found that no unique maximization of the likelihood arises.

It is only when further structure is imposed on the mathematical form of the $\alpha(j, s)$ values (for example $\alpha(j, s) = 1 - (1 - \alpha)^j$ for some π) that a unique maximization process is possible in this example. However, our very aim is to avoid imposing any particular structure. Note that even imposing a reasonable extrinsic requirement such as $\alpha(1, s) \leqslant \cdots \leqslant \alpha(s, s)$ does not to lead to a unique value of $\hat{p}$, since, assuming each $n_j > 0$, (3.4) may be written as

$$\hat{\alpha}(j + 1, s)/\hat{\alpha}(j, s) = (n_{j+1}/n_j)\{(s - j)/(j + 1)\}(1 - \hat{p})/\hat{p} , \qquad (3.5)$$

and thus the requirement $\alpha(j + 1, s)/\alpha(j, s) \geqslant 1$ will be met uniformly in j by choosing $\hat{p}$ less than p_0, where p_0 is the minimum (over j) of $[1 + (n_j/n_{j+1})(j + 1)/(s - j)]^{-1}$.

We have carried through the analysis above assuming that all families in the sample are of the same size s, but the conclusion and the difficulties found above still arise when this restriction no longer applies. Thus the approach to using unrestricted ascertainment probabilities as just described initially appears not to be useful. Nevertheless we will continue to investigate this approach, since we will find that in cases where the data are more complex than in the example just given, the problems just described can be overcome.

4. Complex segregation analysis

As we will note shortly, the real reason why the attempt to use general ascertainment parameters $\alpha(r, s)$ in the model of the previous section failed is that the genetical model is very simple, the only data value for each family being the number of affected children. In more complex cases each family provides several data points, perhaps both genotypic and phenotypic, and in such cases, as we now see, it is possible to use the general ascertainment parameters introduced above to resolve the ascertainment problem, that is to obtain an estimation procedure which will give asymptotically unbiased estimates of genetic parameters no matter what the true (but unknown) ascertainment procedure might be.

We consider first the theoretical aspects of this approach. Any family will exhibit certain data, which we denote D, and which we assume is reasonably substantial, certainly more so than simply the number of affected children. Now not all these data will influence ascertainment: for example, genetic markers having no outward effect, and not directly associated with the disease in question, would presumably have no bearing on the ascertainment probability of the family. We thus divide the data D conceptually into two parts, D_1 and D_2, where we assume that only the D_1 part of the data affects the ascertainment probability. Thus D_1 would presumably include at least (and, in the cases considered above, only) the number of affected children in each family, whereas D_2 would presumably include the genetic marker information just referred to, as well as all information concerning the parents in cases where ascertainment is solely through the children.

We now suppose that in a family of s children, having ascertainment data D_1, there is some probability $\alpha(D_1, s)$ that the family will in fact be ascertained. This probability is not assumed to take any specified mathematical form and is regarded as an entirely free parameter, just as was the parameter $\alpha(r, s)$ introduced in the previous section.

The likelihood contribution of an ascertained family having s children and data $D = (D_1, D_2)$ is the conditional probability of these data given the fact of ascertainment of the family. This is

$$\Pr\{(D_1, D_2) \mid \text{family ascertained}\} = \alpha(D_1, s) Q_s(D_1, D_2) \bigg/ \left[\sum_j \alpha(j, s) Q_s(j) \right],$$
(4.1)

where the summation is over all possible values j of the data relevant to ascertainment, $Q_s(j)$ is the population probability that a family of size s has data j relevant to ascertainment, and $Q_s(D_1, D_2)$ is the population probability that a family of size s has data (D_1, D_2). The likelihood from which parameters are estimated is the product, over families, of contributions of the form (4.1). Note that the ascertainment parameters are determined only up to their ratios, so that as above some further condition (for example that they add to unity over D_1 for each s) must be imposed.

Differentation of the likelihood computed from (4.1) with respect to $\alpha(D_1, s)$ leads to the equations

$$Q_s(D_1) \, \hat{\alpha}(D_1, s) = \frac{n(D_1, s)}{n(s)} \left[\sum_i Q_s(i) \, \hat{\alpha}(i, s) \right],$$
(4.2)

which are analogous to (3.4), with $n(D_1, s)$ being the number of families in the sample having s children and data D_1 and $n(s)$ being the number of families in the sample having s children. It may be shown from (4.2) (Ewens and Shute, 1986), and the equations found from differentiating the likelihood with respect to genetic parameters, that these parameters may then be estimated by using as likelihood the product over all families of

$$Q_s(D_1, D_2) / Q_s(D_1),$$
(4.3)

that is of the conditional probability of the data D_2, given the data D_1. We now make the key observation that this likelihood is entirely free of the unknown ascertainment parameters (which are therefore never actually estimated) and can be written down entirely in terms of the genetic parameters whose estimation is our major aim. This leads to an immediate, and in many ways simple, MLE procedure, in which the estimates of genetic parameters are free from any form of ascertainment assumption other than that ascertainment is only through the D_1 part of the data.

It can perhaps be seen intuitively that estimates of genetic parameters found from conditional probabilities such as (4.3) will be free of ascertainment assumptions. The data D_1 form the link between the ascertainment process and the entire data set in the family, and if we condition on D_1 we are severing this link and in effect using in our estimation procedure only that part of the data unaffected by the ascertainment procedure. We therefore describe parameter estimation using (4.3) as being 'ascertainment-assumption-free', or AAF, estimation. There are clear analogues between the choice between using an AAF procedure and one which assumes a specific ascertainment scheme on the one hand, and between a non-parametric and a normal theory hypothesis testing procedure on the other, although in our case the focus is on estimating parameters rather than on hypothesis testing.

The above calculation makes it clear why no estimation of genetic parameters was possible, in the example of the previous section, when general ascertainment parameters $\alpha(r, s)$ were used. Here the only piece of data from each family is the number of affected children, and this is also D_1, the data relevant to ascertainment. Thus the conditional probability (4.3) is unity, D_2 being empty, and so no useful likelihood emerges. More generally, the approach we have outlined is most useful when the D_2 part of the data is substantial, since this is the only part of the data used in the estimation procedure.

Before giving an example of the use of the approach using general ascertainment parameters, several general comments are in order. First, if the true ascertainment scheme is known, (4.3) should not be used, but rather whatever is the appropriate likelihood derived directly from (2.2), using whatever the true ascertainment procedure might be. Use of (4.3) will not lead to systematically biased estimators in this case, but rather to estimators having larger variances (since they in effect do not use what is now a part of the data carrying useful information, namely D_1) than the variances of the estimators found from (2.2). One can pursue this argument in a different direction also. Suppose the ascertainment scheme is not known, so that a likelihood deriving from (4.3) is appropriate. One could also use as likelihood an expression deriving from conditioning, in each family, not only on the data D_1, but also on the data D_3, where D_3 is part of D_2. Thus, if we write $D_2 = (D_3, D_4)$, asymptotically unbiased estimation arises by replacing (4.3) by

$$Q(D_1, D_2)/Q(D_1, D_3) \, . \tag{4.4}$$

However, the variances of estimators derived from (4.4) can be shown to exceed those found from (4.3), and so one should not use (4.4) unless, perhaps, it were felt that the data D_3 are also possibly relevant to ascertainment.

In particular, ascertainment problems can always be overcome by conditioning on all the phenotypic (i.e., affectedness) information in each family, since presumably these data must always contain that part of the total data relevant to ascertainment. We show later that large standard errors of the estimates of some parameters will arise from this form of conditioning. In practice, of course, there might be cases where it is not clear-cut which part of the data is relevant to ascertainment, and conditioning on all phenotypic information might be chosen as the safest policy, despite these large standard errors. We return to this point later.

A further advantage of the use of (4.3) is that it allows amalgamation of data from different studies, even though different ascertainment schemes may have been used in these studies. (4.3) is also free of assumptions such as that the 'probandship' probability π is the same for all affected individuals in all families, and indeed is free of the very concept of a proband. These arguments point to the practical advantages of use of (4.3).

It is also clear that the arguments which lead to (4.3) can be generalized to cover pedigrees. We take this generalization up in a later section.

5. Estimation of bias and standard error

As noted above, the gain in using an explicit ascertainment model and proceding from (2.2) is that estimates with the highest possible precision are obtained; use of (4.3) leads to estimates with decreased precision. The loss in using an explicit ascertainment model is that biased estimates arise if an incorrect ascertainment scheme is assumed, while this does not occur with use of (4.3) (at least asymptotically). In order to make a reasoned choice between the two approaches, some indication of the sizes of the potential asymptotic biases, and the differences between the variances of the estimators, is desirable.

The approach we use is to assume some set of values for all genetic parameters, some specific family size distribution, and some specific ascertainment scheme. We can then calculate the probability of any specific data D in any family. We now assume 'deterministic' or 'perfect' data: that is, we assume that the number of families in the sample which have any particular data configuration is proportional to the sample probability of that configuration. Using these 'data', we now estimate genetic parameters by ML using either one or other specific ascertainment scheme or else the AAF likelihood found from (4.3), and estimate the variances of unbiased parameter estimates using the information matrix, calculated numerically from the log likelihood function. The advantage of this approach is that it is relatively fast: the disadvantages are that it only provides us with the asymptotic mean values of parameter estimates, and that the information matrix may not give extremely accurate values for the variances. We should more exactly sample a large number of simulated data sets, but computer time problems make this a difficult task and the approach outlined above is more than adequate to establish the points which we wish to make.

6. Numerical example

We use for a numerical example the case of disease for which susceptibility is determined by the genes at a 'susceptibility' locus S closely linked to a marker locus for which all unrelated individuals can be assumed to have different heterozygous genotypes. (An example of such a marker is, in effect, the HLA complex (Tiwari and Terasaki, 1985), the genotype consisting of two HLA haplotypes.) We denote the marker alleles $M_1, M_2, \ldots$. It is assumed, of course, that we cannot establish directly the genotype at the susceptibility locus.

For simplicity, assume that two alleles are possible at the locus S, the susceptibility allele S and the normal allele s, and the penetrances (i.e. probabilities of contracting the disease) for individuals of the three genotypes are

$$
\begin{array}{ccc}
\text{SS} & \text{Ss} & \text{ss} \\
x & \lambda x & 0
\end{array}
$$

Here x and λ are unknown parameters which we wish to estimate; normally $0 \leqslant \lambda \leqslant 1$. We write the population frequency of S as p, also an unknown parameter. Finally, we denote the recombination fraction between the S locus and the marker locus by R, the fourth and final unknown parameter.

The population prevalence of the disease is clearly $p^2 x + 2p(1-p)\lambda x$, which is sometimes effectively known and sometimes unknown. If it is known, parameters are estimated under the constraint that $\hat{p}^2 \hat{x} + 2\hat{p}(1-\hat{p})\hat{\lambda}\hat{x}$ equals the known prevalence, and in this case the number of unknown parameters is in effect reduced to three. We assume in the example below that the prevalence is known.

The data in each family is a triplet (r, g, i). Here, r is the number of affected children and g the number of affected parents. The symbol i describes the 'shared marker genotype' between affected sibs in the following way.

The mother and father in any family are heterozygous for different marker alleles, and this implies that two affected sibs can share two, one or no marker alleles. We denote these three possibilities ac/ac, ac/ad, ac/bd. There are four essentially different sharing patterns among three affected sibs, namely ac/ac/ac, ac/ac/ad, ac/ac/bd and ac/ad/bd. More complicated possibilities arise for four or more affected sibs. The possible patterns are listed by Ethier and Hodge (1985), and the third data point i for each family is simply an index number indicating which of these patterns applies for the affected sibs within that family.

We assume that families are ascertained through the children, so that the data D_1 relevant to ascertainment is simply the number r of affected children in the family. Similarly D_2, the data not affecting the ascertainment probability, is the pair (g, i). We now calculate, for a family of size s, the population probabilities

$Q_s(r)$ and $Q_s(r, g, i)$: the formulae are quite complex and are not reproduced here (see, for example, equation (2.24) in Ewens and Shute, 1986). Equation (4.3) shows that, under the AAF approach, the likelihood contribution for a family of size s with data (r, g, i) is

$$Q_s(r, g, i)/Q_s(r). \tag{6.1}$$

If complete ascertainment had been assumed, the likelihood contribution for this family would be

$$Q_s(r, g, i)\bigg/\bigg[\sum_{r \geqslant 1} Q_s(r)\bigg], \tag{6.2}$$

while if single ascertainment had been assumed, the appropriate expression is

$$Q_s(r, g, i)\bigg/\bigg[\sum_{r \geqslant 1} rQ_s(r)\bigg]. \tag{6.3}$$

We have noted earlier that asymptotically unbiased parameter estimation will arise if we condition this likelihood not only on r (as in (6.1)), but on further data as well. In some cases it might be thought that the number of affected parents might partly determine the ascertainment probability, and if this were so we would use as likelihood contribution

$$Q_s(r, g, i)/Q(r, g). \tag{6.4}$$

(This might also be the appropriate form of conditioning for reasons unconnected with ascertainment, for example if fertility differences exist between affected and unaffected parents.) Our aim is to establish bias and standard error properties of estimators derived from (6.1), (6.2), (6.3) and (6.4), which we describe respectively as AAF, AC (assuming complete ascertainment), AS (assuming single ascertainment), and CAP (conditional on all phenotypic information) estimators respectively.

We next turn to the 'data'. For any specific choice of p, x, λ and R and family size distribution there will exist, in a sample of size n arrived by a complete ascertainment sampling process, some calculable mean value for the number of families having each possible (s, r, i, g) combination. Indeed, under this form of sampling, each mean value is proportional to the population frequency of the appropriate combination. We now assume (in conformity with the discussion of the previous section) that the data are precisely at these mean values, and then find respectively the AC, AS, AAF and CAP estimators of the unknown parameters, together with their standard errors (subject to the constraint provided by the known prevalence). Suppose, for example, that the family size distribution is as depicted in Table 1. Then in a sample of n families, the mean number of families having each possible (s, r, i, g) combination are found by multiplying the various

W. J. Ewens

Table 1

Value of s	2	3	4
Probability	0.600	0.275	0.124

$p = 0.04$, $x = 0.50$, $\lambda = 0.08$, $R = 0.025$

values in Table 2 by n. (We emphasize that these mean values assume a complete ascertainment sampling procedure.) We now use these means as 'data' values, form likelihoods using (6.1), (6.2), (6.3) and (6.4) respectively and thus find MLE's of the four parameters (subject to the constraint imposed by the known prevalence value). It is found that the estimates are as shown in Table 3.

Table 2
Mean proportion of families having each (s, r, i, g) combination, assuming the numerical values in Table 1 and complete ascertainment

			g		
s	r	i	0	1	2
2	2	1	0.16085	0.01775	0.00048
		2	0.09649	0.01426	0.00047
		3	0.00758	0.00251	0.00011
3	2	1	0.19484	0.02027	0.00050
		2	0.11943	0.01606	0.00044
		3	0.00916	0.00283	0.00009
3	3	1	0.00672	0.00084	0.00003
		2	0.00590	0.00146	0.00007
		3	0.00027	0.00015	0.00001
		4	0.00073	0.00031	0.00003
4	2	1	0.15678	0.01546	0.00035
		2	0.09843	0.01215	0.00029
		3	0.00739	0.00217	0.00005
4	3	1	0.01042	0.00124	0.00004
		2	0.00906	0.00203	0.00008
		3	0.00040	0.00020	0.00002
		4	0.00108	0.00042	0.00003
4	4	1	0.00036	0.00005	0.00000
		2	0.00036	0.00010	0.00001
		3	0.00001	0.00001	0.00000
		4	0.00004	0.00004	0.00001
		5	0.00010	0.00006	0.00000
		6	0.00000	0.00000	0.00000
		7	0.00007	0.00003	0.00000
		8	0.00000	0.00000	0.00000

Table 3

Estimation procedure	Likelihood from	$\hat{p}$	$\hat{x}$	$\hat{\lambda}$	$\hat{R}$
AC	(6.2)	0.040	0.500	0.080	0.025
AS	(6.3)	0.045	0.401	0.090	0.017
AAF	(6.1)	0.040	0.500	0.080	0.025
CAP	(6.4)	0.040	0.500	0.080	0.025

As we expect, all parameters are estimated correctly under the AC (assuming complete ascertainment) procedure, since the 'data' are at 'perfect' complete ascertainment values. The estimates derived when single ascertainment is assumed are all biased: this again is as we expect since we know that if we make an incorrect assumption about the nature of the sampling scheme, estimates of genetic parameters will be biased. Finally, AAF and CAP estimates are exactly at the correct values, as theory predicts.

The comments in the previous paragraph concern the potential bias in estimators. It is equally important to evaluate the standard errors of unbiased estimators. The information matrix approximation to these standard errors ($\times \sqrt{n}$) is depicted in Table 4.

Table 4

Estimation procedure	$\hat{p}$	$\hat{x}$	$\hat{\lambda}$	$\hat{R}$
AC	0.164	2.033	0.269	0.300
AAF	0.181	2.798	0.316	0.332
CAP	0.699	13.524	0.742	0.728

We note that, as theory predicts, the AC estimates uniformly have the smallest standard errors. However the AAF method leads to estimators with standard errors not substantially in excess of these. On the other hand the CAP method leads to estimates with quite large standard errors: the reason for this is that in conditioning on r and g, a very large proportion of the information in the data concerning the parameters, and in particular concerning x, is discarded.

We have assumed above that the 'data' derive from a complete ascertainment sampling procedure. We now suppose that the 'data' derive from a single ascertainment procedure. Probabilities for each (s, r, i, g) combination under this assumption are given in Table 5, and from these we obtain, in a sample of n families, the mean number of families having each (s, r, i, g) combination.

The likelihood is now formed using the values in Table 5 as 'data' and, respectively, (6.2), (6.3), (6.1) and (6.4). The estimates are given in Table 6.

46 　　　　　　　　　　　　　　　　*W. J. Ewens*

Table 5
Mean proportion of families having each (s, r, i, g) combination, assuming the numerical values in Table 1 and single ascertainment

			g		
s	r	i	0	1	2
2	2	1	0.15738	0.01737	0.00047
		2	0.09442	0.01396	0.00046
		3	0.00742	0.00245	0.00011
3	2	1	0.19065	0.01983	0.00049
		2	0.11686	0.01571	0.00043
		3	0.00896	0.00277	0.00009
3	3	1	0.00986	0.00124	0.00004
		2	0.00865	0.00125	0.00010
		3	0.00040	0.00022	0.00002
		4	0.00107	0.0046	0.00004
4	2	1	0.15340	0.01513	0.00034
		2	0.09631	0.01189	0.00028
		3	0.00723	0.00212	0.00005
4	3	1	0.01529	0.00182	0.00005
		2	0.01329	0.00298	0.00012
		3	0.00058	0.00029	0.00002
		4	0.00159	0.00062	0.00005
4	4	1	0.00070	0.00009	0.00000
		2	0.00071	0.00020	0.00001
		3	0.00003	0.00002	0.00000
		4	0.00009	0.00007	0.00001
		5	0.00019	0.00011	0.00001
		6	0.00000	0.00000	0.00000
		7	0.00013	0.00006	0.00001
		8	0.00001	0.00001	0.00000

Table 6

Estimation procedure	$\hat{p}$	$\hat{x}$	$\hat{\lambda}$	$\hat{R}$
AC	0.036	0.630	0.069	0.034
AS	0.040	0.500	0.080	0.025
AAF	0.040	0.500	0.080	0.025
CAP	0.040	0.500	0.080	0.025

We note now that, as expected, the AS (assuming single) method gives unbiased estimates and the AC (assuming complete) method leads to biased estimates. (Note also the almost symmetrical bias structure in Tables 3 and 6.) Both the AAF and CAP methods lead to unbiased estimates, as theory requires, verifying so far the claim that these methods apply irrespective of the ascertainment scheme.

Turning to standard errors, the values for unbiased estimators ($\times \sqrt{n}$) are shown in Table 7. Essentially the same comments concerning standard error comparisons apply as were made from Table 4.

Table 7

Estimation procedure	$\hat{p}$	$\hat{x}$	$\hat{\lambda}$	$\hat{R}$
AC	0.162	1.865	0.262	0.292
AAF	0.180	2.760	0.316	0.332
CAP	0.622	12.014	0.678	0.678

Finally we suppose that the data arose from a quadratic ascertainment procedure, that is one in which the probability of ascertainment of a family having r affected sibs is proportional to r^2. (We discuss below a real-world circumstance in which this form of sampling might arise.) Once again, appropriate 'data' points may be calculated (from Table 8), and likelihoods formed using these 'data' and, respectively, (6.2), (6.3), (6.1) and (6.4). The estimators are depicted in Table 9.

There are two points of interest in Table 9. First, both AC and AS estimators are biased, both in the same direction away from the true value, and with AC estimates having approximately twice the bias as AS estimators. The second observation, and one which is central to the main theme, is that no matter what the nature of the ascertainment process, AAF and CAP estimators continue to provide asymptotically unbiased estimators.

The standard errors of the AAF and CAP estimates ($\times \sqrt{n}$) are shown in Table 10. The calculations therein give a picture of the likely sizes of biases, and of the standard error comparisons, for one set of parameter combinations. In order to obtain a more complete picture, the same procedure should be carried out for a wide and representative set of arameter combinations. This is done by Shute and Ewens (1988a), and the broad conclusions described above continue to hold. These calculations allow an objective choice of ascertainment assumption to be made and used in the estimation procedure. If the investigator is quite certain of the nature of the ascertainment procedure, then of course that procedure is assumed in the estimation process and estimators having optimal properties will arise. If the investigator is less certain of the ascertainment procedure, he can assess the likely biases arising when an incorrect ascertainment assumption is made and perhaps decide to use the AAF approach. In acting this

Table 8
Mean number of families having each (s, r, i, g) combination, assuming the numerical values in Table 1 and quadratic ascertainment

			g		
s	r	i	0	1	2
2	2	1	0.15242	0.01682	0.00047
		2	0.09143	0.01351	0.00045
		3	0.00718	0.00238	0.00010
3	2	1	0.18463	0.01921	0.00047
		2	0.11317	0.01522	0.00042
		3	0.00868	0.00268	0.00009
3	3	1	0.01433	0.00179	0.00006
		2	0.001258	0.00311	0.00015
		3	0.00058	0.00032	0.00002
		4	0.00156	0.00066	0.00006
4	2	1	0.14856	0.01465	0.00033
		2	0.09327	0.01151	0.00027
		3	0.00700	0.00206	0.00005
4	3	1	0.02222	0.00264	0.00008
		2	0.01932	0.00433	0.00017
		3	0.00008	0.00043	0.00004
		4	0.00230	0.00090	0.00006
4	4	1	0.00136	0.00019	0.00000
		2	0.00136	0.00038	0.00004
		3	0.00004	0.00004	0.00000
		4	0.00015	0.00015	0.00004
		5	0.00038	0.00023	0.00000
		6	0.00000	0.00000	0.00000
		7	0.00027	0.00011	0.00000
		8	0.00000	0.00000	0.00000

Table 9

Estimation procedure	$\hat{p}$	$\hat{x}$	$\hat{\lambda}$	$\hat{R}$
AC	0.034	0.820	0.054	0.053
AS	0.036	0.645	0.068	0.037
AAF	0.040	0.500	0.080	0.025
CAP	0.040	0.500	0.080	0.025

Table 10

Estimation procedure	$\hat{p}$	$\hat{x}$	$\hat{\lambda}$	$\hat{R}$
AAF	0.194	2.705	0.321	0.337
CAP	0.626	10.336	0.607	0.628

way, he would be in a position analogous to a person deciding to use a non-parametric statistical procedure rather than a procedure which assumes a normal distribution for his data.

We have described in a previous section the 'multiple ascertainment' model, and shown that complete and single ascertainment may be regarded as limiting, or boundary, cases of this model. Thus if the investigator feels that this model is appropriate, then estimates of parameters found by assuming respectively complete and single ascertainment will provide bounds bracketing the multiple ascertainment estimates. However we have stressed also that the complete and single cases, while they do form bounds in the 'multiple ascertainment' model, by no means form bounds in general, and that in particular quadratic ascertainment lies outside the range defined by complete and single ascertainment. This is confirmed by the estimates in Table 9. It is thus a matter of some importance to determine how often, in practice, as ascertainment scheme such as the quadratic might arise. The following example provides a case strongly suggesting approximately quadratic ascertainment.

A series of Genetic Analysis Workshops has been held, more or less annually, using first artificial 'data' (such as that discussed above) and secondly real data to assess the validity of, and then conclusions deriving from, various computer packages designed to estimate genetic parameters in ascertainment sampling processes. In particular, the 1985 workshop (Bishop et al., 1986) was designed in part to assess the genetic basis of IDDM (insulin dependent diabetes mellitus), which is known to be determined in significant part by a susceptibility locus closely linked to HLA. The data for this workshop were derived from various investigators who obtained the data from physicians. It is quite possible, indeed probable, that different ascertainment processes were used first in the process of families contacting physicians, and secondly in the process of physicians transmitting the data for the workshop analysis. With these comments in mind, it is interesting to note the estimates of the parameters p, x, λ and R described above derived from the workshop data using AC, AS and AAF procedures respectively in Table 11.

The comparison with Table 9 is striking, since apart from the estimates of λ, which are effectively equal for all three estimation procedures, the patterns of the estimates of the remaining parameters are the same in the two cases. Thus, it is at least plausible that, for the Workshop data, the true ascertainment procedure was approximately quadratic and that complete and single ascertainment estimates do not bracket the true parameter values. In practice, a quadratic ascertain-

Table 11

Estimation procedure	$\hat{p}$	$\hat{x}$	$\hat{\lambda}$	$\hat{R}$
AC	0.05	0.87	0.01	0.19
AS	0.08	0.61	0.00	0.18
AAF	0.11	0.32	0.00	0.16

ment scheme could arise by an approximately single ascertainment process between families and physicians, followed by a further approximately single ascertainment scheme between physicians and workshop. It is indeed known that various physicians did deliberately select families with large numbers of affected children as being of interest for workshop purposes. This example agrees with the observation made from the estimators in Table 11. This at least suggests the value of an AAF procedure in a complex and largely unknown ascertainment process such as that described above.

We conclude this section with a comment on the 'single ascertainment' likelihood contribution (6.3). The denominator in (6.3) is the mean number of affected children in any family, which in turn is proportional to the probability that an individual taken at random is affected. This has several interesting implications. First, if ascertainment were through parents as well as children, but single ascertainment continued to apply, the denominator in (6.3) would still be appropriate. This indicates a greater generality for the use of (6.3) than its initial derivation might suggest. The same comment applies for pedigrees (Cannings and Thompson, 1977). Second, in the simple model described in an earlier section, the probability that any 'at risk' person is affected is a constant value p, so that use of (6.3) leads to a likelihood contribution for each family in which the simple random sampling likelihood is divided by p. This is observed in (2.5). In general, use of (6.3) leads to corrections only marginally more complex than this: for example, in the case considered earlier in this section the divisor is simply the population prevalence $p^2 x + 2p(1-p)\lambda x$. This leads to significant computational simplification in the calculation of AS parameter estimates.

7. Further comments for nuclear families

In this section we take up a number of points associated with the ascertainment of nuclear families.

The first point concerns the 'data relevant to ascertainment'. In the preceding sections it was assumed that ascertainment was through children, so that the 'data relevant to ascertainment' is the number of affected children in the family. If, on the contrary, ascertainment is through parents, then the data relevant to ascertainment becomes either the number of affected parents or, in cases where it is felt that maternal and paternal affectedness affect the probability of ascertainment differently, the pair $\{m, f\}$, where m describes the mother's affectedness status and f the father's. In the latter case the likelihood contribution from each family is

$$P(\text{data})/P\{m, f\}\,. \tag{7.1}$$

Complications arise when ascertainment is felt to depend not only on the affectedness status of parents, but also on the number of children affected. The AAF likelihood contribution from each family would now be

$$P(\text{data})/P\{m, f, r\}\,, \tag{7.2}$$

with r, as above, the number of affected children in the family.

What is the corresponding likelihood when the concept of a proband is defined and formulae such as (2.16) and (2.17) are used? At least one computer program [POINTER, Lalouel and Morton (1981)] uses a likelihood of the form

$$P(\text{data})/[P\{m, f\} \cdot \{1 - (1 - \pi p)^s\}]\,. \tag{7.3}$$

However, it is easy to see that a formula such as (7.3) is applicable only if the probability of ascertainment of a family is of the form

$$\alpha(m, f)[1 - (1 - \pi)^r]\,, \tag{7.4}$$

where $\alpha(m, f)$ is an unspecified probability depending on maternal and paternal affectedness status and r is the number of affected sibs. But (7.4) is an unnatural probability for this case, being in the form of a product: when ascertainment is potentially through parents or children, it is natural to replace (7.4) by a formula more in the form of a sum. It is easy to produce realistic ascertainment schemes for which use of (7.3), implying the assumption (7.4), leads to substantially biased estimation.

The second point also concerns the 'data relevant to ascertainment'. It is assumed throughout this paper that this is simply the number of affected children in each family. In practice this might be an oversimplification. For example, birth order effects might be relevant (Stene, 1978), and in such cases ascertainment corrections more complex than those developed here will be necessary.

The third point concerns the 'final use' of the data. The data might well initially have been collected by a well-defined ascertainment procedure, for example complete ascertainment, but then only those sampled families with, say, two or more affected children used in the data analysis. This might be the case where shared HLA haplotype information of affected sibs forms an important part of the data, as in the example of the previous section. Here the ascertainment correction must take this 'final use' into account: it is not sufficient simply to make a correction only for the initial complete ascertainment process.

The fourth point concerns the concept of the 'proband sampling frame' (PSF) of Elston and Sobel (1979), namely the set of all individuals who could be probands (in cases where the proband concept applies). This is very close to 'the data relevant to ascertainment'. Now it is possible that the PSF is misspecified by the investigator. Elston and Sobel show that, under single ascertainment, this will not lead to biased estimation. The reason for this may be seen by considering (2.5). Here correcting for ascertainment is carried out simply by dividing the 'simple-random-sampling' likelihood contribution by the probability that an individual is affected, and this procedure is the same whatever the PSF.

The next point concerns the biases that can arise in using (2.8) or (2.9) in inappropriate circumstances, as calculated using (2.12) and (2.15). Davie (1979) has found an estimator which, in the 'multiple ascertainment' model (2.16), is asymptotically unbiased whatever the value of π. Assuming a sample of n families, each having s children, this estimator is

$$\hat{p} = (R - J)/(ns - J), \tag{7.5}$$

where J is the number of families with one proband and R the number of affected children in the sample. From a statistical point of view this is an interesting estimator, since (2.19) shows that R and Q (where Q is the total number of probands) are jointly sufficient statistics for p and π, and (7.5) does not use these. Thus we should be able to improve on (7.5) by using, instead of J, its expected value given R and Q. However the resulting estimator is very complex and not useful in practice.

Finally, in the example of the previous section, parental affectedness information was used as part of the data. If possible, this information should be used, as the gain in precision of the estimators is substantial (Ewens and Spielman, 1985, Table 3).

8. Continuous measurements

So far we have assumed that the phenotypic observation made an each person is the discrete character affected/not affected. In some cases, however, the character observed is some continuous quantitative measure such as blood pressure. Here a more complex theory for ascertainment correction is needed.

We suppose some specified threshold T is given, and assume that a family can be ascertained only if at least one sib has measurement exceeding T. Complete ascertainment applies if the probability of ascertainment of a family is a constant, irrespective of the number of sibs having measurement exceeding T (providing that at least one sib does). Single ascertainment arises when this probability is proportional to the number of children having measurement exceeding T. Other ascertainment rules can of course apply and further, as in the discrete character case, the ascertainment process might not be well defined and an ascertainment-assumption-free procedure might be thought desirable. We establish below what this procedure is.

For convenience, we suppose all families in the sample have s sibs, with respective measurements $x_1, x_2, \ldots, x_s$ for the character of interest. (The more realistic case, where family sizes are not identical, is easily handled using obvious extensions of the theory given below.) We suppose that, in the population at large, these observations have joint density function $f(x_1, x_2, \ldots, x_s)$, and that the measurement for any individual sib has density function $f(x)$, assumed to be the same for each sib.

Suppose for simplicity that the parental measurements are not available and that there is no genetic information: in other words, the data from which estimates are to made consists only of the values $(x_1, x_2, \ldots, x_s)$.

The most important observation to make is that the likelihood from which parameters are estimated, under any ascertainment procedure, is still (2.2). Thus, the likelihood contribution from any given family is

$$f(x_1, \ldots, x_s)/\Pr(A), \tag{8.1}$$

where $x_1, \ldots, x_s$ are the measurements in this family and $\Pr(A)$ is the probability of ascertainment of a family having s children.

$\Pr(A)$ will depend on the nature of the ascertainment process. Under complete ascertainment, $\Pr(A)$ is simply the probability that at least one measurement exceeds T, and so the appropriate likelihood is

$$f(x_1, \ldots, x_s)/\Pr(\text{at least one } x_i \text{ exceeds } T). \tag{8.2}$$

The denominator is found by a (possibly complicated) multiple integration. Under single ascertainment, however, $\Pr(A)$ is proportional to the number of measurements exceeding T. This leads to

$$\Pr(A) = \text{const} \sum_i i \Pr(i \text{ measurements exceed } T\}$$

$$= \text{const} \{\text{mean number of measurements exceeding } T\}$$

$$= \text{const} \{\Pr(x > T)\}, \tag{8.3}$$

where x is the measurement of a randomly chosen sib. This leads to the single ascertainment likelihood contribution

$$f(x_1, \ldots, x_s) \bigg/ \left\{ \int_T^\infty f(x)\,\mathrm{d}x \right\}. \tag{8.4}$$

Under the AAF approach, theory analogous to that for discrete characters shows that for a family where exactly r sibs have measurement exceeding T, the likelihood contribution becomes

$$f(x_1, \ldots, x_s)/\Pr\{r \text{ measurements exceed } T\}. \tag{8.5}$$

As with (8.2), the denominator is found from a possibly complicated integration.

The likelihood contributions (8.2), (8.4) and (8.5) are all obtained by conditioning the data in each family on the fact of ascertainment of each family, using the basic likelihood contribution (2.2). They therefore lead to optimal estimation under the respective ascertainment schemes assumed. For continuous data, in contrast to discrete affected/not affected data, conditioning on observed values of measurements is also possible. Thus under single ascertainment, the likelihood

$$f(x_1, \ldots, x_s)/f(x_1) \tag{8.6}$$

also leads to asymptotically unbiased estimates (Rao et al., 1988).

Here, if single ascertainment is viewed as the limiting case of the 'multiple ascertainment' process as $\pi \to 0$, so that an ascertained family will have exactly one proband, x_1 is the measurement of that proband. However, the definition of single ascertainment is broader than this: single ascertainment simply assumes that the probability of ascertainment of a family is proportional to the number of sibs having measurement exceeding T, and the concept of a proband need not apply. Here x_1 can be taken as the measurement of an arbitrarily chosen sib whose measurement exceeds T.

What is the relation between (8.6) and (8.4)? We may write (8.4) as

$$[f(x_1, \ldots, x_s)/f(x_1)] \cdot \left[f(x_1) \middle/ \left\{ \int_T^\infty \{f(x)\,dx\} \right\} \right]. \tag{8.7}$$

The second factor is simply the density function of x_1. From this it is easy to see (Rao et al., 1988) that the asymptotic variance of estimators using (8.6) exceeds that of estimators using (8.4). Note that no consistent statement can be made about bias properties, and under single ascertainment both (8.4) and (8.6) lead to asymptotically unbiased estimates.

Despite the variance comparison just noted, one might prefer to use (8.6) rather than (8.4) in practice, since in practice the threshold T might not be defined precisely and in such a case biases will arise in using (8.4). This has been pointed out by Young et al. (1988), and their argument is strengthened by the fact that even when T is specified exactly, the increase in the variance of estimators in using (8.6) rather than (8.4) is usually quite small.

Note that (8.6) is appropriate only for single ascertainment and that its use in other cases leads to biased estimation. Further, apart from use of (8.6) under single ascertainment, no form of ascertainment correction which conditions only on measured values is valid, and in particular conditioning on all measurements $x_1, \ldots, x_r$ in any family which exceed T, for which the likelihood contribution is

$$f(x_1, \ldots, x_s)/f(x_1, \ldots, x_r), \tag{8.8}$$

is never valid for any ascertainment procedure.

On the other hand there is one form of conditioning which is rather similar to (8.8) and which always leads to asymptotically unbiased estimates for any ascertainment procedure. The AAF likelihood contribution (8.5) conditions on the number r of measurements in each family which exceed T, and assuming r is the 'data relevant to ascertainment', this leads to asymptotically unbiased estimation whatever the ascertainment procedure. If we condition not only on r, but also on the actual measurements $x_1, \ldots, x_r$ exceeding T, asymptotically unbiased estimation will continue to hold, as the comments preceding esuation (4.4) imply. The likelihood contribution for this form of conditioning is

$$f(x_1, \ldots, x_s) \Big/ \left[\int \cdots \int f(x_1, \ldots, x_s) \, dx_{r+1} \cdots dx_s \right]. \tag{8.9}$$

Here the $(s - r)$-fold integration in the denominator is over all values of $x_{r+1} \ldots x_s$ less than T. Since more information is conditioned on in using (8.9) rather than (8.5), less is subsequently available for estimation, so that estimators deriving from (8.9) have higher variances than those deriving from (8.5).

It is interesting to check the theory given above by a numerical example. The following is taken from Ewens and Green (1988). Suppose that all families in a sample have two children ($s = 2$) and that the joint density function of the measurements of the children in any family is

$$f(x_1, x_2) = \tfrac{1}{2}\theta^3 (x_1 + x_2) \exp[-\theta(x_1 + x_2)], \tag{8.10}$$

where $x_1, x_2 > 0$ and θ is an unknown parameter. For this density function the two marginal density functions are identical, although x_1 and x_2 are not independent; this is what we would expect in practice. Apart from these considerations, (8.10) is chosen purely for computational simplicity.

It is easy, by using standard probabilistic methods, to generate a large number of pairs of observations, each pair having density function (8.10), for any assumed value of θ. It is then possible to carry out an ascertainment process on each pair, given a predetermined threshold T, and then use as data only those observations from families which are 'ascertained'. The parameter θ can then be estimated by maximum likelihood, using likelihoods derived from (8.2), (8.4), (8.5), (8.6), (8.8) and (8.9). We expect (8.5) and (8.9) to give essentially unbiased estimates no matter what the true ascertainment process, (8.2) to give essentially unbiased estimators only if the true ascertainment process is complete ascertainment, (8.4) and (8.6) to give essentially unbiased estimates only for single ascertainment, and (8.8) never to give unbiased estimates. Further, we expect various inequalities between standard errors for asymptotically unbiased estimates; for example, the standard errors of estimates when (8.2) or (8.4) is appropriately used should be less than standard errors arising from (8.5), standard errors of estimation derived from use of (8.4) should be less than those deriving from (8.6) and those deriving from (8.5) less than those deriving from (8.9).

Three 'true' ascertainment schemes were used to generate the data, namely complete, single and quadratic ascertainment. There are thus eighteen combinations of three 'true' ascertainment schemes and six estimation methods.

Table 12

Empirical estimates ($\pm$ one standard error) of unbiased estimators of the parameter θ when $T = 2$. Data from three different ascertainment procedures (complete, single and quadratic) are used (number of families as indicated), and estimates from six different likelihoods are considered

	True ascertainment process		
Likelihood from	Complete (91 561 families)	Single (54 502 families)	quadratic (29 290 families)
(8.2)	0.9997 (± 0.0020)	0.9199	0.8153
(8.4)	1.0815	1.0005 (± 0.0025)	0.8939
(8.5)	1.0005 (± 0.0022)	1.0003 (± 0.0031)	0.9835 (± 0.0036)
(8.6)	1.1762	1.0004 (± 0.0035)	0.8189
(8.8)	1.5479	1.5508	1.5456
(8.9)	1.0040 (± 0.0023)	1.0100 (± 0.0035)	0.9997 (± 0.0078)

Table 12 gives the estimates of θ (when the true value is 1) arising from the large-scale simulation described by Ewens and Green (1988) for all eighteen combinations, together with the number of families from which each estimate was made and the estimated standard error of asymptotically unbiased estimates.

It will be noted that all the theoretical predictions made above are verified by the values in this table, especially the fact that (8.8) never leads to unbiased estimates while (8.5) and (8.9) do whatever the ascertainment process.

The density function (8.10) was chosen purely for mathematical convenience, and it is appropriate to mention briefly the bivariate normal case. For simplicity again, suppose that all families in the sample have two children whose measurements have a bivariate normal distribution with means μ, variances σ^2 and correlation ρ. Our aim is to estimate μ using data from ascertained families where at least one sib has measurement exceeding T.

Perhaps the case of most interest is that of single ascertainment, where we can compare the variances of estimators deriving respectively from (8.4) and (8.6). Using (8.4), the likelihood contribution from any family is

$$\text{const} \exp[-\tfrac{1}{2}[u^2 - 2\rho uv + v^2]/(1 - \rho^2)]/N\{(T - \mu)/\sigma\}, \qquad (8.11)$$

where $u = (x_1 - \mu)/\sigma$, $v = (x_2 - \mu)/\sigma$, and

$$N(z) = \int_z^\infty n(z)\, dz, \qquad (8.12)$$

$$n(z) = (1/\sqrt{2\pi}) \exp(-\tfrac{1}{2}z^2). \qquad (8.13)$$

Using (8.6), the likelihood contribution (using standard results for conditional normal distribution) is

$$\text{const} \exp[-\tfrac{1}{2}[x_2 - \mu - \rho(x_1 - \mu)]^2/(1 - \rho^2)\sigma^2]. \qquad (8.14)$$

The information (second derivative log likelihood) provided by (8.11) is

$$(1 - \rho)/[(1 + \rho)\sigma^2] + g[(T - \mu)/\sigma]\sigma^2 , \qquad (8.15)$$

and that provided by (8.14) is

$$(1 - \rho)/[(1 + \rho)\sigma^2] , \qquad (8.16)$$

where

$$g(z) = 1 + zn(z)/N(z) - [n(z)/N(z)]^2 . \qquad (8.17)$$

Numerical values of $g(z)$ are given in Table 13, and these show that $g(z)$ is always positive. Thus, as theory has predicted, smaller estimation variances arise using (8.11) rather than (8.14). (On the other hand, the remarks of Young et al. (1988) referred to above should be kept in mind in making this comparison.) Note that $g(z) \to 0$ as $z \to \infty$, showing that for large T the two methods of estimation have asymptotically equivalent variance properties: this occurs because for large T, $x_1 > T$ increasingly implies $x_1 \approx T$. Note also that as $z \to -\infty$, $g(z) \to 1$. Thus, the expression (8.15) approaches $2/[(1 + \rho)\sigma^2]$, confirming the fact that with no ascertainment requirement, the variance of $\bar{x}$ is $\frac{1}{2}(1 + \rho)\sigma^2$.

Table 13
Values of $g(z)$ for various z (see equation (8.17) for definitions)

z	$g(z)$
$\to -\infty$	$\to 1$
0	0.364
1	0.199
2	0.114
3	0.071
4	0.047
$\to +\infty$	$\to 0$

9. Pedigrees

We conclude with some brief remarks about pedigrees.

In some cases, pedigrees are ascertained in an ascertainment process, rather than simply nuclear families. Ascertainment corrections are in theory carried out in exactly the same way as for nuclear families: thus, (2.2) and (2.3) apply equally for pedigrees if we simply replace 'family' by 'pedigree' in the surrounding sentence. Similarly, if we can identify the data D_1 relevant to ascertainment in each pedigree, then use of the likelihood contribution (4.3) will lead to ascertainment-assumption-free estimation for pedigrees. provided we re-interpret the likelihood Q to apply for pedigrees rather than families.

In practice, however, pedigrees present problems far more formidable than do families. The calculations are usually far more complicated than for families. But the main difficulty is more a conceptual one: problems arise for pedigrees which do not arise for families, and here we focus on those connected with ascertainment-assumption-free estimation.

The starting point in AAF estimation is that we can identify the data D_1 relevant to ascertainment, and this will often in practice be a far more difficult task for pedigrees than for families. For families we have assumed that ascertainment is through the children; for pedigrees, with several generations and several branches represented in each pedigree, and with individuals in the same generation but in different branches being of possibly quite different ages, the concept of 'the children in the pedigree' might not be clear out. Another problem arises with geographical location. Only part of a pedigree might live in the sampling area and the data relevant to ascertainment can only comprise appropriate individuals in those parts of the pedigree who do live in that area.

Shute and Ewens (1988b) take up various aspects of this 'geographical' problem. Suppose, for example, that all branches of the pedigree live in the sampled area but the investigator (or his computer program) assumes that only certain branches of the pedigree do so. Then the data relevant to ascertainment is the total number of affected children in the pedigree, but the investigator would incorrectly assume that only the number of affected children in a part of the pedigree is relevant. Thus in conditioning on these data he conditions on only a proportion of the total data relevant to ascertainment, and the theory given above then shows that the AAF procedure will no longer lead to unbiased estimation.

On the other hand, it might be the case that only part of the pedigree lives in the sampled area, but the investigator assumes that all the pedigree does. Using the AAF procedure, he will condition on the data relevant to ascertainment as well as further data and thus the discussion surrounding (4.4) shows that asymptotically unbiased estimation will arise. (Note that it makes a difference whether he conditions on the set of numbers of affected children in the various families in the pedigree, or simply the total number over all families in the pedigree. In the latter case the data relevant to ascertainment is not a simple subset of the data used and biased estimation will arise. This matter is discussed in Shute and Ewens, 1988b.) In the case of single ascertainment, the discussion concerning the proband sampling frame shows that asymptotically unbiased estimation arises even when the investigator incorrectly assumes all the pedigree lives in the sampled area.

We now turn the question of the biases that can arise when an incorrect assumption is made about the nature of the ascertainment procedure. In assessing these biases we assume that the investigator has correctly identified those families in the pedigree living in the sampled area. Assuming one or other specific form of ascertainment, a likelihood deriving from (2.3) will be used, with whatever is the appropriate formula for $\Pr(A_i)$. Under the AAF approach a likelihood deriving from (4.3) will be calculated. As is the case for nuclear families, biases will arise if a specific form of ascertainment is assumed incorrectly, while asymptotically unbiased estimates are obtained under the AAF approach. However the biases are somewhat less than those applying for nuclear families, the reason being that for pedigrees, a higher fraction of the data used is not directly connected with the ascertainment process. For the same reason the increase in standard error incurred by using the AAF procedure is smaller for pedigrees than for nuclear families.

10. Final note

The literature on ascertainment sampling is very large, and we have touched only a small proportion on all of it here, since we focus on procedures which attempt to overcome biases arising through an incorrect specification of the ascertainment procedure. The References list a wider range of papers than those directly referred to above, and give a more balanced view of the entire ascertainment literature.

Acknowledgement

I am very grateful for the good advice and many useful suggestions offered by Robert Elston.

References

Bailey, N. T. J. (1951a). The estimation of the frequencies of recessives with incomplete multiple selection. *Ann. Eugen.* **16**, 215–222.

Bailey, N. T. J. (1951b). A classification of methods of ascertainment and analysis in estimating frequencies of recessives in man. *Ann. Eugen.* **16**, 223–225.

Berger, A. and Gold, R. L. (1967). On estimating recessive frequencies from truncated samples. *Biometrics* **23**, 356–360.

Bishop, D. T., Falk, C. T. and MacCluer, J. W. (1985). Proceedings of Genetic Analysis Workshop IV. *Genet. Epidem.* **1**, 1–406.

Boehnke, M. and Greenberg, D. A. (1984). The effects on conditioning on probands to correct for multiple ascertainment. *Am. J. Hum. Genet.* **36**, 1298–1308.

Cannings, C. and Thompson, E. A. (1977). Ascertainment in the sequential sampling of pedigrees. *Clinical Genetics* **12**, 208–212.

Davie, A. M. (1979). The 'singles' method for segregation analysis under incomplete ascertainment. *Ann. Hum. Genet.* **42**, 507–512.

Elandt-Johnson, E. A. (1971). *Probability Models and Statistical Methods in Genetics*. Wiley, New York.

Elston, D. C. and Sobel, E. (1979). Sampling considerations in the gathering and analysis of pedigree data. *Am. J. Hum. Genet.* **31**, 62–69.

Elston, R. C. and Stewart, J. (1971). A general model for the genetic analysis of pedigree data. *Hum. Hered.* **21**, 523–542.

Ethier, S. N. and Hodge, S. E. (1985). Identity-by-descent analysis of sibship configurations. *Am. J. Hum. Genet.* **22**, 263–272.

Ewens, W. J. (1982). Aspects of parameter estimation in ascertainment sampling schemes. *Am. J. Hum. Genet.* **34**, 853–865.

Ewens, W. J. (1984). Conditional and unconditional likelihood solutions. *Am. J. Hum. Genet.* **35**, 232–233.

Ewens, W. J. and Asaba, B. (1984). Estimating parameters in ascertainment sampling schemes: numerical results. *Biometrics* **40**, 367–374.

Ewens, W. J., Hodge, S. E. and Foo Hooi Ping (1986). The effect of a known family-size distribution on the estimation of genetic parameters. *Am. J. Hum. Genet.* **38**, 555–566.

Ewens, W. J. and Shute, N. C. E. (1986). A resolution of the ascertainment sampling problem. I. Theory. *Theoret. Pop. Biol.* **30**, 388–412.

Ewens, W. J. and Spielman, R. S. (1985). Statistical properties of macimum likelihood estimators of parameters of HLA-linked diseases. *Am. J. Hum. Genet.* **37**, 1172–1191.

Ewens, W. J. and Green, R. M. (1988). A resolution of the ascertainment sampling problem. IV. Continuous phenotypes. *Genet. Epidem.* **5**, 433–444.

Fisher, R. A. (1934). The effect of methods of ascertainment upon the estimation of frequencies. *Ann. Eugen.* **6**, 13–25.

Gart, J. J. (1968). A simple, nearly efficient alternative to the simple sib method in the complete ascertainment case. *Ann. Hum. Genet.* **31**, 283–291.

Greenberg, D. A. (1986). The effect of proband designation on segregation analysis. *Am. J. Hum. Genet.* **39**, 329–339.

Haldane, J. B. S. (1932). A method for investigating recessive characters in man. *J. Genet.* **25**, 251–255.

Haldane, J. B. S. (1938). The estimation of the frequencies of recessive characters in man. *Ann. Eugen.* **8**, 255–262.

Hodge, S. E. (1985). Family-size distribution and Ewens' equivalence theorem. *Am. J. Hum. Genet.* **37**, 166–177.

Hodge, S. E. (1988). Conditioning on subsets on the data applications to ascertainment and other genetic problems. *Am. J. Hum. Genet.* **43**, 364–373.

Lalouel, J. M. and Morton, N. E. (1981). Complex segregation analysis with pointers. *Human Heredity* **31**, 312–321.

Li, C. C. and Mantel, N. (1968). A simple method for estimating the segregation ratio under complete ascertainment. *Am. J. Hum. Genet.* **20**, 61–81.

Meyers, D. A. and Murphy, E. A. (1979). A unified theory of ascertainment bias. *Am. J. Hum. Genet.* **31**, 139A.

Morton, N. E. (1959). Genetic tests under incomplete ascertainment. *Am. J. Hum. Genet.* **11**, 1–16.

Morton, N. E. (1984). Trials of segregation analysis by deterministic and macro simulation. In: A. Chakravarti, ed., *Human Population Genetics: The Pittsburgh Symposium*. Van Nostrand, New York.

Rao, D. C., Wette, R. and Ewens, W. J. (1988). Multifunctional analysis of family data ascertained through truncation: A comparative evaluation of two methods of statistical inference. *Am. J. Hum. Genet.* **42**, 506–515.

Risch, N. (1984). Segregation analysis incorporating linkage markers. Part I: Single-locus models with an application to Type 1 diabetics. *Am. J. Hum. Genet.* **36**, 363–386.

Shute, N. C. E. (1988). Statistical and ascertainment problems in human genetics. Ph.D. Thesis, Monash University.

Shute, N. C. E. and Ewens, W. J. (1988a). A resolution of the ascertainment sampling problem. Part II: Generalizations and numerical results. *Am. J. Hum. Genet.* **43**, 374–386.

Shute, N. C. E. and Ewens, W. J. (1988b). A resolution of the ascertainment sampling problem. Part III: Pedigrees. *Am. J. Hum. Genet.* **43**, 387–395.

Stene, J. (1970). Analysis of segregation patterns between sibships within families ascertained in different ways. *Ann. Hum. Genet.* **33**, 261–283.

Stene, J. (1977). Assumptions for different ascertainment models in human genetics. *Biometrics* **33**, 523–527.

Stene, J. (1978). Choice of ascertainment model. Part I: Discrimination between single-proband models by means of birth order data. *Ann. Hum. Genet.* **42**, 219–229.

Stene, J. (1979). Choice of ascertainment model. Part II: Discrimination between multi-proband models by means of birth order data. *Ann. Hum. Genet.* **42**, 493–505.

Stene, J. (1989). The incomplete, multiple ascertainment model: Assumptions, applications and alternative models. *Genet. Epidem.* **6**, 247–251.

Thompson, E. A. and Cannings, C. (1979). Sampling schemes and ascertainment. In: C. F. Sing and M. Skolnick, eds., *Genetic Analysis of Common Diseases: Applications to Predictive Factors in Coronary Disease*. Liss, New York.

Thompson, E. A. and Edwards, A. W. F. (1984). The non-equivalence of likelihood and conditional-likelihood solutions in multinomial sampling. *Am. J. Hum. Genet.* **35**, 229–232.

Tiwari, J. L. and Terasaki, P. I. (1985). *HLA and Disease Association*. Springer-Verlag, Berlin.

Weinberg, W. (1912a). Further contributions to the theory of heredity. Part 4: On methods and sources of error in studies on Mendelian ratios in man. *Archiv. für Rassen- und Gesellschaftsbiologie* **9**, 165–174.

Weinberg, W. (1912b). Further contributions to the theory of heredity. Part 5: On the inheritance of the predisposition to blood disease with methodological supplements to my sibship method. *Archiv. für Rassen- und Gesellschaftsbiologie* **9**, 694–709.

Weinberg, W. (1928). Mathematical foundations of the proband method. *Zeitschrift für Induktive Abstammungs- und Vererbungslehre* **48**, 179–228.

Winter, R. M. (1980). The estimation of phenotype distributions from pedigree data. *Am. J. Hum. Genet.* **7**, 537–542.

Young, M. R., Boehnke, M. and Moll, P. P. (1988). Correcting for single ascertainment by truncation for a quantitative trait. *Am. J. Hum. Genet.* **43**, 105–108.

C. R. Rao and R. Chakraborty, eds., *Handbook of Statistics, Vol. 8*
© Elsevier Science Publishers B.V. (1991) 63–80

3

Statistical Considerations in Applications of Path Analysis in Genetic Epidemiology

D. C. Rao

1. Introduction

The method of path analysis, developed by the pioneer population geneticist Sewall Wright over 60 years ago, is a form of structural linear regression analysis of standardized variables whose purpose is two-fold: to explain the interrelationships among a given set of variables, and to evaluate the relative importance of varying causes influencing a certain variable of interest (Wright, 1921, 1968, 1978; Li, 1975). Although path analysis can be pursued strictly through structural equations, path diagrams, specifying the proposed structural relationships schematically, are often found more helpful in determining the internal consistency of a given model and its limitations. Under a specific model, correlations among the random variables can be derived either by taking mathematical expectations of products of (standardized) variables, or directly from the path diagram following a simple set of tracing rules (e.g., Li, 1975). The gist of the method of path analysis consists of a comparison of these model-based correlations with the actual data. Such model fitting needs appropriate statistical methods for obtaining consistent and efficient estimates of the parameters of the model, as well as for enabling tests of specific null hypotheses.

Although path analysis failed to attract much attention of the statistical community, perhaps due to the lack of a rigorous mathematical framework and due to some degree of subjective judgment involved, it has found two major areas of application: social sciences (e.g., Goldberger and Duncan, 1973) and genetic epidemiology (e.g., Rao et al., 1984). Whereas parameter estimation under preconceived structural models dominated the applications in social sciences, both hypothesis testing and estimation of parameters are emphasized in genetic epidemiology applications. In typical social science applications (e.g., Hope, 1984), causal relations among variables are often uncertain, and overinterpretation of the data based on presumed relationships have led to much, often justified,

This work was partly supported by N.I.H. and N.I.M.H. Grants GM-28719, HL-33973, and MH-31302. The author is grateful to Dr. Reimut Wette for performing the simulations.

criticism (e.g., Freedman, 1987). Although inappropriate applications are cause for much concern, this should not lead to an undue criticism of the method itself as was offered by Freedman (1987). It is possible to circumvent some of the limitations in such studies by considering a series of alternative causal structures and by falsifying some structures as being inconsistent with the data (Rao et al., 1977). It appears that lack of reliance on hypothesis tests in social science applications has attracted much of the criticism.

Although uncertainties pertain to some of the causal relationships considered in genetic epidemiology applications, fortunately, certain standard principles do exist enabling unambiguous causal relationships in other parts of the model. For example, genes 'cause' a certain variable like height, and parental genes determine those of children, and not vice versa. Such standard principles provide a relatively strong foundation for application of path analysis in genetic epidemiology. Development of path models in genetic epidemiology has evolved with progressive refinements based on extensions of theory as well as experience with earlier applications (Cloninger et al., 1983). We shall confine our discussion to a recent model encapturing the standard features of most models, that has also been found to be sufficient in most recent applications.

2. A contemporary path model

The primary goal of path analysis in genetic epidemiology is a resolution of genetic and environmental effects on the given variable, and tests of hypotheses pertaining to the model.

2.1. The basic model for an individual

First consider the basic linear additive model for a single individual involving three causal variables each with zero mean:

$$P^* = G^* + C^* + R^*,\tag{2.1}$$

where P^* is the variable of interest, called the phenotype (e.g., blood pressure), G^* is the relevant genotype, assumed to be polygenic with additive gene action, C^* is the relevant 'familial environment', also called the 'transmissible environment', and R^* is the residual. All three causes, assumed to be uncorrelated with each other, act additively to produce the phenotype. The total phenotypic variance is given by

$$\sigma_p^2 = \sigma_g^2 + \sigma_c^2 + \sigma_r^2\tag{2.2}$$

where σ_x^2 denotes the variance of X. The components of (2.2) are called variance components. Dividing both sides of (2.1) by σ_p yields the basic structural equation

$$P = hG + cC + rR,\tag{2.3}$$

where P, G, C, and R denote the standardized variables, and the standardized partial regression coefficients h, c, and r, also called *path coefficients*, are given by

$$h = \sigma_g/\sigma_p, \qquad c = \sigma_c/\sigma_p, \qquad r = \sigma_r/\sigma_p. \qquad (2.4)$$

Variance of P, obtained from (2.3), defines the *equation for complete determination*

$$h^2 + c^2 + r^2 = 1, \qquad (2.5)$$

the components of which provide the variance components relative to the total phenotypic variance. *Genetic heritability* (h^2) is defined as the proportion of the phenotypic variance explained by the genotype. Likewise, we define *cultural heritability* (c^2) as the proportion of the phenotypic variance explained by the familial environment. In this basic model, h^2 and c^2 are the two unknown parameters, since r^2 is readily obtained from (2.5). Finally, the path coefficients in (2.3) may be distinguished for different classes of individuals, such as parents and children, as considered in Section 2.3.

2.2. Environmental index

Extension of the basic model even to simple nuclear families consisting of parents and children will entail additional parameters. The number of parameters aside, the two basic parameters h^2 and c^2 cannot be resolved using phenotypic data alone in nuclear families. Therefore, so long as one wishes to fit such models to data on simple sampling units such as nuclear families, there is a need to generate additional data on the same family members. Keeping (2.3) in mind, it becomes necessary to generate some information on G or C, neither of which is directly measurable. For this reason, an *environmental index* (I) was introduced as an estimate of the familial environment (C), thus defining a second 'observation' on every family member (Morton, 1974; Rao et al., 1974, 1982). Creation of the index, however, requires actual observations on a set of environmental variables $(X_1, X_2, \ldots, X_n)$ relevant for the phenotype under study. For example, if blood pressure is the phenotype of interest, then relevant environmental variables would seem to be physical activity, dietary habits, smoking, alcohol consumption etc. Since the index is meant to be an estimate of C as it relates to P, I may be created by regressing P on the index variables (X):

$$P = \alpha + \sum_{i=1}^{n} \beta_i X_i + \varepsilon, \qquad (2.6)$$

and then by defining $I = \hat{P} = \hat{\alpha} + \sum \hat{\beta}_i X_i$, where $\hat{\alpha}$ and $\hat{\beta}$ are estimates, and only the significant terms may be retained. Therefore, creation of the environmental index for each individual requires that the X-variables be observed on every individual. In order not to distort the inference, it is necessary to measure as many *X*-variables as are known to be pertinent to the phenotype under study. However, as shown by the following structural equation, *I* is not assumed to be a perfect estimate of *C*:

$$I = iC + sS,\qquad(2.7)$$

where *C* and an uncorrelated residual *S* jointly determine *I*, with $i^2 + s^2 = 1$ (equation for complete determination of the index). Generation of an index for every member of a family introduces only one additional parameter *i*, but gives rise to many more correlations among 'observed' variables, thus making it possible to fit comprehensive models to nuclear family data. Finally, the path coefficient *i* may be distinguished among parents and children.

2.3. Contemporary path model for nuclear families

The basic model of Section 2.1 can be extended to any family structure, which may entail additional modelling assumptions. One particular model found to be sufficient for many studies of disease-related traits in nuclear families is presented in Figure 1. Since all residuals, those of phenotypes and indices, are assumed to be uncorrelated, they are not shown in the path diagram. The model involves a

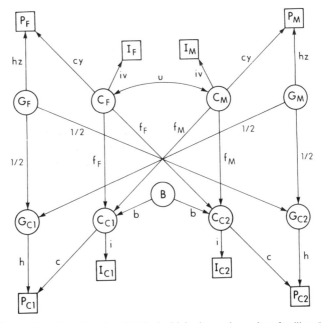

Fig. 1. Path diagram showing cultural and biological inheritance in nuclear families. *P* is phenotype, *G* is genotype, *C* is transmissible environment with index *I*, and *B* is non-transmitted common sibship environment. The subscripts F, M, C1, and C2 denote father, mother, and two children, respectively. Residuals are assumed to be uncorrelated, and are therefore not shown here.

total of ten parameters: three basic parameters for children (h, c, i), their analogues for parents (hz, cy, iv, where $z = y = v = 1$ will make the two sets of parameters equal), and four additional parameters for modelling similarities among the familial environments (C) of parents and children (u, f_F, f_M, and b). It is assumed that a parent is equally correlated with every child, and that all children are equally correlated pairwise. This assumption can, however, be relaxed if necessary.

Resemblance between the two parents is assumed to arise only on account of their correlated (familial) environments (u). Heritabilities are allowed to be different in the two generations: whereas h^2 and c^2 are the genetic and cultural heritabilities in children, they are h^2z^2 and c^2y^2 in parents. *Cultural transmission*, from parental C's to those of their children, is postulated so that fathers (f_F) and mothers (f_M) may contribute differently to their children. The familial environments of children within a family (C_{C1} and C_{C2}) are correlated partly due to cultural transmission from parents to children (f_F, f_M), and partly due to the effect (b) of a non-transmitted common sibling environment (B). Finally, the 'precision' of I as an estimate of C is allowed to be different in children (i) and parents (iv).

The model may be specified by the following set of structural equations for children (and similar ones for parents), the last components in each corresponding to the residuals:

$$P_{C1} (= P_{C2}) = hG_{C1} + cC_{C1} + r_1R_1 , \qquad (2.8a)$$

$$I_{C1} (= I_{C2}) = iC_{C1} + r_2R_2 , \qquad (2.8b)$$

$$G_{C1} (= G_{C2}) = \tfrac{1}{2}G_F + \tfrac{1}{2}G_M + r_3R_3 , \qquad (2.8c)$$

$$C_{C1} (= C_{C2}) = f_F C_F + f_M C_M + bB + r_4R_4 , \qquad (2.8d)$$

with $\text{Cov}(C_F, C_M) = u$. Since all variables are standardized, taking variances of these equations give rise to a set of equations for complete determination, such as the one in (2.5). Since the residual variances, such as r^2 in (2.5), have to be non-negative, these equations, in turn, give rise to a set of inequality constraints among the parameters:

$$h^2 + c^2 \leqslant 1, \qquad h^2z^2 + c^2y^2 \leqslant 1,$$

$$i^2 \leqslant 1, \qquad i^2v^2 \leqslant 1,$$

$$b^2 + f_F^2 + f_M^2 + 2uf_Ff_M \leqslant 1. \qquad (2.9)$$

Therefore, when fitting the model to data, one should seek only those parameter estimates that satisfy the inequality constraints in (2.9). Solutions in violation of the constraints are not valid.

For fitting the model to data, one needs to derive correlations, predicted from the model, among all 'observed' variables (P and I on each member). These may be derived by taking expectations of products of pairs of observed variables. For example, the correlation between P and I of a child may be derived from (2.8) as

$$E(P_{C1} \cdot I_{C1}) = E[(hG_{C1} + cC_{C1} + r_1 R_1)(iC_{C1} + r_2 R_2)]$$
$$= E[(cC_{C1})(iC_{C1})] + 0$$
$$= ci\, E(C_{C1}^2)$$
$$= ci \quad \text{since } V(C_{C1}) = E(C_{C1}^2) = 1\,.$$

Note that all the covariance terms on the right hand side above are zero. Alternatively, the correlations may be derived directly from Figure 1 by following a set of tracing rules (e.g., Li, 1975). Table 1 presents the 16 distinct correlations. Statistical methods necessary for fitting the model to data will be discussed in Section 3.

Table 1

Correlations for nuclear families predicted under the model (Figure 1)[a]

Pair of variables	Correlation (ρ)
(P_F, P_M)	$uc^2 y^2$
(I_F, I_M)	$ui^2 v^2$
(P_F, I_M) or (P_M, I_F)	$ucyiv$
(P_F, I_F) or (P_M, I_M)	$cyiv$
(P_{C1}, P_{C2})	$\frac{1}{2}h^2 + c^2\psi$
(I_{C1}, I_{C2})	ψi^2
(P_C, I_C)	ci
(P_{C1}, I_{C2}) or (P_{C2}, I_{C1})	$c\psi i$
(P_F, P_C)	$\frac{1}{2}h^2 z + c^2 y(f_F + uf_M)$
(I_F, P_C)	$civ(f_F + uf_M)$
(P_F, I_C)	$cyi(f_F + uf_M)$
(I_F, I_C)	$i^2 v(f_F + uf_M)$
(P_M, P_C)	$\frac{1}{2}h^2 z + c^2 y(f_M + uf_F)$
(I_M, P_C)	$civ(f_M + uf_F)$
(P_M, I_C)	$cyi(f_M + uf_F)$
(I_M, I_C)	$i^2 v(f_M + uf_F)$

[a] P = phenotype, I = index and subscripts F, M, and C refer to father, mother, and child, respectively. C1, C2 denote full sibs. $\Psi = b^2 + f_F^2 + f_M^2 + 2uf_F f_M$.

2.4. Temporal trends in family resemblance

Implicit in the path model is the assumption that the genetic (h) and familial environmental (c) effects are constant over time. Although for many phenotypes this is a reasonable assumption, varying gene action and transient environmental effects may be important for some phenotypes. A simple extension of the path model, incorporating time-dependent effects, may be considered for analysis of phenotypes that warrant temporal effects. For example, the basic structural equation in (2.3) may be extended as

$$P(A) = h(A)G + c(A)C + r(A)R , \qquad (2.10)$$

where each variable is made a function of the individual's age. It is possible to describe the temporal variability in $h(A)$ and $c(A)$ as continuous functions of age and the two basic parameters at birth (h and c). An important feature of this extension is that it enables one to test the null hypothesis of no temporal variation. Using such models, Province and Rao (1985b) were able to show significant trends in cultural heritability for systolic blood pressure. Theoretical details of the method can be found in Province and Rao (1985a, 1985b, 1988) and Province et al. (1989).

2.5. Modelling assumptions and possible limitations

Implicit in the path models are several assumptions. Linearity and additivity of effects, such as in (2.3), represent two fundamental assumptions. Cloninger et al. (1983) reviewed some empirical evidence that supports the practical adequacy of the fundamental assumptions as first-order approximations in many situations. Choice of scale is important for these assumptions to hold, and such a scale may not even exist for some variables.

Secondary assumptions include the absence of gene–gene (dominance and epistasis) and gene–environment interactions. In principle, it is possible to test for these assumptions, however, special study designs and large samples are required (Lathrop et al., 1984; McGue et al., 1989). It is assumed that the familial environment (C) partly determines the phenotype (P) and not vice versa. Likewise, it is also assumed that parental environments (C_F and C_M) determine those of the children. Although these assumptions appear to be reasonable for many situations, they are not necessarily adequate for others. The phenotype may also influence the environment, at least in a temporal order. For example, knowledge of high blood pressure may influence an individual to modify the subsequent environment. Likewise, parental phenotypes may influence the environment of the children (e.g., Rao et al., 1976; Fulker, 1988).

Additional modelling assumptions concern the resemblance between parents, and that involving indices. It has been assumed that parents resemble one another only due to their correlated environments (u). Likewise, it is assumed that the index is strictly an estimate of the familial environment only. Alternative assumptions are possible, and that using a wrong set of modelling assumptions can give rise to errors in inference (McGue et al., 1989).

Finally, recent applications of path analysis also make some distributional assumptions for the purposes of maximum likelihood analysis, as discussed below. Tests of hypotheses appear to be sensitive to moderate departures from normality (McGue et al., 1987).

3. Statistical method of analysis

Let us consider a sample of n independent nuclear families. Let X denote the vector of phenotypes and indices on all members of a particular family. The dimension of X varies among families, depending on the number of individuals in the family and on how many individuals the value of P and/or I is missing (unobserved). For a family with two parents and s children, and with no missing data, the dimension of X is $2s + 4$. For likelihood analysis, a certain distributional assumption about X will be necessary, the exact form of which will be discussed in later subsections. For now, let $f_i(X; \theta)$ denote the joint density function of X in the i-th family, with θ being the vector of unknown parameters. Denote the likelihood function by $\{L_i(\theta|X)\}$. Then, the total log-likelihood function for the entire sample of n independent families is

$$\ln L(\theta|X) = \sum_{i=1}^{n} \ln L_i(\theta|X), \tag{3.1}$$

which will be referred to simply as $\ln L$. The unknown parameters (θ) may be estimated by maximizing $\ln L$ numerically, with the aid of numerical optimization packages such as GEMINI and ALMINI (Lalouel, 1979, 1983) and NAG (Numerical Algorithms Group, 1982). Tests of specific null hypotheses on θ may also be performed using the likelihood ratio criterion. For example, let $\ln L_{k+w}$ denote the numerical value of $\ln L$ when $k + w$ parameters are estimated. Let $\ln L_k$ be the value when only k of the $k + w$ parameters are estimated, fixing the w other parameters under a null hypothesis. Then, the likelihood ratio test for testing the null hypothesis (against the alternative hypothesis with $k + w$ parameters) is given by

$$\ln \lambda = 2(\ln L_{k+w} - \ln L_k), \tag{3.2}$$

where $\ln \lambda$ is known to be asymptotically distributed as χ^2 distribution with w degrees of freedom (d.f.). We shall now discuss the specific forms of $f(X; \theta)$ in different situations.

It should be noted that, owing to variable family sizes and missing data, the likelihoods for different families involve different functions of the same parameters giving rise to non-identical parent distributions. Therefore, the regular properties of the ML estimators do not necessarily apply. However, these estimators were empirically shown to be consistent and nearly fully efficient (Wette et al., 1988).

3.1. Random sampling of families

Consider a random sample of n independent nuclear families. Assume that the observations in a given family (X) are jointly distributed according to a suitable multivariate normal distribution. Therefore, the $\ln L_i(\theta|X)$ of (3.1) is given by

$$\ln L_i(\theta|X) = -\tfrac{1}{2}[\ln |\Sigma| + (X - \mu)' \, \Sigma^{-1}(X - \mu)] + \text{constant} \tag{3.3}$$

where μ and Σ are, respectively, the mean vector and covariance matrix of X. Distinct means (μ) and variances (σ^2) may be considered for phenotypes and indices on fathers, mothers, and all children, thus involving a total of six means and six variances. Likewise, the Σ matrix may be written explicitly as a function of the six variances and the 16 distinct correlations presented in Table 1, which are in turn functions of 10 path coefficients (for specifics, see Rao, 1985). Thus, the parameter vector θ consists of six means, six variances, and 10 path coefficients, a total of 22 parameters in all. Likelihood analysis may be performed using (3.1) in conjunction with (3.3). In practice, however, we often fix the six means and four parental variances, corresponding to phenotypes and indices in fathers and mothers, at the sample estimates obtained from the random sample data. This leaves only 12 parameters to be estimated simultaneously by maximum likelihood. It is important to note that, although the practice of fixing the 10 parameters can give rise to theoretical objections, neither the parameter estimates nor the resulting ln L value change noticeably in practical applications whether or not those 10 parameters are estimated simultaneously with the remaining 12 parameters.

3.2. Non-random sampling of families

Non-random sampling is coming into increased use in genetic epidemiological studies. Sometimes, a family is sampled only if the phenotypic value of a certain member, say the father, is above, say, the 95-th percentile in the population. Sometimes a family is sampled only because the phenotypic value of some member is very 'high' or very 'low'. The individual through whom the family is sampled is called a *proband*. There are several alternative methods of analysis depending on how exactly the families are sampled.

3.2.1. Likelihood conditional on the proband's value

This method is applicable to any type of on-random sample so long as each family has a designated proband. Consider a non-random sample of n families, each one sampled through a proband. In a particular family, let X_p denote the proband's phenotypic value, and let $X_{\bar{p}}$ denote the reduced vector of all other variables. Thus, X_p and $X_{\bar{p}}$ comprise X. The appropriate likelihood function for the family is (dropped θ for convenience)

$$L(X_{\bar{p}}|X_p) = f(X)/f(X_p), \tag{3.4}$$

where $f(X)$ is the multivariate normal density function for all the data, and $f(X_p)$ is the univariate normal density function for X_p, whose mean and variance are appropriate elements in μ and Σ. Likelihood analysis proceeds with (3.1) in conjunction with equation (3.4). Details of this method can be found in Hopper and Mathews (1982), Boehnke and Lange (1984), Rao et al. (1988), and Young et al. (1988).

3.2.2. Ascertainment through direct truncation

Direct truncation is defined as the case when probands are ascertained from a certain region of the phenotypic distribution, such as from the 95-th percentile (Rao and Wette, 1987). Consider a sample of n families ascertained this way from the upper tail. With the notation of the preceding section, assume that X has been sampled from a multivariate normal distribution only because $X_p \geqslant T$, where T represents the point of truncation. The likelihood function for a specific family is then given by

$$L(X_p | X_p \geqslant T) = f(X)/Q(X_p \geqslant T),$$ (3.5)

where $Q(X_p \geqslant T) = Q[(T - \mu)/\sigma]$ is the upper tail probability of the (univariate) standard normal distribution corresponding to the normal deviate $Z = (T - \mu)/\sigma$, and μ and σ are appropriate elements of μ and Σ. When the actual value of Q is known (by virtue of the sampling criterion), the denominator in equation (3.5) enters the likelihood function only as a constant and is, therefore, of no further interest. However, the value of Z is then known (as the standard normal deviate corresponding to Q), which fact imposes the linear constraint

$$\mu = T - \sigma Z$$ (3.6)

between μ and σ. Therefore, μ is not estimated as an independent parameter, but only according to equation (3.6). Also, a consistent estimate of T can be obtained externally as the smallest of X_p values of all probands:

$$\hat{T} = \underset{P}{\text{minimum}}\ X_p.$$ (3.7)

Therefore, (3.5) reduces effectively to (3.3), except that, likelihood analysis under this method proceeds with (3.1) and (3.3) subject to (3.6).

3.2.3. Method of moments

Although the two preceding methods provide likelihood approaches for analysis of non-random samples of family data, they are explicitly based on the assumption of multivariate normality. There is an alternative method, based on the method of moments, that can be used with just about any type of non-random sample (Hanis and Chakraborty, 1984). Under this method, first the (distorted) sample estimates of means, variances, and correlations are obtained (say, $\tilde{\beta}$). Then, unbiased estimates ($\hat{\beta}$) are obtained using a set of transformations (T):

$$(\hat{\beta}) = T(\tilde{\beta}).$$ (3.8)

To unbiased estimates of correlations obtained this way, path models can be fit using alternative approaches (see Hanis and Chakraborty, 1984; Rao and Wette, 1989a).

3.3. Data adjustments and normality

It has been conveniently assumed so far that the phenotypes and indices are free from any effects of concomitant variables like age and sex. However, most epidemiological studies do report such concomitant effects on many traits an investigator might be interested in. Such effects are seldom of any direct interest to a genetic epidemiologist, however, they must be addressed to avoid spurious effects on the inference. A prevailing approach is to adjust the phenotypes and indices for such concomitant effects prior to fitting the models discussed earlier. This is done by regressing the variable (phenotype or an index) on the relevant concomitant variables, including applicable interaction terms, and then taking the standardized residual for further analysis. Whereas this approach is straightforward when dealing with random samples, special methods are necessary for non-random samples (see Chakraborty and Hanis, 1987). Alternatively, the means (μ) may be modeled directly as functions of the concomitant variables within the context of the model (Clifford et al., 1984; Moll et al., 1984; Corey et al., 1986). This way, concomitant effects can be investigated simultaneously within the genetic analysis. If heteroscedasticity is present, the variances (σ^2) can also be modeled as functions of the concomitant variables. This concurrent approach to data 'adjustment' and model fitting is possible under the likelihood methods discussed above in Sections 3.1 and 3.2, and is equally applicable to both random and non-random samples.

The likelihood methods discussed above in Sections 3.1 and 3.2 assume joint multivariate normal distribution for the phenotypes and indices within families. Toward this end, a prevailing practice is to convert the standardized residuals into standard normal scores by taking the inverse normal transformation of ranked residuals. Although this does not gurrantee multivariate normality, it has been found to enhance normality quite well. Such normalized variables may then be defined as phenotypes and indices for the purposes of genetic analysis. Alternatively, we may consider a transformation of each X_i (e.g., standardized residual of father's phenotype), denoted by $T_i(X_i)$, and assume that $T(X)$, not X itself, is multivariate normally distributed. The Box–Cox transformation (Box and Cox, 1964), especially as implemented by MacLean et al. (1976), has been found to be useful for this purpose:

$$T_i(X_i) = Y_i = \begin{cases} \dfrac{r_i}{p_i}\left[\left(\dfrac{X_i}{r_i}+1\right)^{p_i}-1\right] & \text{if } p_i \neq 0, \\[2em] r_i \ln\left(\dfrac{X_i}{r_i}+1\right) & \text{if } p_i = 0, \end{cases} \tag{3.9}$$

where r_i is a suitable constant so chosen as to make all values of $(1 + X_i/r_i)$ positive, and the p_i is an unknown parameter to be estimated simultaneously with other model parameters. Usually, $r_i = 6$ suffices since X_i is standardized. As discussed by Andrews et al. (1971), although this transformation does not attain even univariate normality, it enhances normality and possibly improves additivity. Under the assumption that $T(X) = Y$ follows a multivariate normal distribution, the likelihood function for X may be obtained by using the Jacobian of the transformation in (3.9). In practice, only two different transformation parameters (p_i) may be used, one for phenotypes and the other for indices. Finally, so long as the departures from multivariate normality are small to moderate, which can be achieved through transformations of the above type, path analysis appears to be reasonably robust against such departures (Rao et al., 1987; McGue et al., 1987). Hypothesis tests are sensitive to larger departures.

4. Complete and partial environmental indices

Resolution of genetic and cultural inheritance in nuclear families is only possible with the use of environmental indices. Such indices serve two important purposes: they enable analysis of nuclear family data using comprehensive models, and they have been found to increase power for hypothesis testing (McGue et al., 1985). These considerations motivated the development of indices in the first place. Does this mean that we can use any type of index, based on whatever index variables are observed in a study, and still obtain valid inference? It would appear that the validity of inference should depend on the exact type of index used.

Ideally, indices should be based on all the index variables pertinent to the familial environment. In actual studies, however, it may not be known as to what are *all* the relevant variables, and still worse, not even all the *known* variables may have been observed in a given study. This uncertainty led to the use of different types of indices, each based on varying amounts of information, in different studies. It is unclear as to what the consequences are of using indices based on varying amounts of information. This section discusses some potential effects of using different types of indices.

For any investigation on the effects of using partial indices one needs to know what constitutes *all* the information relevant to indices. For this purpose, let us consider a theoretical model where we know all relevant variables on which to base a perfect index, which we call the *complete index*. Consider the 'real' model presented in Figure 2 which postulates a possible scenario. It is assumed that three independent familial environmental components (C_1, C_2, C_3) completely generate the 'familial environment' in path models (C). The contribution of C_i to C is β_i, so that $\sum_{i=1}^{3} \beta_i^2 = 1$. For simplicity of presentation we have assumed the C_i's to be independent, but the essential argument should hold even if they are correlated. We assume that the C_i's are not directly observable, but that they can be estimated by some 'indicator variables', X_i. X_i is completely determined by C_i and an uncorrelated residual. The relative contribution of C_i to X_i is denoted by γ_i. Therefore, $\gamma_i = 1$ if and only if X_i is a 'perfect' estimator of C_i. A certain linear combination of the X_i's is then defined as the 'environmental index', I. Thus, $I = \sum_{i=1}^{3} b_i X_i$. All the variables are assumed to be standardized to zero mean and unit variance. At this stage, it may be useful to give a possible interpretation to

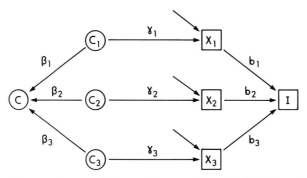

Fig. 2. A path model presenting a possible causal interpretation of the familial environment (*C*), and the environmental index (*I*). The potentially non-observable C_i's completely determine *C*, and *I* is based on the observed index variables (X_i's).

the C_i's and X_i's. For example, C_1 may designate body fat, estimated by a linear function X_1, of obesity and diet. C_2 may denote overall physical fitness, estimated by physical activity (X_2), and C_3 may denote social factors, estimated by a linear function, X_3, of smoking and alcohol intake. Then, γ_1 reflects how good (or bad) an estimate obesity and diet are of body fat, and so on.

Imagine an extension of the path model presented previously (Figure 1) by incorporating Figure 2. For simplicity, assume that $y = z = v = 1$, that the paternal and maternal variables are uncorrelated ($u = 0$), that the 'cultural inheritance' operates from C_i of a parent to C_i of a child (denote the effect by f_i, regardless of the sex of the parent), and that there is no additional sibship common environment ($b = 0$). Further assume that all f_i's are equal ($f_i = f$) and that all γ_i's are equal ($\gamma_i = \gamma$). That is, each parental C_i exerts the same influence on the C_i of a child, and that each C_i is equally precisely estimated by the corresponding X_i. Although these assumptions will necessarily have an effect on the subsequent discussion, they are not central to the simple result presented here.

Table 2
Parameter estimates averaged over 100 replications, resulting from fitting a simple analysis model to data generated by the true model in Figure 2. The same type of index is used on every member.

Index based on	Estimates of			
	h^2	c^2	f	i
X_1	0.55	0.02	0.30	0.72
X_1, X_2	0.52	0.05	0.31	0.70
X_2, X_3	0.51	0.08	0.30	0.71
X_1, X_2, X_3	0.50	0.10	0.30	0.71
True values	0.50	0.10	0.30	0.71

In this setup, the phenotype (P) and all or some of the X_i's are observed on each individual. As discussed before, the index (I) is generated by regressing the P on as many X_i's as are observed. Let us first consider the case of complete index (CI). That is, let us assume that all three X_i's are observed on each individual. Therefore, to generate the index, we should first regress P on all three X_i's: $P = A + \sum_{i=1}^{3} B_i X_i + u$. The expected regression coefficients are thus given by $B_i = \text{Cov}(P, X_i) = c\gamma\beta_i$. Therefore, the standardized regression coefficients, designated by b_i in Figure 2, are

$$b_i = \frac{c\gamma\beta_i}{\sqrt{\sum_{i=1}^{3} c^2 \gamma^2 \beta_i^2}} = \beta_i \quad \text{since} \quad \sum_{i=1}^{3} \beta_i^2 = 1 \,.$$

The complete index is then given by $I = \sum_{i=1}^{3} b_i X_i = \sum_{i=1}^{3} \beta_i X_i$. Thus, data on P, X_1, X_2, and X_3 are reduced to P and I on each individual. To these data 'generated' under the 'true' model, let us now fit a simple 'analysis' model of the usual type given by Figure 1 with the corresponding assumptions (i.e., $y = z = v = 1$, $u = b = 0$, $f_F = f_M = f$). The four parameters to be estimated are h, c, f, and i. Let us designate the estimates resulting from this model fit with a '$\wedge$' ($\hat{c}$, $\hat{f}$, etc.).

The correlation between C and I of an individual, denoted by 'i' in the analysis model, can be readily derived from the true model as $\rho(C, I) = \hat{i} = \sum_{i=1}^{3} \beta_i \gamma b_i = \gamma$ (since $b_i = \beta_i$). Likewise, the correlation between P and I of an individual is: $\rho(P, I) = \hat{c}\,\hat{i} = c \sum_i \beta_i \gamma b_i = c\gamma$. The correlation between C of a parent and C of a child is: $\rho(C_F, C_C) = \hat{f} = \sum_{i=1}^{3} f\beta_i^2 = f$. From similar calculations it can be seen that the model recovers the true parameters (with the notation that $i = \gamma$).

Having observed that complete indices introduce no bias, we can now evaluate the effects of *partial indices*, based on less than complete information; i.e., not all the three X's are used. Let us assume that only one X_i is observed. Then, the standardized regression coefficient of P on X_i is $b_i = 1$. Thus, $I = X_i$. In that case, $\rho(P, I) = \rho(P, X_i) = \hat{c}\,\hat{i} = c\beta_i\gamma$, $\rho(C_F, C_C) = \hat{f} = f$, and $\rho(I_F, I_C) = \rho(X_{iF}, X_{iC}) = \hat{i}^2 \hat{f} = \gamma^2 f$. These imply that $\hat{i} = \gamma$ (implying that i measures how accurately the C_i's are estimated, rather than how many X_i's are observed), $\hat{f}$ is unbiased, but $\hat{c} = c\beta_i \leqslant c$, where $\hat{c}$ designates the actual estimate of c. It is easy to see that, when the index is based on two X's, say X_i and X_j, $\hat{c} = c\sqrt{\beta_i^2 + \beta_j^2} \leqslant c$. In fact, when the complete index, based on all the X_i's, is used, $\hat{c} = c(\sum_{i=1}^{3} \beta_i^2)^{1/2} = c$.

Now, the parent–child phenotypic correlation is $\rho(P_F, P_C) = \frac{1}{2}\hat{h}^2 + \hat{c}^2 \hat{f} = \frac{1}{2}h^2 + c^2 f$. Therefore

$$\hat{h}^2 = h^2 + 2f(c^2 - \hat{c}^2) = h^2 + 2f[c^2 - c^2 \sum_i \beta_i^2] \,,$$

where $\sum_i \beta_i^2$ depends on which of the X_i's the index is based on. In general, so long as the same type of index is used on each member, the following solution appears to hold:

$$\hat{c}^2 = c^2(\Sigma_i \beta_i^2),$$

$$\hat{h}^2 = h^2 + 2c^2 f(1 - \Sigma_i \beta_i^2), \tag{4.1}$$

$$\hat{f} = f, \quad \hat{i} = \gamma.$$

Here, $\Sigma_i \beta_i^2$ should only include terms corresponding to the X_i's used in the index. Therefore, $\hat{h}^2$ and $\hat{c}^2$ are both biased unless all the index data are used. Thus, the effect of partial indices is to underestimate the 'cultural heritability' (c^2) and overestimate the genetic heritability (h^2). The overestimation of h^2 should, however, be only slight, as seen in equation (4.1). It may be pointed out that the estimated proportion of phenotypic variance attributable to familial causes, $\hat{h}^2 + \hat{c}^2$, also involves small bias. The expected bias is $c^2(2f - 1)(1 - \Sigma_i \beta_i^2)$, which should often be negative since f seldom exceeds 0.5.

To get a numerical feel for the magnitude of the bias, we have simulated data on P, X_1, X_2, and X_3 on each member of 1000 nuclear families, each with two parents and two children, with $h^2 = 0.5$, $c^2 = 0.1$, $f = 0.3$, $\gamma^2 = 0.5$, $\beta_1^2 = 0.2$, $\beta_2^2 = 0.3$, and $\beta_3^2 = 0.5$. Thus, C is derived least from C_1 and most from C_3. Then we generated seven types of indices on everybody, depending on which of the X_i's are used. Six of these indices are based on at most two X's, and one is based on all three. The indices were generated exactly as discussed before, by actually regressing P on the appropriate X's. We fitted the path model to these data seven times, each time using a different index, using the statistical method discussed in Section 3. This experiment was replicated 100 times, and the averages of the parameter estimates are reported in Table 2 for four representative cases.

Only 20% of the variability in the familial environment (C) is explained by C_1 (since $\beta_1^2 = 0.2$), the latter estimated imperfectly by X_1. Thus, the first partial index, based on X_1 alone, is highly incomplete. The next partial index, based on X_1 and X_2, corresponds to 50% variability in C, and thus may be regarded as moderate. The third one, based on X_2 and X_3, corresponds to 80% of the variability in C, and may accordingly be regarded as a good (partial) index. The last index presented is a complete index based on all three X's. It is clear that complete indices give rise to no bias, and 'good' (partial) indices give rise to small bias in h^2 and c^2 (the root-mean-squared errors over 100 replications, analogous to standard errors for unbiased estimates, are about 0.03 for h^2 and 0.02 for c^2 for all the indices). When moderately partial indices are used, the bias in the estimate of h^2 appears to be tolerably small, but that in c^2 may be regarded as significantly negative (i.e., underestimation). Only highly partial indices appear to generate considerable bias in c^2, and still not quite so much in h^2. Other parameter estimates are little affected.

Despite the assumptions made, especially those of $f_i = f$ and $\gamma_i = \gamma$, this investigation has shed considerable light on the impact of using different types of indices in path analysis. Even under the simple model, we have seen how partial indices can lead to biased estimates of c^2 and h^2. Under more general models, it is unclear if the other parameter estimators will also be biased. Details of these and other investigations may be found elsewhere (Rao and Wette, 1990).

5. Issues needing further work

Although many of the problems and methods discussed here have been improving over time, two areas received little attention. Progress in these areas is likely to enhance applications of this methodology.

One area concerns the assumptions of 'recursion' and linearity in the path model presented in Figure 1, the relaxation of which would reduce the gap between the model and real data. It was assumed that G and C (together with an uncorrelated residual) produce the phenotype P. It is possible that the phenotype might partly influence C. Such reciprocal interactions would seem to be important. Also, statistical interaction between G and C has not been pursued in any detail.

The statistical methods presented in Section 3 rely on normality assumptions. Although moderate departures from normality may not distort the resulting inference unduly, there is a need to develop alternative statistical approaches to fitting path models to real data. Such methods should possess desirable robustness properties, and may not be based on normality assumptions.

6. Summary

In this chapter we have discussed briefly the reason why application of causal models in social sciences has generated much heat, and the reason why its application in genetic epidemiology is better justified. We have discussed a contemporaty path model much used in actual family studies, and the associated statistical methods. We have presented some new results showing how the estimates of environmental and genetic effects can be distorted by the use of environmental indices based on less than complete data. We conclude the chapter by identifying two problem areas needing further work.

References

Andrews, D. F., Gnanadesikan, R. and Warner, J. L. (1971). Transformations of multivariate data. *Biometrics* **27**, 825–840.

Boehnke, M. and Lange, K. (1984). Ascertainment and goodness of fit of variance component models for pedigree data. In: D. C. Rao, R. C. Elston, L. H. Kuller, M. Feinleib, C. Carter and R. Havlik, eds., *Genetic epidemiology of coronary heart disease: Past, present, and future*, Alan R. Liss, New York, 173–192.

Box, A. E. P. and Cox, D. R. (1964). An analysis of transformations. *J. Roy. Statist. Soc. Ser. B* **26**, 211–252.

Chakraborty, R. and Hanis, C. L. (1987). Nonrandom sampling in human genetics: estimation of familial correlations, model testing, and interpretation. *Statistics in Medicine* **6**, 629–646.

Clifford, C. A., Hopper, J. L., Fulker, D. W. and Murray, R. M. (1984). A genetic and environmental analysis of a twin family study of alcohol use, anxiety, and depression. *Genetic Epidemiology* **1**, 63–79.

Cloninger, C. R., Rao, D. C., Rice, J., Reich, T. and Morton, N. E. (1983). A defense of path analysis in genetic epidemiology. *American Journal of Human Genetics* **35**, 733–756.

Corey, L. A., Eaves, L. J., Mellen, B. G. and Nance, W. E. (1986). Testing for developmental changes in gene expression on resemblance for quantitative traits in kinships of twins: Application to height, weight, and blood pressure. *Genetic Epidemiology* **3**, 73–83.

Freedman, D. A. (1987). As others see us: A case study in path analysis. *Journal of Educational Statistics* **12**, 101–128.

Fulker, D. W. (1988). Genetic and cultural transmission in human behavior. In: B. S. Weir, E. J. Eisen, M. M. Goodman, G. Namkoong, eds., *Proceedings of the 2nd International Conference on Quantitative Genetics*. Sinauer Associates, Inc., Sunderland, MA, 318–340.

Goldberger, A. S. and Duncan, O. D. (1973). *Structural Equation Models in the Social Sciences*. Seminar Press, New York.

Hanis, C. L. and Chakraborty, R. (1984). Nonrandom sampling in Human Genetics: Familial Correlations. *IMA J. Mathematics Applied in Medicine and Biology* **1**, 193–213.

Hope, K. (1984). *As others see us: Schooling and social mobility in Scotland and the United States*. Cambridge University Press, New York.

Hopper, J. L. and Mathews, J. D. (1982). Extensions to multivariate normal models for pedigree analysis. *Ann. Human Genetics* **46**, 373–383.

Lalouel, J. M. (1979). GEMINI-A computer program for optimization of general nonlinear functions. University of Utah Technical report No. 16, December 10, 1979, Salt Lake City, Utah.

Lalouel, J. M. (1983). ALMINI-A computer program for optimization of nonlinear functions subject to nonlinear constraints. In: N. E. Morton, D. C. Rao and J. M. Lalouel, eds., *Methods in Genetic Epidemiology*, S. Karger, New York.

Lathrop, G. M., Lalouel, J. M. and Jacquard, A. (1984). Path analysis of family resemblance and gene-environment in interactions. *Biometrics* **40**, 611–625.

Li, C. C. (1975). *Path Analysis—A Primer*. Boxwood Press, Pacific Grove, CA.

MacLean, C. J., Morton, N. E., Elston, R. C. and Yee, S. (1976). Skewness in commingled distributions. *Biometrics* **32**, 695–699.

McGue, M., Rao, D. C., Iselius, L. and Russell, J. M. (1985). Resolution of genetic and cultural inheritance in twin families by path analysis: Application to HDL-cholesterol. *American Journal of Human Genetics* **37**, 998–1014.

McGue, M., Wette, R. and Rao, D. C. (1987). A Monte Carlo evaluation of three statistical methods used in path analysis. *Genetic Epidemiology* **4**, 129–155.

McGue, M., Wette, R. and Rao, D. C. (1989). Path analysis under generalized marital resemblance: Evaluation of the mixed homogamy model by the Monte Carlo method. *Genetic Epidemiology* **6**, 373–388.

Moll, P. P., Sing, C. F., Lussier-Cacan, S. and Davignon, J. (1984). An application of a model for a genotype-dependent relationship between a concomitant (age) and a quantitative trait (LDL cholesterol) in pedigree data. *Genetic Epidemiology* **1**, 301–314.

Numerical Algorithms Groups (1982). *FORTRAN Subroutine Library Mark 9*. NAG, Oxford.

Province, M. A. and Rao, D. C. (1985a). Path analysis of family resemblance with temporal trends: Applications to height, weight, and quetelet index in Northeastern Brazil. *American Journal of Human Genetics* **37**, 178–192.

Province, M. A. and Rao, D. C. (1985b). A new model for the resolution of cultural and biological inheritance in the presence of temporal trends: Application to systolic blood pressure. *Genetic Epidemiology* **2**, 363–374.

Province, M. A. and Rao, D. C. (1988). Familial aggregation in the presence of temporal trends. *Statistics in Medicine* **7**, 185–198.

Province, M. A., Tishler, P. and Rao, D. C. (1989). A repeated measures model for the investigation of temporal trends using longitudinal family studies: Application to systolic blood pressure. *Genetic Epidemiology* **6**, 333–347.

Rao, D. C. (1985). Application of path analysis in human genetics. In: P. R. Krishnaiah, ed., *Multivariate Analysis*-IV, North-Holland, Amsterdam.

Rao, D. C., Laskarzewski, P. M., Morrison, J. A., Khoury, P., Kelly, K., Wette, R., Russell, J. M. and Glueck, C. J. (1982). The Cincinnati lipid research clinic family study: Cultural and biological determinants of lipids and lipoprotein concentrations. *American Journal of Human Genetics* **34**, 888–903.

Rao, D. C., McGue, M., Wette, R. and Glueck, C. J. (1984). Path analysis in genetic epidemiology. In: A. Chakravarti, ed., *Human Population Genetics: The Pittsburgh Symposium*, Van Nostrand Reinhold, New York.

Rao, D. C., Morton, N. E., Elston, R. C. and Yee, S. (1977). Causal analysis of academic performance. *Behavior Genetics* **7**, 147–159.

Rao, D. C., Morton, N. E. and Yee, S. (1974). Analysis of family resemblance. II. A linear model for familial correlation. *American Journal of Human Genetics* **26**, 331–359.

Rao, D. C., Morton, N. E. and Yee, S. (1976). Resolution of cultural and biological inheritance by path analysis. *American Journal of Human Genetics* **28**, 228–242.

Rao, D. C., Vogler, G. P., Borecki, I. B., Province, M. A. and Russell, J. M. (1987). Robustness of path analysis of family resemblance against deviations from multivariate normality. *Human Heredity* **37**, 107–112.

Rao, D. C. and Wette, R. (1987). Nonrandom sampling in genetic epidemiology: Maximum likelihood methods for multifactorial analysis of quantitative data ascertained through truncation. *Genetic Epidemiology* **4**, 357–376.

Rao, D. C. and Wette, R. (1989). Nonrandom sampling in genetic epidemiology: An implementation of the Hanis–Chakraborty method for multifactorial analysis. *Genetic Epidemiology* **6**, 461–470.

Rao, D. C. and Wette, R. (1990). Environmental index in genetic epidemiology: An investigation of its role, adequacy, and limitations. *American Journal of Human Genetics* **46**, 168–178.

Rao, D. C., Wette, R. and Ewens, W. J. (1988). Multifactorial analysis of family data ascertained through truncation: A comparative evaluation of two methods of statistical inference. *American Journal of Human Genetics* **42**, 506–515.

Wette, R., McGue, M., Rao, D. C. and Cloninger, C. R. (1988). On the properties of maximum likelihood estimators of familial correlations under variable sibship size. *Biometrics* **44**, 717–725.

Wright, S. (1921). Correlation and causation. *Journal of Agricultural Research* **20**, 557–585.

Wright, S. (1968). *Evolution and the Genetics of Populations*, Vol. 1. The University of Chicago Press, Chicago, IL.

Wright, S. (1978). *Evolution and the Genetics of Populations*, Vol. 4. The University of Chicago Press, Chicago, IL.

Young, M. R., Boehnke, M. and Moll, P. P. (1988). Correcting for single ascertainment by truncation for a quantitative trait. *American Journal of Human Genetics* **43**, 705–708.

C. R. Rao and R. Chakraborty, eds., *Handbook of Statistics, Vol. 8*
© Elsevier Science Publishers B.V. (1991) 81–123

4

Statistical Methods for Linkage Analysis

G. M. Lathrop and J. M. Lalouel

1. Introduction

The goal of linkage studies is to assign the genetic factors responsible for phenotype characters to specific chromosome regions by examining their cosegregation with marker loci (i.e., well-characterized genetic systems) exhibiting polymorphic variation in families. Statistical techniques are required to assess evidence of linkage from finite family samples. Once linkage is established, linked markers are tools for further genetic and epidemiological investigations of the phenotype, in genetic risk assessment, and in studies to isolate and characterize the genes responsible for the phenotype.

For many years, the application of the linkage method in human genetics was limited, as only a few polymorphic loci were available as genetic markers. Recent advances in molecular biology, particularly the introduction of methods to study restriction fragment length polymorphisms (RFLPs), have greatly increased the number of linkage applications, by providing abundant polymorphic markers distributed throughout the human genome (see White and Lalouel, 1987). Primary genetic maps or marker loci have now been constructed for all human chromosomes, and localizations have been obtained for the genetic factor responsible for many Mendelian diseases (reviewed in Lathrop and Nakamura, 1989). The power of the linkage approach has been illustrated recently by the isolation of the genes responsible for Duchenne muscular dystrophy (Monaco et al., 1986) and cystic fibrosis (Riordan et al., 1989), after precise chromosomal localizations had been obtained from closely linked markers.

Statistical techniques for linkage investigations in humans were first proposed in the 1930's (reviewed in Ott, 1985). The introduction of the lod (log-odds) approach (Smith, 1953; Morton, 1955) encouraged the wide-spread use of linkage by providing standard methods to report and to analyze results. The introduction of efficient algorithms for likelihood calculations on genetic data (Elston and Stewart, 1971; Lange and Elston, 1975; Cannings et al., 1978), and their implementation in computer programs for linkage analysis (Ott, 1974; Lathrop and Lalouel, 1984) led to routine applications with extended human pedigrees. Recent developments in statistical methods have included techniques for construction of linkage map from multiple marker data obtained from reference panels of families, and methodology for applying linkage maps to determine precise localizations of genes responsible for human genetic diseases (Lathrop and Lalouel, 1988).

In this chapter, we will review statistical methods for linkage analysis, with particular attention to these recent developments.

2. Genetic background

We will assume that readers are familiar with most of the underlying notions and terminology used in human genetics (Cavalli-Sforza and Bodmer, 1971; Vogel and Moltulsky, 1986). This section provides a rapid overview of concepts that are important specifically in linkage studies. Procedures for linkage analysis of loci on the X and Y require slight modifications of the methods discussed here; see Edwards (1971) and Ott (1986) for details. Bailey (1961) presents a classic, and still very useful treatment of many aspects of linkage analysis.

2.1. Mendelian segregation

The goal of most linkage studies is the localization of a gene responsible for a phenotype (trait locus) by the investigation of its cosegregation with a genetic marker. Marker loci are well-characterized Mendelian genetic systems that are easily studied in families, or pedigrees. Codominant marker systems are ideal for linkage studies because the two alleles at the gene on homologous chromosomes can be identified. If the alleles are different, the individual is said to be a heterozygote; otherwise, he or she is a homozygote.

Autosomal genotypes are formed by the combination of two alleles at single locus. In the following, genetic loci will usually be designated by letters (e.g., A, B, C), and alleles will be indicated by the locus designation with a numeric subscript (e.g. A_1, A_2, B_1, B_2). Genotypes are written as allelic combinations, such as A_1A_2 or B_1B_2.

Under Mendelian segregation, the probabilities that a child receives either allele from a parent at a single autosomal locus are equal, and transmissions to different children are independent. Therefore, in a mating of the form $A_1A_2 \times A_3A_4$, children have genotypes A_1A_3, A_1A_4, A_2A_3, or A_2A_4 each with probability $\frac{1}{4}$. The situation is different when the parental alleles are not all distinct. For example, the intercross mating $A_1A_2 \times A_1A_2$, gives rise to genotypes A_1A_2 or A_2A_2 each with probability $\frac{1}{4}$, and A_1A_2 with probability $\frac{1}{2}$. The parental origin of the alleles cannot be determined for heterozygotes issuing from this mating.

2.2. Linkage and recombination

When two loci are located closely on the same chromosome, they usually exhibit non-independent segregation, or linkage, i.e., the gamete transmitted to a child is often of the parental type, or phase. Recombination is said to occur when the child receives alleles inherited from different grandparents. Two loci segregate independently (recombine freely) if the probability of transmission in the parental phase is equal to the probability of transmission in the alternate phase.

Meiotic recombination results from an unequal number of crossing-over events occurring in meiosis between homologous chromosomes in the interval spanning the two loci. In most circumstances, therefore, the frequency of recombination (denoted as θ) increases as a function of the physical distance between the loci until reaching 50% (independent segregation or free recombination). Biological mechanisms underlying recombination and crossing-over have been reviewed recently by White and Lalouel (1987).

2.3. Information for linkage studies

Linkage studies are usually based on the examination of meioses in families or in experimental crosses. We say that a mating is fully informative for linkage between two loci, if the parents are double heterozygotes (e.g. $A_1 B_1/A_2 B_2$), and their genotypes are distinct (e.g. $A_1 B_1/A_2 B_2 \times A_1 B_3/A_3 B_4$). The identity-by-descent relationships can be determined unambiguously for alleles transmitted to the children in such a mating. Note that this is not possible if the mating is an intercross of the form $A_1 A_2 \times A_1 A_2$. If, in addition to being fully-informative, the parental phase is known, all the recombination events between two loci can be directly observed. Otherwise, children are non-recombinant or recombinant only in probability.

The potential information on linkage provided by a genetic locus is a function of the heterozygote frequencies, or heterogzygosity of the marker loci under study. For loci in Hardy–Weinberg equilibrium, the heterogzygosity is $H = 1 - \sum p_i^2$, where p_i is the frequency of the i-th allele in the reference population. If two loci are in linkage equilibrium, the probability that an individual is a double heterozygote is then $H_A H_B$.

Another widely-used measure of information is the polymorphic information content (PIC), defined as the probability that a marker locus will be informative in phase-known meioses segregating for a rare dominant disease (Botstein et al., 1980). Calculation of the mating type probabilities under Hardy–Weinberg equilibrium, and random mating gives $\text{PIC} = H - 2 \sum p_i^2 p_j^2$. The second factor is the probability that a child is a heterozygote issued from an intercross mating, and therefore uninformative for linkage in this situation, as either marker allele could have been transmitted with the disease from the affected parent.

2.4. Recombination at several loci and interference

We first consider linkage between three loci A, B and C. The frequencies of different gametes after meioses are given by the recombination class probabilities R_{000}, R_{100}, R_{001} and R_{010}, where identical (different) subscripts indicate that the alleles at the corresponding loci have been transmitted in the original (recombined) phase. The pairwise recombination rates are:

$$\theta_{AB} = R_{100} + R_{010}, \qquad \theta_{BC} = R_{001} + R_{010}, \qquad \theta_{AC} = R_{100} + R_{001}.$$
$$\tag{2.1}$$

Solving these equations for the recombination class probabilities gives:

$$R_{100} = \tfrac{1}{2}(\theta_{AB} + \theta_{AC} - \theta_{BC}), \qquad R_{001} = \tfrac{1}{2}(\theta_{BC} + \theta_{AC} - \theta_{AB}),$$
$$R_{010} = \tfrac{1}{2}(\theta_{AB} + \theta_{BC} - \theta_{AC}). \tag{2.2}$$

The three-locus situation could also be represented by the 2×2 contingency table, where R_{000}, R_{100} etc. define the cell entries and the θ are marginal frequencies. If recombination is a non-decreasing function of the distance between loci, and the physical order is $A-B-C$, then

$$\theta_{AB} + \theta_{BC} \geqslant \theta_{AC} \geqslant \max(\theta_{AB}, \theta_{BC}), \tag{2.3}$$

where the first inequality follows from the third equation in (2.2). If recombination is independent in each segment, we obtain $\theta_{AC} = \theta_{AB} + \theta_{BC} - 2\theta_{AB}\theta_{BC}$, and the frequency of double recombination is $\theta_{AB}\theta_{BC}$.

Interference, or non-independence of crossing-over in different chromosome segments, has been demonstrated in many organisms by breeding experiments. Typically, the observed frequency of double recombinants is less than the expectations under independence in each segment (positive interference). Coincidence, defined as $c = R_{010}/R_{100}R_{001}$, was introduced by Muller (1916) as a measure of interference for three loci. The condition for positive interference is $c < 1.0$, which implies the additional constraint that

$$\theta_{AB} + \theta_{BC} - 2\theta_{AB}\theta_{BC} \leqslant \theta_{AC}.$$

Although the representation of recombination rates as linear combinations of recombination class frequencies extends to an arbitrary number of loci, estimation of the recombination class probabilities is usually not attempted because their number rapidly becomes very large; instead, lack of interference is usually assumed, and the problem is reduced to determining estimates of recombination for intervals spanned by adjacent loci. Although other formulations for multilocus recombination class probabilities to take account of interference have been suggested (for example: Bailey, 1961, pp. 158–163; Sturt, 1976; Risch and Lange, 1977, 1983; Karlin and Liberman, 1978, 1979), none have been widely used in linkage analysis in humans. One reason for this is the low power to detect deviations from the assumption of independence with the sample sizes available in humans (Lathrop et al., 1985).

2.5. Genetic distance measures

The genetic distance between two loci is defined in terms of the expected number of crossing-overs in the interval that separates them. Many possible functional relationships have been suggested to relate distance to recombination. The two most common distance functions are: Haldane's formula (Haldane, 1919), $x = \frac{1}{2}(1 - \exp(-2\theta))$, and Kosambi's mapping function (Kosambi, 1944), $x = \frac{1}{4}\ln(1 + 2\theta)/(1 - 2\theta)$. The former is compatible with lack of interference. For three loci in the order ABC, Kosambi's mapping function gives:

$$\theta_{AC} = (\theta_{AB} + \theta_{BC})/(1 + 4\theta_{AB}\theta_{BC}). \tag{2.4}$$

Felsenstein (1979) has pointed out that mapping functions are generally of little use for estimation of recombination rates from multiple loci studied in the same family, as most do not provide values for the recombination class frequencies at more than 3 loci. Changes in the mapping functions can give lead to dramatic differences in the genetic distances predicted from the same recombination rates, but only the latter are important in linkage studies, and these are not be greatly affected by the choice of functions (White and Lalouel, 1987).

2.6. Marker loci

DNA polymorphisms provide ideal marker systems, with codominant alleles that can be easily traced in families and extended pedigrees (reviewed in White and Lalouel, 1987). Restriction site polymorphism leading to RFLPs are ubiquitous throughout the genome, and provide marker systems that average 40% heterogzygosity. Other DNA polymorphisms are based on variation in the number of short tandem repeat sequences (Jefferys et al., 1985; Nakamura et al., 1988). These loci exhibit over 70% average heterogzygosity, but may not be distributed uniformly in the genome. Another type of useful DNA markers is based on polymorphisms of short tandem repeats of GT, also with high average heterogzygosity (Weber et al., 1989; Litt et al., 1989). The combination of these systems provide abundant polymorphic markers for linkage studies.

2.7. Genetic maps and reference linkage panels

Genetic maps of marker loci are most often constructed from linkage data on reference panels. In animals and plants, experimental crosses can be designed to provide maximum efficiency for linkage studies. Ideally, experimental panels are formed from parental lines that have been rendered genetically homogeneous by many generations of inbreeding. The strategy is to cross two genetically distinct inbred lines to form a hybrid F_1 population, which can be mated under a variety of designs to create different types of linkage panels. Three frequent designs are backcross, F_2 and recombinant inbred (RI) panels.

A backcross panel is created by crossing the F_1 to one of the original parental lines to produce a single informative, phase-known meiosis for each offspring at any two loci that carry distinct alleles in the parental population. Backcross panels have been particularly useful for mapping studies in the mouse, where they have been constructed from F_1 hybrids formed by laboratory crosses of different mouse subspecies that are genetically very distinct, and that do not interbreed in the wild (Avner et al., 1988). In many plant species, the F_1 generation can be selfed to produce a large panel of F_2 individuals for linkage studies. F_2 panels provide information on two meioses for each individual, but recombinants cannot always be directly scored (see below).

Neither backcross or F_2 panels provide a permanent source of DNA for marker studies. To overcome this drawback, RI strains can be produced by successive selfing in plants, or by successive brother–sister matings in the mouse or other organisms, starting from the F_1 generation. After many generations of inbreeding, alleles at all loci are fixed to one or the other parental types. The origins of different loci provide information on linkage, since they differ only if a recombination event has separated the alleles of the original parental gametes. Once established, the RI panel can be maintained to provide unlimited material for linkage studies. In plants, F_2 panels can be supplemented by large amounts of DNA pooled from a F_3 generation for each F_2 individual; the F_2 genotypes at any locus can be reconstructed by typing the pooled DNA samples.

In man, efficient linkage panels can be created by establishing cell lines to provide a permanent source of DNA on nuclear families with large sibships. Such a panel was initially ascertained in the Utah population (White et al., 1985). Grandparents are included when possible to verify parental genotypes, and to provide phase information. DNA from 40 families (656 potentially informative meioses) selected largely from the Utah panel is distributed by the CEPH (Centre d'Etude du Polymorphism Humain) in Paris so that laboratories working in the field of human linkage studies may have free access to the same material (Dausset et al., 1989). Our group has used a larger panel of 58 families containing 1212 potentially informative meioses, to obtain improved resolution of gene order (Lathrop et al., 1988c).

2.8. Disease phenotypes and quantitative traits

Many phenotypes studied with linkage are binary responses indicating the presence or absence as a disease. In the following, we will denote the disease phenotype as 'd' and the normal phenotype as 'D'. For a two-allele system with complete penetrance, we will often use 'd' and 'D' to refer to the disease and the normal alleles. The locus is recessive if the genotypes DD and Dd have the normal phenotype, and the genotype dd has the disease phenotype; it is dominant if both dD and dd have the disease phenotype, and DD has the normal phenotype. More generally, we can consider penetrance probabilities f_{DD}, f_{Dd} and f_{dd}, representing the probabilities that individuals with the corresponding genotype are affected.

Dichotomous affection status variables are often supplemented by quantitative measurements that are relevant to the disease process. Such variables can increase the power to detect linkage, particularly when they distinguish normal and carrier genotypes in the case of recessive diseases.

2.9. Study designs for linkage analysis of disease phenotypes

The pedigrees or families available for studies of human genetic disease are usually much less informative than the families selected for mapping panels. Pedigrees extending over several generations with multiple occurrence of a rare dominant disease can sometimes be ascertained. A single extended pedigree often contains sufficient meioses to detect linkage, even though marker data will not be available beyond the first two or three generations. Late onset of dominant disease often necessitates the consideration of age-dependent penetrance, and reduces linkage information.

Since recessive diseases are only revealed in matings in which both parents are carriers of a rare allele, nuclear families with at least two affected offspring provide the primary data for most of these diseases. Usually a smaller number of families with three or more affected children are available. Multiple cases of a recessive disease may occur in a single pedigree that contains several consanguineous matings, and such pedigrees are usually more informative than nuclear families for linkage studies. Consanguineous pedigrees, however, are often drawn from small isolates that contain less polymorphism at the marker loci.

3. Introduction to statistical methods

3.1. Likelihood calculations for general pedigrees

For an arbitrary pedigree or any experimental design, the likelihood of a hypothesis can be written formally as:

$$p(X ; \theta, \alpha) = \sum p(X ; \theta, \alpha, G) \, p(G ; \theta, \alpha) \qquad (3.1)$$

$$= \sum \prod p(X_i ; \alpha, G_i) \, p(G ; \theta, \alpha), \qquad (3.2)$$

with the sum taken over G (genotypes), and the product over i (individuals). Here X_i (X) are the phenotypic data for the i-th individual (over all the individuals), G_i (G) is the genotype vector for the i-th individual (over all the individuals), θ is the vector of recombination rates, and α represents the other parameters needed to specify the genetic model, such as penetrance and gene frequencies. The vector G_i consists of genotypes at an arbitrary number of loci. The genotype for the j-th locus, usually denoted as G_{ij}, consists of a pair of alleles, where the first allele has been transmitted by the father and the second allele by the mother.

Equation (3.2) is derived from (3.1) under the assumption that phenotypes for different individuals in the pedigree are independent, conditional on the unknown genotypes and the parameters of the genetic model. Usually, we will also make the assumption that phenotype data X_{ij}, for the i-th individual and the j-th locus, are independent of the phenotypic data and genotypes at other loci, conditional on G_{ij}. This assumption could be violated in the case of multifactorial traits, epistasis, or pleiotropy.

The likelihood (3.1) can be calculated for general pedigrees with a modified Elston–Stewart algorithm (Elston and Stewart, 1971; Lange and Elston, 1975; Cannings et al., 1978). The algorithm is based on the independence of genotypes in children and their descendants, and in ascending and collateral generations, conditional on genotypes in the parents. Application of successive conditioning to the term $P(G; \theta, \alpha)$ allows the likelihood to be calculated recursively based on this independence, and avoids the summation over all possible genotype vectors. For example, consider a pedigree founded by a couple 1 and 2, and denote the phenotypic data and genotypes for a child (i) and all descendants in the branch of the pedigree below i, as Y_i and H_i, respectively, for children $i = 3, \ldots, n$. Then

$$\mathrm{Pr}(G) = \mathrm{Pr}(H_i; G_1, G_1)\,\mathrm{Pr}(G_1)\,\mathrm{Pr}(G_2)\,,$$

and from (3.2) we obtain:

$$\mathrm{Pr}(X) = \sum \mathrm{Pr}(X_1; G_1)\,\mathrm{Pr}(X_2; G_2)\,\mathrm{Pr}(G_1)\,\mathrm{Pr}(G_2)$$
$$\times \prod \left(\sum \mathrm{Pr}(Y_i; H_i)\,\mathrm{Pr}(H_i; G_1, G_2) \right),$$

where the inner sum is taken over H_i, the outer sum is over G_1 and G_2, and the product is over the children. For further details of the Elston–Stewart algorithm in linkage analysis, see Ott (1974).

3.1.1. Pairwise linkage in two- and three-generation nuclear families

Two locus data from meioses from fully-informative matings in nuclear families can be scored as recombinant or non-recombinant, if the grandparents have been studied and completely determine phase. For two loci, the probability of observing r in n meioses is given by the binomial function:

$$p(r; \theta) = b(r; \theta) = n!\,\theta^r(1 - \theta)^{n-r}/r!(n - r)!\,. \tag{3.3}$$

The unrestricted maximum likelihood estimate, $\hat{\theta}$, is r/n. In genetics, the usual practice is to report a restricted maximum likelihood estimate $\hat{\theta} = 0.5$ when $r/n > 0.5$ since θ may be biologically constrained to the interval $0.0 \leqslant \theta \leqslant 0.5$ (although the unrestricted estimate would be preferred for confidence intervals). In phase-known backcross matings, the likelihood also follows (3.3), but only one informative meioses is scored for each child.

When grandparents are not studied, or do not completely determine the phase, but otherwise n informative meioses are scored, the likelihood of the data x is

$$p(x; \theta) = \sum L_{1i}L_{2j}b(r_{ij}; \theta)\,, \tag{3.4}$$

where the sum is taken over all different possible phase combinations in the parents, and r_{ij} is the number of recombinants for this phase combination. The weights L_{1i} and L_{2j} are for the probabilities of the i-th and j-th phases in the parents (labeled 1 and 2) under the assumption of random mating. In general, the weights for each phase depend on information contributed by the grandparents, and linkage disequilibrium between alleles at the two loci. If the latter is negligible, and the grandparents have not been studied, the weights are equal. As an example, the likelihood for a single sibship of size n from a phase-unknown backcross is:

$$p(r;\theta) = \tfrac{1}{2}n!(\theta^r(1-\theta)^{n-r} + \theta^{n-r}(1-\theta)^r)/r!(n-r)!, \qquad (3.5)$$

where r is the number of recombinants in the coupling phase.

The likelihood (3.5) is symmetrical in θ and $1 - \theta$, so that $\hat{\theta}$ can be chosen in the interval from 0.0 to 0.5 whenever a real solution exists to the corresponding likelihood equation. If the solution does not exist, the maximum occurs at $\hat{\theta} = 0.5$ by the monoticity of the function (see the examples in Section 4.1).

Most linkage data in families is a mixture from phase-known and phase-unknown families, and the total likelihood over all families is usually a product of terms with different forms. Other situations that complicate the likelihood functions include missing data, intercrosses, reduced penetrance, extended pedigrees, and consanguinity. Although the equations presented here assume equal male and female recombination rates, they are easily extended to include sex-differences (at the expense of incorporating an additional unknown parameter in the likelihood function).

3.1.2. Experimental designs

Backcross panels lead to a binomial likelihood function as discussed above. RI strains also give rise to a binomial likelihood function as in (3.3), with θ replaced by $R = 2\theta/(1 + 2\theta)$ under selfing and $R = 4\theta/(1 + 6\theta)$ under brother–sister mating; a demonstration of these results is given by Haldane and Waddington (1931).

The likelihood also has a simple form for F_2 data. The four different phenotypic classes in the F_2 generation, and their probabilities in terms of the segregation of gametes from the F_1 parents are shown in Table 1. The likelihood of the data is

$$p(r;\theta) = n!\,\theta^{2r_3 + r_2}(1 - \theta)^{2r_1 + r_2}(\theta^2 + (1 - \theta)^2)^{r_4}/r_1!r_2!r_3!r_4!, \qquad (3.6)$$

where r_i is the number of recombinants observed in the i-th recombination class as defined in Table 1.

3.1.3. Risk calculations

Genetic risk evaluation is one of the most important applications of linkage analysis. The problem is to assign probabilities, or risks of different genotypic status at a trait locus, conditional on marker and other phenotypic data in a

Table 1

Two-locus recombination classes in F_2 offspring derived from inbred paternal lines

Recombination class	Phenotypes	Probability
1. No recombination	$A_1 A_1/B_1 B_1$ $A_2 A_2/B_2 B_2$	$\frac{1}{2}(1 - \theta)^2$
2. Single recombination	$A_1 A_1/B_1 B_2$ $A_1 A_2/B_1 B_1$ $A_2 A_2/B_1 B_2$ $A_1 A_2/B_2 B_2$	$2\theta(1 - \theta)$
3. Double recombination	$A_1 A_1/B_2 B_2$ $A_2 A_2/B_1 B_1$	$\frac{1}{2}\theta^2$
4. Double heterozygote	$A_1 A_2/B_1 B_2$	$\frac{1}{2}(1 - \theta)^2 + \frac{1}{2}\theta^2$

pedigree, and the genetic model. The risks of genotype G_i for individual i is given by

$$\Pr(G_i ; X) = \Pr(Y ; G_i) \Pr(G_i)/\Pr(X), \qquad (3.7)$$

where Y is all of the data except X_i. The terms in this equation are easily calculated using the Elston–Stewart algorithm (often it is necessary to include mutation as part of the genetic model in risk calculations; see Lathrop and Lalouel, 1984).

3.2. Computer implementations

Although analytic formulations are available for some types of linkage data as seen above, computer programs are needed for analysis of general pedigrees, and complex phenotypes often with missing data. Several computer programs are available for linkage analysis in general pedigrees. These include LIPED (Ott, 1974), LINKAGE (Lathrop and Lalouel, 1984) and MENDEL (Lange et al., 1988). Special purpose programs have been developed for specific applications, such as the analysis of X-linkage (Clayton, 1986), and nuclear family data for construction of genetic maps (Lander et al., 1987). Numerical techniques are needed to obtain maximum likelihood estimates. One program widely used for this purpose is GEMINI; it implements a variable metric algorithm (Lalouel, 1979). The EM-algorithm can be applied to some problems in linkage estimation (Smith, 1958; Ott, 1977; Thomson, 1983).

3.3. Monte Carlo methods

Many problems in linkage analysis must be addressed by simulation studies due to the complexity of the distribution of the likelihood estimates and test statistics from general pedigrees. An elegant, new method for simulation studies in general pedigress proposed by Ott (1989) has a surprising relationship to risk evaluation. Ott's method is based on a recursion that follows from the relationship

$$\Pr(G|X) = \Pr(G_1|X)\Pr(G_2|X, G_1)\Pr(G_3|X, G_1, G_2)\cdots.$$

The form of each term in this equation is similar to a risk probability (see Section 3.1.3), and they can be evaluated by any computer program that allows for risk calculation. The simulation scheme consists of sampling genotypes randomly for the first individual based on the risk probabilities; the genotype of this individual is assumed known for a risk calculation, and the sampling of genotypes of the second person, etc. Such sampling is continued until all genotypes are determined; random phenotypes are generated for chosen individuals to simulate a particular pattern of missing and observed marker and trait phenotypes. Many replicates can be drawn for the same pedigree to determine characteristics of estimators and test statistics. Ott's approach has been recently implemented as part of the LINKAGE computer programs (Weeks, Ott and Lathrop, 1990).

Other simulation methods for particular genetic models or pedigree types have also been proposed (Boehnke, 1986; Ploughman and Boehnke, 1989; Lange and Matthyse, 1989).

4. Detection of linkage with pairwise data

4.1. The lod score method

Tests of linkage are usually carried out by the method of lod scores. The lod (logarithm of the odds ratio; see Barnard, 1949) is defined in linkage studies as $z(\theta) = \log_{10} \mathrm{LR}(\theta)$ where $\mathrm{LR}(\theta) = p(x; \theta)/p(x; 0.5)$ (Smith, 1953; Morton, 1955). The hypothesis of no linkage is rejected for large values of $z(\theta)$; clearly, this is equivalent to a test based on $-2\ln(\mathrm{LR})$, but the lod score is retained in linkage analysis for historical reasons. Results from different studies can be combined by summing lod scores when these are presented in the form of tables for selected values of θ.

By convention, $z(\theta) \geqslant 3$ (or $-2\ln\mathrm{LR} \geqslant 13.2$) is usually required to reject the hypothesis of no linkage. The criterion was originally proposed by Morton (1955) in an application of Wald's sequential test, and includes an adjustment of the type I error for the prior probability that two randomly chosen autosomal loci are linked in man. From Wald (1947), approximate stopping bounds on the likelihood ratio statistic for the sequential test of no linkage against a fixed alternative $\theta = \theta_1$ are $A = (1 - \beta)/\alpha$ to reject $\theta = 0.5$, and $B = \beta/(1 - \alpha)$ to reject $\theta = \theta_1$, where α and β are the probabilities of type I and type II error. For $\alpha = 0.001$ and $\beta = 0.01$, Wald's stopping rules give $A \simeq 1000$ and $B \simeq 0.01$, or $\log_{10} A = 3.0$ and $\log_{10} B = -2$. Smith (1959) argued the criterion of $\alpha = 0.001$ leads to a test with approximately 0.02 specificity, i.e. 2% of significant results identify unlinked loci residing on different chromosomes, since the prior probability that two randomly selected autosomal loci are synthenic (reside on the same chromosome) is approximately $\frac{1}{22}$ in humans.

The choice of $z(\theta) > 3$ as a critical value for linkage studies has been of great practical benefit in restricting the number of spurious linkage reports (Rao et al., 1978), but its theoretical basis has engendered considerable disagreement. Chotai (1984) has pointed out that while Wald's stopping rules apply to the test of a fixed alternative, the test statistic for linkage is usually evaluated at the maximum likelihood recombination estimate. Other criticisms have been made, due to the non-sequential nature of most linkage tests, the choice of prior distributions for autosomal linkage, multiple comparisons with a series of unlinked marker loci, or non-random selection of the loci compared. For example, a lod score of 3 is likely to be overly stringent for a test of linkage of a disease to a candidate gene locus thought to be involved in its etiology.

Chotai (1984) presents a cogent analysis of many of the issues involved, and suggest adopting the viewpoint of fixed-sample testing without consideration of the prior probability of linkage. A one-sided test of linkage is then appropriate for the restriction $\theta \leqslant 0.5$. When the likelihood is given by equation (3.3), the probability in large samples that $\hat{\theta} = 0.5$ is $\frac{1}{2}$ under the null hypothesis (i.e. $\Pr(r/n \geqslant 0.5) = \frac{1}{2}$). Therefore, the asymptotic distribution of the likelihood ratio test statistic is $\Pr(-2 \ln \mathrm{LR} > x) = 0.5 \Pr(\chi_1^2 > x)$, which gives $p = 0.0001$ approximately for a lod score of 3.

4.1.1. Phase-unknown families

$\Pr(\hat{\theta} = 0.5) = 0.5$ has been assumed implicitly for other simple family types, and mixtures of phase-known and phase-unknown data, apparently due to the symmetry of the likelihood function (cf. Chotai, 1984). The following argument verifies the assumption for data from a set of phase-unknown backcross, or fully-informative families with two or three children. The likelihood for two informative meioses from a single family is given by (3.6) with $n = 2$, and $r = 0$ (sibs completely identical or completely distinct for both alleles from an informative parent), or $r = 1$ (sibs different for one informative allele and similar for the other). The term contributed to the likelihood is $\theta^2 + (1 - \theta)^2$ for $r = 0$ and $2\theta(1 - \theta)$ for $r = 1$.

Let n_0 and n_1 be the number of pairs with $r = 0$ and $r = 1$, respectively, with $n = n_0 + n_1$. The likelihood is a binomial function with

$$p = 2\theta(1 - \theta), \qquad (4.1)$$

leading to the unrestricted maximum likelihood estimate $\hat{p} = n_1/n$. For $n_1/n < 0.5$, the estimate is $\hat{\theta} = 0.5(1 - \sqrt{1 - 2n_1/n})$; otherwise (4.1) has no real solution, and the maximum is obtained at $\theta = 0.5$ since the likelihood function (3.6) is unimodal. Under the null hypothesis, the condition $n_1/(n_1 + n_0) < 0.5$ obtains with probability $\frac{1}{2}$, which gives the appropriate asymptotic one-sided test. This example is also discussed by Chotai (1984).

Phase-unknown data with three informative meioses leads to a binomial likelihood function where $p = 3\theta(1 - \theta)$ is the term contributed to the likelihood when $r = 1$ or $r = 2$. By the same considerations as above, the maximum likelihood estimate is:

$$\theta = \begin{cases} 0.5(1 - \sqrt{1 - 4n_1/3n}) & \text{for } n_1/n < \tfrac{3}{4}, \\ 0.5 & \text{otherwise}, \end{cases}$$

where n_1 is the total number of families with $r = 1$ or $r = 2$. Since $p = 0.25$ when $\theta = 0.5$, the first condition obtains asymptotically with probability $\tfrac{1}{2}$, against leading to the appropriate one-sided test. Numerical studies suggest that $\Pr(\hat{\theta} = 0.5) = \tfrac{1}{2}$ is satisfied under the null hypothesis for other phase-unknown distributions, and mixtures of phase-known and phase-unknown data.

The adequacy of the one-sided χ^2 approximation is evaluated in Section 4.3.1.

4.1.2. Multiple comparisons

Often a single trait locus will be compared for linkage to many different marker loci. Kidd and Ott (1984) and Thompson (1984) have considered a correction to the significance of the maximum lod score, to account for the number of tests made. The correction factors are easily calculated from the large sample χ^2 approximation, assuming independence of the tests. However, as Ott (1985, p. 79) points out, multiple tests also increases the prior probability that the trait locus is linked to one of the markers. Since the original reason for adopting $z(\theta) > 3$ as a criterion of linkage was to correct for this prior probability, Ott concludes that the critical values should not be adjusted for independent tests. He provides a mathematical arguement to show that 3 can be maintained as the appropriate critical value from this viewpoint, when the number of comparisons does not exceed 100.

4.2. Confidence intervals and exclusion maps

The normal approximation to the distribution of the maximum likelihood estimate provides a means of calculating a confidence interval for θ. The approximation may not be adequate for small values of θ; this is reflected by the non-symmetry of the likelihood curve in this situation. Various transformations can be applied to obtain a closer approximation to normality for small values of θ (Ott, 1985, p. 90), but the usual practice in linkage studies is to report a non-symmetrical confidence interval with limits θ_L and θ_U defined by $z(\hat{\theta}) - z(\theta_L) = z(\hat{\theta}) - z(\theta_U) = 1$ (a 1-lod-unit interval). The 1-lod-unit interval has been shown to provide a good approximation to a 95% confidence interval for linkage data (Ott, personal communication).

Intervals of a different type are also calculated in exclusion mapping, in which recombination is used to eliminate the possibility that a gene lies in certain regions of the genome. Once a region has been excluded, no further markers are types within it. By elimination, the search area is progressively narrowed until linkage is discovered. The exclusion interval is usually defined as the region about a marker for which $\mathrm{Lod}(\theta_\mathrm{E}) \leqslant 2$. This criterion originates in the sequential approach to linkage analysis (Morton, 1955), as described above. A review of exclusion mapping procedure is given in Edwards (1987a).

4.3. Sample sizes needed to detect and exclude linkage

An important application of statistical methods is to evaluate the power to detect linkage with available family material prior to embarking on the collection of marker data. Although asymptotic power results based on the normality of the maximum likelihood estimates can be obtained for some family types, finite sample calculations are useful since the number of informative meioses studies is often limited. The effect of finite sample sizes on the bias of recombination estimates has been considered by a number of researchers (see the discussion in Ott, 1985, p. 37).

Similation methods are usually needed for power calculation if markers are not fully informative, data is missing, dominance or incomplete penetrance occurs, and the families include extended pedigrees. Exact calculations of the expected lod scores are sometimes possible, and can be used as an alternative to the power for evaluating different study designs. The expected number of families needed to achieve a significant result is easily calculated, since the lod scores are additive (examples are given in Thompson et al., 1978). However, for many pedigree structures the evaluation of the expected lod score also requires simulation.

Guidelines for the sample sizes needed to detected linkage are presented for specific cases in a number of different publications (e.g., Ott, 1985; Silver and Buckler, 1986; etc.). The purpose of this section is to collect together a number of such results, to guide in the selection of families or experimental crosses for widely-used studies designs. The power calculations are based on exact numerical calculations for simple mating types and codominant loci, and on simulation studies with the method of Ott (1989) in more complicated situations. In each simulation, 10 000 replicate families were generated, and analyzed as groups of 10, 25, etc. to obtain the power results.

4.3.1. Codominant loci in 2- and 3-generation families

Table 2 provides exact numerical calculations of the power and size of the one-sided likelihood ratio test of linkage for codominant marker data from human families with fully-informative phase-known matings, and phase-unknown families with 2, 3 or 4 informative meioses, based on equations derived above. These results confirm the adequacy of the one-sided χ^2 approximation for $\alpha = 0.0001$ (lod score of 3), for which the test of the linkage is generally conservative for the sample ranges considered here. However, the theoretical size is usually exceeded in the one-sided test when $\alpha = 0.05$, a result that prompted Chotai (1984) to propose that the two-sided significance levels would be more appropriate for judging the test statistic. Ott (1989) has suggested that Monte Carlo resampling

Table 2
Power to detect linkage (lod score > 3), or exclude linkage (lod score < -2), and type I error of the likelihood ratio test for different recombination rates and sample sizes for matings seen in human genetics. The matings are: (1) phase-known, fully-informative or backcross data; (2) phase-unknown families containing two informative meioses; (3) phase-unknown families containing three informative meioses; and (4) phase-unknown families containing four informative meioses. The actual type I error of the test is given with critical values in the lod scale of 3.00 (theoretical type I error = 0.0001) and 0.59 (theoretical type I error = 0.05) for a one-sided test. Sample size is the number of informative meioses for fully-informative or B_1 data, and the number of families otherwise

Mating type	Sample size	Recombination				Type I error	
		0.05	0.1	0.2	0.3	> 3.00	> 0.59
1	10	0.599	0.349	0.107	0.028	0.0010	0.055
		(0.205)	(0.000)	(0.000)	(0.000)		
	25	0.966	0.764	0.234	0.033	<0.0001	0.054
		(0.478)	(0.446)	(0.155)	(0.000)		
	50	1.000	0.999	0.814	0.223	0.0002	0.059
		(0.444)	(0.441)	(0.343)	(0.000)		
2	10	0.369	0.137	0.021	0.004	<0.0001	0.055
		(0.000)	(0.000)	(0.000)	(0.000)		
	25	0.791	0.317	0.021	0.001	<0.0001	0.022
		(0.446)	(0.155)	(0.000)	(0.000)		
	50	0.994	0.719	0.044	0.01	<0.0001	0.016
		(0.441)	(0.384)	(0.000)	(0.000)		
3	10	0.572	0.202	0.015	0.001	<0.0001	0.078
		(0.922)	(0.526)	(0.244)	(0.000)		
	25	0.999	0.890	0.156	0.001	<0.0001	0.030
		(0.997)	(0.970)	(0.378)	(0.000)		
	50	1.000	1.000	0.665	0.041	<0.0001	0.055
		(0.441)	(0.384)	(0.000)	(0.000)		
4	10	0.956	0.657	0.099	0.005	<0.0001	0.037
		(0.973)	(0.831)	(0.130)	(0.000)		
	25	1.000	1.000	0.857	0.031	<0.0001	0.050
		(1.000)	(0.996)	(0.718)	(0.014)		
	50	1.000	1.000	0.943	0.127	<0.0001	0.045
		(1.000)	(1.000)	(0.958)	(0.221)		

schemes should be applied routinely for assigning significance, since only a small number of meioses may be informative for any single marker.

Small sizes on the order of 25–50 informative phase-known meioses, or phase-unknown families provide reasonable power (>0.75) to detect linkage at 10% recombination distance. Phase-unknown families containing two informative meioses are an exception that requires larger samples to achieve this power. (When interpreting these results, it must be recalled that the potential number of informative meioses in phase-unknown families is twice that shown in Table 2 as each parent provides independent information in fully informative matings.)

In addition to the power, Table 2 also shows the probability of obtaining an exclusion interval of different sizes when the markers are unlinked. Even with 50 informative phase-known meioses, the probability of excluding linkage at 5% recombination is under 50%. This shows that large numbers of families will generally be necessary for efficient exclusion mapping based on pair-wise linkage tests.

4.3.2. Experimental designs

Table 3 presents similar results for F_2 and RI panels. A comparison to backcross panels can be obtained from the first study design in Table 1. Although F_2 individuals are potentially informative for two meioses, the results show that F_2 and B_1 panels have roughly equivalent power to detect linkage for an identical number of individuals studied. Small RI panels have considerably potential for the detection of linkage at the recombination distances studies here, particularly when they have been constructed by sib-mating. As the available RI panels in studies of the mouse contain fewer than 30 animals, the lod score criteria of 3 will be obtained with less than 50% probability for markers at > 0.1 recombination distance.

Table 3
Power to detect linkage (lod score > 3), or exclude linkage (lod score < -2), and type I error of the likelihood ratio test for various recombination rates and sample sizes in different mapping panels. The panels are: (1) F_2 generation; (2) RI under selfing; and (3) RI under sib-mating. The actual type I error of the test is given with critical values in the lod scale of 3.00 (type I error = 0.0001) and 0.59 (type I error = 0.05) for a one-sided test

Mating type	Sample size	Recombination				Type I error	
		0.05	0.1	0.2	0.3	> 3.00	> 0.59
1	10	0.551	0.269	0.051	0.001	0.0002	0.068
		(0.884)	(0.668)	(0.051)	(0.001)		
	25	0.997	0.917	0.358	0.048	0.0001	0.056
		(0.996)	(0.968)	(0.679)	(0.158)		
	50	1.000	1.000	0.864	0.227	0.0001	0.048
		(1.000)	(0.999)	(0.939)	(0.501)		
2	10	0.385	0.162	0.035	0.009	0.0001	0.054
		(0.000)	(0.000)	(0.000)	(0.000)		
	25	0.812	0.381	0.046	0.000	< 0.0001	0.053
		(0.478)	(0.385)	(0.288)	(0.000)		
	50	0.499	0.883	0.193	0.15	< 0.0001	0.032
		(0.443)	(0.411)	(0.000)	(0.000)		
3	10	0.188	0.056	0.011	0.004	0.001	0.011
		(0.000)	(0.000)	(0.000)	(0.000)		
	25	0.237	0.321	0.001	0.000	< 0.0001	0.021
		(0.230)	(0.000)	(0.000)	(0.000)		
	50	0.768	0.164	0.038	0.000	< 0.0001	0.016
		(0.319)	(0.175)	(0.000)	(0.000)		

Silver and Buckler (1986) have adopted a Bayesian approach to the problem of detecting linkage in RI strains, under the assumption of a uniform prior distribution of genetic distance obtained from Kosambi's formula. They conclude that linkage studies could be conducted with the ordinary lod score test with critical levels between 0.001–0.004 to obtain a type I error probability of 0.05. The critical level of 0.004 is appropriate if one of the markers is known to lie on the largest mouse chromosome; 0.001 is chosen for the smallest chromosome, or when chromosomal localizations are unknown for both markers. This often provides a more powerful test of linkage, since the 0.001 critical value is usually less stringent than $z(\theta) > 3$. For example, linkage is significant with 4 or less recombinants in 25 RI lines under the former criteria, but a lod scores > 3 is obtained only with 3 or less recombinants. The former criteria give a 75% probability for detecting linkage at $\theta = 0.1$ with the critical level at 0.001, instead of the 32% probability with the criterion of $z(\theta) > 3$ given in Table 3. However, an empirical investigation of a large data base of marker genotypes for loci with known chromosomal localizations that have been characterized on a mouse panel of 26 RI lines suggests that criteria proposed by Silver and Buckler are not sufficiently stringent, as they lead to the inference of linkage between several marker groups known to be on different chromosomes (Julier and Lathrop, unpublished results).

4.3.3. Disease studies

As polymorphic DNA markers covering most of the human genome are now available, with distances of 10–20% recombination between adjacent makers, the limiting factor in most investigations of Mendelian disease is the number of families available for study. Usually genetic markers are studied in a set of families until a significant linkage (lod score > 3) is found. Non-significant positive lod scores (e.g., > 1 or > 2) often serve as a guide to chromosome regions that require further investigation with other markers. We have estimated the power to achieve lod scores greater than these values with simulation methods for selected study designs. In addition, we give the expected lod scores observed from the simulation studies, although these could also be obtained numerically by enumerating the possible outcomes in a single family, and using the fact that the results are additive over all families.

4.3.3.1. Recessive diseases.

Rare recessive diseases are frequently studied in families containing 2 or 3 affected offspring, and unaffected siblings are often not investigated. The simulation results in Table 4 indicate that a lod score > 3 can be achieved at $\theta = 0.1$ if 20 families containing two affected sibs are studied with a completely informative marker. Approximately 50 families are needed to achieve the same power if the marker locus has two, equally frequent alleles. (Exact power calculations can be made in the fully-informative situation, as transmission from each parent is equivalent to a phase-unknown family with two informative meioses; see Table 2.) Roughly 50% fewer families are needed when each contains 3 affected children.

Table 4

Expected lod score, and probabilities of obtaining a lod score >1, >2 or >3 in the study of a rare recessive disease with $\theta = 0.1$. The family types considered are: (1) two affected children; (2) three affected children; and (3) one affected and three non-affected children. The marker locus has either two alleles with equal gene frequency (marker 1), or is fully-informative in all families (marker 2). The parental genotypes are assumed known

Family type	Number of families	Marker 1				Marker 2			
		>1	>2	>3	E(lod)	>1	>2	>3	E(lod)
1	10	0.43	0.15	0.02	1.0	0.86	0.49	0.28	2.1
	25	0.82	0.50	0.25	2.2	1.00	0.97	0.92	5.0
	50	1.00	0.88	0.71	4.0	1.00	1.00	1.00	9.8
2	10	0.79	0.43	0.21	2.0	0.99	0.93	0.86	4.6
	25	0.99	0.95	0.81	4.6	1.00	1.00	1.00	11.1
	50	1.00	1.00	1.00	9.2	1.00	1.00	1.00	22.2
3	10	0.16	0.00	0.00	0.5	0.25	0.20	0.05	1.1
	25	0.30	0.10	0.02	0.9	0.80	0.43	0.21	2.0
	50	0.64	0.28	0.10	1.5	1.00	0.96	0.83	4.7

Table 4 addresses another question frequently raised in the design of linkage studies: Should non-affected children be included in the study of recessive disease when both parents are known carriers of the mutation? Although 50 families with 1 affected and 3 non-affected children would be sufficient to detect linkage with power >0.8 when using completely informative markers, these families contribute much less otherwise (power <0.1 for 50 families with marker 1 in Table 4). Therefore, these families will be useful only if extremely informative marker loci, such as VNTRs or micro-satellites are used for linkage investigations.

4.3.3.2. Dominant diseases. Table 5 shows results for study of a dominant disease. If the disease exhibits complete penetrance, samples on the order of 20–50 families (depending on the marker locus) with 4 offspring allow detection of linkage with power >0.9. A larger number of families is required for detection of linkage with incomplete penetrance when each family contains two affected and two non-affected offspring.

Often dominant disease are studied in extended pedigrees to maximize information on linkage. Simulation programs can be used to determine if a single extended pedigree, or a set of extended pedigrees provides sufficient power to warrant such a study.

4.4. Heterogeneity

Family samples will be heterogeneous for linkage between a phenotype and marker locus if genetic variants at different loci give the same trait phenotype. Typically, a single mutation responsible for the phenotype will be segregating in

Table 5
Expected lod score, and probabilities of obtaining a lod score > 1, > 2 or > 3 in the study of a rare dominant disease with $\theta = 0.1$. The parental genotypes are assumed known, and one parent is affected. The family types considered are: (1) two children (complete penetrance); (2) four children (complete penetrance); (3) two affected and two non-affected children (penetrance 0.9); and (4) two affected and two non-affected children (penetrance 0.5). The marker locus has either two alleles with equal gene frequency (marker 1), or is fully-informative in all families (marker 2)

Family type	Number of families	Marker 1				Marker 2			
		> 1	> 2	> 3	E(lod)	> 1	> 2	> 3	E(lod)
1	10	0.18	0.00	0.00	0.5	0.45	0.15	0.15	1.2
	25	0.43	0.12	0.03	1.0	0.92	0.71	0.29	2.6
	50	0.72	0.40	0.17	1.9	1.00	0.97	0.90	5.1
2	10	0.56	0.22	0.10	1.4	0.99	0.87	0.60	3.8
	25	0.92	0.73	0.49	3.3	1.00	1.00	1.00	9.3
	50	1.00	0.96	0.93	5.8	1.00	1.00	1.00	18.2
3	10	0.46	0.16	0.03	1.1	0.93	0.70	0.42	2.9
	25	0.84	0.56	0.25	2.4	1.00	0.99	0.95	6.8
	50	0.99	0.95	0.75	4.5	1.00	1.00	1.00	13.4
4	10	0.25	0.01	0.00	0.6	0.83	0.53	0.28	2.3
	25	0.50	0.23	0.06	1.2	0.99	0.92	0.76	4.4
	50	0.75	0.48	0.22	2.1	1.00	1.00	1.00	10.6

each family or pedigree selected by the presence of a rare disease. Clinical data may provide a natural classification for subdividing the families, and undertaking separate tests of linkage in each subgroup; usually an overall test is also performed since different mutations at the same locus could also be responsible for clinical heterogeneity.

When testing linkage to a single marker locus in the absence of a clinical subdivision, the sample can be considered to be a mixture of linked and unlinked families, with an unknown mixture proportion. A test of heterogenity in this situation was introduced by Morton (1956). In this test, the likelihood under the null hypothesis of a single recombination rate in all families is compared to the likelihood separate under the hypothesis of separate recombination rates in all families; the likelihood ratio statistic is compared to a χ^2 distribution with $n - 1$ degrees of freedom, where n is the number of families studied, for a test of the null hypothesis of homogeneity. The usual large sample properties of the likelihood ratio test do not apply, since regularity conditions are violated by the introduction of one new parameter to estimate with each family studied.

4.4.1. Example: Detection of outliers
Morton's approach to heterogenity is useful for detecting outlying families in which data errors may be responsible for apparent recombination events. For example, a single error in the genotype of a grandparent may lead to an incorrect

phase assignment in a parent, and the introduction of a cluster of apparent recombination events in the children. Without statistical analysis, examination of the data for errors based on the distribution of recombinants in different families can be misleading, as the number of informative meioses may be quite variable from one family to another. A statistical search for outliers can be made by calculating $LR_{fam} = z(\theta_{fam}) - z(\hat{\theta})$, where θ_{fam} is the maximum likelihood estimate of recombination within the family, and $\hat{\theta}$ is the usual estimate over all families. Large values of $|LR_{fam}|$ identify families in which the genotypes should be verified.

4.4.2. An alternative heterogeneity test

Morton's approach has been largely supplanted by a test in which the likelihood of the data under the alternative hypothesis consists of mixture with two components, assuming an identical recombination fraction in the subset of linked families (Smith, 1963; Ott, 1983). The alternative hypothesis leads to the estimation of two unknown parameters: the recombination rate, θ, and the mixture proportion, α. It has often been assumed that a heterogeneity test will be applied, only after significant linkage is found. The likelihood ratio statistic is judged as a one-sided χ^2 distribution with one degree-of-freedom, although the conditional nature of the test will affects its distribution under the null hypothesis of a single linked group of families.

It seems appropriate to include the estimate of α in the initial test for linkage if heterogeneity is suspected. Risch (1989a) compared the two approaches, and concluded that the latter is more powerful for detection of linkage only for dominant disease in large, extended pedigrees with the proportion of unlinked families less than 40%. To compare the two tests, Risch took a critical value of 17 for the likelihood ratio statistic under the hypothesis of heterogeneity, as this gives $p = 0.0001$ for a χ^2 distribution with two degrees-of-freedom (2 parameters are estimated in the heterogeneity test). However, as the test statistic does not depend on α if $\theta = 0.5$, the usual asymptotic theory cannot be applied to the likelihood ratio. In fact, the asymptotic distribution is not χ^2 under the null hypothesis of no linkage. Davis (1977) provides a method for calculating asymptotic significance levels for tests in which the null hypothesis does not depend on a nuisance parameter that appears under the alternative (Shoukri and Lathrop, submitted).

4.4.3. Sample sizes needed to detect linkage under heterogeneity

Cavalli-Sforza and King (1986) have provided useful tables showing sample sizes needed to achieve an expected lod score of 3 when heterogeneity is present, for various family types and either dominant or recessive inheritance of the disease. We have simulated data to estimate the probability to obtain a lod score > 3 with various samples sizes under 50% heterogeneity with a marker at 0.1 recombination distance from a recessive or dominant disease locus (Table 6). The test statistic reported in Table 6 has been calculated without inclusion of the heterogeneity parameter. When this is incorporated directly into the test, the probability of obtaining a lod score > 3 is little changed, although the type I error increases. With either test statistic, it is apparent that a large number of families are needed to detect linkage if substantial heterogeneity is present.

Estimates are also available of the number of families needed to detect heterogeneity with a linked marker (Cavalli-Sforza and King, 1986; Ott, 1986). It is usual to give a confidence interval for the estimate of α when reporting linkage results, because of the importance of evaluating the possibility of locus hetero-

Table 6
Expected lod score, and probabilities of obtaining a lod score > 1, > 2 or > 3 in the study of disease locus with $\theta = 0.1$ and 50% heterogeneity. The parental marker genotypes are assumed known, and trait is either a rare recessive disease with complete penetrance (family type 1), or a rare dominant disease with complete penetrance (family types 2 and 3). In the study of a recessive disease, the type 1 families are assumed to contain two affected children. In the study of a dominant disease, the family contains a single affected parent, with either two affected children (type 2), or two affected and two non-affected children (type 3). Complete penetrance is assumed at the disease locus. The marker locus has either two alleles with equal gene frequency (marker 1), or is fully-informative in all families (marker 2)

Family type	Number of families	Marker 1				Marker 2			
		> 1	> 2	> 3	$E(\text{lod})$	> 1	> 2	> 3	$E(\text{lod})$
1	25	0.26	0.07	0.01	0.7	0.56	0.21	0.07	1.3
	50	0.47	0.14	0.04	1.1	0.86	0.55	0.35	2.5
	100	0.80	0.46	0.19	2.0	0.99	0.94	0.83	4.7
2	25	0.13	0.02	0.00	0.4	0.34	0.10	0.01	0.8
	50	0.24	0.05	0.00	0.6	0.54	0.24	0.09	1.4
	100	0.37	0.11	0.04	1.0	0.90	0.60	0.28	0.5
3	25	0.43	0.15	0.04	1.0	0.90	0.62	0.39	2.9
	50	0.67	0.36	0.19	1.8	1.00	0.97	0.97	5.5
	100	0.98	0.73	0.48	3.4	1.00	0.96	1.00	10.7

geneity of a genetic disease. Although many family samples show no evidence of heterogeneity, the confidence interval for α may be very large.

Most studies of the power of the heterogeneity test have not taken account of the fact that it may be applied conditionally, after significant evidence of linkage has been obtained in the sample. Chakravarti et al. (1987) have examined this issue for a two stage test with sib-pair data, where the heterogeneity test is applied only after the detection of linkage.

4.5. Complex disease phenotypes

Analysis of complex disease phenotypes often requires consideration of reduced penetrance (perhaps dependent on age or other variables), extended pedigrees, consanguinity, missing data, possible locus heterogeneity, epistasis and various ascertainment criteria that enter into the selection of pedigrees. In principal, these factors can be accounted for in the calculation of the likelihood by the Elston–Stewart algorithm. As we have seen above, incomplete penetrance has the effect of reducing the information for detection of linkage, as the genotype status of unaffected members of the pedigree cannot be precisely determined.

Estimation of gene frequencies, penetrance and the degree of heterogeneity is usually difficult because the families selected for linkage studies are rarely random samples. Therefore, tests of linkage should be repeated under different assumptions to determine the effects of these unknown variables on the inference. An extensive literature is available on the effects of model mis-specification on linkage analysis (see, for example, Hodge and Spence, 1981; Clerget-Darpoux, 1982; Clerget-Darpoux and Bonaiti-Pellié, 1983).

4.6. Affected sib-pair and affected pedigree member methods

An alternative approach to studies of disease linkage is the affected sib-pair method, first introduced by Penrose (1933). Under this method, the frequency with which affected siblings share alleles at a marker locus is compared to Mendelian segregation. The marker and disease are said to be linked if significant distortion of the segregation is observed, i.e. if affected siblings share alleles that are identical-by-descent more frequently than expected under random segregation.

The affected sib-pair method is suited to the study of certain traits, for example disease of late onset or low penetrance, where more extensive family material is difficult to obtain. The method is easy to apply for testing the hypothesis of the involvement of a candidate gene in the disease process, under the assumption of complete linkage. The most fruitful area of application has been in the study of HLA-associated disease, for which the marker (HLA) is sufficiently informative that marker allele identity is equivalent to identity-by-descent. The test requires only the comparison of observed and expected frequencies of sharing 2, 1, or 0 HLA halotypes. The expected frequencies under the null hypothesis are 0.25, 0.5 and 0.25, respectively.

As the sib-pair approach requires no assumptions on the mode of inheritance of the disease, it should be robust to unknown factors, such as discussed above. Limitations of the classical sib-pair method include:

(1) the need for unambiguous assignment of parental origins (identity-by-descent relationships) at the marker loci;

(2) other affected relative-pairs, and sibships containing more than 2 affected children are excluded from the analysis;

(3) marker information from non-affected relatives is not included;

(4) analysis is restricted to a single marker locus.

Lange (1986a,b), Weeks and Lange (1988b), Amos and Elston (1989) and Amos et al. (1989) have suggested modifications to relax some of the assumptions in (1), (2) and (3).

Tests of non-random segregation, such as those proposed by Lange (1986a,b), Cantor and Rotter (1987), Weeks and Lange (1988b), rely on the asymptotic normal approximation under the null hypothesis to some statistic, usually a weighted sum of the number alleles shared by affected siblings, or other affected-pedigree members. Different types of affected relatives can be included in the test, as the probabilities of identity by descent are specified only under the null hypothesis. Despite this attraction, all of the presently available methods suffer the drawback that marker information on intervening relatives is ignored, even though this may provide definitive information on identity-by-descent relations in the affected members (for example, when the genotypes of the intervening relatives show that the affected pedigree-members have received no alleles from a common ancestor).

Several methods have been developed to extend affected sib-pair methods to the study of linkage with a quantitative trait (Haseman and Elston, 1972; Cockerman and Weir, 1983; Amos and Elston, 1989; Amos et al., 1989). Power calculations on the number of affected sib-pairs needed to detect linkage are given by Goldin and Gerhson (1988), who have applied a method due to Blackwelder and Elston (1985). Their studies show that 50–200 sib-pairs needed to detect linkage at a recombination distance of $\theta = 0.1$ with a completely informative marker for common diseases and 50% heterogeneity, if the significance level is 0.05 (see Chakravarti et al., 1987, for related results). Much larger samples would be needed to achieve the usual levels of significance required in linkage tests.

4.6.1. Linkage and affected sib-pair analysis

Although the affected sib-pair method has been most often employed for the study of polymorphism at a candidate genetic locus (i.e. with $\theta = 0$), it can also be used with a linked marker. If $q = \{q_2, q_1, q_0\}$ stand for the identity-by-descent probabilities for a disease determinant at distance θ from the marker locus, we have:

$$p_2 = (1 - r)^2 q_2 + r(1 - r)q_1 + r^2 q_0,$$
$$p_1 = 2r(1 - r)q_1 + ((1 - r)^2 + r^2)q_1 + 2r(1 - r)q_0,$$
$$p_0 = r^2 q_2 + r(1 - r)q_1 + (1 - r)^2 q_0, \tag{4.2}$$

for the identity-by-descent relationships at the marker locus, where $r = 2(1 - \theta)\theta$. The parameters in this model cannot by estimated from sib-pair data is θ is unknown, as only two degrees-of-freedom are available.

The probabilities q_2, q_1, q_0 are readily expressed as a function of the gene frequencies, mating probabilities, penetrances, and genetic heterogeneity under a specific genetic model. In the presence of linkage, the identity-by-descent probabilities deviate from Mendelian proportions. Two genetic models that give the same identical identity-by-descent probabilities at the marker loci, possibly at different recombination rates, will produce identical test results for linkage. Determination of the maximum lod scores under a sufficient number of genetic models to assure wide-coverage of possible identity-by-descent probabilities usually leads to identical conclusions as sib-pair analysis (Risch, 1989b).

4.7. Quantitative phenotypes

Linkage to quantitative phenotypes can be studied with the generalized single locus model, in which the distribution of the phenotype, or a suitably transformed variable, is usually assumed to follow a mixture of normal distributions with different genotype-specific means, and common variance. The properties of segregation analysis, in which the parameters are estimated without linked markers, have been extensively investigated under this model by different researchers. In addition to the effect of a major gene, residual familial correlations due to polygenic and common environmental effects can be accounted (see, for example, Elston and Stewart, 1971; Morton and MacLean, 1974; Lalouel et al., 1983). Although these models extend to linked markers (Hasstedt, 1982), few theoretical investigations have yet been made of the effect of residual correlations on the detection of linkage. Recently, Bonney et al. (1988) introduced an extension of the regressive model approach to segregation analysis (see Bonney, 1984) to incorporate linked markers.

Both the classical and regressive approaches to linkage analysis have proved useful in applications (Leppert et al., 1986, 1988). Linkage analysis of quantitative trait loci in animal or plant genetics requires only a simple application of existing methods used in human genetics, particularly when the crosses are constructed from inbred lines. An application to a study design to detect linkage in cattle pedigrees from an outbred population is given in Georges et al. (1989). For an example of a somewhat different approach to the problem of experimental crosses, see Lander and Botstein (1989).

5. Construction of genetic maps

The accumulation of pairwise linkage results in the form of lod score tables over a number of years led to the detection of linkage groups, prior to systematic use of DNA polymorphisms for mapping studies (Keats et al., 1979). The accumulation of these data stimulated attempts to construct linkage maps of human chromosomes in the 1970's (Robson et al., 1973; Cook et al., 1974; Sturt, 1975; Rao et al., 1979). Methodologies used in these investigations have been reviewed recently in White and Lalouel (1987), and they will not be considered in detail here.

Most recent studies have relied on reference panels as an efficient means to generate linkage maps. In a reference panel, the same meioses are studied for many different loci. Therefore, multilocus statistical techniques have been introduced for map construction. By the nature of the reference panels, the analysis can often be restricted to three-generation nuclear families in humans, or experimentally designed crosses derived from inbred lines.

The construction of a genetic map usually proceeds through several steps. The first problem is to determine linkage groups by the evaluation of two-locus lod scores. Chromosomal assignments of the linkage groups can be obtained by the use of anchor point markers that have been physically localized, by in situ hybridization or in somatic cell hybrids. Gene order and recombination distances are then established from the linkage data within each group.

Several approaches have been proposed for odering loci:

(1) non-parametric or other statistical methods can be used to obtain a linear order from two-locus recombination estimates;

(2) a locus order can be chosen based on minimization of crossover counts; and

(3) likelihood methods can be used to compare different orders.

We often call methods (1) and (2) 'trial map procedures', since they can provide initial or trial orders for maximum likelihood methods.

5.1. Methods based on two-locus recombination rates

Gene order and genetic distances can be estimated through the fit of a mapping function to pairwise recombination data (Rao et al., 1979). The assumptions necessary in the choice of mapping functions can be avoided by adopting non-parametric procedures. Most are based only on the assumption that recombination is an increasing function of the distance between loci. Therefore, a reasonable method for ordering loci is to seek to minimize incompatibilities between the estimated two-locus recombination rates and the chosen gene order. Lalouel (1977) proposed a non-parametric scaling method for this problem. Buetow and Chakravarti (1987) applied a seriation algorithm to the recombination rates to obtain a gene order; their approach does not take into account different variances of the estimates. Weeks and Lange (1988a) suggested that maximization of the sum of adjacent lod scores could be used to obtain a trial order; they provide a numerical algorithm for maximizing this function by 'simulated annealing'.

Wilson (1988) proposed the determination of gene order by minimizing the product of the recombination estimates between adjacent loci, without taking into account other information such as the variance of the estimates. Although she hypothesized that this would be equivalent to maximum likelihood estimation of gene order, counterexamples show that these procedures are generally not equivalent (see Section 5.5.1 for an example).

5.2. Minimization of cross-over counts

Edwards (1987b) has recently reviewed procedures for determining locus order based on the hypothesis of a minimal number of recombination events, an ordering method that has long been used in genetics (Sturtevant, 1913). If n loci from a single, fully-informative meioses are arranged along a chromosome, the recombinant and non-recombinant intervals between adjacent loci can be deduced from the pattern of recombination events between loci pairs. A change in the hypothesized locus order may increase, or decrease the number of recombinant intervals. Since the probability of recombination is less than 0.5, it is reasonable to infer a locus order that minimizes the number of recombinant intervals in the meioses studied.

Thompson (1987) demonstrated the consistency of the minimum crossover estimate of gene order, under certain broad conditions on the spacing of loci and interference, and describes a 'branch-and-bound' algorithm for determining the estimate. For fully-informative, closely linked loci, she also shows equivalence to the maximum likelihood order. In practice, as most data in human genetics are only partially informative, the minimum crossover estimate may differ from the maximum likelihood order.

5.3. Maximum likelihood methods for multiple loci

The maximum likelihood approach requires the evaluation of the likelihood of the multilocus data under a specific gene order, and comparison of different orders in terms of their support, or relative likelihoods. Although the multilocus likelihood from a 3-generation family under a fixed gene order is given by a multilocus extension of equation (3.5), its direct application is rendered computationally difficult as the number of possible phase combinations may increase exponentially with the number of informative loci, or loci with missing parental data; in a phase-unknown mating with n_1 informative loci in the first parent, and n_2 informative loci in the second parent, the number of terms in the outer summation of the multilocus equivalent to (3.4) is $2^{n_1 + n_2}$.

Under the assumption of lack of interference, recombination is independent in each segment, and the multilocus recombination probabilities can be expressed as a function of the recombination rates between adjacent loci under a hypothesized gene order. For data from a fully-informative, phase-known mating, the intervals between adjacent locus can be scored as recombinant or non-recombinant in each gamete as described in the previous section. Due to independence, the probability of the observation is the product of terms of the form θ_i (for recombinant intervals) and $1 - \theta_i$ (for non-recombinant intervals), where θ_i is the recombination rate for the i-th interval. Thus the total likelihood of the data for all gametes for a single mating is

$$p(r ; \theta) = \prod p_i(r_i ; \theta_i) , \qquad (5.1)$$

where r_i is the number of recombination events in the i-th interval, and $p_i(r_i ; \theta_i)$ is the likelihood for the data in the i-th interval; in this case the binomial probability function. The multilocus likelihood also has the form (5.1) for other study designs and mating types, for example, phase-known or phase-unknown backcross data (see Lathrop et al., 1986 for the case of a phase-unknown backcross).

When equation (5.1), or an equivalent relationships holds, the multilocus recombination estimates between adjacent loci are equal to the pairwise estimates. Recently, Morton (1989) has suggested that this relation holds for all data from three-generation nuclear families. Figure 1 shows an example of a family with complete data that contradicts this assertion, and other examples in which (5.1) is violated are readily constructed (Lathrop et al., 1986). Therefore, more sophisti-

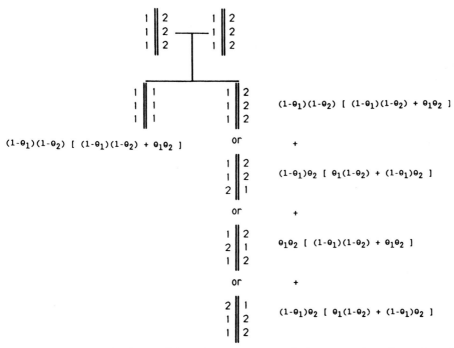

Fig. 1. The likelihood for 3-locus linkage data cannot generally be expressed as the product of likelihoods between adjacent loci. The data is assumed to be from a phase-known mating. The likelihood is the product of two terms, one for each of the two. The likelihood term contributed by the second offspring is the sum of the probabilities of each of the possible multilocus genotypes. The total likelihood cannot be factorized into one term that depends on only on the recombination probability in the first segment (θ_1), and a second that depends only on the recombination probability in the second segment (θ_2).

cated algorithms are generally needed for obtaining the recombination estimates. Missing data, common in most genetic studies, also result in different estimates for the recombination rates.

Several methods are now available for rapid calculation of the multilocus likelihood for codominant markers in three-generation nuclear families when equation (5.1) does not hold. Lathrop et al. (1986) and Lathrop and Lalouel (1988) provide techniques for obtained a partial factorization of the likelihood for a single family, in which terms may depend on the recombination rates in several intervals. Lander and Green (1987) describe a different algorithm, also leading to simple likelihood calculations.

5.4. Experimental designs

Studies of B_1 or F_2 panels lead to a likelihood equation of the form (5.1) if no genotypic data is missing; the maximum of the likelihood for a given order is then the product of the pairwise maxima defined by the equations discussed above. A formal demonstration of (5.1) for a F_2 panel can be obtained from the following argument. The phenotypes from a single F_2 individual resulting from segregation at loci $1, \ldots, i - 1$ $(A_i = \{P_1, \ldots, P_{i-1}\})$ and loci $i + 1, \ldots, n$ $(B_i = \{P_{i+1}, \ldots, P_n\})$ are independent conditional on G_i, the ordered genotype at the i-th locus. Thus:

$$
\begin{aligned}
\Pr(P) = \sum \Pr(A_{i-1} ; G_i) \Pr(B_{i+2} ; G_{i+1}) \\
\times \Pr(P_i ; G_i) \Pr(P_{i+1} ; G_{i+1}) \Pr(G_i, G_{i+1}),
\end{aligned}
\tag{5.2}
$$

where the sum is taken over G_i and G_{i+1}, and $\Pr(G_i, G_{i+1})$ is obtained from Table 1. Furthermore, the symmetry of the cross assures that

$$
\Pr(A_i ; G_i = \{p_1, p_2\}) = \Pr(A_i ; G_i = \{p_2, p_1\}),
$$

and

$$
\Pr(B_i ; G_i = \{p_1, p_2\}) = \Pr(B_i ; G_i = \{p_2, p_1\}).
$$

Application of these equalities to (5.2) gives the desired result by induction on the number of loci. For example, if $P_i = P_{i+1} = (p_1, p_2)$ for the unordered phenotype (i.e. the genotypes are either $p_1 p_1 / p_2 p_2$ or $p_1 p_2 / p_2 p_1$), we obtain:

$$
\begin{aligned}
\Pr(P) &= \Pr(A_{i-1} ; G_i = \{p_1, p_2\}) \Pr(B_{i+2} ; G_{i+1} = \{p_1, p_2\})(1 - \theta)^2 \\
&+ \Pr(A_{i-1} ; G_i = \{p_1, p_2\}) \Pr(B_{i+2} ; G_{i+1} = \{p_2, p_1\})\theta^2 \\
&+ \Pr(A_{i-1} ; G_i = \{p_2, p_1\}) \Pr(B_{i+2} ; G_{i+1} = \{p_1, p_2\})\theta^2 \\
&+ \Pr(A_{i-1} ; G_i = \{p_2, p_1\}) \Pr(B_{i+2} ; G_{i+1} = \{p_2, p_1\})(1 - \theta)^2 \\
&= \Pr(A_{i-1} ; G_i = \{p_1, p_2\}) \Pr(B_{i+2} ; G_{i+1} = \{p_1, p_2\})((1 - \theta)^2 + \theta^2).
\end{aligned}
$$

In most studies individuals will have data missing at some loci. Equation (5.2) then applies to the calculation of the likelihood contribution for a single individual where the recombination rates are between adjacent markers after removal of loci with missing data, but the maximum of the total likelihood is not the product of the maxima of adjacent pairs.

5.4.1. Example: Three-locus data in RI strains

Morton (1988) hypotheses on the validity of (5.1) in RI panels under lack of interference. A 3-locus result, which follows from the discussion of RI in Haldane and Waddington (1931), shows that (5.1) does not hold in this case. The 3-locus recombination class frequencies are obtained by replacing the θ's in equation (2.2) by the appropriate two-locus recombination frequencies, for selfing or sib-mating schemes as given in Section 3.1.2. To obtain conditions under which (5.1) holds, we then solve the equation $R_{01} \cdot R_{01} = R_{010}$ to obtain θ_{AC} in terms of θ_{AB} and θ_{BC}. For selfing, this leads to

$$
\theta_{AC} = (\theta_{AB} + \theta_{BC})/(1 + 4\theta_{AB}\theta_{BC}),
$$

which is the condition for Kosambi level interference [see equation (2.4)]. Under sib-mating, the equivalent condition for independence is

$$\theta_{AC} = (\theta_{AB} + \theta_{BC} + 4\theta_{AB}\theta_{BC})/(1 + 12\theta_{AB}\theta_{BC}).$$

Lack of interference, on the other hand, leads to the inequality $R_{01} \cdot R_{01} < R_{010}$, which is analogous to positive interference in single generation recombination classes.

The consequences of assuming equation (5.1) with > 3 loci merits further investigation (see also Chakravarti and Cox, 1987).

5.5. Comparison of gene orders

Maximum likelihood inference under the assumption of lack of interference requires the comparison of the likelihoods of different orders. The ratio of the likelihoods at the maximum likelihood estimates of recobmination under two orders (the relative odds) is often used as a measure of the evidence in favour of one order, and against the other (Leppert et al., 1986a). With few markers, the likelihoods of all orders can be compared to determine the maximum likelihood order. For moderate size maps, the search must be restricted due to the large number of possible orders (> 1.8 million for 10 loci). Although heuristic search schemes can be envisaged, convergence to the maximum likelihood order is not guaranteed (Lathrop and Lalouel, 1988).

We have adopted a heuristic approach containing the following steps:

Step 1. A trial map algorithm is used to obtain an initial gene order, and maximum likelihood estimates of the multilocus recombination distances are obtained. The loci are divided into subgroups based on the estimated recombination frequencies; the first division is made in the interval of largest recombination, the second in the interval of next largest recombination, etc.; the division is continued until all subgroups contain no more than a pre-determined number of loci.

Step 2. Likelihoods are obtained for the all orders of loci within each subgroup. The within-subgroup likelihoods are compared; those orders with odds less than $\frac{1}{100}$ compared to the maximum likelihood order are rejected, and not considered in further calculations. (The rejection criterion of can be replaced by more stringent values such as $\frac{1}{1000}$.)

Step 3. The within-subgroup loci are fixed at the maximum likelihood order. The order of the subgroups is permuted and their orientations are varied; the likelihood is evaluated for each new order examined. The subgroup permutations may be limited to certain regions of the map; for example, a condition may be imposed to limit investigations to permutations of adjacent subgroup triplets. The orders are ranked by likelihood, and a rejection criterion is applied to eliminate a subset. The remaining permutations are evaluated under non-rejected within-subgroup orders obtained in Step 2.

Step 4. The best-supported order is chosen as a new trial map to restart at Step 1. Iterations are continued until convergence.

The approach is illustrated in the following example.

5.5.1. Example: A linkage map of a region of chromosome 14

Nakamura et al. (1988) have published results of a linkage study of 11 loci from a region of chromosome 14. Genotypic data were obtained for the 59 families of the extended reference panel described above. Preliminary analysis showed that obligate recombinants or non-recombinants could be scored between at least two informative loci for 686 of the 1212 potentially informative meioses. The linear order obtained from the two-locus recombination rates by the seriation algorithm of Buetow and Chakravarti (1987) is 7−9−3−10−1−6−4−5−8−2−11 (see Figure 2a for the marker loci corresponding to the locus numbers). Convergence was obtained after two iterations of the heuristic maximization algorithm (Figure 2a), which gave the final gene order as 7−10−3−9−1−6−4−8−5−2−11. The odds of the first trial order relative to the order at convergence ate 724 : 1. Recombination estimates and physical mapping data for these markers are shown in Figure 2b.

The converged order also exhibits the smallest number of obligate recombination events. Of the 686 chromosomes informative for at least two loci, 33 contained a single recombinant under this order; 2 contained double recombinants. However, when statistical information on the most probable phase under this order is taken into account (as in maximum likelihood analysis), 958 of the chromosomes are informative for at least two loci, and the number of single and double recombinants are 94 and 15, respectively. No chromosomes exhibit obligate or probable recombination in more than two intervals.

The order minimizing the product of the two-locus recombination probabilities between adjacent loci is 7−9−3−10−1−6−8−4−5−2−11. Maximum likelihood

```
    Trial Map : [7] [9 3 10] [1 6 4 5 8] [2] [11]  -2874.42

1st iteration : [7] [10 9 3] [1 6 4 5 8] [2] [11]  -2887.61

2nd iteration : [7] [10 3 9] [1 6 4 8 5] [2] '11]  -2898.20
```

Fig. 2a. Results from the analysis of data from 11 loci on chromosome 14. Ten loci are detected by DNA probes (1 = D14S1; 2 = IGHC; 3 = D14S13; 4 = D14S16; 5 = D14S19; 6 = D14S17; 8 = D14S23; 9 = D14S21; 10 = D14S18; 11 = D14S20); one is a protein polymorphism (7 = PI). The loci were initially ordered from the pairwise recombination rates by seriation, and then divided into subgroups of no more than 5 loci based on the recombination estimates. Orders of the loci within each subgroup were ranked by likelihood, and those with relative odds of $<\frac{1}{100}$ of the maximum likelihood order were rejected from further consideration. The maximum likelihood subgroup orders were used in tests of alternative permutations and orientations of the subgroups (between-subgroup orders and orientations). Only those between-subgroup orders involving interchanges of adjacent subgroup triplets were considered. The tested orders were again ranked by likelihood, and those with relative odds $<\frac{1}{100}$ compared to the best-supported order were examined under the non-rejected between subgroup orders. The best-supported was chosen for a new trial map. Convergence was obtained after two iterative cycles.

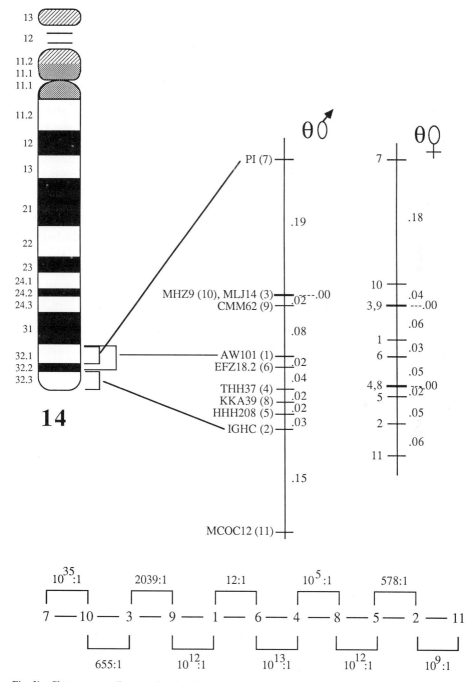

Fig. 2b. Chromosome diagram showing the distance between markers, physical data on their localizations, and odds against inversions of adjacent loci. (Reprinted by permission from Nakamura et al., 1989.)

analysis shows that the odds against this order are $> 10^{13} : 1$ compared to the best-supported order found by the heuristic algorithm.

5.6. Statistical tests of gene order

An advantage of the maximum likelihood method is that different orders can be compared by reporting the relative odds, or the likelihood ratio at the maximum likelihood recombination estimates. Thompson (1984) has considered the probability of correct inference of gene order from completely informative 3-locus data based on the likelihood ratio criteria. She employed a normal approximation to compare this approach to inference of order from independent pairwise data for the three loci, and found that the former is considerably more efficient. Bishop (1985) gives similar small sample results. Both these studies assume completely informative data, for which inference of gene order from the comparison of the pairwise recombination estimates will give the same results as the likelihood ratio criteria.

The likelihood ratio can also be considered as a test statistic, and we can attempt to assign a significance to odds in favour of one order against an alternative. In addition to inference to gene order, this approach would lead to the rejection of an alternative order for small values of $-2 \ln LR$, where LR is the likelihood ratio. However, the distribution of the likelihood ratio statistic cannot be derived from standard likelihood theory, because non-nested compound hypotheses are compared when testing different gene orders. Elsewhere, we have examined the application of isotonic inference (Barlow et al., 1972; Chotai, 1983) to this problem for simple cases of 3-locus data from fully informative matings (Lathrop et al., 1987). Maximum likelihood estimates are found after imposing restrictions on the recombination class probabilities valid for a particular order, such as $R_{010} \leqslant \min(R_{100}, R_{001})$ under the order $A-B-C$, i.e. double recombinants are less frequent than single recombinants. Other possible constraints include positive interference [equation (2.3) above], or lack of interference; see Lathrop et al. (1987) for details.

The significance of the LR to reject the order $A-B-C$ can be calculated for the asymptotically least-favourable alternative with $R_{000} = R_{100} = R_{001} = R_{010} = 0.25$, or equivalently, $\theta_{AB} = \theta_{BC} = \theta_{AC} = 0.5$. The least favourable values of θ (denoted θ_L) are chosen so that

$$\Pr(-2 \ln LR < x ; \theta_L) < \Pr(-2 \ln LR < x ; \theta),$$

for all values of θ when $A-B-C$ is the true gene order. Therefore, rejecting $A-B-C$ based on the significance of the likelihood ratio statistic calculated under the least-favourable alternative is conservative, since the same or greater significance would be assigned for other values of the unknown parameters.

The least-favourable alternative is extremely conservative for recombination < 0.25. We have proposed an 'adaptive' approach in which the least-favourable significance value is replaced by that obtained from the distribution of LR at the maximum likelihood estimates under the order-constrained solution. Exact numerical calculations show that the 'adaptive' test is generally conservative, and has greater power than the least-favourable approach.

The tests outlined here can be adapted to mixed mating types, or multiple loci under lack of interference. Generally, it is not possible to calculate exact significance levels, but a resampling scheme, in which the marker genotypes in parents and grandparents are maintained and genotypes in the children are randomly generated, can be used to approximate the distribution of the test statistics.

5.7. *Sex differences in recombination rates*

Sex differences in recombination rates within the chromosome region spanned by a linkage map can be assessed in two different ways (O'Connell et al., 1987). First, the significance of the average sex effect can be determined by comparing the likelihood of the model with no sex difference, to that of a model with a constant ratio, k, of female–male genetic distance, i.e. $k = k_i = f(\theta_{fi})/f(\theta_{mi})$ for all intervals i, where f is a mapping function, usually Haldane's or Kosambi's function (see Section 2.6). The likelihood ratio test for $k = 1$ is asymptotically distributed as a χ_1^2 under the null hypothesis, which provides a test of no sex difference. Female genetic distances are typically 2–3 fold superior to male genetic distances throughout many chromosome regions (White and Lalouel, 1987).

Second, a model in which the sex-effect may be different in each interval spanned by adjacent loci can be investigated by letting k_i vary in each interval. The likelihood ratio test of the variable sex-effect vs. the constant sex-effect models has $n - 1$ degrees of freedom, where n is the number of intervals between adjacent markers with non-zero recombination frequency. Donis-Keller et al. (1987) have used a bootstrap scheme to assign significance levels in small samples.

5.8. *Interference*

The assumption of lack of interference seems a reasonable first-order simplification for determination of order in genetic linkage maps. Even the largest data samples presently available in humans have little power to detect expected levels of interference at three loci. The most information will be provided by meioses that are fully informative at three loci, for which detection of interference reduces to a test of non-independence in a two-way contingency table. The normal approximation for the large sample distribution of the log relative risk for Kosambi level interference shows the sample sizes of more than > 490 triply-informative phase-known meioses are needed to achieve a power of 0.9 to detect interference in the most optimal situation, at 18% recombination between adjacent markers, and more than 2000 are needed if the phase is unknown (Lathrop et al., 1985; Bishop and Thompson, 1988). Recombination estimates obtained under the assumption of lack of interference often have smaller mean square error even when interference is present (Lathrop et al., 1985).

Most studies of interference in man with > 3 loci have relied on fit of mapping functions to pairwise recombination estimates, although several investigators have suggested methods of specifying multilocus recombination class probabilities for investigations of interference with > 3 loci as discussed in Section 2.5. Morton and MacLean (1984) have suggested a general method of approximation to the recombination class frequencies from an arbitrary map function. Although none of these methods have yet been widely tested on data, it is likely that significant investigations of interference in man will require larger data samples than those that are presently available.

6. Multiple marker loci and disease phenotypes

Linkage studies of Mendelian disease usually occur in two phases. The first consists of finding an initial chromosomal localization(s) for the gene(s) responsible for the disease. The second is to determine a precise localization, and to locate markers at a close distance on each side of the disease locus, so as to define of interval spanning approximately 1% recombination containing the gene. A region of this size is amenable to other approaches to isolate the gene; well-characterized families in which the disease shows recombination with one of the closest flanking markers are helpful for eliminating expressed sequences in the interval as candidates for the gene responsible for the disease.

Initial genetic localizations have most often been obtained by pairwise analysis with a single marker locus. However, as maps of linked markers are now available, detection of linkage is also based on data from several marker loci on the same chromosome. An efficient strategy to obtain a refined placement for a gene of interest is to isolate RFLPs from the region of its initial localization, and construct a map from genotypes obtained in a reference panel of families. RFLPs that are localized to the region of the gene are characterized in informative families, and the placement of the disease is determined with respect to the fixed map of markers by maximum likelihood analysis. See Lathrop et al. (1988c) for an example of this approach applied to the localization of the gene responsible for cystic fibrosis.

6.1. Detection of linkage with several marker loci

In this section, we attempt to clarify certain statistical issues associated with tests of linkage between a disease locus, and multiple marker loci by considering data from phase-known, fully-informative meioses with two markers (A and B), and a dominant fully-penetrant allele at the disease locus (D).

6.1.1. Theoretical argument

Under the assumption of no interference, the maximum likelihood gene order is found by comparing the likelihoods of each of 3 possible orders as obtained from equation (5.1). The likelihood of the hypothesis of no linkage between the disease and marker loci (denoted D, $A-B$) is obtained by fixing $\theta_{AD} = \theta_{BD} = 0.5$, and estimating θ_{AB} under the order $D-A-B$ or $A-B-D$. Usually, we will not be interested in the test of linkage between the marker loci.

In analogy with the two locus approach, linkage is said to be significant if the likelihood ratio exceeds 3.0 on the lod scale. The size and power of this multilocus test have not examined in detail, but the following argument demonstrates that the type I error associated with the 2- and 3-locus tests are different. Suppose that the null hypothesis D, $A-C$ holds. From the consistency of the order estimates, $A-D-B$ must be rejected in large samples, and the test of linkage will reduce to a comparing the likelihoods of the hypotheses $D-A-B$ and $A-B-D$ to D, $A-B$. From equation (5.1), the likelihood ratio test statistic of D, $A-B$ vs. either $D-A-B$ or $A-B-D$ is simply the lod score from the corresponding pairwise test of linkage between $D-A$ or $D-B$, respectively.

If the linkage between $A-B$ is complete ($\theta_{AB} = 0$), the two lod scores must be identical, and the multilocus test will have the properties of a single pairwise test as described in Section 4. On the other hand, if $\theta_{AB} = 0.5$, the lod scores are independent, and the multilocus test is equivalent to two pairwise tests. Therefore, we may argue that the significance level, α, of the multilocus test in large samples should be adjusted by a factor of 2 to account for multiple comparisons (see Section 4.2 for an alternative viewpoint). If the recombination between A and B lies between 0.0 and 0.5, the two lod scores are positively correlated, and the significance level is between α and 2α.

6.1.2. Numerical results

Table 7 shows the type I error and power of the 3-locus test of linkage for fully-informative, phase-known data with critical values corresponding to the significance levels of 0.0001 and 0.05 from the χ_1^2 distribution. In addition to various recombination rates, we have also considered the effect of different sample sizes on both power and type I error. The results were obtained by comparing the maximum likelihood under each of the three possible gene orders to the likelihood under D, $A-B$, and the largest value of the likelihood ratio was chosen as the test statistic. For the purposes of comparing 3-locus and 2-locus tests, we also report the power to detect linkage with a single marker using 2.66 as a critical value for the lod score (to adjust the theoretical type I error to be approximately 0.0002 for a single test, or 0.0001 for two independent pairwise test).

As predicted by the theoretical argument, the type I error is always less than twice the corresponding value from Table 2 for the pair-wise test of linkage for an equivalent number of phase-known meioses. The power of the 3-locus test under the true order $A-D-B$ is greater than the equivalent pairwise test, especially for large recombination distances, where 2- to 3-fold increases in the power are seen. Since the pairwise test for each locus is independent, the power to detect linkage considering each marker locus separately can be obtained from as $1 - (1 - \beta)^2$, where $1 - \beta$ is the power of the pairwise test statistic from Table 2. For example, at 0.3 recombination between the disease and the markers, the two

Table 7

Power and type I error of the test of linkage between two marker loci (A and B) and a dominant disease (D), under the assumption of fully-informative data, and complete penetrance. For the order $A-D-B$, we give the power to detect linkage with the criteria: (1) multilocus lod score > 3 for the analysis of two marker loci (top line of each entry); (2) pairwise lod score > 2.66 for the analysis of a single marker locus (bottom line of each entry). Under the order $D-A-B$, the type I error is calculated for the 3-locus test with D unlinked to the marker loci, for the critical values corresponding to $\alpha = 0.0001$ (top line) and $\alpha = 0.05$ bottom line from the χ_1^2 distribution

Order	Sample size	Recombination between markers (between disease and markers ADB)			
		0.095 (0.050)	0.180 (0.100)	0.320 (0.200)	420 (0.300)
$A-D-B$ (D equal	10	0.84	0.58	0.20	0.06
distance from A		(0.60)	(0.35)	(0.11)	(0.03)
and B)	25	1.000	0.99	0.68	0.16
		(0.99)	(0.90)	(0.42)	(0.09)
	50	1.000	1.000	0.99	0.61
		(1.000)	(1.000)	(0.81)	(0.22)
$D-A-B$ (D not linked	10	0.016	0.018	0.019	0.019
to A and B)		(0.078)	(0.091)	(0.108)	(0.115)
	25	0.0001	0.0002	0.0002	0.0003
		(0.079)	(0.089)	(0.100)	(0.108)
	50	0.0003	0.0003	0.0003	0.0004
		(0.087)	(0.097)	(0.108)	(0.116)

pairwise test give a power of 0.41 to detect linkage with 50 meioses, while the 3-locus test has a power of 0.61.

The results presented in this section illustrate statistical differences between pairwise and multilocus test of linkage. In practice, multilocus tests will be further complicated by phase-unknown data, reduced penetrance and extended pedigrees. Although each marker locus will be informative only in portion of families, when several markers are characterized from a single chromosome nearly all families will be informative for linkage with at least one marker flanking each chromosome segment that is tested. On average, therefore, we may expect multilocus test to show considerably more power than pairwise tests for the detection of linkage. The above results show that a positive lod score < 3 could be accepted as evidence of linkage within an interval without increasing the type I error, as consistency of the order estimate assures that this event has very low probability in large samples under the null hypothesis. As the significance level depends on the number of meioses studied, it may be useful to adopt Monte Carlo procedures for the evaluation of test.

6.2. Gene order and the location score method

Multilocus linkage analysis is now widely employed to determine the placement of the disease locus with respect to the markers. Usually, the families available for disease tudies are limited, and may not provide sufficient information for precise estimates of marker order and recombination distances between marker loci. Information on the markers obtained in reference families can be incorporated by including these data in the analysis.

If the reference panel is large, the recombination and order estimates will be little modified by the addition of the disease families. The order and distances between markers can then be considered fix, and the placement of the disease is the only parameter that need be estimated. This procedure is called the location score method. The location score is defined to be the $-2 \ln LR$, where LR is the ratio of the likelihood of the data for a given placement of the disease locus to its likelihood with the disease locus unlinked to the markers (Lathrop et al., 1984). The method leads to a graphical representation of the results, in which the location scores are given for points along the map of the marker loci; an example is shown in Figure 3. A similar graphical representation for pairwise data was introduced by Cook et al. (1974).

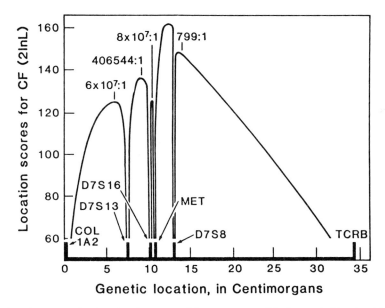

Fig. 3. Location score showing the placement of the gene responsible for cystic fibrosis with respect to the map of six marker loci from chromosome 7q. The horizontal axis shows the placement of the markers, and the vertical axis gives the value of the location score obtained by the analysis of a large number of multi-case families (Lathrop et al., 1988c). These data show that most likely placement of the gene is between the MET and D7S8 loci. This localization has been confirmed by isolation of the gene (Riordan et al., 1989). Odds are reported against the maximum likelihood locations under alternative placements of the gene. (Reprinted by permission from Lathrop et al., 1988c.)

6.3. Other applications of the location score method

Edwards (1987a) has pointed out the increased power obtained from multiple loci with the location score approach for exclusion mapping. Location scores are also potentially useful for the detection of locus heterogeneity. An example showing the detection of heterogeneity in X-linked spastic paraplegia is given in Keppen et al. (1987).

A recent study by Martinez and Goldin (1989) has examined the power of the location score test for the detection of linkage with heterogeneity, but only when the test statistic is calculated under the assumption of no heterogeneity. In contrast to the pairwise situation described in Section 4.4, it may be necessary to incorporate heterogeneity directly into the location score calculation for the detection of linkage; otherwise, families that exhibit no linkage to the markers will apparently contain multiple recombination events throughout chromosome regions, and give evidence against all interior placements of the disease locus.

7. Conclusion

Linkage analysis has become a standard method for addressing many problems in human genetics, and in other species. The advent of new DNA technology has led to rapid development of the statistical techniques for map construction, and the analysis of multiple marker loci. Although many statistical issues, particularly in the area of significance testing with multiple loci, remain to be addressed, the presently available statistical techniques have proved their usefulness in many practical applications. It is likely that the realm of these applications will continue to expand, as the biological tools for linkage studies improve over the next few years.

References

Amos, C. I. and Elston, R. C. (1989). Robust methods for the detection of genetic linkage for quantitative data from pedigrees. *Genet. Epidemiol.* **6**, 349–360.

Amos, C. I., Elston, R. C., Wilson, A. F. and Bailey-Wilson, J. E. (1989). A more powerful robust sib-pair test of linkage for quantitative traits. *Genet. Epidemiol.* **6**, 435–450.

Avner, P., Amar, L., Dandolo, L. and Guenet, J. L. (1988). The impact of interspecific *Spretus* crosses on knowledge of the mouse X chromosome and autosomes. *Trends in Genetics* **4**, 18–23.

Bailey, N. T. J. (1961). *Introduction to the Mathematical Theory of Genetic Linkage.* Clarendon Press, Oxford.

Barlow, R. E., Bartholomew, D. J., Bremner, J. M. and Brunk, H. (1973). *Statistical Inference Under Order Restrictions.* Wiley, New York.

Barnard, G. (1949). Statistical inference. *J. Roy. Statist. Soc. B* **11**, 115–149.

Buetow, K. and Chakravarti, A. (1987). Multipoint mapping using seriation. *Am. J. Hum. Genet.* **41**, 180–201.

Bishop, D. T. (1985). The information content of phase-known matings for ordering genetic loci. *Genet. Epidemiol.* **2**, 349.

Bishop, D. T. and Thompson, E. A. (1988). Linkage information and bias in the presence of interference. *Genet. Epidemiol.* **5**, 75–80.

Boehnke, M. (1986). Estimating the power of a proposed linkage study: A practical computer simulation approach. *Am. J. Hum. Genet.* **39**, 513–527.

Blackwelder, W. C. and Elston, R. C. (1985). A comparison of sib-pair linkage tests for disease susceptibility loci. *Genet. Epidemiol.* **2**, 85–97.

Bonney, G. E. (1984). On the statistical determination of major gene mechanisms in continuous human traits: regressive models. *Am. J. Med. Genet.* **18**, 731–749.

Bonney, G., Lathrop, G. M. and Lalouel, J. M. (1988). Combined linkage and segregation analysis using regressive models. *Am. J. Hum. Genet.* **43**, 29–37.

Botstein, D., White, R., Skolnick, M. and Davis, R. (1980). Construction of a genetic linkage map in man using restriction fragment length polymorphisms. *Am. J. Hum. Genet.* **32**, 314.

Cannings, C., Thompson, E. A. and Skolnick, M. H. (1976). Calculations of likelihoods in pedigrees of arbitrary complexity. *Adv. App. Prob.* **8**, 622–625.

Cantor, R. M. and Rotter, J. (1987). Marker concordance in pairs of distant relatives: A new method of linkage analysis for common diseases. *Am. J. Hum. Genet.* **41**, A252.

Cavalli-Sforza, L. L. and Bodmer, W. F. (1977). *The Genetics of Human Populations.* Freeman, San Francisco.

Cavalli-Sforza, L. L. and King, M. C. (1986). Detecting linkage for genetically heterogeneous diseases and detecting heterogeneity with linkage data. *Am. J. Hum. Genet.* **5**, 599–616.

Chakravarti, A. and Cox, T. (1987). Multipoint mapping using recombinant inbred (RI) strains. Human gene mapping 9. *Cytogen Cell Genetics* **46**, 591.

Chakravarti, A., Badner, J. A. and Li, C. C. (1987). Tests of linkage and heterogeneity in Mendelian disease using identity by descent scores. *Genet. Epidemiol.* **4**, 255–266.

Chotai, J. (1983). Isotonic inference for populations related to the uniform distribution. *Communications in statistics: Theory and methods* **12**, 2109–2118.

Chotai, J. (1984). On the lod score method in linkage analysis. *Ann. Hum. Genet.* **48**, 359–378.

Clayton, J. (1986). A multipoint linkage analysis program for X-linked disorders, with the example of Duchenne muscular dystrophy and seven DNA probes. *Hum. Genet.* **73**, 68–72.

Clerget-Darpoux, F. M. (1982). Bias of the estimated recombination fraction and lod score due to an association between a disease gene and a marker gene. *Ann. Hum. Genet.* **44**, 195–204.

Clerget-Darpoux, F. M., Bonaiti-Pellie, C. (1983). Possible misuse of the lod score method. *Am. J. Hum. Genet.* **35**, 195A.

Cockerman, C. C. and Weir, B. S. (1983). Linkage between a marker locus and a quantitative trait of sibs. *Am. J. Hum. Genet.* **35**, 263–273.

Cook, P. J. L., Robson, E. B., Buckton, K. E., Jacobs, P. A. and Polani, P. E. (1974) Segregation of genetic markers in families with chromosome polymorphism and structural rearrangements involving chromosome 1. *Ann. Hum. Genet.* **37**, 261–274.

Dausset, J., Cann, H., Cohen, D., Lathrop, G. M., Lalouel, J. M. and White, R. (1990). Centre d'Etude du Polymorphisme Humain (CEPH): Collaborative genetic mapping of the human genome. *Genomics* **6**, 575–577.

Davis, R. B. (1977). Hypothesis testing when a nuisance parameter is present only under the alternative. *Biometrika* **64**, 247–254.

Donis-Keller, H., Green, P., Helms, C., Cartinhour, S., Weiffenbach, B. and 28 others. (1987). A genetic linkage map of the human genome. *Cell* **51**, 319–337.

Edwards, H. J. (1971). The analysis of X linkage. *Ann. Hum. Genet.* **34**, 229–250.

Edwards, H. J. (1987a). Exclusion mapping. *J. Med. Genet.* **24**, 539–543.

Edwards, J. H. (1987b). The locus ordering problem. *Ann. Hum. Genet.* **51**, 251–258.

Elston, R. C. and Stewart, J. (1971). A general model for the genetic analysis of pedigree data. *Hum. Hered.* **21**, 523.

Felsenstein, J. (1979). A mathematically tractable family of genetic mapping functions with different amounts of interference. *Genetics* **91**, 769–775.

Georges, M., Lathrop, M., Hilbert, P., Marcotte, A., Schwers, A., Swillens, S., Vassart, G. and Hanset, R. (1989). On the use of DNA fingerprints in mapping the bovine 'muscular hypertrophy' gene. *Genomics*, to appear.

Goldin, L. and Gershon, E. S. (1988). Power of the affected-sib-pair method for heterogeneous disorders. *Genet. Epidemiol.* **5**, 35–42.

Haldane, J. B. S. (1919). The combination of linkage values, and the calculation of distance between the loci of linked factors. *J. Genet.* **8**, 299–309.

Haldane, J. B. S. and Waddington, C. H. (1931). Inbreeding and linkage. *Genetics* **16**, 357–374.

Haseman, J. K. and Elston, R. C. (1972). The investigation of linkage between a quantitative trait and a marker locus. *Behav. Genet.* **2**, 3–19.

Hasstedt, S. J. (1982). Linkage analysis using mixed, major gene with generalized penetrance or three-locus models. Hum. Gene Mapping 6. *Cytogenet. Cell Genetics* **18**, 284–285.

Hodge, S. and Spence, A. (1981). Some epistatic two-locus models of disease. II. The confounding of linkage and association. *Am. J. Genet.* **33**, 396–406.

Jeffreys, A., Wilson, V. and Thein, S. (1985). Hypervariable 'minisatellite' regions in human DNA. *Nature* **314**, 67–73.

Karlin, S. and Liberman, U. (1978). Classifications and comparisons of multilocus recombination distributions. *Proc. Natl. Acad. Sci. USA* **75**, 6332–6336.

Karlin, S. and Liberman, U. (1979). A natural class of multilocus recombination processes and related measures of crossover interference. *Adv. Appl. Prob.* **II**, 479–501.

Keats, B. J. B., Morton, N. E., Rao, D. C., Williams, W. (1979). *A source book for linkage in man.* Johns Hopkins University Press, Baltimore.

Keppen, L. D., Leppert, M., O'Connell, P., Nakamura, Y., Stauffer, D., Lathrop, M., Lalouel, J. M. and White, R. (1987). Etiological heterogeneity in X-linked spastic paraplegia. *Am. J. Hum. Genet.* **41**, 933–943.

Kidd, K. K. and Ott, J. (1984). Power and sample size in linkage studies. Human Gene Mapping 7. *Cytogenet. Cell Genet.* **37**, 510–511.

Kosambi, D. D. (1944). The estimation of map distance from recombination values. *Ann. Eugen.* **12**, 172–175.

Lalouel, . M. (1977). Linkage mapping from pair-wise recombination data. *Heredity* **38**, 61–77.

Lalouel, J. M. (1979). GEMINI: A computer program for optimization of general non-linear functions. University of Utah. Department of Medical Biophysics and Computing. Technical Report 14. Salt Lake City, Utah.

Lalouel, J. M., Rao, D. C., Morton, N. E. and Elston, R. C. (1983). A unified model for complex segregation analysis of quantitative traits. *Am. J. Hum. Genet.* **35**, 816–826.

Lander, E. S. and Green, P. (1987). Construction of multilocus linkage maps in humans. *Proc. Natl. Acad. Sci. USA* **84**, 2363–2367.

Lander, E. S., Green, P., Abrahamson, J., Barlow, A., Daly, M. J., Lincoln, S. L. and Newbers, L. (1987). MAPMAKER: An interactive computer package for constructing primary genetic linkage maps of experimental populations. *Genomics* **1**, 182–195.

Lander, E. S. and Botstein, D. (1989). Mapping mendelian factors underlying quantitative traits using RFLP linkage map. *Genetics* **121**, 185–199.

Lange, K. and Elston, R. C. (1975). Extensions to pedigree analysis. I. Likelihood calculations for simple and complex pedigrees. *Hum. Hered.* **25**, 95–105.

Lange, K. (1986a). The affected sib-pair method using identity by state relations. *Am. J. Hum. Genet.* **39**, 148–150.

Lange, K. (1986b). A test statistic for the affected-sib-set method. *Ann. Hum. Genet.* **50**, 283–290.

Lange, K., Weeks, D. and Boehnke, M. (1988). Programs for pedigree analysis: MENDEL, FISHER, and dGENE. *Genet. Epidemiol.* **5**, 471–462.

Lange, K. and Matthyse, S. (1989). Simulation of pedigree genotypes by random walks. Submitted for publication.

Lathrop, G. M. and Lalouel, J. M. (1984). Simple calculation of lod-scores on small computers. *Am. J. Hum. Genet.* **36**, 460–465.

Lathrop, G. M. and Lalouel, J. M. (1988). Efficient computations in multilocus linkage analysis. *Am. J. Hum. Genet.* **42**, 498–505.

Lathrop, G. M., Lalouel, J. M., Julier, C. and Ott, J. (1984). Strategies for multilocus linkage analysis in humans. *Proc. Natl. Acad. Sci. USA* **881**, 3443–3446.

Lathrop, G. M., Lalouel, J. M., Julier, C. and Ott, J. (1985). Multilocus linkage analysis in humans: detection of linkage and estimation of recombination. *Am. J. Hum. Genet.* **37**, 482–498.

Lathrop, G. M., Lalouel, J. M. and White, R. L. (1986). Calculation of human linkage maps: likelihood calculations for multilocus linkage analysis. *Genet. Epidemiol.* **3**, 39–52.

Lathrop, G. M., Chotai, J., Ott, J. and Lalouel, J. (1987). Tests of gene order from three-locus linkage data. *Ann. Hum. Genet.* **51**, 235–249.

Lathrop, G. M., Nakamura, Y., Cartwright, P., O'Connell, P., Leppert, M., Jones, C., Tateishi, H., Bragg, T., Lalouel, J. M. and White, R. (1988a). A primary genetic map of markers for human chromosome 10. *Genomics* **2**, 157–164.

Lathrop, G. M., Lalouel, J. M. and White, R. L. (1988b). The number of meioses needed to resolve gene order in a 1% linkage map. Human Gene Mapping 10. *Cytogenet. Cell Genet.* **46**, 643.

Lathrop, G. M., Farrall, M., O-Connell, P., Wainwright, B., Leppert, M., Nakamura, Y., Lench, N., Kruyer, H., Dean, M., Park, M., Vande Woude, G., Lalouel, G. M., Williamson, R. and White, R. (1988c). Refined linkage map of chromosome 7 in the region of the cystic fibrosis gene. *Am. J. Genet.* **42**, 38–44.

Lathrop, G. M. and Nakamura, Y. (1989). Mapping the human genome. In: J. D. Karand, L. Chao and G. W. Wan, eds., *Methods in nucleic acids*, CRC Press, Boca Raton, 157–180.

Leppert, M., Cavanee, W., Callahan, P., Holm, T., O'Connell, P., Thompson, K., Lathrop, G. M., Lalouel, J. M. and White, R. L. (1986a). A genetic linkage map for chromosome 13. *Am. J. Hum. Genet.* **39**, 425–437.

Leppert, M., Hasstedt, S. J., Holm, T., O'Connell, P., Wu, L., Ash, O., Williams, R. and White, R. (1986b). A DNA probe for the LDL receptor gene is tightly linked to hypercholesterolemia in a pedigree with early coronary disease. *Am. J. Hum. Genet.* **39**, 300–306.

Leppert, M., Breslow, J., Wu, L., Hasstedt, S. J., O'Connell, P., Lathrop, G. M., Williams, R., White, R. and Lalouel, J. M. (1988). Inference of a molecular defect of apolipoprotein *B* in a hypobeta-lipoproteinemia by linkage analysis in a large kindred. *J. Clin. Invest.* **82**, 847–851.

Litt, M. and Luty, J. A. (1989). A hypervariable microsatellite revealed by in vitro amplification of a dinucleotide repeat within the cardiac muscle actin gene. *Am. J. Hum. Genet.* **44**, 397–401.

Martinez, M. and Goldin, L. (1989). The detection of linkage and heterogeneity in nuclear families for complex disorders: One versus two marker loci. *Am. J. Hum. Genet.* **44**, 552–559.

Monaco, A., Neve, R., Colletti-Feener, C., Bertelson, C., Kurnit, D. and Kunkel, L. (1986). Isolation of candidate cDNAs for portions of the Duchenne muscular dystrophy gene. *Nature* **323**, 646–650.

Morton, N. E. (1955). Sequential tests for the detection of linkage. *Am. J. Hum. Genet.* **7**, 277–318.

Morton, N. E. (1956). The detection and estimation of linkage between the genes for elliptocytosis and the Rh blood types. *Am. J. Hum. Genet.* **8**, 80–96.

Morton, N. E. and MacLean, C. J. (1984). Multilocus recombination frequencies. *Genet. Res.* **44**, 99–108.

Morton, N. E. (1988). Multipoint mapping and the emperor's clothes. *Ann. Hum. Genet.* **52**, 309–318.

Muller, H. J. (1916). The mechanism of crossing-over. *Am. Nat.* **50**, 193–194.

Nakamura, Y., Leppert, M., O'Connell, P., Wolff, R., Holm, T., Culver, M., Martin, C., Fujimoto, E., Hoff, M., Kumlin, E. and White, R. (1987). Variable number of tandem repeat (VNTR) markers for human gene mapping. *Science* **235**, 1616–1622.

Nakamura, Y., Lathrop, G. M., O'Connell, P., Leppert, M., Lalouel, J. M. and White, R. (1989). A small region of human chromosome 14 shows a high frequency of recombination. *Genomics* **4**, 76–81.

O'Connell, P., Lathrop, G. M., Law, M., Leppert, M., Nakamura, Y., Hoff, M., Kumlik, E., Thomas, W., Elsner, T., Ballard, L., Goodman, P., Azen, E., Sadler, J. E., Lai, G. Y., Lalouel, J. M. and White, R. (1987). A primary genetic linkage map for human chromosome 12. *Genomics* **1**, 93–102.

Ott, J. (1974). Estimation of the recombination fraction in human pedigrees: efficient computation of the likelihood for human linkage studies. *Am. J. Hum. Genet.* **26**, 588–597.

Ott, J. (1977). Counting methods (EM algorithm) in human pedigree analysis: Linkage and segregation analysis. *Ann. Hum. Genet.* **40**, 443–454.

Ott, J. (1983). Linkage analysis and family classification under heterogeneity. *Ann. Hum. Genet.* **42**, 353–364.

Ott, J. (1985). *Analysis of Human Genetic Linkage*. Johns Hopkins University Press, Baltimore.

Ott, J. (1986). *Y* linkage and pseudoautosomal linkage. *Am. J. Hum. Genet.* **38**, 891–897.

Ott, J. and Lathrop, G. M. (1987). Goodness of fit test for locus order in 3-point mapping. *Genet. Epidemiol.* **4**, 51–58.

Ott, J. (1989). Computer-simulation methods in human linkage analysis. *Proc. Natl. Acad. Sci. USA* **86**, 4175–4178.

Penrose, L.S. (1935). The detection of autosomal linkage in data which consists of pairs of brothers and sisters of unspecified parentage. *Ann. Eugen* **6**, 133–138.

Ploughman, L. M. and Boehnke, M. (1989). Estimating the power of a proposed linkage study for a complex genetic trait. *Am. J. Hum. Genet.* **44**, 513–527.

Rao, D. C., Keats, B. J. B., Morton, N. E., Yee, S. and Lew, R. (1978). Variability of human linkage data. *Am. J. Hum. Genet.* **30**, 516–529.

Rao, D. C., Keats, B. J. B., Lalouel, J. M., Morton, N. E. and Yee, S. (1979). A maximum likelihood map of chromosome 1. *Am. J. Hum. Genet.* **31**, 680–696.

Riordan, J. R., Rommens, J. M., Kerem, B., Alon, N., Rozmahel, R., Grzelczk, Z., Zielinski, J., Lok, S., Plavsic, N., Chou, J. L., Drumm, M., Iannuzzi, M. C., Collins, F. S. and Tsui, L. C. (1989). Identification of the cystic fibrosis gene: cloning and characterization of complementary DNA. *Science* **245**, 1066–1072.

Risch, N. and Lange, K. (1977). An alternative model of recombination and interference. *Ann. Hum. Genet.* **43**, 61–70.

Risch, N. and Lange, K. (1983). Statistical analysis of multilocus recombination. *Biometrics* **39**, 949–963, 1983.

Risch, N. (1987). Assessing the role of HLA-linked and unlinked determinants of disease. *Am. J. Hum. Genet.* **40**, 1–14.

Risch, N. (1989a). Linkage detection tests under heterogeneity. *Genet. Epidemiol.* **6**, 473–480.

Risch, N. (1989b). Genetics of IDDM: Evidence for complex inheritance with HLA-DR. *Genet. Epidemiol.* **6**, 143–148.

Robson, E. B., Cook, P. J. L., Corney, G., Hopkinson, D. A., Noades, J. and Cleghorn, T. E. (1973). Linkage data on Rh, PGM1, PGD, pepsidae C and Fy from family studies. *Ann. Hum. Genet.* **36**, 393–399.

Silver, J. and Buckler, C. E. (1986). Statistical considerations for linkage analysis using recombinant inbred strains and backcrosses. *Proc. Natl. Acad. Sci. USA* **83**, 1423–1427.

Smith, C. A. B. (1953). The detection of linkage in human genetics (with discussion). *J. R. Stat. Soc.* B **15**, 153–192.

Smith, C. A. B. (1958). Counting methods in genetical statistics. *Ann. Eugen.* **21**, 254–276.

Smith, C. A. B. (1959). Some comments on the statistical methods used in linkage investigations. *Am. J. Hum. Genet.* **11**, 289–304.

Smith, C. A. B. (1963). Testing for heterogeneity of recombination values in human genetics. *Am. Hum. Genet.* **27**, 175–182.

Sturt, E. (1975). The use of LOD score for the determination of the order of loci on a chromosome. *Ann. Hum. Genet.* **39**, 255–260.

Sturtevant, A. H. (1913). The linear association of six sex-linked factors in Drosophila, as shown by their mode of association. *J. Exp. Zool.* **14**, 43–59.

Thompson, E. A., Kravitz, K., Hill, J. and Skolnick, M. (1978). Linkage and the power of a pedigree structure. In: N. E. Morton and C. S. Chung, eds., *Genetic Epidemiology.* Academic Press, New York, 247–253.

Thompson, E. A. (1983). Optimal ampling for pedigree analysis: parameter estimates and genotype uncertainty. *Theor. Pop. Biol.* **24**, 39–58.

Thompson, E. A. (1984a). Information gain in joint linkage analysis. *IMA J. Math. Appl. Med. Biol.* **1**, 31.

Thompson, E. A. (1984b). Interpretation of LOD scores with a set of marker loci. *Genet. Epidemiol.* **1**, 357–362.

Thompson, E. A. (1987). Crossover counts and likelihood in multipoint linkage analysis. *IMA J. Math. App. Med. Biol.* **4**, 93–108.

Vogel, F. and Moltulsky, A. G. (1979). *Human Genetics.* Springer, New York.

Wald, A. (1947). *Sequential Analysis.* Wiley, New York.

Weber, J. L. and May, P. E. (1989). Abundant class of human DNA polymorphism which can be types using polymerase chain reaction. *Am. J. Hum. Genet.* **44**, 388–396.

Weeks, D. and Lange, K. (1988a). Preliminary ranking procedures for multilocus ordering. *Genomics* **3**, 236–242.

Weeks, D. and Lange, K. (1988b). The affected-pedigree-member method of linkage analysis. *Am. J. Hum. Genet.* **42**, 315–326.

Weeks, D., Ott, J. and Lathrop, G. M. (1990) SLINK: A general simulation program for linkage analysis. *Am. J. Hum. Genet.* **47**, A204.

White, R., Leppert, M., Bishop, T., Barker, D., Berkowitz, J., Brown, C., Callahan, P., Holm, T. and Jerominski, L. (1985). Construction of linkage maps with DNA markers for human chromosomes. *Nature* **313**, 101.

White, R. and Lalouel, J. M. (1987). Investigations of genetic linkage in human families. In: H. Harris and K. Hirschhorm, eds., *Advances in Human Genetics*, Vol. 16., Plenum, New York, 121–128.

Wilson, S. (1988). A major simplification in the preliminary ordering of linked loci. *Genet. Epidemiol.* **5**, 75–80.

C. R. Rao and R. Chakraborty, eds., *Handbook of Statistics, Vol. 8*
© Elsevier Science Publishers B.V. (1991) 125–144

5

Statistical Design and Analysis of Epidemiologic Studies: Some Directions of Current Research

Norman Breslow

1. Introduction

Epidemiology is the branch of public health science concerned with the occurrence of disease in populations and how that occurrence is related to genetic and environmental risk factors. Epidemiologic research influences many health promotion and disease prevention practices, such as the personal decision to quit smoking or the societal decision to regulate emissions of ionizing radiation. Statisticians have played a major role in the development of epidemiology as a science, and many of its fundamental concepts and methods rest heavily on those of statistics. This paper reviews several recent developments in statistical methodology that were either motivated by, or have major applications to, problems in epidemiology. It is divided into two sections, Design and Analysis.

2. Design

Classical epidemiologic designs for studying disease/risk factor associations at the level of the individual are the cohort design and the case-control design. With the cohort design, groups of subjects defined often on the basis of a particular exposure are followed forward in time to ascertain disease specific incidence or mortality rates. A famous example is the cohort of survivors of the atom bomb blasts in Hiroshima and Nagasaki (Beebe, Ishida and Jablon, 1962). With the case-control design, cases of a particular disease are ascertained as they arise in a defined population or seek hospital treatment and their exposure histories, obtained retrospectively by interview or other means, are compared to those of control subjects who are sampled from the population 'at risk'. Statistical methods for the design and analysis of cohort and case-control studies are discussed at length by Breslow and Day (1980, 1987).

Supported in part by USPHS Research Grant No. CA40644. This work is based on an invited address at the *Symposium on the Future of Theory and Methods in Epidemiology*, held during the 1987 meeting of the International Epidemiological Association in Helsinki.

At the *1981 Symposium on Epidemiology in Occupational Health*, Miettinen (1982, 1985) put forward two new design concepts for epidemiologic research that tended to blur the distinction between cohort and case-control studies and that were remarkably prophetic of future developments. First, as an alternative to the usual sample of disease-free controls, he proposed a sample drawn drom the study 'base' without regard to the disease outcome. Second, for situations where expensive data acquisition remains to be done after ascertainment of disease and exposure status, he suggested that subsamples be selected with regard to *both* characteristics. In the intervening years, considerable work has been done on these two design concepts and associated methods of data analysis, so that today they are ready for full exploitation. Both should see substantial use in future years.

2.1. A numerical example

Table 1 presents a ficticious numerical example that will help to fix ideas. Part A shows the exposure distribution of a cohort of 10 200 subjects identified as to whether or not they develop the disease of interest during the study period. For concreteness, and also to emphasize that we have in mind a follow-up study of persons who are initially disease-free, we may think of asbestos insulation workers kept under surveillance for lung cancer for 10 years starting from the 20-th anniversary of their employment. The odds ratios indicate that moderately exposed workers have lung cancer rates about 3 times those of the lightly exposed, whereas for the heavily exposed this ratio is closer to 9. These ratios could be interpreted as incidence rate ratios if the entries in the row labeled 'non-cases' were exactly proportional to the 'person-years' of observation contributed by members of the cohort and likewise if the control sample in the case-control study was drawn using 'incidence density' sampling (Greenland and Thomas, 1982; Miettinen, 1976). However, this refinement is unnecessarily complicated for present purposes.

Table 1(B) shows the expected exposure distribution of controls, selected as a 10% random sample of the non-cases, that are to be used with all available cases in a classical case-control study. Such designs are well known to produce unbiased estimates of the odds ratios that would be observed if the entire cohort were followed up (Cornfield, 1951). Multiple logistic regression analysis is used to estimate exposure odds ratios that are adjusted for covariables (Breslow and Day, 1980). Case-control sampling from cohort data files has been suggested as a means of reducing the computational burden of the analysis, or reducing the number of subjects for whom covariable information, for example on smoking, needs to be collected (Mantel, 1983). For a time-dependent analysis, a few controls are selected randomly from among those who are disease-free at the time each case is diagnosed. 'Risk sets' consisting of a case and its time matched controls are analyzed by conditional logistic regression so as to approximate the results that would be obtained from a proportional hazards regression analysis of data from the entire cohort (Breslow and Day, 1980; Liddel et al., 1977; Prentice and Pyke, 1979).

Table 1
Design options in epidemiology[a]

Exposure	LO	MED	HI	TOT
(A) Full cohort (base)				
Cases	300	200	50	550
Non-cases	7700	1800	150	9650
Total	8000	2000	200	10200
Odds ratio	1.0	2.9	8.6	
(B) Case-control sample				
Cases	300	200	50	550
Controls[b]	770	180	50	965
Odds ratio[b]	1.0	2.9	8.6	
(C) Case-cohort (case-base) sample				
Cases[b]	240 + 60	160 + 40	40 + 10	440 + 110
Controls[b]	1540	360	300	1930
Sub-cohort[b]	1600	400	40	2040
Odds ratio[b]	1.0	2.9	8.6	
(D) 'Balanced' two-stage sample (selectivity in both series)				
Cases[b]	50	50	50	150
Controls[b]	150	150	150	450
Odds ratio[b]	1.0	1.0	1.0	

[a] After Miettinen (1982).
[b] Expected values.

2.2. The case-cohort design

Table 1(C) shows the expected exposure distributions for an alternative type of study design termed by Miettinen (1982) a 'case-base' design and by Prentice (1986) a 'case-cohort' design. Here, a subcohort selected without regard to the disease outcome is used to estimate the exposure distribution in the cohort and thus the denominators of the disease rates at each level of exposure. Returning to the example, among those in the 20% subcohort sampled at random from the 10200 insulation workers we expect 40 to be heavily exposed to asbestos. Ten will develop lung cancer. Forty additional lung cancer cases will be ascertained from among the 160 heavily exposed workers who are not in the subcohort. Unbiased estimates of the exposure odds ratios result from a comparison of exposure levels for the non-cases in the subcohort with those of the cases in the entire cohort.

Table 1(D) illustrates a 'balanced two-stage' design where both disease and exposure data are used to select the subsamples. All the cases and all of the heavily exposed are selected in order to increase the information available about the risk of high exposure. Statistical analyses of data from this design clearly need to be adjusted for the biased sampling, however, in order to estimate the odds ratios of interest. We return to this design later.

2.3. Advantages of the case-cohort design

The case-cohort design offers certain advantages over the case-control design as a technique for reduction of the number of subjects for whom covariable data are collected or processed. For time-matched analyses, a subcohort member is included in the risk set for all cases who are diagnosed while he is under observation. Thus, he may serve as a control for several cases. Consequently, the sizes of the risk sets are generally larger than for a time-matched case-control study (Breslow and Day, 1980), even though the total number of subjects are the same. The regression coefficients of interest are estimated with greater precision. However, the statistical analysis is more complicated due to the need to account for correlations between the individual risk set contributions to the estimating equations (Prentice, 1986). Furthermore, if it turns out that only a few subcohort members are still under observation at the later times when most cases of disease occur, the risk sets for such cases may be reduced in size with a consequent loss of efficiency. (See note added in proof.)

Since no knowledge of their disease outcome is required, subcohort members may be selected at the outset of the study. When implemented in the context of a dietary intervention trial, for example, this means that biochemical assays of serum samples may be carried out at once for the subcohort in order to monitor the effectiveness of and compliance with the intervention. Similar assays are made of the stored serum for all cases as they arise and the resulting data are combined with those from the subcohort for statistical analysis. Clearly, one needs to be very sure that assay results for samples stored several years are equivalent to those conducted with fresh material; otherwise, the results may be biased.

Table 2 illustrates a case-cohort design used to study a possible association between external radiation exposure and lung cancer in nuclear energy workers (Peterson et al., 1987). The investigators identified more than 5000 white male operations employees who had been monitored for external radiation for at least three years and who had terminated employment in 1965 or later, when questions about tobacco use became a routine part of the periodic medical exam. Eighty-six lung cancer deaths were ascertained through records of the Social Security Administration and state and national death registers. The subcohort consisted of 455 men, stratified by year of birth in rough proportion to the numbers of lung cancer deaths. Detailed time-dependent smoking histories were abstracted from the medical records. In spite of this effort, it turned out that adjustment for tobacco use had little effect on the patterns of lung cancer risk across radiation exposure groups. There was a slightly elevated lung cancer risk among the moderately exposed, but a non-significant negative trend overall.

Table 2
Case-cohort study of lung cancer and radiation exposure in nuclear workers[a]

Period of birth	Lung cancer deaths	Subcohort members		
			Selected	
		Eligible	All	Lung cancer
1900–1904	19	413	95	6
1905–1909	18	688	95	3
1910–1914	27	949	135	4
1915–1919	13	938	70	0
1920–1924	6	1012	30	0
1925–1929	2	922	10	0
1930–1934	1	523	10	0
1900–1934	86	5445	455	13

[a] *Source:* Peterson, Gilbert, Stevens and Buchanan (1987).

2.4. The two-stage design

One serious drawback to both case-cohort and case-control designs is that they ignore data on exposure and disease for subjects not included in the subsamples selected for covariable ascertainment. In the preceding example, radiation dosimetries were available but not used for 4917 nuclear energy workers who did not die from lung cancer and who were not selected for the subcohort. Such designs also may be inefficient because they fail to use all the subjects who have experienced a rare exposure (see Table 1). One needs 'selectivity in both series' so that all 'rare' subjects, whether cases or exposed, are used in the analysis (Miettinen, 1982).

White (1982) and Walker (1982) published separate accounts of how to combine the 'first stage' data on exposure and disease, i.e. data available from the entire cohort (Table 1(A)), with the covariable data collected from a balanced second stage subsample (Table 1(D)) so as to estimate covariable adjusted odds ratios relating disease and exposure. White also derived a variance for the adjusted odds ratio that correctly accounts for sampling errors in the first stage data. However, both papers were limited to the simple situation of sampling from a fourfold table (case-control vs. exposed/nonexposed) and to covariables having a small number of discrete values.

Breslow and Cain (1988) recently developed the modifications of logistic regression needed to analyze data from samples that are stratified according to disease and any number of exposure levels, using model equations containing discrete and continuous exposure variables, covariables and interaction terms. Theirs is an extension of the work of econometricians Manski and McFadden (1981) on 'choice based', i.e. outcome selective, sampling for binary response analysis. Breslow and Cain show how to adjust output from standard logistic regression analysis of the second stage data so as to account both for the biased sampling and for the extra data available from the first stage sample. For the situations considered by White and Walker, in which exposure is represented as a single binary variable, one simply multiplies the adjusted odds ratio from the logistic analysis by the reciprocal of the cross-product ratio of the sampling fractions to adjust for the bias. More precisely, let us denote by N_{11} and N_{10} the numbers of exposed and non-exposed cases in the cohort as a whole and by N_{01} and N_{00} the corresponding numbers of non cases. Suppose that the sizes of the subsamples drawn from each of these four categories are n_{11}, n_{10}, n_{01}, and n_{00}, and further that the regression coefficient for the binary exposure variable in the logistic regression model fitted to the subsampled data is $\hat{\beta}$. Then the adjusted estimate of the log odds ratio is

$$\hat{\beta}_{\text{adj}} = \hat{\beta} + \log\left(\frac{N_{11}N_{00}n_{10}n_{01}}{N_{10}N_{01}n_{11}n_{00}}\right).$$

Similarly, if $\hat{\sigma}^2$ is the standard variance estimate for $\hat{\beta}$ obtained from the program, then

$$\hat{\sigma}^2_{\text{adj}} = \hat{\sigma}^2 - \left[\left(\frac{1}{n_{11}} - \frac{1}{N_{11}}\right) + \left(\frac{1}{n_{10}} - \frac{1}{N_{10}}\right) \right.$$
$$\left. + \left(\frac{1}{n_{01}} - \frac{1}{N_{01}}\right) + \left(\frac{1}{n_{00}} - \frac{1}{N_{00}}\right)\right]$$

is the variance estimate for $\hat{\beta}_{\text{adj}}$ that accounts for the first stage data. When the exposure effects are modelled continuously, the adjustment involves more complicated matrix manipulations (Breslow and Cain, 1988).

2.5. Analyses of the Framingham data

Tables 3–6 illustrate the application of two stage logistic regression methodology to 18 years follow-up data from Framingham (Kannel and Gorton, 1974), the cohort study of a small Massachusetts town that first identified several important risk factors of coronary heart disease (CHD). A public use tape kindly supplied by the National Center for Health Statistics contained data for 5052 individuals with no prior CHD diagnosis who had known values recorded at the initial examination for height, weight, systolic blood pressure, age and cigarette smoking. Their distribution by age and sex is shown in Table 3(A). For purposes of this illustrative exercise, age and sex were regarded as the 'exposure' whereas relative weight, blood pressure, etc. were regarded as the 'covariables'.

Several different sampling schemes were used to reduce the size of the dataset for efficient computer analysis. First, two samples of 673 controls were selected for comparison with the 673 cases of CHD, one a simple random sample and the

Table 3
Sampling from the Framingham cohort

	Males			Females			Total
Age:	30 –	40 –	50 –	30 –	40 –	50 –	
			(A) Entire cohort				
CHD	73	158	189	24	75	154	673
No CHD	742	603	481	996	870	687	4379
Total	815	761	670	1020	945	841	5052
			(B) Control samples: all available cases				
Random	113	80	76	151	136	11	673
Stratified	73	158	189	24	75	154	673
			(C) Standard case-control sample				
Cases	18	25	45	5	19	37	149
Controls	27	25	16	34	22	25	149
			(D) Balanced two stage sample				
Cases	25	25	25	24	25	25	149
Controls	25	25	25	24	25	25	149

other stratified by age and sex (Table 3(B)). Additional simple random samples of 149 cases and 149 controls were drawn from the cases and non cases (Table 3(C)) to illustrate the problems with a 'small' case-control study. Finally, in hopes of demonstrating the increased efficiency of the balanced design, two more samples of 149 cases and 149 controls were selected using stratification by age and sex (Table 3(D)).

Logistic regression models were fit to the entire cohort of 5052 men and women in order to establish a reference point for comparison with results obtained from the various case-control samples (Table 4). Except for the constant term and a binary indicator of female sex, all variables were continuous. Most were centered about modal values and then scaled so that the regression coefficients would be of comparable magnitude. Note that the regression coefficient for age is reduced by inclusion of the covariables, indicating that some of the age effect is mediated by the association of age with high blood pressure and other risk factors. Measurements of serum cholesterol were missing for over one-third of the initial cohort, and a binary indicator of serum availability was added to the model shown in Table 4(C) to account for this. Those for whom bloods were taken had a lower CHD risk. The examples below use only the five variable model (Table 4(B)) so as to avoid the complication of missing values. Results for the seven variable model were similar.

Table 4
Logistic regression fits to the entire cohort[a]

Variable	Coefficient	Standard error
(A) Two-variable model		
CONSTANT	− 1.129	0.058
SEX (F)	− 0.894	0.087
(AGE-50)/10	0.784	0.052
(B) Five-variable model		
CONSTANT	− 1.420	0.095
SEX (F)	− 0.848	0.097
(AGE-50)/10	0.677	0.057
(FRW-100)/100	0.950	0.266
(SBP-142.6)/100	1.558	0.181
CIGS (PACK)	0.381	0.073
(C) Seven-variable model		
CONSTANT	− 1.217	0.095
SEX (F)	− 0.906	0.098
(AGE-50)/10	0.651	0.057
(FRW-100)/100	0.930	0.270
(SBP-142.6)/100	1.543	0.185
CIGS (PACK)	0.381	0.073
SC AVAILABLE	− 0.330	0.090
(SC-227.3)/100	0.634	0.122

[a] Key: FRW = Framingham relative weight, SBP = systolic blood pressure (mmHg), and SC = Serum cholesterol (mg/100 cc).

Table 5 presents logistic regression fits to data for all 673 CHD cases and an equal number of controls. The 'unadjusted' entries are standard program output for the 1346 observations; the 'adjusted' coefficients and standard errors account for the extra data shown in Table 3(A). No adjustment is needed for the random case-control sample so far as validity of the relative risk estimates is concerned (Breslow and Day, 1980). By taking account of the Table 3(A) data, however, we are able to correct the constant term and to make the estimates of age and sex effects more precise. Note that the adjusted standard errors for the 'exposure' variables are quite close to those for the entire cohort (Table 4(B)) and that the coefficient for sex, at least, is moved substantially closer to its cohort value of − 0.85. The adjustment has negligible effect on the coefficients and standard errors for the covariables, information about which is exclusively contained in the case-control sample. (Had age and sex been analyzed as discrete variables with interactions, the adjustment would have no effect at all on the covariables (Breslow and Cain, 1988).) On the other hand, adjustment is clearly needed for the stratified case-control sample (Table 5(B)) in order to account for the fact that

Table 5
Logistic regression fits to the second stage data: all available cases and an equal number of controls
(N = 1346)[a]

Variable	Unadjusted		Adjusted	
	Coefficient	Standard error	Coefficient	Standard error
(A) Random control sample				
CONSTANT	0.51	0.12	− 1.39	0.09
SEX	− 1.00	0.14	− 0.91	0.11
AGE	0.69	0.08	0.73	0.07
FRW	1.25	0.38	1.26	0.38
SBP	1.49	0.27	1.50	0.27
CIGS	0.36	0.11	0.36	0.11
(B) Stratified control sample				
CONSTANT	− 0.27	0.10	− 1.46	0.08
SEX	0.05	0.13	− 0.79	0.11
AGE	0.00	0.08	0.65	0.07
FRW	0.68	0.36	0.81	0.36
SBP	1.50	0.25	1.57	0.25
CIGS	0.38	0.10	0.42	0.10

[a] See Table 4 for key. Variables are centered and scaled as shown there.

cases and controls were balanced on age and sex. The change in the covariable coefficients is more pronounced here also.

Table 6 presents analogous results for the smaller case-control samples (Tables 3(C) and 3(D)). None of the 'covariable' effects are well estimated here and those for relative weight and smoking even fail to demonstrate statistical significance. The standard errors for age and sex are reduced substantially by the adjustment, but do not now approach those obtained from the full cohort analysis (Table 4(B)). Nevertheless, the age and sex effects are reasonably well estimated by the adjusted coefficients.

2.6. Relative efficiency

A somewhat surprising feature of Table 6 is that the adjusted standard errors from analysis of the balanced sample are little if any smaller than those for the random sample. The anticipated gains from stratification are not realized in this example because there are no 'rare' categories of 'exposure'. The two sexes are equally represented and there is sufficient variation in age so that the linear age term is well determined even from the random subsamples.

Table 7 presents relative efficiencies, i.e. ratios of asymptotic variances, for regression coefficients estimated from balanced vs. unbalanced samples for a simple model with one binary exposure, one binary covariable that may confound

Table 6
Logistic regression fits to the second stage data: case-control samples (N = 298)[a]

Variable	Unadjusted		Adjusted	
	Coefficient	Standard error	Coefficient	Standard error
(A) Random samples of cases and controls				
CONSTANT	0.58	0.25	−1.18	0.18
SEX	−0.69	0.28	−0.95	0.16
AGE	0.70	0.16	0.70	0.09
FRW	0.36	0.80	0.33	0.81
SBP	2.11	0.72	2.15	0.71
CIGS	0.14	0.23	0.19	0.23
(B) Balanced samples of cases and controls				
CONSTANT	−0.18	0.23	−1.28	0.16
SEX	0.14	0.26	−0.92	0.16
AGE	−0.14	0.16	0.68	0.11
FRW	−0.12	0.78	−0.07	0.78
SBP	2.58	0.61	2.55	0.62
CIGS	0.26	0.23	0.27	0.23

[a] See Table 4 for key. Variables are centered and scaled as shown there.

the disease/exposure association and their interaction (Breslow and Cain, 1988). It is assumed that only 5% of the population is exposed, that exposure carries a relative risk of 2, and that the first stage sample is so much larger than the second that only sampling errors from the latter need be accounted for. The table entries show that, when the exposure is rare, the balanced design offers substantial advantages for estimation of exposure effects in the presence of a strong confounder, at the cost of a slight reduction in efficiency for estimation of the confounder effect. The balanced design is much more efficient for estimation of interaction effects.

2.7. Summary

In summary, two new strategies for subsampling and analysis of data from epidemiologic studies have been developed that should see increasing use in future years. The case-cohort design is particularly well suited to prospective intervention trials where it is desired to monitor covariables on a sample while storing biological specimens or other materials from the remainder of the cohort for later processing. The two-stage design is especially appropriate for retrospective cohort studies where data on primary exposures and disease outcome are already available and it remains to collect the covariable information.

Table 7
Relative efficiencies of balanced vs. 'case-control' sampling[a]

RR of confounder[b]	Odds ratio linking exposure & confounder	Relative efficiency for estimating		
		Exposure effect	Confounder effect	Interaction
(A) 'Common' disease: cases and controls balanced on exposure				
	0.2	1.4	0.7	4.4
0.2	1.0	4.4	1.0	3.5
	5.0	3.6	1.5	3.3
	0.2	0.7	0.7	5.8
1.0	1.0	1.0	1.0	4.1
	5.0	1.1	1.1	3.9
	0.2	1.8	0.8	6.5
5.0	1.0	3.7	1.0	4.1
	5.0	1.4	0.8	3.4
(B) 'Rare' disease: all available cases used				
	0.2	1.0	0.8	1.4
0.2	1.0	1.2	0.9	1.5
	5.0	1.3	0.9	1.7
	0.2	1.0	0.7	2.3
1.0	1.0	1.0	0.8	2.1
	5.0	1.0	0.8	2.1
	0.2	1.1	0.8	2.9
5.0	1.0	1.2	0.8	1.9
	5.0	1.0	0.8	1.7

[a] *Source:* Breslow and Cain (1988).
[b] Relative risk (RR) for a confounder that affects 30% of the population.

3. Analysis

The computer developments of the past decade have revolutionized the methods of data analysis used by professional statisticians. Powerful desktop workstations are now available at relatively low cost that facilitate real time graphical exploration of moderately large databases. Formal inference protocols, where test statistics derived from regression coefficients are referred to tables of the chi-square distribution, are giving way to nonparametric estimation procedures in which statistical significance is evaluated by repeated computer sampling from the database itself. These developments are starting to impact data analysis in epidemiology and in other fields where statistical methods play an important role. Two examples will serve to illustrate current trends.

3.1. The generalized additive model

Hastle and Tibshirani (1986, 1987) developed a generalization of linear regression models of the form

$$y = \alpha + \beta_1 x_1 + \beta_2 x_2 + \cdots + \beta_p x_p + \varepsilon,$$

in which the regression variables x with unknown coefficients β are replaced by smooth functions $s = s(x)$. Thus, the model equation is given as

$$y = \alpha + s_1(x_1) + s_2(x_2) + \cdots + s_p(x_p) + \varepsilon.$$

This frees the statistician from the constraints of a linear dependence of y on each x. However, the contributions to the regression equation from different x's are still additive. Formerly, it was possible to relax the linearity assumption by replacing each linear term with a low degree polynomial in x or a parametric non-linear function. The advantage of the new methods is that no such specification of functional form is needed. The goal is to let the data 'speak for themselves'.

Various 'smoothers' are available to estimate the s functions. Hastie and Tibshirani recommend a local linear smoother. If only a single x variable is involved, this means estimating s at x by the linear regression of y on x using only observations (y_i, x_i) for which x_i is close to x. Multiple regression variables are accounted for in the usual fashion by regressing the residuals involving those terms that are already in the model on each added x variable. The width of the neighborhoods are determined by a parameter known as the span. For a span of 0.1, the linear regression at x uses only 10% of the data points, the 5% with nearest x_i on either side of x. A span of 1.0 corresponds to simple linear regression. By varying the span in the range from 0 to 1, one can control the degree of smoothness. Objective selection of the 'optimal' span for each x is possible using the criterion of 'cross-validation'. One minimizes the prediction error when each y observation is estimated from its associated x's, using a regression equation determined from the rest of the data (Stone, 1974). In practice, however, it is sometimes desirable to consider a variety of different spans and to pick one that appears to give a 'reasonable' picture of the relationship. There is the usual tradeoff between excess variance if the span is too small so that the estimated regression function is too wiggly, and excess bias if the span is too large so that the true curvature in the relationship is lost.

Hastie and Tibshirani have extended this approach for use with generalized linear models (McCullagh and Nelder, 1983) fitted by maximum likelihood, of which the paradigm is logistic regression. Using their program, we fit models with varying spans to the 673 cases and 673 randomly sampled controls from Framingham (Table 3(B)). Separate fits were made for males and females that included age, relative weight, blood pressure and cigarettes as regression variables (see Table 4(B)). Figures 1 and 2 contrast the 'smooths' estimated using spans of 0.1, 0.3 and 0.5 with the lines whose slopes were determined by the usual logistic

AGE   SYSTOLIC BLOOD PRESSURE

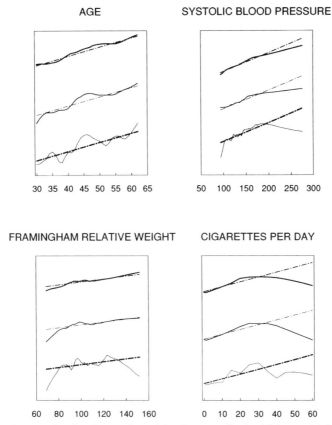

FRAMINGHAM RELATIVE WEIGHT   CIGARETTES PER DAY

Fig. 1. Smooth regression functions estimated by the generalized additive model for Framingham males with spans of 0.1 (lowest curve in each panel), 0.3 (middle curve) and 0.5 (upper curve).

regression approach. Both are normalized so that the estimated values, $s(x)$ or βx, sum to zero over all subjects. For most variables, there is remarkably little evidence for a departure from linearity. A possible exception is the relation of CHD risk to cigarette consumption. There is a slight decline in risk from non-smokers to very light smokers, especially for females, with a gradual rise in risk thereafter. For men, there is a suggestion that the CHD risk may stop rising after 1–2 packs/day.

Substantial work remains to be done on statistical inference procedures in connection with these new methods. How are we to decide, for example, whether the simple linear model for cigarettes provides an adequate description of the data in comparison with the generalized additive fit with span = 0.1? Hastie and Tibshirani provide only a partial answer. Using the 'deviance' (twice the log likelihood ratio) as a measure of the discrepancy between the fitted model and the observed data, they note that this usually has a chi-square distribution with degrees of freedom equal to $n - p$, the number of subjects minus the number of

AGE   SYSTOLIC BLOOD PRESSURE

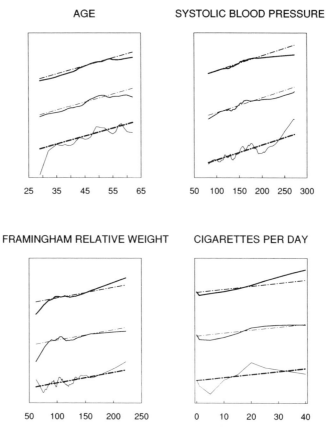

FRAMINGHAM RELATIVE WEIGHT   CIGARETTES PER DAY

Fig. 2. Smooth regression functions estimated by the generalized additive model for Framingham females with spans of 0.1 (lowest curve in each panel), 0.3 (middle curve) and 0.5 (upper curve).

fitted parameters. For the generalized additive model, they estimate an 'effective number' of parameters from the trace of the 'projection' matrix at the final stage of iteration. Table 8 summarizes the results of an 'analysis of deviance' based on this concept. Both theoretical and empirical calculations suggest that the expected difference in deviances for two nested models is approximately equal to the difference in the corresponding values of p. However, the *variability* of the deviance difference is often greater than twice the p difference. One therefore should resist the temptation to refer such deviance differences to tables of chi-square. It usually will be necessary to resort to resampling methods (Efron and Tibshirani, 1986) in order to accurately assess the significance of the observed results. In fact, the greatest danger in these new methods lies in the temptation to overinterpret insignificant features of the fitted curves. While they offer great power and flexibility, they also pose a trap for the unwary.

Table 8
Analysis of deviance: goodness-of-fit statistics for the models shown in Figures 1 and 2

Span	Males		Females	
	Deviance	p [a]	Deviance	p
0.1	770.5	47.1	702.2	47.1
0.3	812.5	17.3	732.8	17.0
0.5	824.0	11.8	744.7	11.5
1.0 (linear)	833.1	5	755.2	5

[1] p = effective number of parameters used in fitting model.

3.2. Nonparametric estimation of the SMR

Our second example of graphical analysis of epidemiologic data concerns a procedure for nonparametric estimation of the standardized mortality ratio (SMR) as a continuous function of age or time (Andersen et al., 1985; Breslow and Day, 1987; Breslow and Langholz, 1987). In occupational mortality studies, it is useful to calculate the SMR (defined as the ratio of the observed number of cases to the number expected from standard rates) according to time since initial employment, time since termination of employment, duration of employment or other time-varying factors. One hopes thereby to answer questions about the latent interval between exposure and disease, about the attenuation of risk following cessation of exposure, or even about the biological mechanisms involved in the disease process. Similarly, by examining changes in the SMR according to age, calendar time or other variables involved in its definition, one can examine whether the assumption of a constant risk ratio that implicitly underlies its use as a summary measure is well founded (Breslow and Day, 1985).

As an example, Table 9 presents observed and expected numbers of respiratory cancer deaths, broken down according to several fixed and time-varying factors for a cohort of 8014 Montana smelter workers exposed to airborne arsenic (Breslow, 1985; Lee-Feldstein, 1983). Workers were first employed between 1885 and 1956. The follow-up period was 1938–1977 and a worker was entered on study if still employed in 1938 or upon the completion of one year of employment if hired thereafter. Expected rates were based on U.S. national mortality rates for white males, specific for age and calendar year of follow-up. Workers employed prior to 1925, when changes in the smelting process substantially reduced the concentrations of airborne arsenic, had respiratory mortality rates approximately 3.6 times expected. For those employed later the ratio was only 1.6. There were substantial variations in the SMR according to birthplace, duration of time since first employment, level of arsenic exposure and calendar year of follow-up (Breslow, 1985).

Interpretation of these results, however, should take account of the strong correlations between the risk factors. Since follow-up did not commence until

Table 9
Variations in respiratory cancer SMR's for Montana smelter workers[a]

Factor	Level	Number of deaths		SMR
		Observed	Expected[b]	
Year of	1885–1924	115	31.8	3.62
employment	1925–1955	161	98.2	1.64
Birthplace	U.S.	198	110.0	1.80
	Foreign	80	21.0	3.81
Years since	1–14	101	61.2	1.65
first	15–29	59	31.9	1.85
employed	30 +	116	36.8	3.15
Arsenic	Light	153	95.7	1.60
exposure	Moderate	91	26.9	3.39
	Heavy	32	7.4	4.34
Year of	1938–1949	34	8.4	4.03
follow-up	1950–1959	65	22.1	2.94
	1960–1969	94	44.6	2.11
	1970–1977	83	54.9	1.51

[a] *Source:* Breslow (1985).
[b] From U.S. national rates for white males, accounting for age and calendar year of follow-up.

1938, for example, nearly everyone hired prior to 1925 contributed person-years of observation only to the later categories of years since initial employment. When the change in the SMR according to time since hire is adjusted for the effect of period of hire in a Poisson regression analysis, it is no longer statistically significant (Breslow, 1985).

Recent developments in techniques of survival analysis facilitate the estimation of the SMR as a continuous function of a continuous time variable, without the need for arbitrary grouping of each factor into discrete intervals. Briefly, suppose that $\lambda(t)$ denotes the death rate for a randomly selected cohort member at time t, and that $\lambda^*(t)$ denotes the expected rate based on his age and the calendar year at t. The function we want to estimate is the ratio $\theta(t) = \lambda(t)/\lambda^*(t)$. For technical reasons, it is easiest first to estimate the cumulative SMR $\Theta(t) = \int_0^t \theta(s)\, ds$ using the formula

$$\hat{\Theta}(t) = \sum_{t_i < t} \frac{d_i}{\sum_j Y_j(t_i)\lambda_j^*(t_i)},$$

where d_i denotes the number of deaths that occur at t_i and $Y_j(t_i)$ denotes whether ($Y = 1$) or not ($Y = 0$) the j-th cohort member is under observation at that time. Thus, $\hat{\Theta}$ is analogous to the standard nonparametric estimator of cumulative mortality, except that the denominator summands, usually the total numbers of deaths at t_j, are replaced by the expected numbers based on standard rates.

The graph of $\hat{\Theta}(t)$ for the Montana cohort, with t equals time since initial employment, is shown in Figure 3. Since most interest is in the slope of the curve, which appears to increase markedly some 30–40 years since initial employment, this graph is difficult to interpret in quantitative terms. Using 'kernel' methods (Ramlau-Hansen, 1983; Yandell, 1983), it is possible to derive a smooth estimator $\hat{\theta}(t)$ from the cumulative. The degree of smoothness depends on a bandwidth b such that jumps in $\hat{\Theta}$ at observed times of death t_i in the interval $(t - b, t + b)$ contribute to the estimation of θ at t. Figure 4 shows two smoothed estimates

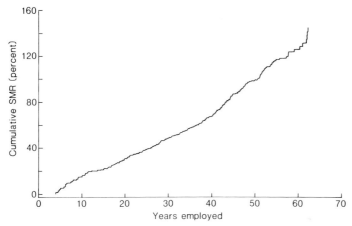

Fig. 3. Cumulative respiratory cancer SMR by years since initial employment for Montana smelter workers (reproduced from Breslow and Langholz (1987)).

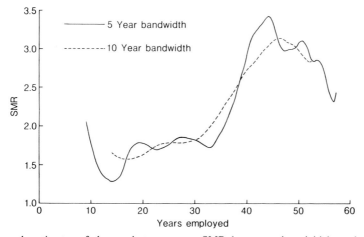

Fig. 4. Smooth estimates of the respiratory cancer SMR by years since initial employment for Montana smelter workers: comparison of five- and ten-year bandwidths (reproduced from Breslow and Langholz (1987)).

derived from Figure 3, with bandwidths of 5 and 10 years. These confirm that the SMR rises to a peak of nearly 3.5 some 40 years or so from the date of first employment.

Grouped data analysis demonstrated that the effect on the SMR of time since initial employment is largely due to confounding by period of employment (Breslow, 1985). A similar covariable adjustment of the nonparametric estimate may be developed from an extension of the basic model equation, viz.

$$\lambda(t \,|\, z(t)) = \theta(t)\,\lambda^*(t)\,\exp\{\beta z(t)\}\,,$$

where z is a vector of (possibly time-dependent) covariables and β is a vector of regression coefficients. This has the same structure as Cox's (Cox, 1972) famous proportional hazards model, except that the unknown 'background' rates are expressed as the product $\theta(t)\,\lambda^*(t)$ of known standard rates times a nonparametric relative risk function (Andersen et al., 1985). By incorporating $\log\lambda^*(t) = z_0(t)$ into the regression equation as a covariable with known coefficient $\beta_0 = 1$, standard programs may be used to obtain estimates of β and $\theta(t)$.

Figure 5 presents a comparison of the unadjusted estimate of $\beta(t)$,[1] and an estimate adjusted via the proportional hazards model for period of employment, birthplace and number of years worked in areas of the smelter with potential for heavy or moderate arsenic exposure. The difference between the two is striking. Instead of providing evidence for a possible latent interval between onset of exposure and respiratory cancer effect, as might be suggested by Figure 4, the covariable adjusted analysis shows a constant or even slightly declining SMR. For cohort members born in the U.S., hired in 1925 or later and who were accumulat-

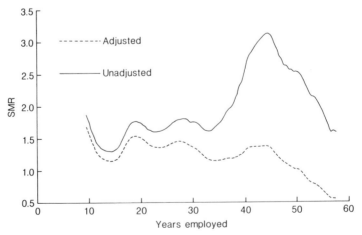

Fig. 5. Respiratory cancer SMR by year since initial employment for Montana smelter workers with and without covariable adjustment (reproduced from Breslow and Langholz (1987)).

---

[1] Obtained using a slightly different smoothing technique than that described earlier.

ing no heavy or moderate arsenic exposure, there is little evidence that respiratory cancer rates were elevated over background.

3.3. Summary

In summary, computationally intensive, graphical methods of data analysis are now available that free the epidemiologist from the constraints imposed by linear models and by the grouping of data into arbitrary categories. Much work remains to be done on associated statistical inference procedures, however, so that apparently interesting features of the graphs that are due partly or wholly to sampling errors are not overinterpreted.

Note added in proof

Recent work by Langholz, B. and Thomas, D. C. (1990, Nested case-control and case-cohort methods of sampling from a cohort: a critical comparison. *Am. J. Epidemiol.* **131**, 169–176) suggests that the efficiency advantages of the case-cohort design mentioned in Section 2.3 have been overstated. For studies with random censoring and staggered entry, the case-cohort design may be substantially *less* efficient due to high correlations between the individual risk set contributions to the pseudo-likelihood analysis.

References

Andersen, P. K., Borch-Johnsen, K., Deckert, T. et al. (1985). A Cox regression model for the relative mortality and its application to diabetes mellitus survival data. *Biometrics* **41**, 921–932.

Beebe, G. W., Ishida, M. and Jablon, S. (1962). Studies of the mortality of A-bomb survivors. I. Plan of study and mortality in the medical subsample (selection 1), 1950–1958. *Radiation Research* **16**, 253–280.

Breslow, N. E. (1985). Multivariate cohort analysis. *Natl. Cancer Inst. Monograph* **67**, 149–56.

Breslow, N. E. and Cain, K. C. (1988). Logistic regression for two stage case-control studies. *Biometrics* **44**, 891–899.

Breslow, N. E. and Day, N. E. (1980). *Statistical Methods in Cancer Research I: The Analysis of Case-Control Studies*. IARC Scientific Publication No. 32, International Agency for Research on Cancer, Lyon.

Breslow, N. E. and Day, N. E. (1985). The standardized mortality ratio. In: P. K. Sen, ed., *Biostatistics: Statistics in Biomedical, Public Health and Environmental Scienc.*, Springer-Verlag, New York, 55–74.

Breslow, N. E. and Day, N. E. (1987). *Statistical Methods in Cancer Research II: Design and Analysis of Cohort Studies*. IARC Scientific Publication No. 82, International Agency for Research on Cancer, Lyon.

Breslow, N. E. and Langholz, B. (1987). Nonparametric estimation of relative mortality functions. *J. Chronic Dis.* **40**, 895–995.

Cornfield, J. A. (1951). A method of estimating comparative rates from clinical data. *J. Natl. Cancer Inst.* **11**, 1269–1275.

Cox, D. R. (1972). Regression models and life tables (with discussion). *J. R. Stat. Soc. B* **34**, 187–220.

Efron, B. and Tibshirani, R. (1986). Bootstrap methods for standard errors, confidence intervals, and other measures of statistical accuracy. *Stat. Sci.* **1**, 54–77.

Greenland, S. and Thomas, D. C. (1982). On the need for the rare disease assumption in case-control studies. *Am. J. Epidemiol.* **116**, 547–553.

Hastie, T. and Tibshirani, R. (1986). Generalized additive models (with discussion). *Stat. Sci.* **1**, 297–318.

Hastie, T. and Tibshirani, R. (1987). Generalized additive models: some applications. *J. Am. Stat. Assoc.* **82**, 371–386.

Hsieh, D. A., Manski, C. F. and McFadden, D. (1985). Estimation of response probability from augmented retrospective observations. *J. Am. Stat. Assoc.* **80**, 651–662.

Kannel, W. B. and Gorton, T. (ed.) (1974). *The Framingham Study.* DHEW Publication No. (NIH) 74-559. US GPO, Washington, DC.

Lee-Feldstein, A. (1983). Arsenic and respiratory cancer in humans: Follow-up of copper smelter employees in Montana. *J. Natl. Cancer Inst.* **70**, 601–609.

Liddel, F. D. K., McDonald, J. D. and Thomas, D. C. (1977). Methods for cohort analysis: Appraisal by application to asbestos mining (with discussion). *J. R. Stat. Soc. A* **140**, 469–490.

Manski, C. F. and McFadden, D. (1981). Alternative estimators and sample designs for discrete choice analysis. In: C. F. Manski and D. McFadden, eds., *Structural Analysis of Discrete Data.* The MIT Press, Cambridge, MA, 2–50.

Mantel, N. (1983). Synthetic retrospective studies and related topics. *Biometrics* **29**, 479–486.

McCullagh, P. and Nelder, J. A. (1983). *Generalized Linear Models.* Chapman and Hall, London.

Miettinen, O. (1976). Estimability and estimation in case-referent studies. *Am. J. Epidemiol.* **103**, 226–235.

Miettinen, O. (1982). Design options in epidemiologic research: an update. *Scand. J. Work Environ. Health* **8** (Suppl. 1), 7–14.

Miettinen, O. (1985). *Theoretical Epidemiology.* Wiley, New York.

Peterson, G. R., Gilbert, E. S., Stevens, R. G. and Buchanan, J. A. (1987). A case-cohort study of lung cancer among males based on smoking data obtained from industrial medical records. (Submitted for publication.)

Prentice, R. L. (1986). A case-cohort design for epidemiologic cohort studies and disease prevention trials. *Biometrika* **73**, 1–11.

Prentice, R. L. and Pyke, R. L. (1979). Logistic disease incidence models and case-control studies. *Biometrika* **66**, 403–411.

Ramlau-Hansen, H. (1983). Smoothing counting process intensities by means of kernel functions. *Ann. Stat.* **11**, 453–466.

Stone, M. (1974). Cross-validatory choice and assessment of statistical predictions (with discussion). *J. R. Stat. Soc. B* **36**, 111–147.

Walker, A. M. (1982). Anamorphic analysis: sampling and estimation for covariable effects when both exposure and disease are known. *Biometrics* **38**, 1025–1032.

White, J. E. (1982). A two stage design for the study of the relationship between a rare exposure and a rare disease. *Am. J. Epidemiol.* **115**, 119–128.

Yandell, B. S. (1983). Nonparametric inference for rates with censored survival data. *Ann. Stat.* **11**, 1119–1135.

C. R. Rao and R. Chakraborty, eds., *Handbook of Statistics, Vol. 8*
© Elsevier Science Publishers B.V. (1991) 145–202

6

Robust Classification Procedures and Their Applications to Anthropometry

N. Balakrishnan and R. S. Ambagaspitiya

1. Introduction

The problem of classifying a given individual as a member of one of two (or more than two) groups to which that particular individual can possibly belong has been of great interest to statisticians, anthropologists, taxonomists, psychologists, sociologists and many others. Various contributions have appeared on this subject from the beginning of this century. Under the assumption that the two groups have multivariate normal distributions with different mean vectors but a common dispersion matrix, one would simply use the linear classification procedure introduced by Fisher (1936) for this purpose. One may also refer to Welch (1939), Rao (1947, 1948a,b) and Smith (1947) for some other derivations and interpretations of this procedure. Under the assumption that the two groups have multivariate normal distributions with different mean vectors and unequal dispersion matrices, one would use the quadratic classification procedure; see, for example, Anderson (1958) and Srivastava and Khatri (1979).

In this chapter, we primarily use the data obtained from an elaborative and extensive anthropometric survey conducted by Majumdar (1941) in the United Provinces of India in which about twelve to fourteen different physical measurements were made on each individual from 23 different castes and tribes. This data has earlier been statistically analyzed by Mahalanobis and Rao (1948) and also utilized by Rao (1948b) for illustrating the usefulness of the classical classification procedure in his extremely interesting, stimulating and pioneering paper that was read before the Research Section of the Royal Statistical Society. In his article, Rao has emphasized the main aim of anthropological classifications as, in the words of Morant (1939), "to unravel the course of human evolution, and it may be taken for granted to-day that the proper study of the natural history of man is concerned essentially with the mode and path of his descent".

We discuss in this chapter some robust univariate and multivariate two-way classification procedures based on Tiku's (1967, 1970, 1980) MML (modified maximum likelihood) estimators based on symmetrically type-II censored samples from the normal distribution and show that they are more efficient than the respective classical two-way classification procedures, when one of the two errors of misclassification is at a fixed level. We also show that these procedures have a smaller average error rate than the corresponding classical procedures and a considerably smaller average error rate than some non-parametric classification procedures like the nearest neighbour method and the method based on density estimates, when both the errors of misclassification are allowed to float. We use the anthropometric data given by Majumdar (1941) in order to illustrate the various robust classification procedures considered in this chapter.

In Section 2 we present briefly the derivation of the MML estimators in the univariate normal case as given by Tiku (1967) and its multivariate extension due to Tiku (1988) and Tiku and Balakrishnan (1988c) and some properties of these estimators. In Section 3 we discuss the classical univariate two-way linear classification procedure and an analogous procedure based on the MML estimators. By comparing their performance under normal and various non-normal models, we show that the latter is quite efficient and robust to departures from normality. In Section 4 we present the multivariate two-way classification procedure based on the MML estimators and by applying it to Majumdar's anthropometric data we demonstrate that this procedure is overall more efficient than the classical multivariate two-way linear classification procedure, when we fix one of the two errors of misclassification at a specified level (equal to 0.05). The superiority of the robust procedure in this case is quite possibly due to the presence of some outliers in the data from each group. We also demonstrate that the robust procedure has a much smaller average error rate than some of the non-parametric or distribution-free classification procedures like the nearest neighbour method and the method based on Loftsgaarden and Quesenberry's density estimate. In Section 5 we discuss the classical univariate two-way quadratic classification procedure and an analogous procedure based on the MML estimators. By comparing their performance under normal and various non-normal models, we show that the latter is quite efficient and robust to departures from normality. We also propose a transformed linear classification procedure and a robust analogue of it based on the MML estimators for the classification problem considered in this section. In Section 6 we present the multivariate two-way quadratic classification procedure based on the MML estimators and show it to be overall more efficient than the classical multivariate two-way quadratic classification procedure. We also present bivariate extensions of the classical and the robust transformed linear classification procedures described earlier in Section 5 and compare their performance under normal and several non-normal models. We present some illustrative examples by using Majumdar's anthropometric data. In Section 7 we discuss the classical univariate two-way classification procedure based on a dichotomous and a normal variable given by Chang and Afifi (1974) and a robust analogue of it based on the MML estimators. We show through Monte Carlo simulations that the robust procedure is more efficient than the classical procedure under normal and various non-normal models, when one of the two errors of misclassification is at a fixed level. We present some examples to illustrate the methods of classification discussed in this section. In Section 8 we present the robust multivariate two-way classification procedure based on a dichotomous and a multivariate normal variable and show that it is more efficient and robust to departures from normality as compared to the corresponding classical classification procedure. We present some examples to illustrate the methods of classification discussed in this section. Finally, in Section 9 we briefly describe some related problems of research which need further investigation and point out their potential applications.

Finally, by recalling the following statement of Karl Pearson (1898), "No scientific investigation can be final; it merely represents the most probable conclusion which can be drawn from the data at the disposal of the writer. A wide range of facts, or more refined analysis, experiment, and observations will lead to new formulae and new theories. This is the essence of scientific progress", we hope our effort in proposing and studying these robust classification procedures is worthwhile and will generate some interest and possibly some other procedures for the problems that are considered in this chapter.

2. Modified maximum likelihood estimation

In this section we first give a brief description of the MML estimators for the univariate normal case as derived by Tiku (1967) and then present its extension to the multivariate normal case as given by Tiku (1988) and Tiku and Balakrishnan (1988c).

2.1. Univariate MML estimation

Let $x_{(1)} \leqslant x_{(2)} \leqslant \cdots \leqslant x_{(n)}$ be the order statistics obtained from a sample of size n from the normal $N(\mu, \sigma^2)$ population. Then, by considering the symmetrically type-II censored sample $x_{(r+1)} \leqslant x_{(r+2)} \leqslant \cdots \leqslant x_{(n-r)}$ and writing down the likelihood function based on it as

$$L = \frac{n!}{(r!)^2} \, \sigma^{-(n-2r)} \exp\left\{ -\tfrac{1}{2} \sum_{i=r+1}^{n-r} z_{(i)}^2 \right\} \{\Phi(z_{(r+1)})\}^r \{1 - \Phi(z_{(n-r)})\}^r,$$

(2.1)

where $z_{(i)} = (x_{(i)} - \mu)/\sigma$, $\Phi(z) = \int_{-\infty}^z \phi(t)\,\mathrm{d}t$, and $\phi(t) = (1/\sqrt{2\pi})\,\mathrm{e}^{-t^2/2}$, Tiku (1967) has noted that the maximum likelihood equations

$$\partial \ln L / \partial \mu = 0 \quad \text{and} \quad \partial \ln L / \partial \sigma = 0,$$

involve the functions

$$\psi_1(z_{(r+1)}) = \phi(z_{(r+1)})/\Phi(z_{(r+1)})$$

and

$$\psi_2(z_{(n-r)}) = \phi(z_{(n-r)})/[1 - \Phi(z_{(n-r)})],$$

and hence do not admit explicit solutions. Tiku (1967), therefore, has suggested the use of linear approximations

$$\psi_1(z_{(r+1)}) \approx \alpha - \beta z_{(r+1)} \quad \text{and} \quad \psi_2(z_{(n-r)}) \approx \alpha + \beta z_{(n-r)} \tag{2.2}$$

in order to simplify the maximum likelihood equations. The resulting equations

$$\partial \ln L^*/\partial \mu = 0 \quad \text{and} \quad \partial \ln L^*/\partial \sigma = 0,$$

are referred to as the MML (modified maximum likelihood) equations and they do admit explicit solutions. These solutions are referred to as the MML estimators of μ and σ and are given by

$$\hat{\mu} = \frac{1}{m} \left[\sum_{i=r+1}^{n-r} x_{(i)} + r\beta\{x_{(r+1)} + x_{(n-r)}\} \right], \tag{2.3}$$

and

$$\hat{\sigma} = [B + (B^2 + 4AC)^{1/2}]/2\{A(A-1)\}^{1/2}, \tag{2.4}$$

where

$$A = n - 2r,$$

$$m = n - 2r + 2r\beta,$$

$$B = r\alpha\{x_{(n-r)} - x_{(r+1)}\},$$

$$C = \sum_{i=r+1}^{n-r} x_{(i)}^2 + r\beta\{x_{(r+1)}^2 + x_{(n-r)}^2\} - m\hat{\mu}^2. \tag{2.5}$$

The coefficients α and β depend only on the proportion of censoring, $q = r/n$, and are obtained from the equations

$$\beta = -\frac{\phi(t)\{t - \phi(t)/q\}}{q} \quad \text{and} \quad \alpha = \frac{\phi(t)}{q} - \beta t, \tag{2.6}$$

where $t = \Phi^{-1}(1-q)$. Asymptotically, (2.2) are strict equalities (Tiku, 1968; Bhattacharyya, 1985) in which case $\hat{\mu}$ and $\hat{\sigma}$ in (2.3) and (2.4) are identical to the MLE of μ and σ, respectively.

The distributions of $\hat{\mu}$ and $\hat{\sigma}$ have been studied by Tiku (1980) in the sampling theory framework and by Tan (1985), Tan and Balakrishnan (1986, 1987), Balakrishnan (1988), and Balakrishnan and Tan (1988) from a Bayesian viewpoint. A book length account of the efficiency properties of $\hat{\mu}$ and $\hat{\sigma}$, and the robustness features of several inference procedures based on $\hat{\mu}$ and $\hat{\sigma}$, is available in Tiku, Tan and Balakrishnan (1986).

2.2. Multivariate MML estimation

Let x_i, $i = 1, 2, \ldots, n$, be a random sample from a p-variate normal distribution $N_p(\mu, \Sigma)$. Then the likelihood function is given by

$$L = \left\{ \frac{1}{(2\pi)^{p/2} |\Sigma|^{1/2}} \right\}^n \exp\left\{ -\tfrac{1}{2} \sum_{i=1}^{n} (x_i - \mu)^T \Sigma^{-1} (x_i - \mu) \right\}$$

$$= \left\{ \frac{1}{(2\pi)^{p/2} |\Sigma|^{1/2}} \right\}^n \exp\left\{ -\frac{1}{2|R|} \sum_{k=1}^{p} R_{kk} \sum_{i=1}^{n} \frac{(x_{ik} - \mu_k)^2}{\sigma_k^2} \right.$$

$$\left. - \frac{1}{2|R|} \sum_{k=1}^{p} \sum_{\substack{l=1 \\ l \neq k}}^{p} R_{kl} \sum_{i=1}^{n} \frac{(x_{ik} - \mu_k)(x_{il} - \mu_l)}{\sigma_k \sigma_l} \right\},$$

$$(2.7)$$

where R is the $p \times p$ correlation matrix and R_{kl} is the co-factor of the (k, l)-th element in R. Let us now denote $x_{(1)k} \leqslant x_{(2)k} \leqslant \cdots \leqslant x_{(n)k}$ $(k = 1, 2, \ldots, p)$ for the order statistics of the x_{ik} $(1 \leqslant i \leqslant n, 1 \leqslant k \leqslant p)$ observations. Then we have

$$\frac{1}{\sigma_k^2} \sum_{i=1}^{n} (x_{ik} - \mu_k)^2 = \frac{1}{\sigma_k^2} \sum_{i=1}^{n} (x_{(i)k} - \hat{\mu}_k + \hat{\mu}_k - \mu_k)^2$$

$$= \frac{1}{\sigma_k^2} \{ C_{kk} + m(\hat{\mu}_k - \mu_k)^2 \} + D_{kk},$$

$$(2.8)$$

where C_{kk} has exactly the same expression as C in (2.5) with $\hat{\mu}$ replaced by $\hat{\mu}_k$, and

$$D_{kk} = \sum_{i=1}^{r} \hat{z}_{(i)k}^2 + \sum_{i=n-r+1}^{n} \hat{z}_{(i)k}^2 - r\beta \{ \hat{z}_{(r+1)k}^2 + \hat{z}_{(n-r)k}^2 \}$$

$$+ \frac{2n}{\sigma_k^2} (\hat{\mu}_k - \mu_k)(\bar{x}_k - \hat{\mu}_k) + \frac{(n-m)}{\sigma_k^2} (\hat{\mu}_k - \mu_k)^2,$$

$$(2.9)$$

with $\hat{z}_{(i)k} = (x_{(i)k} - \hat{\mu}_k)/\sigma_k$. Tiku (1988) has recently shown that, for large n, D_{kk}/n converges to its expected value $2q(1 + \alpha t) + (1 - n/m)/n$.

Let us similarly write $w_{ik, l} = (x_{ik} - \mu_k)(x_{il} - \mu_l)$, $\hat{w}_{ik, l} = (x_{ik} - \hat{\mu}_k)(x_{il} - \hat{\mu}_l)$, and $\hat{w}_{(i)k, l}$ for the order statistics of $\hat{w}_{ik, l}$ $(1 \leqslant i \leqslant n; 1 \leqslant k, l \leqslant p, k \neq l)$. Then have

$$\frac{1}{\sigma_k \sigma_l} \sum_{i=1}^{n} w_{ik, l} = \frac{1}{\sigma_k \sigma_l} \sum_{i=1}^{n} (x_{ik} - \hat{\mu}_k + \hat{\mu}_k - \mu_k)(x_{il} - \hat{\mu}_l + \hat{\mu}_l - \mu_l)$$

$$= \frac{1}{\sigma_k \sigma_l} \{ C_{kl} + m(\hat{\mu}_k - \mu_k)(\hat{\mu}_l - \mu_l) \} + D_{kl},$$

$$(2.10)$$

where

$$C_{kl} = \sum_{i=r+1}^{n-r} \hat{w}_{(i)k, l} + r\beta\{\hat{w}_{(r+1)k, l} + \hat{w}_{(n-r)k, l}\} \qquad (2.11)$$

and

$$D_{kl} = \sum_{i=1}^{r} \hat{u}_{(i)k, l} + \sum_{i=n-r+1}^{n} \hat{u}_{(i)k, l} - r\beta\{\hat{u}_{(r+1)k, l} + \hat{u}_{(n-r)k, l}\}$$

$$+ \frac{n}{\sigma_k \sigma_l}\{(\hat{\mu}_k - \mu_k)(\bar{x}_l - \hat{\mu}_l) + (\hat{\mu}_l - \mu_l)(\bar{x}_k - \hat{\mu}_k)$$

$$+ (1 - m/n)(\hat{\mu}_k - \mu_k)(\hat{\mu}_l - \mu_l)\}, \qquad (2.12)$$

where $\hat{u}_{(i)k, l} = \hat{w}_{(i)k, l}/\sigma_k \sigma_l$, $i = 1, 2, \ldots, n$. Tiku (1988) has shown that, for large n, D_{kl}/n converges to its expected value $\rho_{kl}\{2q(1 + \alpha t) + (1 - n/m)/n\}$.

The modified likelihood function L^* is derived on lines exactly similar to the ones discussed by Tiku (1967) and Persson and Rootzen (1977) in the univariate case, by replacing D_{kk} and D_{kl} by their expected values, for large n (see Tiku and Balakrishnan, 1988c). Thus,

$$L \approx L^*,$$

where

$$L^* = \left\{\frac{1}{(2\pi)^{p/2}|\Sigma|^{1/2}}\right\}^n \exp\left\{-\frac{m}{2|R|}\right.$$

$$\times\left[\sum_{k=1}^{p} R_{kk}\frac{(\hat{\mu}_k - \mu_k)^2}{\sigma_k^2} + \sum_{k=1}^{p}\sum_{\substack{l=1\\l\neq k}}^{p} R_{kl}\frac{(\hat{\mu}_k - \mu_k)(\hat{\mu}_l - \mu_l)}{\sigma_k \sigma_l}\right]$$

$$-\frac{1}{2|R|}\left[\sum_{k=1}^{p}\frac{R_{kk}C_{kk}}{\sigma_k^2} + \sum_{k=1}^{p}\sum_{\substack{l=1\\l\neq k}}^{p}\frac{R_{kl}C_{kl}}{\sigma_k \sigma_l}\right]$$

$$\left.-\frac{1}{2}p\left[2r(1 + \alpha t) + \left(1 - \frac{n}{m}\right)\right]\right\}$$

$$\propto\left\{\frac{1}{(2\pi)^{p/2}|\Sigma|^{1/2}}\right\}^n \exp\left\{-\frac{m}{2|R|}\right.$$

$$\times\left[\sum_{k=1}^{p} R_{kk}\frac{(\hat{\mu}_k - \mu_k)^2}{\sigma_k^2} + \sum_{k=1}^{p}\sum_{\substack{l=1\\l\neq k}}^{p} R_{kl}\frac{(\hat{\mu}_k - \mu_k)(\hat{\mu}_l - \mu_l)}{\sigma_k \sigma_l}\right]$$

$$\left.-\frac{1}{2|R|}\left[\sum_{k=1}^{p}\frac{R_{kk}C_{kk}}{\sigma_k^2} + \sum_{k=1}^{p}\sum_{\substack{l=1\\l\neq k}}^{p}\frac{R_{kl}C_{kl}}{\sigma_k \sigma_l}\right]\right\}, \qquad (2.13)$$

since the term $\frac{1}{2}p[2r(1 + \alpha t) + (1 - n/m)]$ is free of the parameters.

It is of interest to mention here that $[\ln L - \ln L^*]/n \to 0$ as $n \to \infty$. The MML estimators of μ_k is $\hat{\mu}_k$ and of $\sigma_k^2 = C_{kk}/A$ which is corrected for bias and taken as $\hat{\sigma}_k^2 = C_{kk}/(A - 1)$. The MML estimator of ρ_{kl} is obtained from (2.13) as $C_{kl}/\sqrt{C_{kk}C_{ll}}$ which is asymptotically unbiased. If $x_{ik} = ax_{il}$ (for positive a), $i = 1, 2, \ldots, n$, then we see that $C_{kl}/\sqrt{C_{kk}C_{ll}} = 1$ as it should. Since for large n, $\hat{\sigma}_k^2 \approx C_{kk}/(A - 1)$, Tiku (1988) has defined a slightly amended estimator of ρ_{kl} as

$$\hat{\rho}_{kl} = \frac{1}{A - 1} \frac{C_{kl}}{\hat{\sigma}_k \hat{\sigma}_l} = \frac{m}{A} \frac{\hat{\sigma}_{kl}}{\hat{\sigma}_k \hat{\sigma}_l}, \qquad (2.14)$$

where $\hat{\sigma}_{kl} = [A/(A - 1)] (C_{kl}/m)$. Interestingly enough, this is exactly the same estimator as was originally proposed by Tiku and Balakrishnan (1986) as the MML estimator of the location parameter of $(x_k - \mu_k)(x_l - \mu_l)$, and subsequently used by Tiku and Balakrishnan (1988a,b) and Tiku, Gill and Balakrishnan (1989) to develop some robust multivariate procedures.

Since the modified likelihood function L^* is almost identical to the likelihood function L, the large-sample variances and covariances of $\hat{\mu}_k$, $\hat{\sigma}_k$ and $\hat{\rho}_{kl}$ may be obtained as the elements of the matrix (Tiku and Balakrishnan, 1988c)

$$((J_{kl}))^{-1},$$

where $J_{11} = E(-\partial^2 \ln L^*/\partial\mu_1^2)$, $J_{12} = E(-\partial^2 \ln L^*/\partial\mu_1\partial\mu_2)$, etc... We note here that these expected values of the partial derivatives of all orders of $\ln L^*$ are exactly similar to those of $\ln L$. In the context of robustness, r is chosen to be the integer value $r = [0.5 + 0.1n]$; see, for example, Tiku, Tan and Balakrishnan (1986). In this case, we have $q = r/n = 0.1$, $\beta = 0.8309$, $m = n - 2r + 2r\beta = n(1 - 2q + 2q\beta)$ and $A = n - 2r = n(1 - 2q)$, and hence $\sqrt{m/A} \approx 1.1$.

3. Robust univariate classification procedures and applications

In this section we first discuss the classical univariate two-way linear classification procedure and then describe the analogous robust linear classification procedure based on the MML estimators as proposed by Balakrishnan, Tiku and El Shaarawi (1985). By comparing these two procedures, when one of the two errors of misclassification is fixed at a specified level, under normal and various non-normal models, we show that the robust procedure is quite efficient and robust to departures from normality. When both the errors of misclassification are allowed to float, we also show that the robust procedure has a smaller average error rate than the classical procedure and also several distribution-free procedures. We present some illustrative examples by using Majumdar's (1941) anthropometric data.

3.1. Classical linear classification procedure

Let x_{1i}, $i = 1, 2, \ldots, n_1$, and x_{2i}, $i = 1, 2, \ldots, n_2$, be independent random samples from populations Π_1 and Π_2 assumed to be normal $N(\mu_1, \sigma^2)$ and $N(\mu_2, \sigma^2)$, respectively. Let us denote the sample means by

$$\bar{x}_1 = \frac{1}{n_1} \sum_{i=1}^{n_1} x_{1i} \quad \text{and} \quad \bar{x}_2 = \frac{1}{n_2} \sum_{i=1}^{n_2} x_{2i},$$

and the sample variances by

$$s_1^2 = \frac{1}{n_1 - 1} \sum_{i=1}^{n_1} (x_{1i} - \bar{x}_1)^2 \quad \text{and} \quad s_2^2 = \frac{1}{n_2 - 1} \sum_{i=1}^{n_2} (x_{2i} - \bar{x}_2)^2,$$

and the pooled sample variance by

$$s^2 = \frac{(n_1 - 1)s_1^2 + (n_2 - 1)s_2^2}{n_1 + n_2 - 2}.$$

Then, to classify a new independent observation x_0, the classical linear classification procedure is to classify x_0 into Π_1 or Π_2 according as (Anderson, 1951)

$$V = \{x_0 - \tfrac{1}{2}(\bar{x}_1 + \bar{x}_2)\}(\bar{x}_1 - \bar{x}_2)/s^2 \gtrless C, \tag{3.1}$$

where the cut-off point C is chosen such that one of the two errors of misclassification is at a pre-specified level.

Let us now denote the probability of wrongly classifying x_0 into Π_2 when x_0 actually belongs to Π_1 by e_{12} and similarly the probability of wrongly classifying x_0 into Π_1 when x_0 actually belongs to Π_2 by e_{21}. Note that $1 - e_{21}$ is simply the probability of correctly classifying x_0 into Π_2 when x_0 actually belongs to Π_2.

Even though $\bar{x}_1$ and $\bar{x}_2$ are independently distributed as $N(\mu_1, \sigma^2/n_1)$ and $N(\mu_2, \sigma^2/n_2)$, and $(n_1 + n_2 - 2)s^2/\sigma^2$ is independently distributed as χ^2 with $n_1 + n_2 - 2$ degrees of freedom, the distribution of V in (3.1) does not assume a manageable form; see, for example, Sinha and Giri (1975). It is, therefore, difficult to determine the cut-off point C in (3.1) for a specified e_{12}, for example. But, by using the first four moments of V worked out by Balakrishnan, Tiku and El Shaarawi (1975) one may fit a four-moment Pearson curve for the distribution of V by making use of the tables of Johnson et al. (1963) and determine an approximate value for the cut-off point C. Alternatively, one may use the following asymptotic argument to approximate C.

Asymptotically (as $n_1, n_2 \to \infty$), $\bar{x}_1$, $\bar{x}_2$ and s tend to μ_1, μ_2 and σ, respectively, and hence the distribution of V in (3.1) is the same as that of Fisher's linear discriminant function

$$U = \{x_0 - \tfrac{1}{2}(\mu_1 + \mu_2)\}(\mu_1 - \mu_2)/\sigma^2 . \tag{3.2}$$

The distribution of U is normal with mean

$$E(U|x_0 \in \Pi_1) = \tfrac{1}{2}\delta^2 = -E(U|x_0 \in \Pi_2), \tag{3.3}$$

and variance

$$\text{Var}(U|x_0 \in \Pi_1) = \text{Var}(U|x_0 \in \Pi_2) = \delta^2 , \tag{3.4}$$

where $\delta^2 = (\mu_1 - \mu_2)^2/\sigma^2$ is the standardized squared distance between the two populations Π_1 and Π_2. Therefore, with e_{12} fixed as 0.05, for example, under the assumption of normality the cut-off point C in (3.1) may be determined approximately as

$$C = \tfrac{1}{2}\delta^2 - 1.64|\delta| . \tag{3.5}$$

Since δ is unknown, C can not be determined from (3.5); however, we may use an estimator of δ^2 given by (Lachenbruch and Mickey, 1968)

$$\hat{\delta}^2 = \left(\frac{n_1 + n_2 - 4}{n_1 + n_2 - 2}\right) \frac{(\bar{x}_1 - \bar{x}_2)^2}{s^2} \tag{3.6}$$

to obtain a suitable estimator of C from (3.5) to be

$$\hat{C} = \tfrac{1}{2}\hat{\delta}^2 - 1.64|\hat{\delta}| . \tag{3.7}$$

Then, the classical two-way linear classification procedure with e_{12} fixed as 0.05, is to classify the observation x_0 into Π_1 or Π_2 according as

$$V = \{x_0 - \tfrac{1}{2}(\bar{x}_1 + \bar{x}_2)\}(\bar{x}_1 - \bar{x}_2)/s^2 \gtrless \hat{C} , \tag{3.8}$$

where the cut-off point $\hat{C}$ is as given in (3.7). It should be pointed out here that the estimator $\hat{\delta}^2$ in (3.6) is a biased estimator of δ^2 and the unbiased estimator of δ^2 given by $\hat{\delta}^2 - (1/n_1 + 1/n_2)$ may take on inadmissible negative values.

3.2. Robust linear classification procedure

A robust linear classification procedure analogous to the classical one in (3.8) has been obtained by Balakrishnan, Tiku and El Shaarawi (1985) by replacing $\bar{x}_1, \bar{x}_2$ and s by the corresponding MML estimators $\hat{\mu}_1, \hat{\mu}_2$ and $\hat{\sigma}$; here $\hat{\mu}_1, \hat{\sigma}_1$ are the MML estimators of μ_1 and σ obtained from the sample x_{1i}, $i = 1, 2, \ldots, n_1$, and $\hat{\mu}_2$, $\hat{\sigma}_2$ are the MML estimators of μ_2 and σ obtained from the sample x_{2i}, $i = 1, 2, \ldots, n_2$, and $\hat{\sigma}^2$ is the pooled MML estimator of σ^2 given by

$$\hat{\sigma}^2 = \frac{(A_1 - 1)\,\hat{\sigma}_1^2 + (A_2 - 1)\,\hat{\sigma}_2^2}{A_1 + A_2 - 2}. \tag{3.9}$$

Then the robust two-way linear classification procedure, with e_{12} fixed as 0.05, is to classify the observation x_0 into Π_1 or Π_2 according as

$$V_R = \{x_0 - \tfrac{1}{2}(\hat{\mu}_1 + \hat{\mu}_2)\}(\hat{\mu}_1 - \hat{\mu}_2)/\hat{\sigma}^2 \gtrless \hat{C}_R, \tag{3.10}$$

where the cut-off point $\hat{C}_R$ is given by

$$\hat{C} = \tfrac{1}{2}\hat{\delta}_R^2 - 1.64|\,\hat{\delta}_R|, \tag{3.11}$$

with $\hat{\delta}_R^2$ being the robust estimator of δ^2 given by

$$\hat{\delta}_R^2 = \left(\frac{A_1 + A_2 - 4}{A_1 + A_2 - 2}\right)\frac{(\hat{\mu}_1 - \hat{\mu}_2)^2}{\hat{\sigma}^2}. \tag{3.12}$$

Since the MML estimators $\hat{\mu}$ and $\hat{\sigma}^2$ have (asymptotically) exactly the same distributions as $\bar{x}$ and s^2, viz. $\sqrt{m}(\hat{\mu} - \mu)/\sigma$ and $(A - 1)\hat{\sigma}^2/\sigma^2$ are independently distributed as standard normal and χ^2 with $(A - 1)$ degrees of freedom (see Tiku, 1982; Tiku, Tan and Balakrishnan, 1986), respectively, the distribution of V_R is the same as that of V (for large n_1 and n_2) under the assumption of normality. Hence, for large n_1 and n_2, the classical and robust linear classification procedures will both have the same $1 - e_{21}$ (the probability of correctly classifying x_0 into Π_2 when x_0 actually belongs to Π_2) for a pre-fixed e_{12}, under the assumption of normality.

In order to examine the robustness features of the two linear classification procedures, n pseudo-random observations were simulated for the x_1 and x_2 samples each from the following models:

(1) Normal;
(2) Logistic;
(3) Outlier-normal model $(n - 1)N(0, 1)$ and $1N(0, 10)$;
(4) Outlier-normal model $(n - 1)N(0, 1)$ and $1N(0, 100)$;
(5) Outlier-normal model $(n - 2)N(0, 1)$ and $2N(0, 16)$; and
(6) Mixture-normal model $0.9N(0, 1) + 0.1N(0, 16)$. $\tag{3.13}$

A constant δ was added to the x_2 sample in order to create a location shift between the two populations Π_1 and Π_2. Then, with e_{12} fixed as 0.05, the values of e_{12} and $1 - e_{21}$ were simulated (based on 1000 Monte Carlo runs) for both the procedures under the six models given in (3.13). These values are presented in Table 3.1 for $\delta = 0.5$, 1.5 and 3.0 and $n_1 = n_2 = n = 20$. It is quite clear from Table 3.1 that the robust procedure has more stable e_{12} values than the corresponding classical procedure. It is also very clear from Table 3.1 that the robust linear procedure has considerably larger $1 - e_{21}$ values than the corresponding classical linear procedure under the non-normal models in (3.13).

Table 3.1
Simulated values of e_{12} and $1 - e_{21}$ for the classical and robust procedures for the models in (3.13) when $n_1 = n_2 = n = 20$

Model	e_{12} values						$1 - e_{21}$ values					
	$\delta = 0.5$		$\delta = 1.5$		$\delta = 3.0$		$\delta = 0.5$		$\delta = 1.5$		$\delta = 3.0$	
	V	V_R	V	V_R	V	V_R	V	V_R	V	V_R	V	V_R
(1)	0.057	0.056	0.056	0.055	0.051	0.042	0.145	0.156	0.453	0.458	0.918	0.915
(2)	0.055	0.067	0.052	0.060	0.046	0.052	0.138	0.169	0.448	0.484	0.916	0.924
(3)	0.035	0.049	0.032	0.044	0.031	0.038	0.101	0.135	0.357	0.440	0.858	0.905
(4)	0.011	0.058	0.011	0.047	0.005	0.036	0.021	0.122	0.107	0.406	0.378	0.894
(5)	0.020	0.049	0.018	0.042	0.017	0.037	0.047	0.109	0.198	0.381	0.647	0.859
(6)	0.037	0.065	0.039	0.064	0.039	0.060	0.092	0.164	0.252	0.395	0.653	0.794

3.3. Comparison with distribution-free procedures

We have examined the performance of the classical and the robust linear classification procedures in Section 3.2 when the probability of error of misclassification e_{12} is at a pre-fixed level. In certain instances, however, it may not be of interest to restrict e_{12} to assume a pre-fixed level; that is, one may be interested in allowing both e_{12} and e_{21} to float. In this case, the classical linear classification procedure is to classify the independent observation x_0 into Π_1, Π_2, or arbitrarily into Π_1 or Π_2, according as

$$V^* = \{x_0 - \tfrac{1}{2}(\bar{x}_1 + \bar{x}_2)\}(\bar{x}_1 - \bar{x}_2) \gtrless = 0 , \tag{3.14}$$

and the analogous robust linear classification procedure is to classify the observation x_0 into Π_1, Π_2, or arbitrarily into Π_1 or Π_2, according as

$$V_R^* = \{x_0 - \tfrac{1}{2}(\hat{\mu}_1 + \hat{\mu}_2)\}(\hat{\mu}_1 - \hat{\mu}_2) \gtrless = 0 . \tag{3.15}$$

The robustness features of the linear classification procedure in (3.14) have been examined by several authors. Lachenbruch, Sneeringer and Revo (1973) have studies the probabilities of misclassification for a variety of non-normal models through Monte Carlo simulations. On the other hand, by adopting a method similar to the one given by John (1960a,b) in the normal case, Ching'anda

(1976), Subrahmaniam and Ching'anda (1978), Ching'anda and Subrahmaniam (1979), Balakrishnan and Kocherlakota (1985), and Kocherlakota, Balakrishnan and Kocherlakota (1987) have examined the behaviour of the two errors of misclassification through their distribution functions and expected values for a wide range of non-normal models including Edgeworth series, Johnson's, mixture-normal, and truncated normal distributions. Recently, Kocherlakota, Kocherlakota and Balakrishnan (1987) have given a unified treatment of all the above developments on the asymptotic expansions of the distribution functions and the expected values of the errors of misclassification.

It is of interest to compare the performance of the above two linear procedures with some distribution-free classification procedures available in statistical literature. The ones that are included in this comparative study are briefly described below.

1. Procedures based on Parzen's kernel density estimator: Parzen (1962) has proposed a kernel density estimator to be

$$\hat{f}_m(z) = \frac{1}{mh(m)} \sum_{j=1}^{m} K\left(\frac{z - x_j}{h(m)}\right),$$

(3.16)

where $h(m) = m^{-1/8}$ and $K(w) = (1/\sqrt{2\pi})\exp(-\frac{1}{2}w^2)$. Now, let $\hat{f}_{n_1}(x_0)$ and $\hat{f}_{n_2}(x_0)$ be the kernel density estimates in (3.16) obtained from the samples x_{1i}, $i = 1, 2, \ldots, n_1$, and x_{2i}, $i = 1, 2, \ldots, n_2$, respectively. Then the classification procedure is to classify the observation x_0 into Π_1, Π_2, or arbitrarily into Π_1 or Π_2, according as

$$\frac{\hat{f}_{n_1}(x_0)}{\hat{f}_{n_2}(x_0)} \gtrless = 1$$

(3.17)

2. Procedure based on Loftsgaarden–Quesenberry's density estimator: The density estimator proposed by Loftsgaarden and Quesenberry (1965) is given by

$$\hat{f}_m(x) = (k(m) - 1)/2mr_{k(m)},$$

(3.18)

where $r_{k(m)}$ is the distance from z to the $k(m)$-th closest x_i to z as determined by the Euclidean distance; $k(m)$ is chosen to be nearer to $\sqrt{m}$ as suggested by Loftsgaarden and Quesenberry (1965). Now, let $\hat{f}_{n_1}(x_0)$ and $\hat{f}_{n_2}(x_0)$ be the density estimators in (3.18) obtained from the samples x_{1i}, $i = 1, 2, \ldots, n_1$, and x_{2i}, $i = 1, 2, \ldots, n_2$, respectively. Once again, the classification procedure is to classify the new observation x_0 into Π_1, Π_2, or arbitrarily into Π_1 or Π_2, exactly according to (3.17).

3. Hudimoto's Mann–Whitney type procedure: Hudimoto (1964) has defined Mann–Whitney type statistics, namely, $p_1 = U_{n_1}/n_1$ and $p_2 = 1 - U_{n_2}/n_2$, where

U_{n_1} = Number of x_{1i} ($i = 1, 2, \ldots, n_1$) less than x_0, and U_{n_2} = Number of x_{2i} ($i = 1, 2, \ldots, n_2$) less than x_0. Then, Hudimoto (1964) has proposed to classify the new observation x_0 into Π_1 if and only if $p_1 < p_2$.

4. Gupta–Kim's procedure based on Parzen's kernel density estimator: Gupta and Kim (1978) have used Parzen's kernel density estimator to propose a procedure similar to that of Hudimoto (1964). By defining

$$p_1^* = \frac{1}{n_1} \sum_{j=1}^{n_1} \Phi\left(\frac{x_0 - x_{1j}}{h(n_1)}\right) \quad \text{and} \quad p_2^* = \frac{1}{n_2} \sum_{j=1}^{n_2} \Phi\left(\frac{x_0 - x_{2j}}{h(n_2)}\right),$$

where $h(m) = m^{-1/8}$ and $\Phi(\cdot)$ is the cumulative distribution function of a standard normal variable, Gupta and Kim (1978) have proposed to classify x_0 into Π_1 if and only if $p_1^* < p_2^*$.

5. Das Gupta's Wilcoxon type procedure: Das Gupta (1964) has looked into classification procedures based on Wilcoxon type statistic. By defining $U = 1/n_1$ (Number of x_{1i} greater than x_0) and $V = 1/n_2$ (Number of x_{2i} greater than x_0), Das Gupta (1964) has proposed to classify the observation x_0 into Π_1 if and only if $|U - \frac{1}{2}| < |V - \frac{1}{2}|$.

6. Das Gupta's Kolmogorov type procedure: Das Gupta (1964) has also proposed a Kolmogorov type classification procedure as follows. By defining S_1 and S_2 to be the empirical distribution functions based on the samples from Π_1 and Π_2 and d_{0i} to be the Kolmogorov distance between x_0 and S_i Das Gupta (1964) has proposed to classify the observation x_0 into Π_i if $d_{0i} = \min(d_{01}, d_{02})$.

7. Fix–Hodges' nearest-neighbour procedure: Fix and Hodges (1951) have proposed the following heuristic nearest-neighbour procedure. First, calculate the Euclidean distances d_i ($i = 1, 2, \ldots, n_1 + n_2$), where $d_i = d(x_0, x_{1i})$, $i = 1, 2, \ldots, n_1$ and $d_i = d(x_0, x_{2i})$, $i = n_1 + 1, n_1 + 2, \ldots, n_1 + n_2$. Among the smallest m of these d_i's, let m_1 correspond to observations from the x_1 sample and m_2 correspond to observations from the x_2 sample. Then, Fix and Hodges have proposed to classify the observation x_0 into Π_1 if $m_1 > m_2$, Π_2 if $m_1 < m_2$, and arbitrarily into Π_1 or Π_2 if $m_1 = m_2$; refer to Gessaman and Gessaman (1972) for more details. Unfortunately, the choice of m seems to be very crucial for this procedure. We have used $m = \sqrt{n_1 + n_2}$ in this comparative study; see Section 4 for more details regarding this point.

8. Stoller's procedure: Let $S_1(x)$ and $S_2(x)$ be the empirical distribution functions based on the x_1 and x_2 samples, respectively. By defining the function

$$T(x) = \frac{[n_1 S_1(x) + n_2(1 - S_2(x))]}{(n_1 + n_2)}, \tag{3.19}$$

Stoller (1954) has proposed to classify the observation x_0 into Π_1 if and only if

$x_0 < x^*$, where the cut-off point x^* is that value of x which maximizes the function $T(x)$ in (3.19).

In order to compare the performance of all the classification procedures described above, n pseudo-random observations were simulated for the x_1 and x_2 samples each from all the six models given in (3.13) and, in addition, from

(7) Student's t distribution with 4 degrees of freedom.   (3.20)

Once again, a constant δ was added to the x_2 sample in order to create a location shift between the two populations Π_1 and Π_2. Then, the values of the average error rate $\frac{1}{2}(e_{12} + e_{21})$ were simulated (based on 1000 Monte Carlo runs) for all the classification procedures described above. These values are presented in Table 3.2 for $\delta = 2$ and 3 and $n_1 = n_2 = n = 20$. It is quite clear from this table that the robust linear classification procedure based on V_R^* in (3.15) is superior overall to all the other procedures. It should be pointed out here, however, that even though the classification procedure based on V_R^* in (3.15) has a slightly smaller average error than the classification procedure based on V^* in (3.14), they both will have exactly the same average error rate asymptotically. This is evident from the fact that both V^* and V_R^* will have the same distribution, asymptotically, as the statistic

$$Z = \{x_0 - \tfrac{1}{2}(\mu_1 + \mu_2)\}(\mu_1 - \mu_2).$$

By considering some more non-normal models and some larger sample sizes, Balakrishnan, Tiku and El Shaarawi (1985) have made more extensive comparisons of these procedures.

Table 3.2
Simulated values of the average error rate $\frac{1}{2}(e_{12} + e_{21})$ for the classical, the robust and various distribution-free classification procedures for the models in (3.13) and (3.20) when $n_1 = n_2 = n = 20$

δ	Model	V^*	V_R^*	P	LQ	H	GK	W	K	FH	S
2	(1)	0.151	0.150	0.160	0.167	0.157	0.157	0.184	0.186	0.157	0.168
	(2)	0.147	0.147	0.150	0.154	0.149	0.147	0.172	0.171	0.155	0.162
	(3)	0.153	0.153	0.161	0.167	0.153	0.157	0.187	0.191	0.155	0.171
	(4)	0.180	0.155	0.159	0.169	0.152	0.159	0.189	0.184	0.156	0.169
	(5)	0.154	0.155	0.163	0.165	0.158	0.164	0.187	0.190	0.163	0.171
	(6)	0.196	0.190	0.200	0.202	0.193	0.186	0.229	0.231	0.198	0.209
	(7)	0.263	0.256	0.275	0.279	0.270	0.265	0.297	0.290	0.275	0.285
3	(1)	0.055	0.054	0.059	0.062	0.072	0.060	0.102	0.098	0.062	0.078
	(2)	0.071	0.070	0.071	0.069	0.076	0.069	0.089	0.099	0.072	0.078
	(3)	0.060	0.055	0.058	0.064	0.073	0.061	0.102	0.102	0.057	0.077
	(4)	0.072	0.056	0.060	0.064	0.077	0.061	0.104	0.095	0.057	0.077
	(5)	0.060	0.057	0.059	0.064	0.070	0.065	0.099	0.092	0.057	0.083
	(6)	0.124	0.122	0.125	0.121	0.130	0.123	0.163	0.159	0.124	0.125
	(7)	0.179	0.172	0.185	0.189	0.183	0.177	0.196	0.260	0.180	0.197

3.4. *Illustrative examples*

In this section we consider two sets of data taken from the anthropometric data collected by Majumdar (1941) and illustrate the classical linear and the robust linear classification procedures described in Sections 3.1 and 3.2, respectively.

3.4.1. *Example 1*

In Tables 3.3 and 3.4, we have given the measurements on nasal breadth of 106 individuals from the Agharia group and 180 individuals from the Bhil group, respectively.

Table 3.3
Measurements on nasal breadth of $n_1 = 106$ individuals from the Agharia group

i	x_{1i}	i	x_{1i}	i	x_{1i}	i	x_{1i}	i	x_{1i}	i	x_{1i}	i	x_{1i}
1	1683	17	1620	33	1683	49	1683	65	1671	81	1578	97	1748
2	1674	18	1655	34	1733	50	1558	66	1732	82	1707	98	1677
3	1663	19	1609	35	1738	51	1560	67	1645	83	1687	99	1750
4	1802	20	1645	36	1712	52	1705	68	1684	84	1647	100	1637
5	1765	21	1746	37	1664	53	1675	69	1599	85	1546	101	1715
6	1606	22	1690	38	1690	54	1674	70	1698	86	1730	102	1602
7	1709	23	1662	39	1673	55	1689	71	1539	87	1725	103	1698
8	1678	24	1613	40	1684	56	1650	72	1620	88	1674	104	1659
9	1647	25	1738	41	1624	57	1680	73	1713	89	1718	105	1530
10	1643	26	1730	42	1655	58	1687	74	1800	90	1748	106	1679
11	1754	27	1667	43	1743	59	1652	75	1770	91	1715		
12	1625	28	1674	44	1694	60	1808	76	1768	92	1607		
13	1590	29	1662	45	1637	61	1622	77	1585	93	1624		
14	1650	30	1666	46	1745	62	1636	78	1661	94	1814		
15	1630	31	1675	47	1699	63	1653	79	1735	95	1585		
16	1676	32	1632	48	1646	64	1689	80	1689	96	1680		

From these two samples, we find:

$$\bar{x}_1 = 1674.6981, \quad \bar{x}_2 = 1630.1611, \quad s_1^2 = 3336.8381, \quad s_2^2 = 3060.0225,$$

$$\hat{\mu}_1 = 1675.2760, \quad \hat{\mu}_2 = 1629.9518, \quad \hat{\sigma}_1^2 = 3087.2207, \quad \hat{\sigma}_2^2 = 3151.6277,$$

$$\hat{C} = -0.98286 \qquad \text{and} \qquad \hat{C}_R = -0.99769.$$

So, with e_{12} (the probability of wrongly classifying an individual from the Agharia group into the Bhil group) fixed as 0.05, by applying the classical linear classification procedure in (3.8) and reclassifying each individual in Tables 3.3 and 3.4, we obtain estimates of e_{12} and $1 - e_{21}$ (the probability of correctly classifying an individual from the Bhil group into itself) to be 0.047 and 0.194, respectively. Similarly, by applying the robust linear classification procedure in (3.10) and reclassifying each individual in Tables 3.3 and 3.4, we obtain estimates of e_{12} and $1 - e_{21}$ to be 0.047 and 0.439, respectively.

Table 3.4
Measurements on nasal breadth of $n_2 = 180$ individuals from the Bhil group

i	x_{2i}	i	x_{2i}	i	x_{2i}	i	x_{2i}	i	x_{2i}	i	x_{2i}	i	x_{2i}
1	1668	27	1678	53	1611	79	1670	105	1699	131	1716	157	1590
2	1597	28	1706	54	1640	80	1570	106	1616	132	1644	158	1617
3	1729	29	1620	55	1604	81	1562	107	1579	133	1614	159	1644
4	1606	30	1656	56	1543	82	1523	108	1546	134	1575	160	1632
5	1713	31	1760	57	1708	83	1591	109	1587	135	1664	161	1593
6	1686	32	1656	58	1552	84	1663	110	1600	136	1595	162	1762
7	1607	33	1697	59	1634	85	1618	111	1628	137	1631	163	1651
8	1522	34	1637	60	1687	86	1566	112	1636	138	1585	164	1619
9	1674	35	1628	61	1608	87	1580	113	1607	139	1682	165	1594
10	1610	36	1563	62	1673	88	1600	114	1645	140	1653	166	1634
11	1706	37	1653	63	1662	89	1632	115	1607	141	1656	167	1665
12	1608	38	1616	64	1511	90	1592	116	1642	142	1570	168	1613
13	1664	39	1657	65	1676	91	1700	117	1662	143	1708	169	1592
14	1567	40	1642	66	1598	92	1622	118	1574	144	1647	170	1646
15	1590	41	1656	67	1707	93	1631	119	1696	145	1624	171	1698
16	1711	42	1557	68	1690	94	1651	120	1695	146	1621	172	1592
17	1643	43	1707	69	1654	95	1510	121	1735	147	1626	173	1480
18	1603	44	1630	70	1615	96	1776	122	1608	148	1605	174	1672
19	1558	45	1619	71	1540	97	1588	123	1668	149	1605	175	1587
20	1755	46	1652	72	1654	98	1638	124	1540	150	1654	176	1615
21	1625	47	1628	73	1690	99	1560	125	1662	151	1713	177	1568
22	1675	48	1621	74	1605	100	1575	126	1644	152	1578	178	1580
23	1634	49	1540	75	1584	101	1624	127	1740	153	1635	179	1696
24	1668	50	1626	76	1565	102	1663	128	1739	154	1715	180	1586
25	1570	51	1555	77	1575	103	1541	129	1660	155	1608		
26	1563	52	1558	78	1760	104	1605	130	1616	156	1681		

3.4.2. Example 2

In Tables 3.5 and 3.6, we have given the measurements on nasal breadth of 85 individuals from the Basti–Brahmin group and 98 individuals from the Oraon group, respectively.

From these two samples, we find:

$$\bar{x}_1 = 1646.0706, \quad \bar{x}_2 = 1615.7041, \quad s_1^2 = 2610.4761, \quad s_2^2 = 3507.9587,$$

$$\hat{\mu}_1 = 1647.6534, \quad \hat{\mu}_2 = 1617.0022, \quad \hat{\sigma}_1^2 = 2439.1426, \quad \hat{\sigma}_2^2 = 3396.6770,$$

$$\hat{C} = -0.74323 \qquad \text{and} \qquad \hat{C}_R = -0.76152.$$

So, with e_{12} (the probability of wrongly classifying an individual from the Basti–Brahmin group into the Oraon group) fixed as 0.05, by applying the classical linear classification procedure in (3.8) and reclassifying each individual in Tables 3.5 and 3.6, we obtain estimates of e_{12} and $1 - e_{21}$ (the probability of correctly classifying an individual from the Oraon group into itself) to be 0.035

Table 3.5
Measurements on nasal breadth of $n_1 = 85$ individuals from the Bhasti–Brahmin group

i	x_{1i}	i	x_{1i}	i	x_{1i}	i	x_{1i}	i	x_{1i}	i	x_{1i}	i	x_{1i}
1	1564	14	1610	27	1617	40	1677	53	1584	66	1645	79	1690
2	1710	15	1675	28	1542	41	1681	54	1670	67	1578	80	1653
3	1639	16	1604	29	1589	42	1644	55	1575	68	1667	81	1580
4	1668	17	1717	30	1699	43	1700	56	1625	69	1622	82	1680
5	1675	18	1713	31	1599	44	1605	57	1695	70	1700	83	1652
6	1679	19	1703	32	1690	45	1732	58	1670	71	1666	84	1614
7	1598	20	1605	33	1694	46	1602	59	1716	72	1605	85	1742
8	1672	21	1637	34	1640	47	1717	60	1631	73	1688		
9	1620	22	1525	35	1678	48	1642	61	1565	74	1674		
10	1674	23	1609	36	1682	49	1642	62	1618	75	1678		
11	1652	24	1576	37	1593	50	1655	63	1597	76	1728		
12	1620	25	1630	38	1695	51	1622	64	1650	77	1668		
13	1639	26	1456	39	1692	52	1689	65	1667	78	1635		

Table 3.6
Measurements on nasal breadth of $n_1 = 98$ individuals from the Oraon group

i	x_{2i}	i	x_{2i}	i	x_{2i}	i	x_{2i}	i	x_{2i}	i	x_{2i}	i	x_{2i}
1	1655	15	1632	29	1627	43	1623	57	1588	71	1620	85	1572
2	1617	16	1600	30	1682	44	1644	58	1666	72	1619	86	1460
3	1553	17	1609	31	1572	45	1562	59	1592	73	1550	87	1628
4	1635	18	1670	32	1558	46	1580	60	1700	74	1657	88	1552
5	1652	19	1678	33	1533	47	1737	61	1560	75	1692	89	1588
6	1694	20	1536	34	1694	48	1702	62	1644	76	1557	90	1606
7	1688	21	1681	35	1611	49	1599	63	1608	77	1596	91	1638
8	1641	22	1647	36	1467	50	1563	64	1567	78	1540	92	1582
9	1520	23	1778	37	1591	51	1515	65	1635	79	1620	93	1566
10	1596	24	1666	38	1644	52	1532	66	1490	80	1590	94	1667
11	1646	25	1656	39	1660	53	1577	67	1640	81	1629	95	1689
12	1685	26	1516	40	1620	54	1700	68	1625	82	1600	96	1616
13	1614	27	1565	41	1670	55	1728	69	1630	83	1614	97	1654
14	1617	28	1627	42	1630	56	1640	70	1670	84	1480	98	1607

and 0.143, respectively. Similarly, by applying the robust linear classification procedure in (3.10) and reclassifying each individual in Tables 3.5 and 3.6, we obtain estimates of e_{12} and $1 - e_{21}$ to be 0.035 and 0.163, respectively.

4. Robust multivariate linear classification procedure and applications

In this section we first present the classical multivariate two-way linear classification procedure and then develop the analogous robust multivariate linear classification procedure based on the MML estimators. By applying these two procedures on Majumdar's anthropometric data, we demonstrate that the robust procedure developed here performs overall more efficiently than the classical procedure, when one of the two errors of misclassification is fixed at a specified level. In the case when both the errors of misclassification are allowed to float, we also show that the robust procedure has a smaller average error rate than some non-parametric classification procedures like the nearest neighbour method and the method based on density estimates.

4.1. Classical linear classification procedure

Let us assume x_{1k} ($k = 1, 2, \ldots, n_1$) and x_{2k} ($k = 1, 2, \ldots, n_2$) to be independent random samples from populations Π_1 and Π_2 which are taken to be $N_p(\mu_1, \Sigma)$ and $N_p(\mu_2, \Sigma)$, respectively. Let us denote the sample means by

$$\bar{x}_1 = \frac{1}{n_1} \sum_{k=1}^{n_1} x_{1k} \quad \text{and} \quad \bar{x}_2 = \frac{1}{n_2} \sum_{k=1}^{n_2} x_{2k}$$

and the sample variance–covariance matrices by

$$S_1 = \frac{1}{n_1 - 1} \sum_{k=1}^{n_1} (x_{1k} - \bar{x}_1)(x_{1k} - \bar{x}_1)^{\mathrm{T}},$$

and

$$S_2 = \frac{1}{n_2 - 1} \sum_{k=1}^{n_2} (x_{2k} - \bar{x}_2)(x_{2k} - \bar{x}_2)^{\mathrm{T}},$$

and the pooled sample variance–covariance matrix by

$$S = \frac{(n_1 - 1)S_1 + (n_2 - 1)S_2}{n_1 + n_2 - 2}.$$

Then, to classify a new independent observation x_0, the classical linear classification procedure is to classify x_0 into Π_1 or Π_2 according as (Rao, 1948a,b; Anderson, 1951)

$$V = \{x_0 - \tfrac{1}{2}(\bar{x}_1 + \bar{x}_2)\}^{\mathrm{T}} S^{-1}(\bar{x}_1 - \bar{x}_2) \gtrless C, \tag{4.1}$$

where, as before, the cut-off point C is determined such that one of the two errors of misclassification (say e_{12}) is at a pre-specified level.

Although $\bar{x}_1$ and $\bar{x}_2$ are independently distributed as $N_p(\mu_1, (1/n_1)\Sigma)$ and $N_p(\mu_2, (1/n_2)\Sigma)$, respectively, and $(n_1 + n_2 - 2)S$ is independently distributed as Wishart $W_p(n_1 + n_2 - 2, \Sigma)$, the distribution of the statistic V in (4.1) does not assume a manageable form. Several authors have derived different representations and series expansions for the distribution of V (see, for example, Wald, 1944; Anderson, 1951; Sitgreaves, 1952; Bowker, 1960; Kabe, 1963; Okamoto, 1963; Dunn and Varady, 1966; Sinha and Giri, 1975). Hence, it is very difficult to evaluate the cut-off point C exactly for a specified value of e_{12}. In the asymptotic situation, however, $\bar{x}_1$, $\bar{x}_2$ and S tend to their respective expected values μ_1, μ_2 and Σ, and as a result the distribution of V in (4.1) is the same as that of Fisher's linear discriminant function

$$U = \{x_0 - \tfrac{1}{2}(\mu_1 + \mu_2)\}^{\mathrm{T}} \Sigma^{-1} (\mu_1 - \mu_2). \tag{4.2}$$

The statistic U in (4.2) is a linear function of x_0 and hence has a p-variate normal distribution with mean

$$\mathrm{E}(U \mid x_0 \in \Pi_1) = \tfrac{1}{2}\delta^2 = -\mathrm{E}(U \mid x_0 \in \Pi_2), \tag{4.3}$$

and variance

$$\mathrm{Var}(U \mid x_0 \in \Pi_1) = \mathrm{Var}(U \mid x_0 \in \Pi_2) = \delta^2, \tag{4.4}$$

where $\delta^2 = (\mu_1 - \mu_2)^{\mathrm{T}} \Sigma^{-1} (\mu_1 - \mu_2)$ is the standardized squared distance between the two populations Π_1 and Π_2. Therefore, with e_{12} fixed as α, for example, under the assumption of normality the cut-off point C in (4.1) may be determined approximately as

$$C = \tfrac{1}{2}\delta^2 - z_{(\alpha)}|\delta|, \tag{4.5}$$

where $z_{(\alpha)}$ is the upper α percentage point of the standard normal distribution. Since δ^2 is unknown in practice, we replace C in (4.5) by an estimate

$$\hat{C} = \tfrac{1}{2}\overset{\circ}{\delta}{}^2 - z_{(\alpha)}|\overset{\circ}{\delta}|. \tag{4.6}$$

By realizing now that

$$(1/n_1 + 1/n_2)^{-1}(\bar{x}_1 - \bar{x}_2)^{\mathrm{T}} S^{-1}(\bar{x}_1 - \bar{x}_2)$$
$$\rightsquigarrow \text{Hotelling's } T_p^2(n_1 + n_2 - 2, (1/n_1 + 1/n_2)^{-1}\delta^2),$$

we may obtain an unbiased estimator of δ^2 as

$$\overset{\circ}{\delta}{}^2 - (1/n_1 + 1/n_2)p, \tag{4.7}$$

where

$$\overset{\circ}{\delta}{}^2 = \frac{n_1 + n_2 - p - 3}{n_1 + n_2 - 2}(\bar{x}_1 - \bar{x}_2)^{\mathrm{T}} S^{-1}(\bar{x}_1 - \bar{x}_2). \tag{4.8}$$

But as mentioned by Lachenbruch and Mickey (1968), the unbiased estimator of δ^2 in (4.7) can take on inadmissible negative values and hence the biased estimator δ^2 in (4.8) itself serves as a suitable estimator of δ^2. Then, the classical two-way linear classification procedure, with e_{12} fixed as α, is to classify the observation x_0 into Π_1 or Π_2 according as

$$V = \{x_0 - \tfrac{1}{2}(\bar{x}_1 + \bar{x}_2)\}^T S^{-1}(\bar{x}_1 - \bar{x}_2) \gtrless \hat{C},$$ (4.9)

where the cut-off point $\hat{C}$ is as given in (4.6).

4.2. Robust linear classification procedure

A robust linear classification procedure analogous to the classical procedure in (4.9) is developed here by making use of the multivariate MML estimators derived in Section 2.2. Let $(\hat{\mu}_1, \hat{\Sigma}_1)$ and $(\hat{\mu}_2, \hat{\Sigma}_2)$ be the MML estimators of (μ_1, Σ) and (μ_2, Σ) obtained from the samples x_{1k}, $k = 1, 2, \ldots, n_1$, and x_{2k}, $k = 1, 2, \ldots, n_2$, respectively. Let $\hat{\Sigma}$ be the pooled MML estimator of Σ given by

$$\hat{\Sigma} = \frac{[(A_1 - 1)\hat{\Sigma}_1 + (A_2 - 1)\hat{\Sigma}_2]}{(A_1 + A_2 - 2)}.$$ (4.10)

Then the robust classification procedure for classifying a new observation x_0 is to classify it into Π_1 or Π_2 according as

$$V_R = \{x_0 - \tfrac{1}{2}(\hat{\mu}_1 + \hat{\mu}_2)\}^T \hat{\Sigma}^{-1}(\hat{\mu}_1 - \hat{\mu}_2) \gtrless C_R,$$ (4.11)

where the cut-off point C_R is determined so that the error of misclassification e_{12} is fixed as α.

It may be noted from the modified likelihood function L^* in (2.13) that asymptotically $\hat{\mu}_1$ and $\hat{\mu}_2$ have independent normal distributions $N_p(\mu_1, (1/m_1)\Sigma)$ and $N_p(\mu_2, (1/m_2)\Sigma)$, respectively, and $(A_1 + A_2 - 2)\Sigma$ is independently distributed as Wishart $W_p(A_1 + A_2 - 2, \Sigma)$. In the asymptotic situation, $\hat{\mu}_1$, $\hat{\mu}_2$ and $\hat{\Sigma}$ tend to their respective expected values μ_1, μ_2 and Σ and as a result the distribution of V_R in (4.11) is the same as that of Fisher's linear discriminant function U in (4.2). Consequently, we replace the cut-off point C_R in (4.11) by an estimate

$$\hat{C}_R = \tfrac{1}{2}\hat{\delta}_R^2 - z_{(\alpha)}|\hat{\delta}_R|,$$ (4.12)

with e_{12} fixed as α. A robust unbiased estimator of δ^2 is given by

$$\hat{\delta}_R^2 - (1/m_1 + 1/m_2)p,$$ (4.13)

where

$$\hat{\delta}_R^2 = \frac{A_1 + A_2 - p - 3}{A_1 + A_2 - 2}(\hat{\mu}_1 - \hat{\mu}_2)^T \hat{\Sigma}^{-1}(\hat{\mu}_1 - \hat{\mu}_2).$$ (4.14)

As mentioned earlier in Section 4.1, the robust unbiased estimator of δ^2 in (4.13) can take on inadmissible values and hence the biased estimator $\hat{\delta}_R^2$ in (4.14) itself serves as a suitable estimator of δ^2. Then, the robust two-way linear classification procedure, with e_{12} fixed as α, is to classify the observation x_0 into Π_1 or Π_2 according as

$$V_R = \{x_0 - \tfrac{1}{2}(\hat{\mu}_1 + \hat{\mu}_2)\}^T S^{-1}(\hat{\mu}_1 - \hat{\mu}_2) \gtrless \hat{C}_R, \qquad (4.15)$$

where the cut-off point $\hat{C}_R$ is as given in (4.12). It should be mentioned here that Tiku and Balakrishnan (1984) and Tiku (1988a) have proposed and studied a robust linear classification procedure somewhat similar to the one in (4.15) for the bivariate case.

4.3. Comparisons based on Majumdar's anthropometric data

The total number of individual measurements on any individual reported by Majumdar (1941) did not exceed 16, but usually the first 12 measurements of the following 14 were taken:
  (1) Stature,
  (2) Sitting height,
  (3) Head length,
  (4) Head breadth,
  (5) Minimum frontal breadth,
  (6) Nasal length,
  (7) Nasal breadth,
  (8) Nasal depth,
  (9) Total facial length,
  (10) Upper facial length,
  (11) Bizygomatic breadth,
  (12) Bigonial breadth,
  (13) Head height, and
  (14) Weight.
In some cases, measurements were also taken on
  (15) Orbito-nasal breadth,
  (16) Orbito-nasal arc,
  (17) Auricular height, and
  (18) Span.
A brief description of the characters measured may be secured from Majumdar (1941).

The various castes and tribes that have been included in the anthropometric survey of Majumdar (1941) are as follows:
  (1) Basti–Brahmins,
  (2) Other Brahmins,
  (3) Agharias,
  (4) Chattris,

  (5) Muslims,
  (6) Bhatus,
  (7) Habrus,
  (8) Bhils,
  (9) Doms,
 (10) Artisan Ahirs,
 (11) Artisan Kurmis,
 (12) Other Artisans,
 (13) Artisan Kahars,
 (14) Tharu males,
 (15) Chamars,
 (16) Cheros,
 (17) Majhwars,
 (18) Panikas,
 (19) Kharwars,
 (20) Oraons,
 (21) Rajwars,
 (22) Korwas, and
 (23) Tharu females.

A brief report about the various castes and tribes listed above may be had from the works Risley (1891), Roy (1915), Majumdar (1941, 1944), or the supplementary ethnological note by Mahalanobis (1948). Should a more detailed account of the castes and tribes of the United Provinces considered above be required, one may refer to the volumes of Sir William Crooke (1896).

First, in order to assess the multivariate normality of the data from each group, we computed the generalized distance $(x - \bar{x})^{\mathrm{T}} S^{-1}(x - \bar{x})$ for each observation x in the group, where $\bar{x}$ and S are the sample mean and sample variance–covariance matrix of that group, respectively. Under multivariate normality, we know that these generalized distances will have approximately a χ^2 distribution with p degrees of freedom. So, the ordered observed generalized distances were plotted against the expected values of order statistics from the χ^2 distribution with p degrees of freedom. These 'probability plots' or 'Q–Q plots' show departures from a linear relationship between these two sets of values and hence a departure from multivariate normality of the data (see Everitt, 1978; Gnanadesikan, 1977). The required expected values of order statistics from the χ^2 distribution with p degrees of freedom were obtained by using David and Johnson's (1954) approximation; one may refer to David (1981) and Arnold and Balakrishnan (1989) for details on this method of approximation. One would like to believe that the presence of a few outliers in a group of large size may not affect the classical linear classification procedure. But, unfortunately this does not seem to be the case as we will demonstrate in this section that the performance of the classical classification procedure is indeed impaired by the presence of these few outliers in the groups. We will also display that the robust linear classification procedure performs overall much better than the classical procedure.

We applied the classical linear and the robust linear classification procedures in (4.9) and (4.15) to classify each individual in Π_i into either Π_i or Π_j, after fixing the error of misclassification e_{ij} as 0.05, for $1 \leqslant i < j \leqslant 23$. We then obtained estimates of the error of misclassification e_{ij} and $1 - e_{ji}$ (the probability of correctly classifying an individual from Π_j into itself) for $1 \leqslant i < j \leqslant 23$ for both the procedures. From these values, we found the average of e_{ij} $(1 \leqslant i < j \leqslant 23)$ and $1 - e_{ji}$ values to be 0.054 and 0.578, respectively, for the classical linear classification procedure, and 0.061 and 0.635 correspondingly for the robust linear classification procedure. We see immediately that the robust procedure has a slightly larger average of estimated $1 - e_{ji}$ values (the difference being 0.057). This difference may not appear to be significant. However, if we take into account the number of classifications performed in order to obtain these estimates (that is, 31 053 and 33 539 classifications performed to find the average of the estimated e_{ij} and $1 - e_{ji}$ values, respectively), we simply observe that the robust procedure wrongly classifies an individual from Π_i into Π_j $(1 \leqslant i < j \leqslant 23)$ 217 times more than the classical procedure; at the same time, we observe that the robust procedure correctly classifies an individual from Π_j into itself $(1 \leqslant i < j \leqslant 23)$ 1912 times more than the classical procedure.

Furthermore, a careful examination of the estimates of e_{ij} and $1 - e_{ji}$ for the two procedures also revealed the following points:

(1) There seemed to be a considerable overlap between groups 1 and 2, namely, Basti–Brahmins and other Brahmins; for example, the estimate of $1 - e_{21}$ was only 0.326 even for an estimated value of e_{12} as high as 0.105 for the classical procedure; similarly, the estimate of $1 - e_{21}$ was only 0.304 even for an estimated value of e_{12} as high as 0.151 for the robust procedure.

(2) Between groups 10, 11, 12 and 13 (Ahirs, Kurmis, Other and Kahar Artisans), we once again observed a considerable overlap and hence a poor discrimination. For example, among these four groups we had the average of the estimated e_{ij} values as 0.070 and the average of the estimated $1 - e_{ji}$ values as only 0.273 for the classical procedure; similarly, we had the average of the estimated e_{ij} values as 0.091 and the average of the estimated $1 - e_{ji}$ values as only 0.285 for the robust procedure.

(3) Upon comparing groups 14 and 23 (Tharu males and females), we observed a good difference between the two groups and hence a good discrimination. For example, we had the estimate of the error of misclassification $e_{14, 23}$ as 0.047 and the estimate of the probability of correct classification $1 - e_{23, 14}$ as 0.514 for the classical procedure; similarly, we observed the estimate of $e_{14, 23}$ as 0.026 and the estimate of $1 - e_{23, 14}$ to be as large as 0.872 for the robust procedure.

(4) After looking at the values corresponding to group 22 (Korwas), we observed a considerable difference between the Korwas and all the other groups. The remarkable discrimination of the Korwas from all other groups by both the classification procedures may be attributed to their unwritten law prohibiting any inter-tribal intercourse and inter-tribal marriage and also their highly developed sense of solidarity and comradeship; see Majumdar (1944). We also noted that the least discrimination of the Korwas occurred against group 19 (Kharwars) and this seems to agree well with the comment made by Majumdar (1944) that the Kharwars display marked similarities with the Korwas.

Finally, based on the first two points made above, we combined the two Brahmin groups (groups 1 and 2) into one group, and similarly combined the four Artisan groups (groups 10, 11, 12 and 13) into one group. Then, we once again applied the classical linear and the robust linear classification procedures in (4.9) and (4.15) to classify each individual in Π_i^* into either Π_i^* or Π_j^*, after fixing the error of misclassification e_{ij}^* as 0.05, for $1 \leq i < j \leq 19$. We then obtained estimates of the error of misclassification e_{ij}^* and $1 - e_{ji}^*$ (the probability of correctly classifying an individual from Π_j^* into itself) for $1 \leq i < j \leq 19$ for both the procedures. From these values we found the average of e_{ij}^* ($1 \leq i < j \leq 19$) and $1 - e_{ji}^*$ values to be 0.052 and 0.593, respectively, for the classical linear classification procedure, and 0.059 and 0.658 correcpondingly for the robust linear procedure. We note that these are slight improvements over the previous results based on all 23 different groups. We also observe immediately that the robust classification procedure has a slightly larger average of estimated e_{ij}^* values (the difference being only 0.007) and, at the same time, also has a larger average of estimated $1 - e_{ji}^*$ values (the difference being 0.065). Now by taking into account the number of classifications performed in order to obtain these estimates (that is, 27033 and 25815 classifications performed to find the average of the estimated e_{ij}^* and $1 - e_{ji}^*$ values, respectively) we simply observe that the robust procedure wrongly classifies an individual from Π_i^* into Π_j^* ($1 \leq i < j \leq 19$) 189 times more than the classical procedure; at the same time, we observe that the robust procedure correctly classifies an individual from Π_j^* into itself ($1 \leq i < j \leq 19$) 1678 times more than the classical procedure. All these results clearly demonstrate the overall superiority of the robust linear classification procedure in (4.15) over the classical linear classification procedure in (4.9).

4.4. Comparison with distribution-free procedures

Let us now consider the classical classification procedure for classifying an individual into either Π_1 or Π_2, without restricting the error of misclassification e_{12} to assume a pre-fixed level. The procedure for classifying a new observation x_0 is based on the statistic V in (4.1), and is to classify x_0 into Π_1 or Π_2 according as

$$V = \{x_0 - \tfrac{1}{2}(\bar{x}_1 + \bar{x}_2)\}^{\mathrm{T}} S^{-1}(\bar{x}_1 - \bar{x}_2) \gtrless 0 . \tag{4.16}$$

We applied the above classical linear procedure to classify each observation in Π_i into either Π_i or Π_j for $1 \leq i, j \leq 23$, $i \neq j$, and then determined estimates of the errors of misclassification e_{ij} and e_{ji} for $1 \leq i < j \leq 23$. From these values, we found the average of the estimated e_{ij} values as 0.187 and the average of the estimated e_{ji} values as 0.178. We also observed a poor discrimination between groups 1 and 2 (Basti–Brahmins and other Brahmins) as the estimates of e_{12} and e_{21} were as large as 0.384 and 0.413, respectively. Similarly, we observed a considerable overlap between the four Artisan groups (groups 10, 11, 12 and 13) as the averages of the estimated e_{ij} and e_{ji} values were 0.360 and 0.361, respectively. Based on these two points, as done earlier in Section 4.3, we combined the two Brahmin groups and the four Artisan groups and then applied the linear classification procedure in (4.16) in order to classify each individual in Π_i^* into Π_i^* or Π_j^* for $1 \leqslant i, j \leqslant 19$, $i \neq j$. We then determined estimates of the errors of misclassification e_{ij}^* and e_{ji}^* for $1 \leqslant i < j \leqslant 19$, and found from these values the average of the estimated e_{ij}^* values to be 0.177 and the average of the estimated e_{ji}^* values to be 0.168. We observe slight improvement over the previous results based on all 23 groups.

In a similar way, the robust classification procedure for classifying a new observation x_0 is based on the statistic V_R in (4.11) and is to classify x_0 into Π_1 or Π_2 according as

$$V_R = \{x_0 - \tfrac{1}{2}(\hat{\mu}_1 + \hat{\mu}_2)\}^{\mathrm{T}} \hat{\Sigma}^{-1}(\hat{\mu}_1 - \hat{\mu}_2) \gtrless 0 . \qquad (4.17)$$

We applied the above robust linear procedure to classify each observation in Π_i into either Π_i or Π_j for $1 \leqslant i, j \leqslant 23$, $i \neq j$, and then determined estimates of the errors of misclassification e_{ij} and e_{ji} for $1 \leqslant i < j \leqslant 23$. From these values, we found the average of the estimated e_{ij} values as 0.175 and the average of the estimated e_{ji} values as 0.159. We note that these average error rates are smaller than the corresponding values of the classical classification procedure. This means that the robust procedure in (4.17) has 1010 fewer misclassifications than the classical procedure in (4.16). As done earlier, we once again combined the two Brahmin groups (groups 1 and 2) and the four Artisan groups (groups 10, 11, 12 and 13) and then applied the robust procedure in (4.17) to classify each individual in Π_i^* into either Π_i^* or Π_j^* for $1 \leqslant i, j \leqslant 19$, $i \neq j$. We then determined estimates of the errors of misclassification e_{ij}^* and e_{ji}^* for $1 \leqslant i < j \leqslant 19$, and found from these values the average of the estimated e_{ij}^* values to be 0.166 and the average of the estimated e_{ji}^* values to be 0.149, which is a slight improvement over the previous result. Both these average error rates are once again smaller than the corresponding values of the classical classification procedure. This means that the robust procedure in (4.17) has 787 fewer misclassifications than the classical classification procedure in (4.16).

In order to compare the above two procedures with some distribution-free procedures, we first considered the method based on density estimates which is as follows. Let $\hat{f}_{n_1}(x_0)$ and $\hat{f}_{n_2}(x_0)$ be consistent estimators of the density functions of the populations Π_1 and Π_2 at the point x_0. Then the procedure is simply to classify x_0 into Π_1 or Π_2 according as

$$\frac{\hat{f}_{n_1}(x_0)}{\hat{f}_{n_2}(x_0)} \gtrless 1 . \qquad (4.18)$$

The density estimator we used in this study is the one due to Loftsgaarden and Quesenberry (1965) and is given by

$$\hat{f}_m(x_0) = \left\{ \frac{k(m) - 1}{m} \right\} \frac{p\Gamma(p/2)}{2\pi^{p/2} r_{k(m)}^p} , \qquad (4.19)$$

where m is the size of the sample on which the density estimator is based, $r_{k(m)}$ is the distance from x_0 to the $k(m)$-th closest observation in the sample to x_0 as measured by the Euclidean distance function; $k(m)$ was chosen to be nearer to $\sqrt{m}$ and refer to Loftsgaarden and Quesenberry (1965) and Gessaman and Gessaman (1972) for details. It should be mentioned that Tiku and Balakrishnan (1984) have compared the performance of the classification procedure based on the above density estimator in the bivariate case to that based on the bivariate extension of Parzen's density estimator as given by Cacoullos (1964). We applied the above procedure based on (4.18) and (4.19) to classify each observation in Π_i into either Π_i or Π_j for $1 \leqslant i, j \leqslant 23$, $i \neq j$, and determined the estimates of the errors of misclassification e_{ij} and e_{ji} for $1 \leqslant i < j \leqslant 23$. From these values, we first of all noted this classification procedure to be inconsistent in the sense that some of the error rates were greater than 0.5. We also found the average of the estimated e_{ij} values $(1 \leqslant i < j \leqslant 23)$ to be 0.246 and the average of the estimated e_{ji} values to be 0.220. It should be noted that both these values are considerably larger than the corresponding values of the robust classification procedure. This means that the robust procedure in (4.17) has 4251 fewer misclassifications than the distribution-free classification procedure based on Loftsgaarden and Quesenberry's density estimator given above in (4.19).

Next, we considered the nearest neighbour classification procedure proposed by Fix and Hodges (1951) as described in Section 3.3 in its multivariate form. As mentioned earlier in Section 3.3, the choice of m seems to be very crucial for this procedure. In any case, we applied this procedure with various choices of m to classify each observation in Π_i into either Π_i or Π_j for $1 \leqslant i, j \leqslant 23$, $i \neq j$, and determined the estimates of the errors of misclassifications e_{ij} and e_{ji} for $1 \leqslant i < j \leqslant 23$. From these values, we observed the nearest neighbour classification procedure also to be inconsistent as some of the error rates were greater than 0.5. Further, we found the averages of the estimated e_{ij} and e_{ji} values $(1 \leqslant i < j \leqslant 23)$ to be 0.324 and 0.267, 0.358 and 0.298, and 0.295 and 0.247, corresponding to the cases $m = 10\%$ of $(n_i + n_j)$, $m = 20\%$ of $(n_i + n_j)$, and $m = \sqrt{n_i + n_j}$, respectively. From these values and also based on comparisons with some other choices of m, we observed that the choice of $m = \sqrt{n_i + n_j}$ yielded the smallest error rates. However, all these average error rates are larger than the corresponding values of the procedure based on Loftsgaarden and Quesenberry's density estimator which in turn, as mentioned earlier, are considerably larger than the corresponding average error rates of the robust procedure. Thus, for example, in comparison to the nearest neighbour procedure with $m = \sqrt{n_i + n_j}$, the robust procedure in (4.17) has 6677 fewer misclassifications.

Yet another interesting distribution-free procedure based on 'statistically equivalent blocks' due to Anderson (1966) and Gessaman (1970) has not been considered in this comparative study as it requires unrealistically large sample sizes (of the order of 700) to achieve somewhat higher efficiency (or smaller error rates) than the Loftsgaarden and Quesenberry's procedure (see, for example, Gessaman and Gessaman, 1972).

5. Robust univariate quadratic classification procedure and applications

In this section we first present the classical univariate two-way quadratic classification procedure and then develop the analogous robust univariate two-way quadratic classification procedure based on the MML estimators. By comparing these two procedures, when the error of misclassification e_{12} is at a pre-fixed level, under normal and some non-normal models, we show that the robust quadratic procedure is quite efficient and robust to departures from normality. The simulated e_{12} values of these two procedures, however, are usually somewhat larger than the pre-fixed level. So, we develop a linear classification procedure and a robust analogue of it as alternatives to the above quadratic classification procedures; these procedures have simpler asymptotic distribution theory, simulated e_{12} values closer to the presumed level, and $1 - e_{21}$ values only slightly smaller than the corresponding quadratic classification procedures. Finally, we present an illustrative example by using Majumdar's (1941) anthropometric data.

5.1. Classical quadratic classification procedure

Let x_{1i}, $i = 1, 2, \ldots, n_1$, and x_{2i}, $i = 1, 2, \ldots, n_2$, be independent random samples from populations Π_1 and Π_2 assumed to be normal $N(\mu_1, \sigma_1^2)$ and $N(\mu_2, \sigma_2^2)$, respectively. Let us denote the sample means by

$$\bar{x}_1 = \frac{1}{n_1} \sum_{i=1}^{n_1} x_{1i} \quad \text{and} \quad \bar{x}_2 = \frac{1}{n_2} \sum_{i=1}^{n_2} x_{2i},$$

and the sample variances by

$$s_1^2 = \frac{1}{n_1 - 1} \sum_{i=1}^{n_1} (x_{1i} - \bar{x}_1)^2 \quad \text{and} \quad s_2^2 = \frac{1}{n_2 - 1} \sum_{i=1}^{n_2} (x_{2i} - \bar{x}_2)^2.$$

Then, to classify a new independent observation x_0, the classical quadratic classification procedure is to classify x_0 into Π_1 or Π_2 according as (Anderson, 1951; Srivastava and Khatri, 1979)

$$W = \frac{(x_0 - \bar{x}_1)^2}{s_1^2} - \frac{(x_0 - \bar{x}_2)^2}{s_2^2} \lessgtr C, \tag{5.1}$$

where the cut-off point C is chosen such that the error of misclassification e_{12} (for example) is at a pre-fixed level.

To determine the constant C for a specified level of e_{12}, we require the distribution of W which unfortunately does not assume a manageable form. In the asymptotic case, however, as pointed out by Tiku and Balakrishnan (1985, 1989),

$\bar{x}_1$, $\bar{x}_2$, s_1^2 and s_2^2 converge to μ_1, μ_2, σ_1^2 and σ_2^2, respectively, in which case the distribution of the statistic W in (5.1) is the same as that of

$$D = \frac{(x_0 - \mu_1)^2}{\sigma_1^2} - \frac{(x_0 - \mu_2)^2}{\sigma_2^2}. \tag{5.2}$$

Even though the distribution of D, being that of the difference of two correlated χ^2 variables, is intractable, its moments may be worked out without great difficulty; for example, when x_0 is actually from Π_1, the first four moments are as follows:

$$E(D|x_0 \in \Pi_1) = 1 - \phi^2 - \delta^2,$$

$$\text{Var}(D|x_0 \in \Pi_1) = 2(1 - \phi^2)^2 + 4\delta^2\phi^2,$$

$$\mu_3(D|x_0 \in \Pi_1) = 8(1 - \phi^2)^3 + 24\delta^2\phi^2(1 - \phi^2),$$

and

$$\mu_4(D|x_0 \in \Pi_1) = 60(1 - \phi^2)^4 + 240\delta^2\phi^2(1 - \phi^2)^2 + 48\delta^4\phi^4, \tag{5.3}$$

where $\phi^2 = \sigma_1^2/\sigma_2^2$ and $\delta^2 = (\mu_1 - \mu_2)^2/\sigma_2^2$. Since the coefficient of skewness and kurtosis of D are exactly equal to 0 and 3, respectively, when $\phi^2 = \sigma_1^2/\sigma_2^2 = 1$ (the equal variance case), the distribution of W may be approximated reasonably by a normal distribution for large n_1 and n_2. This normal approximation may be suitable only in some range of δ^2 and ϕ^2, and in general a four-moment Pearson curve may be fitted by making use of the tables of Johnson et al. (1963). If we use the normal approximation, the cut-off point C in (5.1) may be approximated by

$$C = 1 - \phi^2 - \delta^2 + 1.64\{2(1 - \phi^2)^2 + 4\delta^2\phi^2\}^{1/2}, \tag{5.4}$$

when we fix the error of misclassification e_{12} to be 0.05. Since ϕ^2 and δ^2 are unknown, we may use the estimators

$$\hat{\phi}^2 = \frac{n_2 - 3}{n_2 - 1} \frac{s_1^2}{s_2^2} \tag{5.5}$$

and

$$\hat{\delta}^2 = \frac{n_2 - 3}{n_2 - 1} \frac{(\bar{x}_1 - \bar{x}_2)^2}{s_2^2}, \tag{5.6}$$

in equation (5.4) to obtain a suitable estimate of C to be

$$\hat{C} = 1 - \hat{\phi}^2 - \hat{\delta}^2 + 1.64\{2(1 - \hat{\phi}^2)^2 + 4\hat{\delta}^2\hat{\phi}^2\}^{1/2}. \tag{5.7}$$

Then, the classical quadratic classification procedure, with e_{12} fixed as 0.05, is to classify x_0 into Π_1 or Π_2 according as

$$W = \frac{(x_0 - \bar{x}_1)^2}{s_1^2} - \frac{(x_0 - \bar{x}_2)^2}{s_2^2} \lessgtr \hat{C},$$
(5.8)

where the cut-off point $\hat{C}$ is as given in (5.7). It should be mentioned here that the estimator $\hat{\phi}^2$ in (5.6) is unbiased for ϕ^2 while the estimator $\hat{\delta}^2$ in (5.6) is biased for δ^2; we use this biased estimator since the unbiased estimator of δ^2 given by

$$\hat{\delta}^2 - \frac{n_2 - 3}{n_2 - 1} \frac{s_1^2}{n_1 s_2^2} - \frac{1}{n_2},$$
(5.9)

may take on inadmissible negative values.

5.2. Robust quadratic classification procedure

A robust quadratic classification procedure analogous to the classical one in (5.8) has been obtained by Tiku and Balakrishnan (1985, 1989) by replacing $\bar{x}_1$, $\bar{x}_2$, s_1^2 and s_2^2 by the corresponding MML estimators $\hat{\mu}_1$, $\hat{\mu}_2$, $\hat{\sigma}_1^2$ and $\hat{\sigma}_2^2$; here $\hat{\mu}_1$ and $\hat{\sigma}_1$ are the MML estimators of μ_1 and σ_1 obtained from the sample x_{1i}, $i = 1, \ldots, n_1$, and $\hat{\mu}_2$ and $\hat{\sigma}_2$ are the MML estimators of μ_2 and σ_2 obtained from the sample x_{2i}, $i = 1, \ldots, n_2$. Then the robust two-way quadratic classification procedure, with e_{12} fixed as 0.05, is to classify the observation x_0 into Π_1 or Π_2 according as

$$W_R = \frac{(x_0 - \hat{\mu}_1)^2}{\hat{\sigma}_1^2} - \frac{(x_0 - \hat{\mu}_2)^2}{\hat{\sigma}_2^2} \lessgtr \hat{C}_R,$$
(5.10)

where the cut-off point $\hat{C}_R$ is given by

$$\hat{C}_R = 1 - \hat{\phi}_R^2 - \hat{\delta}_R^2 + 1.64 \{2(1 - \hat{\phi}_R^2)^2 + 4\hat{\delta}_R^2 \hat{\phi}_R^2\}^{1/2},$$
(5.11)

where $\hat{\phi}_R^2$ and $\hat{\delta}_R^2$ being the robust estimators of ϕ^2 and δ^2 given by

$$\hat{\phi}_R^2 = \frac{A_2 - 3}{A_2 - 1} \frac{\hat{\sigma}_1^2}{\hat{\sigma}_2^2}$$
(5.12)

and

$$\hat{\delta}_R^2 = \frac{A_2 - 3}{A_2 - 1} \frac{(\hat{\mu}_1 - \hat{\mu}_2)^2}{\hat{\sigma}_2^2}.$$
(5.13)

Since the MML estimators $\hat{\mu}$ and $\hat{\sigma}^2$ have asymptotically exactly the same distributions as $\bar{x}$ and s^2, viz., $\sqrt{m}(\hat{\mu} - \mu)/\sigma$ and $(A - 1)\hat{\sigma}^2/\sigma^2$ are independently distributed as standard normal and χ^2 with $A - 1$ degrees of freedom (see Tiku, Tan and Balakrishnan, 1986), respectively, the distribution of W_R in (5.10) is the same as that of W (for large n_1 and n_2) under the assumption of normality. Therefore, for large n_1 and n_2, the classical and the robust quadratic classification procedures will both have the same $1 - e_{21}$ value for a fixed e_{12}, under the assumption of normality.

In order to examine the robustness features of these two procedures, n pseudo-random observations were simulated for the x_1 and x_2 samples from the following models:

(1) Normal,
(2) Outlier-normal model $(n - 1)N(0, 1)$ and $1N(0, 100)$, and
(3) Outlier-normal model $(n - 2)N(0, 1)$ and $2N(0, 100)$.   (5.14)

A constant μ_2 was added to the x_2 sample, after multiplying the x_2 sample by σ_2, in order to create a location shift between the two populations Π_1 and Π_2. Then, with e_{12} fixed as 0.05, the values of e_{12} and $1 - e_{21}$ were simulated (based on 1000 Monte Carlo runs) for both the procedures under the three models given in (5.14). These values are presented in Table 5.1 for $\sigma_2 = 2$ and 3, $\mu_2 = 1$ and 3, and $n_1 = n_2 = n = 50$. It is clear from this table that both the procedures have somewhat larger e_{12} values than the presumed value of 0.05, under normality. The e_{12} values of the robust procedure are however fairly stable from distribution to distribution, whereas the e_{12} values of the classical procedure are seen to be quite sensitive to the presence of outliers. Furthermore, we observe from the table that the robust quadratic procedure has much larger $1 - e_{21}$ value than the classical quadratic procedure whenever there are few outliers in the samples.

Table 5.1
Simulated values of e_{12} and $1 - e_{21}$ for the classical and the robust quadratic classification procedures for the models in (5.14) when $n_1 = n_2 = n = 50$

Model	Procedure	e_{12} values				$1 - e_{21}$ values			
		$\sigma_2 = 2$		$\sigma_2 = 3$		$\sigma_2 = 2$		$\sigma_2 = 3$	
		$\mu_2 = 1$	$\mu_2 = 3$	$\mu_2 = 1$	$\mu_2 = 3$	$\mu_2 = 1$	$\mu_2 = 3$	$\mu_2 = 1$	$\mu_2 = 3$
(1)	W	0.084	0.082	0.086	0.088	0.46	0.80	0.56	0.73
	W_R	0.076	0.072	0.078	0.080	0.45	0.79	0.57	0.73
(2)	W	0.026	0.022	0.028	0.030	0.32	0.58	0.41	0.59
	W_R	0.072	0.070	0.068	0.070	0.45	0.79	0.56	0.72
(3)	W	0.014	0.012	0.014	0.016	0.22	0.32	0.24	0.33
	W_R	0.068	0.066	0.068	0.068	0.44	0.78	0.55	0.72

5.3. Transformed linear classification procedures

Tiku and Balakrishnan (1985, 1989) have made transformations on the observations and proposed transformed linear classification procedures based on these observations, which are based on the statistics

$$V^* = \{x_0/s_1 - \tfrac{1}{2}(\bar{x}_1/s_1 + \bar{x}_2/s_2)\}(\bar{x}_1/s_1 - \bar{x}_2/s_2), \tag{5.15}$$

and

$$V_R^* = \{x_0/\hat{\sigma}_1 - \tfrac{1}{2}(\hat{\mu}_1/\hat{\sigma}_1 + \hat{\mu}_2/\hat{\sigma}_2)\}(\hat{\mu}_1/\hat{\sigma}_1 - \hat{\mu}_2/\hat{\sigma}_2). \tag{5.16}$$

The classical procedure is based on the statistic V^* in (5.15) and is to classify x_0 into Π_1 or Π_2 according as

$$V^* \gtrless \hat{C}^*, \tag{5.17}$$

where the cut-off point $\hat{C}^*$ is derived (by proceeding exactly on the same lines as in Section 3.1), by fixing the error of misclassification e_{12} as 0.05, to be

$$\hat{C}^* = \frac{1}{2}\left\{\frac{a_1 \bar{x}_1}{s_1} - \frac{a_2 \bar{x}_2}{s_2}\right\}^2 - 1.64 \left|\frac{a_1 \bar{x}_1}{s_1} - \frac{a_2 \bar{x}_2}{s_2}\right|; \tag{5.18}$$

here

$$a_1 = \frac{\Gamma(\tfrac{1}{2}(n_1 - 1))}{(\tfrac{1}{2}(n_1 - 1))^{1/2}\,\Gamma(\tfrac{1}{2}(n_1 - 2))} \quad \text{and} \quad a_2 = \frac{\Gamma(\tfrac{1}{2}(n_2 - 1))}{(\tfrac{1}{2}(n_2 - 1))^{1/2}\,\Gamma(\tfrac{1}{2}(n_2 - 2))}.$$

Similarly, the robust procedure is based on the statistic V_R^* in (5.16) and is to classify x_0 into Π_1 or Π_2 according as

$$V_R^* \gtrless \hat{C}_R^*, \tag{5.19}$$

where the cut-off point $\hat{C}_R^*$, with e_{12} fixed as 0.05, is given by

$$\hat{C}_R^* = \frac{1}{2}\left\{\frac{b_1 \hat{\mu}_1}{\hat{\sigma}_1} - \frac{b_2 \hat{\mu}_2}{\hat{\sigma}_2}\right\}^2 - 1.64 \left|\frac{b_1 \hat{\mu}_1}{\hat{\sigma}_1} - \frac{b_2 \hat{\mu}_2}{\hat{\sigma}_2}\right|; \tag{5.20}$$

here,

$$b_1 = \frac{\Gamma(\tfrac{1}{2}(A_1 - 1))}{(\tfrac{1}{2}(A_1 - 1))^{1/2}\,\Gamma(\tfrac{1}{2}(A_1 - 2))} \quad \text{and} \quad b_2 = \frac{\Gamma(\tfrac{1}{2}(A_2 - 1))}{(\tfrac{1}{2}(A_2 - 1))^{1/2}\,\Gamma(\tfrac{1}{2}(A_2 - 2))}.$$

Asymptotically, the distribution of V_R^* is exactly the same as that of V^* under the assumption of normality. Hence, for large n_1 and n_2, the V^* procedure in (5.17) and V_R^* in (5.19) will both have the same $1 - e_{21}$ (the probability of classifying x_0 into Π_2 when x_0 in fact belongs to Π_2) for a pre-fixed e_1, under the assumption of normality.

N. Balakrishnan and R. S. Ambagaspitiya

In order to examine the robustness features of the above two procedures, n pseudo-random observations were simulated for the x_1 and x_2 samples each from the three models in (5.14) and in addition the following four models:

(4) Logistic,
(5) Double exponential,
(6) Mixture-normal model $0.9 \ N(0, 1) + 0.1 \ N(0, 16)$, and
(7) Student's t with 4 degrees of freedom.   (5.21)

After multiplying the x_2 sample by σ_2, a constant μ_2 was added to the x_2 sample in order to create a location shift between the populations Π_1 and Π_2. Then, with e_{12} fixed as 0.05, the values of e_{12} and $1 - e_{21}$ were simulated (based on 1000 Monte Carlo runs) for both the procedures under the three models in (5.14) and the four models in (5.21). These values are presented in Table 5.2 for $\sigma_2 = 2$ and 3, $\mu_2 = 1$ and 3, and $n_1 = n_2 = n = 30$. It is clear from this table that both these procedures have lower e_{12} values (and closer to the presumed level of 0.05) than the corresponding quadratic classification procedures, under normality. We also note that the e_{12} values of the robust procedure are fairly stable from distribution to distribution while those of the classical procedure are seen to be quite sensitive to departures from normality. We further observe from Table 5.2 that the trans- formed robust linear procedure in (5.19) has much larger $1 - e_{21}$ values than the transformed classical linear procedure in (5.17) under the various non-normal models.

Table 5.2
Simulated values of e_{12} and $1 - e_{21}$ for the classical and the robust transformed linear classification procedures for the models in (5.14) and (5.21) when $n_1 = n_2 = n = 30$

Model	Procedure	e_{12} values				$1 - e_{21}$ values			
		$\sigma_2 = 2$		$\sigma_2 = 3$		$\sigma_2 = 2$		$\sigma_2 = 3$	
		$\mu_2 = 1$	$\mu_2 = 3$	$\mu_2 = 1$	$\mu_2 = 3$	$\mu_2 = 1$	$\mu_2 = 3$	$\mu_2 = 1$	$\mu_2 = 3$
(1)	V^*	0.067	0.067	0.066	0.067	0.36	0.76	0.40	0.68
	V_R^*	0.069	0.070	0.070	0.070	0.36	0.76	0.39	0.68
(2)	V^*	0.021	0.019	0.019	0.019	0.17	0.51	0.23	0.48
	V_R^*	0.060	0.060	0.059	0.060	0.34	0.75	0.38	0.66
(3)	V^*	0.010	0.009	0.010	0.010	0.09	0.34	0.13	0.36
	V_R^*	0.059	0.060	0.059	0.060	0.33	0.72	0.37	0.66
(4)	V^*	0.069	0.068	0.070	0.067	0.30	0.70	0.34	0.62
	V_R^*	0.070	0.070	0.071	0.070	0.34	0.74	0.38	0.66
(5)	V^*	0.063	0.054	0.065	0.054	0.27	0.65	0.32	0.61
	V_R^*	0.068	0.060	0.069	0.061	0.33	0.70	0.35	0.65
(6)	V^*	0.038	0.038	0.044	0.038	0.24	0.59	0.28	0.56
	V_R^*	0.060	0.061	0.060	0.061	0.34	0.70	0.36	0.64
(7)	V^*	0.060	0.052	0.061	0.053	0.27	0.62	0.31	0.57
	V_R^*	0.062	0.060	0.062	0.059	0.32	0.68	0.36	0.62

Upon comparing the performance of the classical and the robust transformed linear classification procedures to those of the classical and the robust quadratic classification procedures, we observe the former two to have their simulated e_{12} values to be closer to the pre-fixed value of 0.05 and at the same time they have only slightly smaller $1 - e_{21}$ values, under normality. For example, the simulated values of e_{12} and $1 - e_{21}$ (based on 1000 Monte Carlo runs) of all these four classification procedures are presented in Table 5.3 under the three models in (5.14) for $\mu_2 = 1$, $\sigma_2 = 2$ and $n_1 = n_2 = n = 50$.

Table 5.3
Simulated values of e_{12} and $1 - e_{21}$ for the classical and the robust quadratic procedures and the classical and the robust transformed linear procedures for the models in (5.14) when $\mu_2 = 1$, $\sigma_2 = 2$ and $n_1 = n_2 = n = 50$

Model	e_{12} values				$1 - e_{21}$ values			
	W	W_R	V^*	V_R^*	W	W_R	V^*	V_R^*
(1)	0.084	0.076	0.052	0.054	0.46	0.45	0.42	0.42
(2)	0.026	0.072	0.028	0.054	0.32	0.45	0.33	0.42
(3)	0.014	0.068	0.014	0.054	0.22	0.44	0.24	0.41

From Table 5.3, we observe that the V_R^* procedure has its e_{12} value to be very stable under all three models and also to be very close to the presumed value of 0.05. In addition, we note that it has its $1 - e_{21}$ value to be only slightly smaller than the robust quadratic classification procedure. Furthermore, the V_R^* procedure may also be extended to the multivariate situation as explained in Section 6.2.

5.4. Illustrative example

In this section we consider a data set taken from the anthropometric data collected by Majumdar (1941) and illustrate all the classification procedures described in Sections 5.1, 5.2 and 5.3.

5.4.1. Example 3

In Tables 5.4 and 5.5, we have given the measurements on nasal length of 85 individuals from the Basti–Brahmin group and 158 individuals from the Chamar group, respectively.

From these two samples, we find:

$$\bar{x}_1 = 863.2353, \quad \bar{x}_2 = 804.2342, \quad s_1^2 = 728.3333, \quad s_2^2 = 1292.2803,$$

$$\hat{\mu}_1 = 863.7777, \quad \hat{\mu}_2 = 803.3435, \quad \hat{\sigma}_1^2 = 652.9681, \quad \hat{\sigma}_2^2 = 1463.2822,$$

$$\hat{C} = 1.90462, \quad \hat{C}_R = 1.75105, \quad \hat{C}^* = 30.4502, \quad \hat{C}_R^* = 56.7217.$$

So, with e_{12} (the probability of wrongly classifying an individual from the

Table 5.4
Measurements on nasal length of $n_1 = 85$ individuals from the Basti–Brahmin group

i	x_{1i}	i	x_{1i}	i	x_{1i}	i	x_{1i}	i	x_{1i}	i	x_{1i}	i	x_{1i}
1	870	14	878	27	864	40	885	53	845	66	841	79	901
2	917	15	842	28	804	41	881	54	899	67	841	80	885
3	842	16	853	29	805	42	882	55	857	68	878	81	843
4	847	17	893	30	870	43	870	56	849	69	848	82	884
5	869	18	873	31	822	44	840	57	896	70	899	83	862
6	878	19	901	32	913	45	886	58	866	71	871	84	836
7	813	20	851	33	905	46	846	59	887	72	833	85	854
8	933	21	855	34	884	47	861	60	845	73	895		
9	856	22	831	35	878	48	872	61	818	74	889		
10	881	23	849	36	884	49	858	62	850	75	873		
11	849	24	837	37	850	50	845	63	849	76	919		
12	843	25	841	38	883	51	838	64	815	77	882		
13	845	26	807	39	874	52	869	65	879	78	863		

Table 5.5
Measurements on nasal length of $n_2 = 158$ individuals from the Chamar group

i	x_{2i}	i	x_{2i}	i	x_{2i}	i	x_{2i}	i	x_{2i}	i	x_{2i}	i	x_{2i}
1	802	24	768	47	830	70	808	93	818	116	774	139	793
2	786	25	773	48	848	71	778	94	788	117	750	140	801
3	798	26	781	49	858	72	773	95	798	118	814	141	826
4	812	27	753	50	834	73	828	96	812	119	824	142	826
5	807	28	778	51	825	74	798	97	793	120	870	143	838
6	780	29	758	52	820	75	823	98	712	121	834	144	807
7	766	30	763	53	845	76	760	99	752	122	804	145	802
8	792	31	839	54	772	77	784	100	774	123	779	146	819
9	792	32	780	55	810	78	770	101	772	124	729	147	860
10	807	33	794	56	788	79	775	102	828	125	827	148	890
11	770	34	799	57	757	80	768	103	773	126	786	149	871
12	751	35	806	58	802	81	798	104	748	127	876	150	853
13	813	36	822	59	797	82	803	105	748	128	769	151	864
14	775	37	884	60	827	83	758	106	758	129	854	152	879
15	823	38	875	61	754	84	758	107	810	130	754	153	795
16	803	39	835	62	757	85	798	108	834	131	770	154	853
17	778	40	797	63	802	86	853	109	784	132	760	155	861
18	848	41	815	64	800	87	851	110	787	133	831	156	885
19	851	42	840	65	758	88	760	111	769	134	821	157	861
20	821	43	859	66	766	89	810	112	814	135	850	158	788
21	803	44	790	67	794	90	773	113	774	136	822		
22	778	45	837	68	828	91	840	114	864	137	823		
23	798	46	824	69	789	92	755	115	834	138	825		

Basti–Brahmin group into the Chamar group) fixed as 0.05, by applying the classical quadratic classification procedure in (5.8) and reclassifying each individual in Tables 5.4 and 5.5, we obtain estimates of e_{12} and $1 - e_{21}$ (the probability of correctly classifying an individual from the Chamar group into itself) to be 0.082 and 0.696, respectively. For the robust quadratic classification procedure in (5.10), we obtain similarly estimates of e_{12} and $1 - e_{21}$ to be 0.082 and 0.734, respectively. Next, we applied the classical transformed linear classification procedure in (5.17) and obtained estimates of e_{12} and $1 - e_{21}$ to be 0.071 and 0.639, respectively. In a similar way, we applied the robust transformed linear classification procedure in (5.19) and obtained estimates of e_{12} and $1 - e_{21}$ to be 0.047 and 0.614, respectively.

6. Robust multivariate quadratic classification procedure and applications

In this section we first present the classical multivariate two-way quadratic classification procedure and then give its robust analogue based on the MML estimators. Next, we extend the transformed linear classification procedures developed in Section 5.3 to the bivariate case. Through Monte Carlo simulations we show that the robust procedure is more powerful than the classical procedure, for a pre-fixed e_{12} value. Finally, we present an illustrative example by using Majumdar's (1941) anthropometric data.

6.1. Classical and robust quadratic classification procedures

Let us assume x_{1k} ($k = 1, 2, \ldots, n_1$) and x_{2k} ($k = 1, 2, \ldots, n_2$) to be random samples from populations Π_1 and Π_2 which are taken to be $N_p(\mu_1, \Sigma_1)$ and $N_p(\mu_2, \Sigma_2)$, respectively. Let us denote the sample means by

$$\bar{x}_1 = \frac{1}{n_1} \sum_{k=1}^{n_1} x_{1k} \quad \text{and} \quad \bar{x}_2 = \frac{1}{n_2} \sum_{k=1}^{n_2} x_{2k},$$

and the sample variance–covariance matrices by

$$S_1 = \frac{1}{n_1 - 1} \sum_{k=1}^{n_1} (x_{1k} - \bar{x}_1)(x_{1k} - \bar{x}_1)^{\mathrm{T}},$$

and

$$S_2 = \frac{1}{n_2 - 1} \sum_{k=1}^{n_2} (x_{2k} - \bar{x}_2)(x_{2k} - \bar{x}_2)^{\mathrm{T}}.$$

Let x_0 be the new independent observation that needs to be classified. Then, the classical quadratic classification procedure is to classify x_0 into Π_1 or Π_2 according as

$$W = (x_0 - \bar{x}_1)^{\mathrm{T}} S_1^{-1}(x_0 - \bar{x}_1) - (x_0 - \bar{x}_2)^{\mathrm{T}} S_2^{-1}(x_0 - \bar{x}_2) \lessgtr C, \qquad (6.1)$$

where the cut-off point C is determined such that the error of misclassification e_{12} (for example) is at a pre-fixed level. Of course, if we do not pre-fix one of the errors of misclassification and allow both the errors to float, then the classical quadratic classification procedure is to classify x_0 into Π_1 or Π_2 according as

$$W = (x_0 - \bar{x}_1)^{\mathrm{T}} S_1^{-1}(x_0 - \bar{x}_1) - (x_0 - \bar{x}_2)^{\mathrm{T}} S_2^{-1}(x_0 - \bar{x}_2) \lessgtr 0. \qquad (6.2)$$

Let $(\hat{\mu}_1, \hat{\Sigma}_1)$ and $(\hat{\mu}_2, \hat{\Sigma}_2)$ be the MML estimators of (μ_1, Σ_1) and (μ_2, Σ_2) obtained from the samples x_{1k}, $k = 1, 2, \ldots, n_1$, and x_{2k}, $k = 1, 2, \ldots, n_2$, respectively. Then the robust classification procedure is to classify the new independent observation x_0 into Π_1 or Π_2 according as

$$W_R = (x_0 - \hat{\mu}_1)^{\mathrm{T}} \hat{\Sigma}_1^{-1}(x_0 - \hat{\mu}_1) - (x_0 - \hat{\mu}_2)^{\mathrm{T}} \hat{\Sigma}_2^{-1}(x_0 - \hat{\mu}_2) \lessgtr C_R,$$
$$(6.3)$$

where the cut-off point C_R is determined such that e_{12} is at a pre-specified level. Once again, if we do not fix e_{12} and allow both the errors to float, then the robust quadratic classification procedure is to classify x_0 into Π_1 or Π_2 according as

$$W_R = (x_0 - \hat{\mu}_1)^{\mathrm{T}} \hat{\Sigma}_1^{-1}(x_0 - \hat{\mu}_1) - (x_0 - \hat{\mu}_2)^{\mathrm{T}} \hat{\Sigma}_2^{-1}(x_0 - \hat{\mu}_2) \lessgtr 0. \quad (6.4)$$

Under normality, the distributions of W and W_R do not assume manageable forms even in the asymptotic case. So, the cut-off points C and C_R in (6.1) and (6.3), respectively, can not be determined in this case even approximately for a specified value of the error of misclassification e_{12}.

6.2. Transformed linear classification procedures

In this section we extend the classical and the robust transformed linear classification procedures developed in Section 5.3 to the bivariate case. We display the superiority of the robust procedure over the classical procedure through somewhat limited Monte Carlo simulations. Let

$$\begin{pmatrix} x_{11i} \\ x_{12i} \end{pmatrix}, \quad i = 1, 2, \ldots, n_1, \qquad \text{and} \qquad \begin{pmatrix} x_{21i} \\ x_{22i} \end{pmatrix}, \quad i = 1, 2, \ldots, n_2,$$

be independent random samples from populations Π_1 and Π_2, respectively. Let (x_{01}, x_{02}) be the new observation that needs to be classified into Π_1 or Π_2. First of all, let us denote

$$\bar{x}_{11} = \frac{1}{n_1} \sum_{i=1}^{n_1} x_{11i}, \qquad \bar{x}_{12} = \frac{1}{n_1} \sum_{i=1}^{n_1} x_{12i},$$

$$s_{11}^2 = \frac{1}{n_1 - 1} \sum_{i=1}^{n_1} (x_{11i} - \bar{x}_{11})^2, \qquad s_{12}^2 = \frac{1}{n_1 - 1} \sum_{i=1}^{n_1} (x_{12i} - \bar{x}_{12})^2,$$

$$\bar{x}_{21} = \frac{1}{n_2} \sum_{i=1}^{n_2} x_{21i}, \qquad \bar{x}_{22} = \frac{1}{n_2} \sum_{i=1}^{n_2} x_{22i},$$

$$s_{21}^2 = \frac{1}{n_2 - 1} \sum_{i=1}^{n_2} (x_{21i} - \bar{x}_{21})^2, \qquad s_{22}^2 = \frac{1}{n_2 - 1} \sum_{i=1}^{n_2} (x_{22i} - \bar{x}_{22})^2,$$

$$b_1 = \sum_{i=1}^{n_1} (x_{11i} - \bar{x}_{11})(x_{12i} - \bar{x}_{12}) / [\sum_{i=1}^{n_1} (x_{11i} - \bar{x}_{11})^2],$$

and

$$b_2 = \sum_{i=1}^{n_2} (x_{21i} - \bar{x}_{21})(x_{22i} - \bar{x}_{22}) / [\sum_{i=1}^{n_2} (x_{21i} - \bar{x}_{21})^2].$$

Let us define

$$x_{1,2.1,i} = x_{12i} - b_1 x_{11i}, \quad i = 1, 2, \dots, n_1,$$

and

$$x_{2,2.1,i} = x_{22i} - b_2 x_{21i}, \quad i = 1, 2, \dots, n_2,$$

and denote their means and variances by $\bar{x}_{12.1}$, $\bar{x}_{22.1}$, $s_{12.1}^2$ and $s_{22.1}^2$, respectively. With these quantities, let us now define for $i = 1, 2, \dots, n_1$

$$y_{11i} = \frac{x_{11i}}{s_{11}} \quad \text{and} \quad y_{12i} = \frac{x_{1,2.1,i}}{s_{12.1}}$$

and for $i = 1, 2, \dots, n_2$

$$y_{21i} = \frac{x_{21i}}{s_{21}} \quad \text{and} \quad y_{22i} = \frac{x_{2,2.1,i}}{s_{22.1}},$$

and denote their respective means by $\bar{y}_{11}$, $\bar{y}_{12}$, $\bar{y}_{21}$ and $\bar{y}_{22}$. Then, the classical classification procedure is to classify x_0 into Π_1 or Π_2 according as

$$V^* = \{ y_0 - \tfrac{1}{2}(\bar{y}_1 + \bar{y}_2) \}^T (\bar{y}_1 - \bar{y}_2) \gtrless C, \tag{6.5}$$

where $y_{01} = x_{01}/s_{11}$, $y_{02} = (x_{02} - b_1 x_{01})/s_{12.1}$, and $y_0 = (y_{01}, y_{02})^T$; C is chosen in (6.5) such that the error of misclassification e_{12} is at a pre-fixed level. As Tiku and Balakrishnan (1985, 1989) have shown, the cut-off point C in (6.5) may be estimated, with e_{12} fixed to be 0.05, as

$$\hat{C} = \tfrac{1}{2}\hat{t}^2 - 1.64|\hat{t}|,$$ (6.6)

where

$$\hat{t}^2 = \left\{\frac{\bar{x}_{11}}{s_{11}} - \frac{\bar{x}_{21}}{s_{21}}\right\}^2 + \left\{\frac{\bar{x}_{12.1}}{s_{12.1}} - \frac{\bar{x}_{22.1}}{s_{22.1}}\right\}^2.$$ (6.7)

Let $(\hat{\mu}_{11}, \hat{\sigma}_{11}^2)$, $(\hat{\mu}_{21}, \hat{\sigma}_{21}^2)$, $(\hat{\mu}_{12.1}, \hat{\sigma}_{12.1}^2)$, and $(\hat{\mu}_{22.1}, \hat{\sigma}_{22.1}^2)$ be the MML estimators of (μ, σ^2) calculated from the samples x_{11i} $(i = 1, 2, \ldots, n_1)$, x_{21i} $(i = 1, 2, \ldots, n_2)$, $x_{1,2.1,i}$ $(i = 1, 2, \ldots, n_1)$, and $x_{2,2.1,i}$ $(i = 1, 2, \ldots, n_2)$, respectively. Then, Tiku and Balakrishnan (1985, 1989) have proposed the robust classification procedure as to classify x_0 into Π_1 or Π_2 according as

$$V_R^* = \{y_0^* - \tfrac{1}{2}(\hat{\mu}_1 + \hat{\mu}_2)\}^T(\hat{\mu}_1 - \hat{\mu}_2) \gtrless C_R,$$ (6.8)

where $y_{01}^* = x_{01}/\hat{\sigma}_{11}$, $y_{02}^* = (x_{02} - b_1 x_{01})/\hat{\sigma}_{12.1}$, and $y_0^* = (y_{01}^*, y_{02}^*)^T$, and $\hat{\mu}_1$ and $\hat{\mu}_2$ are the MML extimators determined from the transformed observations $(y_{11i}^* = x_{11i}/\hat{\sigma}_{11},\ y_{12i}^* = x_{1,2.1,i}/\hat{\sigma}_{12.1})$, $i = 1, 2, \ldots, n_1$ and $(y_{21i}^* = x_{21i}/\hat{\sigma}_{21},\ y_{22i}^* = x_{2,2.1,i}/\hat{\sigma}_{22.1})$, $i = 1, 2, \ldots, n_2$; here, C_R is chosen in such a way that the error of misclassification e_{12} is at a pre-fixed level. If e_{12} is fixed as 0.05, then the cut-off point C_R in (6.8) may be estimated by

$$\hat{C}_R = \tfrac{1}{2}\hat{t}_R^2 - 1.64|\hat{t}_R|,$$ (6.9)

where

$$\hat{t}_R^2 = \left\{\frac{\hat{\mu}_{11}}{\hat{\sigma}_{11}} - \frac{\hat{\mu}_{21}}{\hat{\sigma}_{21}}\right\}^2 + \left\{\frac{\hat{\mu}_{12.1}}{\hat{\sigma}_{12.1}} - \frac{\hat{\mu}_{22.1}}{\hat{\sigma}_{22.1}}\right\}^2.$$ (6.10)

Under bivariate normality, the distribution of V_R^* is exactly the same as that of V^* asymptotically; hence, the robust classification procedure will have the same $1 - e_{21}$ value as that of the classical classification procedure for a pre-fixed value of e_{12}, under bivariate normality. But, the robust procedure becomes much more efficient than the classical procedure under non-normal models.

In order to examine the robustness features of these two procedures, n pairs of pseudo-random observations were simulated through the equations

$$x_{11} = (1/\sqrt{2})\{\sqrt{1+\rho}\,z_1 + \sqrt{1-\rho}\,z_2\}$$ (6.11)

and

$$x_{12} = (1/\sqrt{2})\{\sqrt{1+\rho}\,z_1 - \sqrt{1-\rho}\,z_2\},$$ (6.12)

where z_1 and z_2 are pseudo-random observations from each of the seven models given in (5.14) and (5.21). The x_2 sample was obtained by multiplying the pairs of observations by $(\sigma_{21}\ \sigma_{22})^T$ and then adding the vector $(\mu_{21}\ \mu_{22})^T$ in order to create a location shift between the populations Π_1 and Π_2. Then, with e_{12} fixed as 0.05, the values of e_{12} and $1 - e_{21}$ were simulated (based on 1000 Monte Carlo runs) for both the procedures under all the seven models. These values are presented in Table 6.1 for the choices of parameters

$$(A) \qquad \mu_1 = \begin{pmatrix} 0 \\ 0 \end{pmatrix}, \qquad \mu_2 = \begin{pmatrix} 2 \\ 2 \end{pmatrix}, \qquad \Sigma_1 = \begin{pmatrix} 1.0 & 0.5 \\ 0.5 & 1.0 \end{pmatrix}, \qquad \Sigma_2 = \begin{pmatrix} 4 & 2 \\ 2 & 4 \end{pmatrix},$$

and

$$(B) \qquad \mu_1 = \begin{pmatrix} 0 \\ 0 \end{pmatrix}, \qquad \mu_2 = \begin{pmatrix} 3 \\ 3 \end{pmatrix}, \qquad \Sigma_1 = \begin{pmatrix} 1.0 & 0.5 \\ 0.5 & 1.0 \end{pmatrix}, \qquad \Sigma_2 = \begin{pmatrix} 4 & 2 \\ 2 & 4 \end{pmatrix},$$

and $n_1 = n_2 = n = 50$. It is quite clear from this table that the robust procedure in (6.8) has its e_{12} value to be very stable and quite close to the presumed value of 0.05 while that of the classical procedure in (6.5) seems to be very sensitive to departures from bivariate normality. We also observe from Table 6.1 that the robust procedure has its $1 - e_{21}$ values to be exactly the same as that of the classical procedure under normality, and at the same time has its $1 - e_{21}$ values to be much larger than that of the classical classification procedure, under the non-normal models.

Table 6.1
Simulated values of e_{12} and $1 - e_{21}$ for the classical and the robust procedures for the models in (5.14) and (5.21) for the choice of parameters (A) and (B) when $n_1 = n_2 = n = 50$

Model	e_{12} values				$1 - e_{21}$ values			
	(A)		(B)		(A)		(B)	
	V^*	V_R^*	V^*	V_R^*	V^*	V_R^*	V^*	V_R^*
(1)	0.048	0.044	0.048	0.048	0.66	0.66	0.87	0.87
(2)	0.024	0.044	0.016	0.048	0.50	0.62	0.65	0.81
(3)	0.012	0.040	0.012	0.044	0.38	0.62	0.56	0.81
(4)	0.048	0.056	0.048	0.056	0.62	0.64	0.80	0.83
(5)	0.036	0.040	0.032	0.040	0.58	0.60	0.74	0.79
(6)	0.044	0.058	0.040	0.052	0.54	0.64	0.75	0.82
(7)	0.056	0.058	0.060	0.064	0.56	0.60	0.76	0.80

6.3. Illustrative example

In this section we present a data set taken from the anthropometric data collected by Majumdar (1941) and illustrate the classical and the robust classification procedures described in Section 6.2.

6.3.1. Example 4

In Tables 6.2 and 6.3, we have given the measurements on upper facial length and bigonial breadth of 85 individuals from the Basti–Brahmin group and 148 individuals from the Bhatu group, respectively.

Table 6.2
Measurements on upper facial length and bigonial breadth of $n_1 = 85$ individuals from the Basti–Brahmin group

i	x_{1i}^T	i	x_{1i}^T	i	x_{1i}^T	i	x_{1i}^T	i	x_{1i}^T
1	(35 129)	18	(37 111)	35	(35 116)	52	(35 111)	69	(35 117)
2	(30 127)	19	(34 118)	36	(34 111)	53	(36 118)	70	(40 117)
3	(36 116)	20	(36 120)	37	(39 120)	54	(37 130)	71	(35 119)
4	(36 128)	21	(37 118)	38	(31 120)	55	(37 111)	72	(40 116)
5	(33 113)	22	(32 115)	39	(36 122)	56	(38 111)	73	(39 108)
6	(40 123)	23	(42 114)	40	(34 130)	57	(40 125)	74	(41 127)
7	(32 117)	24	(32 112)	41	(38 121)	58	(37 118)	75	(36 130)
8	(39 127)	25	(34 117)	42	(38 115)	59	(37 120)	76	(36 119)
9	(35 106)	26	(35 114)	43	(34 121)	60	(47 109)	77	(39 114)
10	(33 125)	27	(34 117)	44	(39 108)	61	(35 114)	78	(41 112)
11	(39 120)	28	(35 113)	45	(36 126)	62	(39 117)	79	(41 120)
12	(38 115)	29	(36 105)	46	(35 120)	63	(40 116)	80	(38 120)
13	(35 108)	30	(35 117)	47	(35 130)	64	(36 120)	81	(40 118)
14	(36 116)	31	(32 112)	48	(39 129)	65	(34 121)	82	(36 125)
15	(35 127)	32	(39 123)	49	(35 129)	66	(38 120)	83	(37 111)
16	(35 111)	33	(38 117)	50	(42 123)	67	(38 109)	84	(35 106)
17	(36 120)	34	(40 117)	51	(40 117)	68	(35 117)	85	(32 122)

From these two samples, we find

$$\bar{x}_{11} = 36.6000, \quad \bar{x}_{21} = 35.6081, \quad s_{11}^2 = 8.2190, \quad s_{21}^2 = 7.3012,$$

$$\hat{\mu}_{11} = 36.6459, \quad \hat{\mu}_{21} = 35.5280, \quad \hat{\sigma}_{11}^2 = 6.9535, \quad \hat{\sigma}_{21}^2 = 7.3732,$$

$$b_1 = -0.14106, \quad b_2 = -0.17627,$$

$$\bar{x}_{12.1} = 123.2097, \quad \bar{x}_{22.1} = 123.7563, \quad s_{12.1}^2 = 39.0723, \quad s_{22.1}^2 = 36.3102,$$

$$\hat{\mu}_{12.1} = 123.2936, \quad \hat{\mu}_{22.1} = 123.6609, \quad \hat{\sigma}_{12.1}^2 = 44.5747, \quad \hat{\sigma}_{22.1}^2 = 36.6924,$$

$$\hat{C} = -1.0881, \quad \hat{C}_R = -1.2340.$$

So, with e_{12} (the probability of wrongly classifying an individual from the Basti–Brahmin group into the Bhatu group) fixed as 0.05, by applying the classical procedure in (6.5) and reclassifying each individual in Tables 6.2 and 6.3, we obtain estimates of e_{12} and $1 - e_{21}$ (the probability of correctly classifying an individual from the Bhatu group into itself) to be 0.058 and 0.236, respectively. For the robust procedure in (6.8), we obtain similarly estimates of e_{12} and $1 - e_{21}$ to be 0.071 and 0.676, respectively. It is of interest to mention here that the classical and the robust linear classification procedures developed in Sections 4.1 and 4.2 yield estimates of e_{12} and $1 - e_{21}$ in this case to be 0.071 and 0.095, and 0.071 and 0.128, respectively.

Table 6.3
Measurements on upper facial length and bigonial breadth of $n_2 = 148$ individuals from the Bhatu group

i	x_{2i}^{T}	i	x_{2i}^{T}	i	x_{2i}^{T}	i	x_{2i}^{T}	i	x_{2i}^{T}
1	(37 114)	31	(33 124)	61	(36 118)	91	(34 124)	121	(35 113)
2	(35 120)	32	(32 117)	62	(42 118)	92	(40 125)	122	(38 111)
3	(32 118)	33	(37 119)	63	(43 118)	93	(33 114)	123	(38 121)
4	(32 118)	34	(38 114)	64	(34 113)	94	(36 118)	124	(40 112)
5	(39 116)	35	(30 113)	65	(31 113)	95	(35 115)	125	(35 108)
6	(35 127)	36	(37 117)	66	(35 118)	96	(35 119)	126	(36 134)
7	(34 108)	37	(36 115)	67	(32 113)	97	(37 121)	127	(38 109)
8	(35 118)	38	(34 126)	68	(37 120)	98	(31 112)	128	(33 118)
9	(36 113)	39	(34 115)	69	(35 120)	99	(34 112)	129	(35 125)
10	(32 120)	40	(40 127)	70	(35 117)	100	(40 109)	130	(35 125)
11	(35 118)	41	(32 117)	71	(36 116)	101	(38 115)	131	(40 107)
12	(32 116)	42	(35 119)	72	(36 122)	102	(36 121)	132	(33 127)
13	(38 123)	43	(39 108)	73	(38 122)	103	(37 118)	133	(31 107)
14	(36 112)	44	(37 119)	74	(40 105)	104	(35 106)	134	(32 122)
15	(35 112)	45	(35 122)	75	(35 117)	105	(36 120)	135	(37 132)
16	(35 120)	46	(34 111)	76	(34 120)	106	(38 124)	136	(37 122)
17	(35 103)	47	(40 109)	77	(39 110)	107	(37 113)	137	(35 116)
18	(34 120)	48	(38 118)	78	(35 118)	108	(41 124)	138	(33 123)
19	(35 125)	49	(33 113)	79	(30 108)	109	(35 112)	139	(36 105)
20	(35 110)	50	(40 111)	80	(35 115)	110	(38 119)	140	(35 118)
21	(35 120)	51	(30 120)	81	(32 118)	111	(43 113)	141	(35 113)
22	(36 120)	52	(35 129)	82	(35 128)	112	(40 111)	142	(32 131)
23	(35 118)	53	(31 119)	83	(37 128)	113	(36 118)	143	(37 114)
24	(37 115)	54	(32 117)	84	(30 112)	114	(35 121)	144	(33 111)
25	(39 115)	55	(38 109)	85	(39 116)	115	(34 125)	145	(37 125)
26	(33 126)	56	(32 118)	86	(38 109)	116	(37 115)	146	(37 122)
27	(35 118)	57	(37 116)	87	(33 114)	117	(40 113)	147	(35 123)
28	(36 126)	58	(37 113)	88	(36 114)	118	(34 130)	148	(41 120)
29	(37 121)	59	(33 120)	89	(31 122)	119	(35 112)		
30	(38 124)	60	(34 114)	90	(36 117)	120	(35 132)		

7. Robust univariate linear classification procedure based on dichotomous and continuous variables and applications

In this section we first discuss the classical univariate two-way linear classification procedure based on dichotomous and normal variables and then describe the analogous robust linear classification procedure based on the MML estimators as proposed by Balakrishnan and Tiku (1988b). By comparing these two procedures, when one of the two errors of misclassification is at a pre-specified level, under normal and various non-normal models, we show that the robust procedure is quite efficient and robust to departures from normality. We present an example to illustrate both the procedures.

7.1. Classical linear classification procedure based on dichotomous and normal variables

Let X be a Bernoulli variate with probability mass function

$$\Pr(X = x) = \theta^x(1 - \theta)^{1-x}, \quad x = 0, 1,$$

and Y be a continuous variable such that the conditional distribution of Y, given $X = x$, is normal with mean $\mu(x) = \mu + x\delta$ and variance $\sigma^2(x) = \sigma^2 + x\gamma^2$. Let $(x_{1i}\ y_{1i})^T$, $i = 1, 2, \ldots, n_1$, be a random sample of size n_1 from population Π_1 which is the model specified above with θ_1 and $\mu_1(x)$. Similarly, let $(x_{2i}\ y_{2i})^T$, $i = 1, 2, \ldots, n_2$, be a random sample of size n_2 from population Π_2 which is the model specified above with θ_2 and $\mu_2(x)$. Then the classical estimates of the parameters are given by

$$\hat{\theta}_i = \frac{n_i(1)}{n_i}, \quad i = 1, 2,$$

where $n_i(1)$ is the number of observations in the i-th sample that correspond to $x = 1$, and

$$\hat{\mu}_i(x) = \bar{y}_i(x),$$

and

$$\hat{\sigma}^2(x) = s^2(x) = \frac{(n_1(x) - 1)s_1^2(x) + (n_2(x) - 1)s_2^2(x)}{n_1(x) + n_2(x) - 2},$$

for $x = 0$ and 1, and $i = 1$ and 2.

Let $(x_0\ y_0)^T$ be an independent new observation that needs to be classified into either Π_1 or Π_2. Then, as shown by Chang and Afifi (1974), Krzanowski (1975) and Balakrishnan and Tiku (1988b), the classical linear classification procedure is to classify $(x_0\ y_0)^T$ into Π_1 or Π_2 according as

$$V(x_0) = \{y_0 - \tfrac{1}{2}(\bar{y}_1(x_0) + \bar{y}_2(x_0))\} \frac{(\bar{y}_1(x_0) - \bar{y}_2(x_0))}{s^2(x_0)}$$

$$+ x_0 \ln[\hat{\theta}_1/\hat{\theta}_2] + (1 - x_0)\ln[(1 - \hat{\theta}_1)/(1 - \hat{\theta}_2)] \gtrless C, \quad (7.1)$$

where the cut-off point C is determined so that the error of misclassification e_{12} (for example) is at a pre-specified level. By using asymptotic arguments, Balakrishnan and Tiku (1988b) have shown that (with e_{12} fixed as α) the cut-off point C in (7.1) is approximately the solution of the equation

$$(1 - \theta_1)\Phi\left\{\frac{C - \tfrac{1}{2}\eta^2(0) - \ln[(1 - \theta_1)/(1 - \theta_2)]}{|\eta(0)|}\right\}$$

$$+ \theta_1\Phi\left\{\frac{C - \tfrac{1}{2}\eta^2(1) - \ln(\theta_1/\theta_2)}{|\eta(1)|}\right\} = \alpha, \quad (7.2)$$

where $\Phi(\ )$ is the cumulative distribution function of a standard normal variable, and $\eta^2(x) = (\mu_1(x) - \mu_2(x))^2/\sigma^2(x)$ for $x = 0, 1$. Since all the parameters in equation (7.2) are unknown, the cut-off point C can not be determined from (7.2). However, by using the classical estimates of all the parameters, we may obtain an estimate of the cut-off point C as the solution $\hat{C}$ of the equation

$$(1 - \hat{\theta}_1) \Phi \left\{ \frac{\hat{C} - \frac{1}{2}\hat{\eta}^2(0) - \ln[(1 - \hat{\theta}_1)/(1 - \hat{\theta}_2)]}{|\hat{\eta}(0)|} \right\}$$

$$+ \hat{\theta}_1 \Phi \left\{ \frac{\hat{C} - \frac{1}{2}\hat{\eta}^2(1) - \ln(\hat{\theta}_1/\hat{\theta}_2)}{|\hat{\eta}(1)|} \right\} = \alpha, \qquad (7.3)$$

where $\hat{\eta}^2(x) = (\bar{y}_1(x) - \bar{y}_2(x))^2/s^2(x)$, $x = 0, 1$.

In some instances, one may not be interested in pre-fixing the error of mis-classification e_{12} and instead be interested in allowing both the errors of mis-classification to float. In this case, the classical linear classification procedure is to classify the independent observation $(x_0\ y_0)^{\mathrm{T}}$ into Π_1, Π_2 or arbitrarily into Π_1 or Π_2, according as

$$V^*(x_0) = \{ y_0 - \tfrac{1}{2}(\bar{y}_1(x_0) + \bar{y}_2(x_0)) \} \frac{(\bar{y}_1(x_0) - \bar{y}_2(x_0))}{s^2(x_0)}$$

$$+ x_0 \ln[\hat{\theta}_1/\hat{\theta}_2] + (1 - x_0) \ln[(1 - \hat{\theta}_1)/(1 - \hat{\theta}_2)] \gtrless = 0 . \qquad (7.4)$$

By adopting a method similar to the one given by John (1960a,b), Balakrishnan, Kocherlakota and Kocherlakota (1986, 1988) and Kocherlakota, Kocherlakota and Balakrishnan (1987) have examined the behaviour of the two errors of mis-classification through their distribution functions and expected values for the normal as well as different non-normal models including the mixture-normal and truncated normal distributions. Tiku, Balakrishnan and Ambagaspitiya (1989) have also studied the errors of misclassification by using asymptotic arguments under normal and wide range of non-normal models.

7.2. *Robust linear classification procedure based on dichotomous and continuous variables*

Let $\hat{\theta}_i = n_i(1)/n_i$, $i = 1, 2$, where $n_i(1)$ is the number of observations in the i-th sample that correspond to $x = 1$, be the estimate of θ_i as before. Further, let $(\hat{\mu}_1(x), \hat{\sigma}_1^2(x))$ and $(\hat{\mu}_2(x), \hat{\sigma}_2^2(x))$ be the MML estimators of $(\mu_1(x), \sigma_1^2(x))$ and $(\mu_2(x), \sigma_2^2(x))$ obtained from the two samples, respectively. Let $\hat{\sigma}^2(x)$ be the pooled MML estimator of $\sigma^2(x)$ defined by

$$\hat{\sigma}^2(x) = \frac{(A_1(x) - 1)\hat{\sigma}_1^2(x) + (A_2(x) - 1)\hat{\sigma}_2^2(x)}{A_1(x) + A_2(x) - 2},$$

for $x = 0$ and 1. Then, the robust classification procedure as given by Balakrishnan and Tiku (1988b) is to classify the new independent observation $(x_0 \ y_0)^T$ into Π_1 or Π_2 according as

$$V_R = \{ y_0 - \tfrac{1}{2}(\hat{\mu}_1(x_0) + \hat{\mu}_2(x_0)) \} \frac{(\hat{\mu}_1(x_0) - \hat{\mu}_2(x_0))}{\hat{\sigma}^2(x_0)}$$

$$+ x_0 \ln[\hat{\theta}_1/\hat{\theta}_2] + (1 - x_0) \ln[(1 - \hat{\theta}_1)/(1 - \hat{\theta}_2)] \gtrless C_R, \qquad (7.5)$$

where the cut-off point C_R is determined so that the error of misclassification e_{12} is at a pre-fixed level. By using asymptotic arguments, Balakrishnan and Tiku (1988b) have shown that (with e_{12} fixed as α) the cut-off point C_R in (7.5) is approximately the solution of the equation

$$(1 - \theta_1) \Phi \left\{ \frac{C_R - \tfrac{1}{2}\eta^2(0) - \ln[(1 - \theta_1)/(1 - \theta_2)]}{|\eta(0)|} \right\}$$

$$+ \theta_1 \Phi \left\{ \frac{C_R - \tfrac{1}{2}\eta^2(1) - \ln(\theta_1/\theta_2)}{|\eta(1)|} \right\} = \alpha . \qquad (7.6)$$

By using the MML estimates of all the parameters calculated from the two samples, we may obtain an estimate of the cut-off point C_R as the solution $\hat{C}_R$ of the equation

$$(1 - \hat{\theta}_1) \Phi \left\{ \frac{\hat{C}_R - \tfrac{1}{2}\hat{\eta}_R^2(0) - \ln[(1 - \hat{\theta}_1)/(1 - \hat{\theta}_2)]}{|\hat{\eta}_R(0)|} \right\}$$

$$+ \hat{\theta}_1 \Phi \left\{ \frac{\hat{C}_R - \tfrac{1}{2}\hat{\eta}_R^2(1) - \ln(\hat{\theta}_1/\hat{\theta}_2)}{|\hat{\eta}_R(1)|} \right\} = \alpha , \qquad (7.7)$$

where

$$\hat{\eta}_R^2(x) = \frac{(\hat{\mu}_1(x) - \hat{\mu}_2(x))^2}{\hat{\sigma}^2(x)} , \qquad x = 0, 1 .$$

If we let instead both the errors of misclassification to float, then the corresponding robust linear classification procedure is to classify the observation $(x_0 \ y_0)^T$ into Π_1, Π_2 or arbitrarily into Π_1 or Π_2, according as

$$V_R^*(x_0) = \{ y_0 - \tfrac{1}{2}(\hat{\mu}_1(x_0) + \hat{\mu}_2(x_0)) \} \frac{(\hat{\mu}_1(x_0) - \hat{\mu}_2(x_0))}{\hat{\sigma}^2(x)}$$

$$+ x_0 \ln[\hat{\theta}_1/\hat{\theta}_2] + (1 - x_0) \ln[(1 - \hat{\theta}_1)/(1 - \hat{\theta}_2)] \gtrless = 0 . \qquad (7.8)$$

Tiku, Balakrishnan and Ambagaspitiya (1989) have studied the errors of mis-classification of the robust procedure in (7.8) by using asymptotic arguments for the normal as well as wide range of non-normal models.

Due to the asymptotic distributional properties of the MML estimators mentioned earlier, the statistic $V_R(x_0)$ in (7.5) asymptotically has exactly the same distribution as the statistic $V(x_0)$ in (7.1), under normality; hence, the two procedures will have exactly the same $1 - e_{21}$ value, for a pre-fixed value of e_{12}, under normality.

In order to examine the robustness features of the two linear classification procedures, $n_1 = n_2 = n$ pseudo-random observations were simulated for the y observations of the two samples from each of the following seven models:

(1) Normal,
(2) Outlier-normal model $(n - 1)N(0, 1)$ and $1N(0, 100)$,
(3) Outlier-normal model $(n - 2)N(0, 1)$ and $2N(0, 16)$,
(4) Logistic,
(5) Double exponential,
(6) Mixture-normal model $0.9N(0, 1) + 0.1N(0, 16)$, and
(7) Student's t distribution with four degrees of freedom. (7.9)

Then, with e_{12} fixed as 0.05, the values of e_{12} and $1 - e_{21}$ were simulated (based on 1000 Monte Carlo runs) for both the procedures under all seven models given in (7.9). These values are presented in Table 7.1 for $\sigma^2 = \gamma^2 = 1$, $\delta = 1$,

Table 7.1
Simulated values of e_{12} and $1 - e_{21}$ for the classical and the robust linear classification procedures for the models in (7.9) when $\mu_1 = 0$, $\sigma^2 = \gamma^2 = 1$, $\delta = 1$, $\theta_1 = \theta_2 = \theta$ and $n_1 = n_2 = n = 50$

θ	Model	e_{12} values						$1 - e_{21}$ values					
		$\mu_2 = 2.00$		$\mu_2 = 2.25$		$\mu_2 = 2.50$		$\mu_2 = 2.00$		$\mu_2 = 2.25$		$\mu_2 = 2.50$	
		V^*	V_R^*	V^*	V_R^*	V^*	V_R^*	V^*	V_R^*	V^*	V_R^*	V^*	V_R^*
0.1	(1)	0.058	0.060	0.058	0.060	0.058	0.062	0.606	0.616	0.700	0.704	0.760	0.762
	(2)	0.012	0.056	0.012	0.058	0.012	0.058	0.262	0.576	0.340	0.668	0.388	0.724
	(3)	0.030	0.054	0.028	0.056	0.030	0.054	0.438	0.562	0.506	0.650	0.580	0.706
	(4)	0.056	0.064	0.052	0.060	0.056	0.060	0.622	0.672	0.708	0.748	0.774	0.794
	(5)	0.058	0.060	0.056	0.064	0.056	0.060	0.412	0.506	0.482	0.578	0.542	0.664
	(6)	0.026	0.062	0.024	0.058	0.024	0.058	0.334	0.540	0.406	0.622	0.474	0.694
	(7)	0.064	0.068	0.064	0.068	0.060	0.066	0.416	0.461	0.492	0.604	0.562	0.674
0.5	(1)	0.050	0.052	0.050	0.052	0.052	0.052	0.518	0.520	0.590	0.592	0.658	0.664
	(2)	0.012	0.048	0.012	0.048	0.012	0.044	0.184	0.478	0.226	0.556	0.270	0.628
	(3)	0.014	0.040	0.014	0.036	0.014	0.038	0.304	0.464	0.378	0.534	0.426	0.604
	(4)	0.050	0.056	0.050	0.056	0.050	0.054	0.526	0.558	0.604	0.644	0.672	0.714
	(5)	0.056	0.060	0.054	0.060	0.046	0.060	0.332	0.416	0.392	0.488	0.448	0.572
	(6)	0.032	0.056	0.032	0.058	0.032	0.058	0.276	0.444	0.326	0.524	0.374	0.594
	(7)	0.052	0.058	0.052	0.056	0.058	0.062	0.332	0.420	0.386	0.490	0.450	0.554

$\theta_1 = \theta_2 = 0.5$ and 0.1, $\mu_1 = 0$, $\mu_2 = 2.00(0.25)2.50$ and $n_1 = n_2 = 50$. It is quite clear from Table 7.1 that the robust linear procedure has more stable e_{12} values than the corresponding classical linear procedure. It is also very clear from Table 7.1 that the robust linear classification procedure in (7.5) has considerably larger $1 - e_{21}$ values than the corresponding classical linear classification procedure in (7.1) under the non-normal models in (7.9).

7.3. *Illustrative example*

In this section we present a simulated data set and illustrate the classical and the robust classification procedures described in Sections 7.1 and 7.2.

Table 7.2
Measurements on upper facial length of $n_1 = 100$ individuals from Group 1 along with their gender

i	x_{1i}	y_{1i}	i	x_{1i}	y_{1i}	i	x_{1i}	y_{1i}	i	x_{1i}	y_{1i}	i	x_{1i}	y_{1i}
1	0	44.5	21	1	34.1	41	1	34.8	61	0	40.5	81	0	40.8
2	0	23.0	22	1	44.0	42	1	38.5	62	1	43.6	82	0	40.5
3	0	37.6	23	1	37.6	43	0	36.4	63	0	39.0	83	0	28.6
4	0	45.1	24	1	37.7	44	0	41.2	64	0	38.6	84	1	44.0
5	0	42.1	25	0	36.7	45	0	47.8	65	0	34.3	85	0	35.0
6	0	39.2	26	0	40.5	46	0	48.4	66	1	46.3	86	0	46.8
7	1	42.8	27	1	36.4	47	0	40.9	67	1	38.0	87	1	51.2
8	0	43.7	28	1	41.7	48	0	38.6	68	0	43.4	88	0	29.4
9	1	37.4	29	1	41.8	49	0	41.1	69	1	49.9	89	0	39.7
10	1	49.9	30	0	39.2	50	0	43.2	70	0	39.1	90	1	42.7
11	1	33.9	31	0	30.4	51	0	40.8	71	0	41.0	91	0	43.2
12	0	35.4	32	0	38.5	52	1	48.2	72	1	54.5	92	1	44.3
13	0	45.2	33	1	40.4	53	0	35.8	73	0	37.6	93	0	41.5
14	0	40.2	34	0	41.5	54	1	50.5	74	0	45.8	94	0	49.1
15	1	39.2	35	0	36.3	55	1	41.7	75	0	35.3	95	0	39.4
16	1	44.1	36	0	43.7	56	0	35.9	76	1	42.2	96	1	41.9
17	0	44.3	37	0	38.3	57	0	35.5	77	0	39.7	97	1	42.3
18	0	44.7	38	0	40.6	58	1	37.7	78	1	42.2	98	1	43.5
19	1	39.6	39	0	42.9	59	1	45.5	79	1	49.8	99	0	41.9
20	1	44.0	40	0	40.1	60	0	38.1	80	0	42.0	100	0	40.7

7.3.1. *Example 5*
In Tables 7.2 and 7.3, we have given a simulated data set in an anthropometric set-up with x denoting the gender (0 for female, 1 for male) and y the upper facial length for two different groups of people.

Table 7.3
Measurements on upper facial length of $n_2 = 80$ individuals from Group 2 along with their gender

i	x_{2i}	y_{2i}	i	x_{2i}	y_{2i}	i	x_{2i}	y_{2i}	i	x_{2i}	y_{2i}	i	x_{2i}	y_{2i}
1	0	52.8	17	0	38.1	33	1	39.9	49	1	41.6	65	1	35.7
2	0	36.1	18	1	32.2	34	0	31.2	50	1	44.3	66	0	40.6
3	0	24.6	19	0	26.5	35	1	50.0	51	0	29.6	67	0	31.4
4	0	40.7	20	0	35.2	36	0	38.7	52	1	46.4	68	1	47.4
5	0	37.6	21	1	31.8	37	0	34.6	53	1	44.0	69	1	39.6
6	1	35.8	22	1	38.4	38	1	36.5	54	1	41.2	70	1	37.6
7	0	41.5	23	0	33.5	39	0	37.7	55	0	40.5	71	1	38.1
8	0	44.6	24	0	37.9	40	0	37.0	56	0	34.6	72	0	32.6
9	0	36.2	25	0	35.7	41	1	41.0	57	1	36.4	73	1	39.7
10	1	32.2	26	0	38.3	42	1	37.6	58	1	32.9	74	0	39.9
11	1	30.2	27	0	30.8	43	0	36.7	59	1	34.4	75	1	47.3
12	0	33.5	28	1	32.0	44	1	42.7	60	1	38.7	76	0	42.8
13	0	40.7	29	1	48.4	45	1	49.0	61	1	45.0	77	1	36.9
14	1	41.0	30	1	38.7	46	0	29.4	62	0	35.4	78	1	42.6
15	1	37.4	31	0	33.0	47	0	32.0	63	0	40.8	79	1	37.6
16	1	39.6	32	1	43.8	48	0	37.8	64	0	34.6	80	1	43.6

From these two samples, we find:

$$\bar{y}_1(0) = 39.8446, \quad \bar{y}_2(0) = 36.2827, \quad s_1^2(0) = 21.9430, \quad s_2^2(0) = 26.7565,$$

$$\hat{\mu}_1(0) = 40.1439, \quad \hat{\mu}_2(0) = 36.1609, \quad \hat{\sigma}_1^2(0) = 14.8989, \quad \hat{\sigma}_2^2(0) = 18.6869,$$

$$\bar{y}_1(1) = 43.5738, \quad \bar{y}_2(1) = 39.7397, \quad s_1^2(1) = 24.7587, \quad s_2^2(1) = 26.0199,$$

$$\hat{\mu}_1(1) = 42.5738, \quad \hat{\mu}_2(1) = 39.7397, \quad \hat{\sigma}_1^2(1) = 26.7490, \quad \hat{\sigma}_2^2(1) = 33.1631,$$

$$\hat{\theta}_1 = 0.3800, \quad \hat{\theta}_2 = 0.5125, \quad \hat{C} = -0.9024, \quad \hat{C}_R = -0.9796.$$

So, with e_{12} (the probability of wrongly classifying an individual from group 1 into group 2) fixed as 0.05, by applying the classical classification procedure in (7.1) and reclassifying each individual in Tables 7.2 and 7.3, we obtain estimates of e_{12} and $1 - e_{21}$ (the probability of correctly classifying an individual from group 2 into itself) to be 0.060 and 0.150, respectively. Similarly, by applying the robust linear classification procedure in (7.5) and reclassifying each individual in Tables 7.2 and 7.3, we obtain estimates of e_{12} and $1 - e_{21}$ to be 0.060 and 0.200, respectively.

8. Robust multivariate linear classification procedure based on dichotomous and continuous variables and applications

In this section we discuss the extension of the classical and the robust two-way linear classification procedures based on dichotomous and continuous variables presented in Sections 7.1 and 7.2 to the bivariate situation as proposed by Balakrishnan and Tiku (1988b). By comparing these two procedures, when one of the two errors of misclassification is at a pre-specified level, under normal and various non-normal models, we show that the robust procedure is quite efficient and robust to departures from normality. Finally, we present an example to illustrate both these procedures.

8.1. Classical linear classification procedure based on dichotomous and normal variables

Let X be a random Bernoulli variable with probability mass function

$$\Pr(X = x) = \theta^x(1 - \theta)^{1-x}, \quad x = 0, 1,$$

and $Y = (y_1 \ y_2)^{\mathrm{T}}$ be a bivariate continuous variable such that the conditional distribution of Y, given $X = x$, is bivariate normal with mean

$$\mu(x) = \begin{pmatrix} \mu_1(x) \\ \mu_2(x) \end{pmatrix} = \mu + x\delta = \begin{pmatrix} \mu_1 \\ \mu_2 \end{pmatrix} + x \begin{pmatrix} \delta_1 \\ \delta_2 \end{pmatrix},$$

and variance–covariance matrix

$$\Sigma(x) = \Sigma + x\Gamma.$$

Now, let $(x_{1i} \ y_{1i})^{\mathrm{T}}$, $i = 1, 2, \ldots, n_1$, be a random sample of size n_1 from population Π_1 which is the model specified above with θ_1 and $\mu_1(x)$. Similarly, let $(x_{2i} \ y_{2i})^{\mathrm{T}}$, $i = 1, 2, \ldots, n_2$, be a random sample of size n_2 from population Π_2 which is the model specified above with θ_2 and $\mu_2(x)$. Then the classical estimates of the parameters are given by

$$\hat{\theta}_i = n_i(1)/n_i, \quad i = 1, 2,$$

where $n_i(1)$ is the number of observations in the i-th sample that correspond to $x = 1$, and

$$\hat{\mu}_i = \bar{y}_i(x),$$

and

$$\hat{\Sigma}(x) = S(x) = \frac{(n_1(x) - 1)S_1(x) + (n_2(x) - 1)S_2(x)}{n_1(x) + n_2(x) - 2}$$

for $x = 0$ and 1, and $i = 1$ and 2; here, $S_i(x)$ is the sample variance–covariance matrix from the i-th sample and is given by

$$S_i(x) = \frac{1}{n_i(x) - 1} \sum_{l=1}^{n_i(x)} (y_{il}(x) - \bar{y}_i(x))(y_{il}(x) - \bar{y}_i(x))^{\mathrm{T}}.$$

Let $(x_0 \; y_0)^T$ be an independent new observation that needs to be classified into either Π_1 or Π_2. Then, as shown by Chang and Afifi (1974), Krzanowski (1975) and Balakrishnan and Tiku (1988b), the classical linear classification procedure is to classify $(x_0 \; y_0)^T$ into Π_1 or Π_2 according as

$$V(x_0) = \{ y_0 - \tfrac{1}{2}(\bar{y}_1(x_0) + \bar{y}_2(x_0)) \}^T S^{-1}(x_0)(\bar{y}_1(x_0) - \bar{y}_2(x_0))$$
$$+ x_0 \ln [\, \hat{\theta}_1 / \hat{\theta}_2] + (1 - x_0) \ln [(1 - \hat{\theta}_1)/(1 - \hat{\theta}_2)] \gtrless C^*, \quad (8.1)$$

where the cut-off point C^* is determined so that the error of misclassification e_{12} (for example) is at a pre-specified level. Balakrishnan and Tiku (1988b) have shown through asymptotic arguments that an estimate of the cut-off point C^* may be obtained as the solution $\hat{C}^*$ of the equation (with e_{12} fixed as α)

$$(1 - \hat{\theta}_1) \Phi \left\{ \frac{\hat{C}^* - \tfrac{1}{2}\hat{\eta}^2(0) - \ln[(1 - \hat{\theta}_1)/(1 - \hat{\theta}_2)]}{|\hat{\eta}(0)|} \right\}$$
$$+ \hat{\theta}_1 \Phi \left\{ \frac{\hat{C}^* - \tfrac{1}{2}\hat{\eta}^2(1) - \ln(\hat{\theta}_1/\hat{\theta}_2)}{|\hat{\eta}(1)|} \right\} = \alpha, \quad (8.2)$$

where

$$\hat{\eta}^2(x) = (\bar{y}_1(x) - \bar{y}_2(x))^T S^{-1}(x)(\bar{y}_1(x) - \bar{y}_2(x)), \quad x = 0, 1 .$$

In the case when both the errors of misclassification are allowed to float, the classical linear classification procedure is to classify the new independent observation $(x_0 \; y_0)^T$ into Π_1 or Π_2, or arbitrarily into Π_1 or Π_2, according as

$$V(x_0) \gtrless = 0 . \quad (8.3)$$

8.2. Robust linear classification procedure based on dichotomous and continuous variables

Let $b_1(x)$ and $b_2(x)$ be the sample regression coefficients based on the two samples. Then, let us denote the pooled sample regression coefficient by $b(x)$. Now define

$$y_{1, 2.1, j}(x) = y_{12j}(x) - b(x) y_{11j}(x), \quad j = 1, 2, \ldots, n_1(x),$$

and

$$y_{2, 2.1, j}(x) = y_{22j}(x) - b(x) y_{21j}(x), \quad j = 1, 2, \ldots, n_2(x),$$

for $x = 0$ and 1. Let $(\hat{\mu}_{11}(x), \hat{\sigma}_{11}^2(x))$, $(\hat{\mu}_{21}(x), \hat{\sigma}_{21}^2(x))$, $(\hat{\mu}_{12.1}(x), \hat{\sigma}_{12.1}^2(x))$, and $(\hat{\mu}_{22.1}(x), \hat{\sigma}_{22.1}^2(x))$ be the MML estimators of (μ, σ^2) obtained from the samples $y_{11j}(x)$, $j = 1, \ldots, n_1(x)$, $y_{21j}(x)$, $j = 1, \ldots, n_2(x)$, $y_{1, 2.1, j}(x)$, $j = 1, \ldots, n_1(x)$, and $y_{2, 2.1, j}(x)$, $j = 1, \ldots, n_2(x)$. Let $\hat{\sigma}_1^2(x)$ and $\hat{\sigma}_{2.1}^2(x)$ be the pooled MML estimators defined by

$$\hat{\sigma}_1^2(x) = \frac{(A_1(x) - 1)\,\hat{\sigma}_{11}^2(x) + (A_2(x) - 1)\,\hat{\sigma}_{21}^2(x)}{A_1(x) + A_2(x) - 2}$$

and

$$\hat{\sigma}_{2.1}^2(x) = \frac{(A_1(x) - 1)\,\hat{\sigma}_{12.1}^2(x) + (A_2(x) - 1)\,\hat{\sigma}_{22.1}^2(x)}{A_1(x) + A_2(x) - 2}.$$

By using the MML estimators, Balakrishnan and Tiku (1988b) have proposed the robust linear classification procedure as to classify the observation $(x_0 \; y_0)^{\mathrm{T}}$ into Π_1 or Π_2 according as

$$V_R(x_0) = \frac{\{y_{01}(x_0) - \frac{1}{2}(\hat{\mu}_{11}(x_0) + \hat{\mu}_{21}(x_0))\}(\hat{\mu}_{11}(x_0) - \hat{\mu}_{21}(x_0))}{\hat{\sigma}_1^2(x_0)}$$

$$+ \frac{\{y_{02.1}(x_0) - \frac{1}{2}(\hat{\mu}_{12.1}(x_0) + \hat{\mu}_{22.1}(x_0))\}(\hat{\mu}_{12.1}(x_0) - \hat{\mu}_{22.1}(x_0)}{\hat{\sigma}_{2.1}^2(x_0))}$$

$$+ x_0 \ln[\,\hat{\theta}_1/\hat{\theta}_2\,] + (1 - x_0)\ln[(1 - \hat{\theta}_1)/(1 - \hat{\theta}_2)] \gtrless C_R^*\,, \quad (8.4)$$

where the cut-off point C_R^* is determined so that the error of misclassification e_{12} is at a pre-specified level; here, $y_{02.1}(x_0) = y_{02}(x_0) - b(x_0)\,y_{01}(x_0)$. Balakrishnan and Tiku (1988b) have shown through asymptotic arguments that an estimate of the cut-off point C_R^* may be obtained as the solution $\hat{C}_R^*$ of the equation (with e_{12} fixed as α)

$$(1 - \hat{\theta}_1)\,\Phi\left\{\frac{\hat{C}_R^* - \frac{1}{2}\hat{\eta}_R^2(0) - \ln[(1 - \hat{\theta}_1)/(1 - \hat{\theta}_2)]}{|\hat{\eta}_R(0)|}\right\}$$

$$+ \hat{\theta}_1\,\Phi\left\{\frac{\hat{C}_R^* - \frac{1}{2}\hat{\eta}_R^2(1) - \ln(\hat{\theta}_1/\hat{\theta}_2)}{|\hat{\eta}_R(1)|}\right\} = \alpha\,, \quad (8.5)$$

where

$$\hat{\eta}_R^2(x) = \frac{1}{\hat{\sigma}_1^2(x)}(\hat{\mu}_{11}(x) - \hat{\mu}_{21}(x))^2 + \frac{1}{\hat{\sigma}_{2.1}^2(x)}(\hat{\mu}_{12.1}(x) - \hat{\mu}_{22.1}(x))^2\,.$$

$$(8.6)$$

In the case when both the errors of misclassification are allowed to float, the robust linear classification procedure is to classify the new independent observation $(x_0 \; y_0)^{\mathrm{T}}$ into Π_1 or Π_2, or arbitrarily into Π_1 or Π_2, according as

$$V_R(x_0) \gtrless = 0\,, \quad (8.7)$$

where the statistic $V_R(x_0)$ is as given in (8.4).

Due to the asymptotic distributional properties of the MML estimators mentioned earlier, the statistic $V_R(x_0)$ in (8.4) asymptotically has exactly the same distribution as the statistic $V(x_0)$ in (8.1), under normality; hence, the two procedures will have exactly the same $1 - e_{21}$ value, for a pre-assigned value of e_{12}, under normality.

In order to examine the robustness features of these two procedures, n pairs of pseudo-random observations $(y_{i1} \; y_{i2})^T$ were simulated through the equations (6.11) and (6.12) by using the seven models in (7.9). The y_2 sample was obtained simply by adding the vector μ_2, which created a location shift between the two populations Π_1 and Π_2. Then, with e_{12} fixed as 0.05, the values of e_{12} and $1 - e_{21}$ were simulated (based on 1000 Monte Carlo runs) for both the procedures under all seven models in (7.9). These values are presented in Table 8.1 for the choice of parameters

Table 8.1
Simulated values of e_{12} and $1 - e_{21}$ for the classical and robust linear classification procedures for the models in (7.9) when $\theta_1 = \theta_2 = 0.1$, $n_1 = n_2 = 50$; for equations of μ_1, μ_2, δ, Σ, and Γ, see text

Model	e_{12} values						$1 - e_{21}$ values					
	$\mu = 1$		$\mu = 2$		$\mu = 3$		$\mu = 1$		$\mu = 2$		$\mu = 3$	
	V	V_R	V	V_R	V	V_R	V	V_R	V	V_R	V	V_R
(1)	0.052	0.052	0.042	0.050	0.044	0.042	0.334	0.336	0.730	0.734	0.944	0.956
(2)	0.026	0.054	0.022	0.046	0.018	0.042	0.174	0.330	0.470	0.712	0.748	0.918
(3)	0.036	0.046	0.030	0.042	0.028	0.042	0.214	0.304	0.584	0.682	0.858	0.910
(4)	0.048	0.054	0.046	0.050	0.044	0.050	0.316	0.340	0.740	0.772	0.942	0.956
(5)	0.060	0.060	0.066	0.064	0.062	0.064	0.204	0.262	0.506	0.608	0.814	0.886
(6)	0.060	0.064	0.054	0.060	0.040	0.060	0.150	0.258	0.464	0.622	0.766	0.866
(7)	0.064	0.066	0.060	0.062	0.060	0.068	0.168	0.248	0.512	0.596	0.812	0.862

$$\mu_1 = \begin{pmatrix} 0 \\ 0 \end{pmatrix}, \quad \mu_2 = \begin{pmatrix} \mu \\ \mu \end{pmatrix}, \quad \delta = \begin{pmatrix} 1 \\ 1 \end{pmatrix},$$

$$\Sigma = \begin{pmatrix} 1.0 & 0.5 \\ 0.5 & 1.0 \end{pmatrix}, \quad \Gamma = \begin{pmatrix} 1 & 0 \\ 0 & 1 \end{pmatrix},$$

for $\mu = 1(1)3$, $\theta_1 = \theta_2 = 0.1$ and $n_1 = n_2 = 50$. It is quite clear from this table that the robust procedure in (8.4) has its e_{12} value to be very stable and close to the presumed value of 0.05 while that of the classical procedure in (8.1) seems to be very sensitive to departures from bivariate normality. We also observe from Table 8.1 that the robust procedure has its $1 - e_{21}$ value to be exactly the same as that of the classical procedure under normality, and at the same time has its $1 - e_{21}$ value to be much larger than that of the classical classification procedure under all the non-normal models.

N. Balakrishnan and R. S. Ambagaspitiya

8.3. Illustrative example

In this section we present a simulated data set and illustrate the classical and the robust linear classification procedures described in Section 8.1 and 8.2.

8.3.1. Example 6

In Tables 8.2 and 8.3, we have presented a simulated data set in an anthropometric set-up which gives measurements on upper facial length and head length along with the gender for $n_1 = 100$ and $n_2 = 80$ individuals from two different groups.

Table 8.2
Measurements on upper facial length and head length of $n_1 = 100$ individuals from Group 1 along with their gender

i	x_{1i}	y_{1i}^{T}	i	x_{1i}	y_{1i}^{T}	i	x_{1i}	y_{1i}^{T}
1	1	(54.5 38.4)	35	0	(45.9 50.6)	69	1	(39.7 41.7)
2	1	(54.7 46.7)	36	1	(45.1 42.2)	70	0	(39.3 40.5)
3	1	(53.8 42.8)	37	1	(42.9 39.8)	71	1	(42.6 45.1)
4	0	(39.4 46.7)	38	0	(44.0 43.5)	72	1	(39.3 30.1)
5	0	(39.3 47.8)	39	0	(43.5 48.1)	73	1	(41.1 45.7)
6	0	(34.5 41.5)	40	0	(37.6 47.7)	74	1	(36.6 42.0)
7	0	(50.4 50.1)	41	1	(43.7 40.0)	75	1	(28.8 40.5)
8	1	(41.1 38.9)	42	0	(45.1 39.1)	76	0	(49.0 44.7)
9	0	(33.8 31.9)	43	1	(36.4 43.8)	77	0	(38.8 42.2)
10	1	(41.3 45.3)	44	0	(32.7 33.8)	78	0	(33.9 35.8)
11	1	(47.7 59.3)	45	0	(42.7 40.8)	79	0	(37.2 44.3)
12	1	(39.8 35.4)	46	1	(42.0 43.2)	80	1	(39.0 40.8)
13	0	(22.6 32.1)	47	1	(46.6 53.1)	81	0	(40.7 40.5)
14	1	(42.1 46.2)	48	1	(42.1 40.8)	82	1	(44.2 44.7)
15	0	(34.6 47.4)	49	1	(46.3 43.3)	83	1	(44.2 46.2)
16	1	(49.4 48.4)	50	0	(40.0 47.6)	84	1	(42.2 40.0)
17	0	(37.3 38.8)	51	1	(46.8 41.7)	85	1	(40.4 38.2)
18	0	(38.8 37.7)	52	0	(44.6 35.6)	86	1	(36.2 41.6)
19	0	(38.6 39.5)	53	0	(39.7 39.6)	87	1	(38.5 41.1)
20	0	(37.4 44.5)	54	0	(45.3 37.6)	88	0	(37.4 39.5)
21	0	(45.0 35.3)	55	0	(45.4 46.8)	89	0	(30.4 48.4)
22	0	(36.3 42.0)	56	0	(47.3 50.8)	90	1	(38.2 38.0)
23	1	(43.0 42.0)	57	0	(35.3 37.9)	91	0	(42.8 36.9)
24	1	(35.2 46.0)	58	1	(39.0 47.5)	92	0	(40.5 46.9)
25	1	(42.8 44.2)	59	0	(40.9 46.5)	93	1	(35.9 43.3)
26	0	(33.1 36.1)	60	0	(41.1 51.2)	94	0	(41.4 38.0)
27	1	(43.1 46.1)	61	0	(37.7 48.4)	95	0	(46.7 51.5)
28	1	(49.2 52.0)	62	0	(40.4 43.2)	96	0	(43.9 49.0)
29	0	(41.7 35.3)	63	0	(46.0 43.4)	97	1	(50.3 47.4)
30	0	(39.8 43.8)	64	1	(48.6 43.2)	98	1	(52.7 43.8)
31	0	(33.5 34.9)	65	1	(40.4 40.4)	99	1	(44.4 56.1)
32	0	(37.2 36.4)	66	0	(45.0 39.5)	100	0	(35.1 37.5)
33	1	(38.7 43.2)	67	1	(46.3 52.6)			
34	0	(35.9 35.5)	68	1	(44.7 37.5)			

Table 8.3
Measurements on upper facial length and head length of $n_2 = 80$ individuals from Group 2 along with their gender

i	x_{2i}	y_{2i}^T	i	x_{2i}	y_{2i}^T	i	x_{2i}	y_{2i}^T
1	0	(43.5 33.5)	28	0	(33.9 40.3)	55	1	(39.0 37.1)
2	0	(35.0 36.5)	29	1	(35.0 40.7)	56	1	(42.2 41.8)
3	0	(29.0 31.8)	30	1	(39.9 40.3)	57	1	(34.9 42.3)
4	0	(31.4 35.6)	31	1	(29.8 36.0)	58	0	(27.6 33.3)
5	1	(46.9 41.0)	32	0	(25.8 27.9)	59	1	(34.0 35.9)
6	0	(39.5 49.0)	33	0	(31.5 37.3)	60	1	(37.8 33.5)
7	1	(38.4 46.6)	34	0	(30.6 33.4)	61	1	(42.8 46.9)
8	0	(34.7 31.6)	35	1	(42.5 35.5)	62	1	(32.6 39.6)
9	1	(34.9 33.2)	36	1	(33.0 36.2)	63	1	(40.5 47.5)
10	0	(52.7 49.7)	37	0	(36.9 35.6)	64	0	(39.1 44.8)
11	0	(31.1 32.5)	38	0	(30.8 26.4)	65	0	(45.0 46.7)
12	1	(39.3 39.2)	39	1	(51.9 49.6)	66	0	(32.7 38.9)
13	0	(41.7 37.9)	40	1	(26.7 26.4)	67	0	(26.0 37.4)
14	0	(36.4 37.1)	41	1	(34.3 40.6)	68	0	(38.3 38.8)
15	1	(35.5 35.0)	42	0	(35.0 36.7)	69	1	(42.4 42.2)
16	0	(40.8 35.7)	43	0	(43.5 42.7)	70	0	(38.0 36.1)
17	0	(39.5 39.8)	44	1	(45.7 51.7)	71	0	(31.7 42.2)
18	1	(37.5 45.8)	45	0	(37.5 38.1)	72	0	(39.5 39.7)
19	1	(39.1 40.8)	46	0	(32.5 33.3)	73	0	(35.0 30.7)
20	0	(36.2 36.1)	47	1	(41.1 41.8)	74	0	(32.2 36.7)
21	1	(35.0 37.4)	48	1	(40.8 38.6)	75	0	(34.8 39.0)
22	1	(40.3 43.4)	49	0	(28.4 35.3)	76	1	(46.8 40.8)
23	0	(34.6 34.1)	50	0	(42.0 45.1)	77	1	(38.4 41.8)
24	0	(40.5 49.1)	51	0	(32.1 26.5)	78	1	(33.2 38.0)
25	0	(34.0 49.9)	52	0	(35.6 34.7)	79	1	(36.3 35.7)
26	0	(39.7 34.7)	53	0	(31.8 36.9)	80	0	(37.3 38.8)
27	0	(35.3 42.6)	54	0	(36.2 45.0)			

From these two samples, we find:

$$\bar{y}_{11}(0) = 39.7490, \quad \bar{y}_{21}(0) = 35.6773, \quad s_{11}^2(0) = 26.5042, \quad s_{21}^2(0) = 27.4400,$$

$$\hat{\mu}_{11}(0) = 39.8813, \quad \hat{\mu}_{21}(0) = 35.6168, \quad \hat{\sigma}_{11}^2(0) = 24.3128, \quad \hat{\sigma}_{21}^2(0) = 22.2241,$$

$$\bar{y}_{11}(1) = 43.0150, \quad \bar{y}_{21}(1) = 38.4347, \quad s_{11}^2(1) = 28.8526, \quad s_{21}^2(1) = 27.0837,$$

$$\hat{\mu}_{11}(1) = 42.8334, \quad \hat{\mu}_{21}(1) = 38.4516, \quad \hat{\sigma}_{11}^2(1) = 26.5951, \quad \hat{\sigma}_{21}^2(1) = 24.0393,$$

$$\bar{y}_{12.1}(0) = 20.5431, \quad \bar{y}_{22.1}(0) = 18.5423, \quad s_{12.1}^2(0) = 24.6948, \quad s_{22.1}^2(0) = 22.2808,$$

$$\hat{\mu}_{12.1}(0) = 20.4563, \quad \hat{\mu}_{22.1}(0) = 18.4084, \quad \hat{\sigma}_{12.1}^2(0) = 27.1502, \quad \hat{\sigma}_{22.1}^2(0) = 18.5207,$$

$$\bar{y}_{12.1}(1) = 24.5276, \quad \bar{y}_{22.1}(1) = 23.0209, \quad s_{12.1}^2(1) = 24.7171, \quad s_{22.1}^2(1) = 16.2804,$$

$$\hat{\mu}_{12.1}(1) = 24.5416, \quad \hat{\mu}_{22.1}(1) = 23.0833, \quad \hat{\sigma}_{12.1}^2(1) = 17.5973, \quad \hat{\sigma}_{22.1}^2(1) = 18.3230,$$

$$\hat{\theta}_1 = 0.4700, \quad \hat{\theta}_2 = 0.4125, \quad \hat{C} = -1.0827, \quad \hat{C}_R = -1.12169.$$

So, with e_{12} (the probability of wrongly classifying an individual from group 1 into group 2) fixed as 0.05, by applying the classical classification procedure in (8.1) and reclassifying each individual in Tables 8.2 and 8.3, we obtain estimates of e_{12} and $1 - e_{21}$ (the probability of correctly classifying an individual from group 2 into itself) to be 0.040 and 0.525, respectively. Similarly, by applying the robust linear classification procedure in (8.4) and reclassifying each individual in Tables 8.2 and 8.3, we obtain estimates of e_{12} and $1 - e_{21}$ to be 0.040 and 0.563, respectively.

9. Some other problems of research and applications

We have demonstrated that the robust univariate and multivariate classification procedures based on the MML estimators that have been developed and discussed in this paper perform overall more efficiently than the classical classification procedures, when one of the errors of misclassification is at a pre-specified level. Also, in the case when both the errors of misclassification are allowed to float, we have shown that the robust classification procedure has a smaller average error rate than the classical classification procedure and a much smaller average error rate than many distribution-free classification procedures including the nearest-neighbour procedure and the procedures based on density estimates.

Several related problems of research emerge from this study. For example:

(1) we may consider some other robust classification procedures based on the prominent trimmed estimators (Tukey and McLaughlin, 1963; Dixon and Tukey, 1968), Winsorized estimators (Dixon, 1960; Dixon and Tukey, 1968), and M-estimators (Huber, 1981; Maronna, 1976), and compare their performance with the robust procedures developed in this paper;

(2) we may generalize these two-way classification procedures to the case of classifying a new individual into one of k populations;

(3) we may consider the classical and the robust multivariate quadratic classification procedures discussed in Section 6.1 when one of the two errors of misclassification is at a pre-specified level, and compare their performance with those of some distribution-free procedures and the procedure due to Anderson and Bahadur (1966);

(4) we may consider some of the well-known classical methods of estimating the missing values in a multivariate data and develop the robust analogues based on the MML estimators, and then study the performance of these imputing methods in the context of classification; Jackson (1968), Chan (1970), and Chan and Dunn (1972) have compared the classical imputing methods in the context of classification through extensive Monte Carlo studies;

(5) we may consider the classification procedure based on normal and multinomial variables (see Krzanowski, 1975) and develop its robust analogue based on the MML estimators, and compare the performance of these two procedures.

Some work has been done on a few of the problems mentioned above and we hope to report these findings in a future paper.

Acknowledgements

Thanks are due to Miss Yvette Macabuag for helping us in collecting and compiling the anthropometric data from its original source. The first author would also like to thank the Natural Sciences and Engineering Research Council of Canada for funding this project.

References

Anderson, T. W. (1951). Classification by multivariate analysis. *Psychometrika* **16**, 31–50.

Anderson, T. W. (1958). *Introduction to Multivariate Statistical Analysis*, Wiley, New York.

Anderson, T. W. (1966). Some nonparametric multivariate procedures based on statistically equivalent blocks. In: *Proc. of an International Symposium on Multivariate Analysis*, Academic Press, New York.

Anderson, T. W. and Bahadur, R. (1962). Classification into two multivariate normal distributions with different covariance matrices. *Ann. Math. Stat.* **33**, 420–431.

Arnold, B. C. and Balakrishnan, N. (1989). *Relations, Bounds and Approximations for Order Statistics*, Lecture Notes in Statistics No. 53, Springer-Verlag, New York.

Balakrishnan, N. (1988). Bayesian insight into a robust two-sample *t*-test based on asymmetric censored samples. *Biometrical Journal* **30**, 425–440.

Balakrishnan, N. and Kocherlakota, S. (1985). Robustness to nonnormality of the linear discriminant function: Mixtures of normal distributions, *Commun. Stat. – Theor. Meth.* **14**, 465–478.

Balakrishnan, N., Kocherlakota, S. and Kocherlakota, K. (1986). On the errors of misclassification based on dichotomous and normal variables. *Ann. Inst. Stat. Math.* **38**, 529–538.

Balakrishnan, N., Kocherlakota, S. and Kocherlakota, K. (1988). Robustness of the double discriminant function in nonnormal situations. *S. Afr. Stat. J.* **22**, 15–43.

Balakrishnan, N. and Tan, W. Y. (1988). Bayesian insight into a robust test for linear contrast based on asymmetric censored samples. *Biometrical Journal* **30**, 517–532.

Balakrishnan, N. and Tiku, M. L. (1988a). Robust classification procedures. In: H. H. Bock, ed., *Classification and Related Methods of Data Analysis*, North-Holland, Amsterdam, 269–276.

Balakrishnan, N. and Tiku, M. L. (1988b). Robust classification procedures based on dichotomous and continuous variables. *Journal of Classification* **5**, 53–80.

Balakrishnan, N., Tiku, M. L. and El Shaarawi, A. H. (1985). Robust univariate two-way classification. *Biometrical Journal* **27**, 123–138.

Bhattacharyya, G. K. (1985). The asymptotics of maximum likelihood and related estimators based on Type II censored data. *J. Am. Stat. Assoc.* **80**, 398–404.

Bowker, A. H. (1960). A representation of Hotelling's T^2 and Anderson's classification statistic W in terms of simple statistics. In: *Contributions to Probability and Statistics: Essays in Honour of H. Hotelling*, Stanford University Press, Stanford, CA.

Cacoullos, T. (1964). Estimation of a multivariate density. Technical Report No. 40, Department of Statistics, University of Minnesota, Minneapolis.

Chan, L. S. (1970). The treatment of missing values in discriminant analysis. Doctoral Dissertation, University of California, Los Angeles.

Chan, L. S. and Dunn, O. J. (1972). The treatment of missing values in discriminant analysis-I. The sampling experiment. *J. Am. Stat. Assoc.* **67**, 473–477.

Chang, P. C. and Afifi, A. A. (1974). Classification based on dichotomous and continuous variables. *J. Am. Stat. Assoc.* **69**, 336–339.

Ching'anda, E. F. (1976). Misclassification errors and their distributions. M.Sc. Thesis, University of Manitoba, Winnipeg.

Ching'anda, E. F. and Subrahmaniam, K. (1979). Robustness of the linear discriminant function to nonnormality: Johnson's system. *J. Stat. Plann. Inf.* **3**, 69–77.

Crooke, W. (1896). *Castes and Tribes of North Western Provinces and Oudh*, Vols. 1–6, Lucknow, India.

Das Gupta, S. (1964). Nonparametric classification rules, *Sankhya, Ser. A* **26**, 25–30.

David, F. N. and Johnson, N. L. (1954). Statistical treatment of censored data. I. Fundamental formulae. *Biometrika* **41**, 228–240.

David, H. A. (1981). *Order Statistics*, 2nd edition, Wiley, New York.

Dixon, W. J. (1960). Simplified estimation from censored normal samples. *Ann. Math. Stat.* **31**, 385–391.

Dixon, W. J. and Tukey, J. W. (1968). Approximate behavior of the distribution of Winsorized *t* (trimming/Winsorization 2). *Technometrics* **10**, 83–98.

Dunn, O. J. and Varady, P. D. (1966). Probabilities of correct classification in discriminant analysis. *Biometrics* **22**, 908–924.

Everitt, B. S. (1978). *Graphical Techniques in Multivariate Analysis*, Heinemann Educational Books, London.

Fisher, R. A. (1936). The use of multiple measurements in taxonomic problems. *Annals of Eugenics* **7**, 179–188.

Fix, E. and Hodges, J. L., Jr. (1951). Discriminating analysis; Nonparametric discrimination; Consistency properties. USAF School of Aviation Medicine, Report No. 4.

Gessaman, M. P. (1970). A consistent nonparametric multivariate density estimator based on statistically equivalent blocks. *Ann. Math. Stat.* **41**, 1344–1346.

Gessaman, M. P. and Gessaman, P. H. (1972). A comparison of some multivariate discrimination procedures. *J. Am. Stat. Assoc.* **67**, 468–472.

Gnanadesikan, R. (1977). *Methods for Statistical Data Analysis of Multivariate Observations*, Wiley, New York.

Gupta, A. K. and Kim, B. K. (1978). On a distribution-free discriminant analysis. *Biometrical Journal* **20**, 729–736.

Huber, P. J. (1981). *Robust Statistics*, Wiley, New York.

Hudimoto, H. (1964). On a distribution-free two-way classification. *Ann. Inst. Stat. Math.* **16**, 247–253.

Jackson, E. C. (1968). Missing values in linear multiple discriminant analysis. *Biometrics* **24**, 835–844.

John, S. (1960a). On some classification problems I. *Sankhya* **22**, 301–308.

John, S. (1960b). On some classification problems. *Sankhya* **22**, 309–316.

Johnson, N. L., Nixon, E., Amos, D. E. and Pearson, E. S. (1963). Tables of percentage points of Pearson curves. *Biometrika* **50**, 459–498.

Kabe, D. G. (1963). Some results on the distribution of two random matrices used in classification procedures. *Ann. Math. Stat.* **34**, 181–185.

Kocherlakota, S., Balakrishnan, N. and Kocherlakota, K. (1987). The linear discriminant function: Sampling from the truncated normal distribution. *Biometrical Journal* **29**, 131–139.

Kocherlakota, S., Kocherlakota, K. and Balakrishnan, N. (1987). Asymptotic expansions for errors of misclassification: Nonnormal situations. In: A. K. Gupta, ed., *Advances in Multivariate Statistical Analysis*, D. Reidel Publishing Company, Dordrecht, 191–211.

Krzanowski, W. J. (1975). Discrimination and classification using both binary and continuous variables. *J. Am. Stat. Assoc.* **70**, 782–790.

Lachenbruch, P. A. and Mickey, M. R. (1968). Estimation of error rates in discriminant analysis. *Technometrics* **10**, 1–11.

Lachenbruch, P. A., Sneeringer, C. and Revo, L. T. (1973). Robustness of the linear and quadratic discriminant functions to certain types of nonnormality. *Commun. Stat.* **1**, 39–57.

Loftsgaarden, D. O. and Quesenberry, C. P. (1965). A non-parametric estimate of a multivariate density function. *Ann. Math. Stat.* **36**, 1049–1051.

Mahalanobis, P. C. (1948). Anthropological observations. Supplement: Ethnological notes. *Sankhya* **9**, 181–236.

Mahalanobis, P. C. and Rao, C. R. (1948). Statistical analysis. *Sankhya* **9**, 111–180.

Majumdar, D. N. (1941). *Collection of U.P. Anthropometric Data in the 1941 Survey*, India.

Majumdar, D. N. (1944). *The Fortunes of Primitive Tribes*, Lucknow, India.

Maronna, R. A. (1976). Robust M-estimators of multivariate location and scatter. *Ann. Stat.* **4**, 51–67.

Morant, G. M. (1939). The use of statistical methods in the investigation of problems of classification in anthropology. *Biometrika* **31**, 72–98.

Okamoto, M. (1963). An asymptotic expansion for the distribution of linear discriminant function. *Ann. Math. Stat.* **34**, 1286–1301. Correction **39**, 1358–1359.

Parzen, E. (1962). On estimation of a probability density and mode. *Ann. Math. Stat.* **33**, 1065–1076.

Pearson, K. (1898). Mathematical contributions to the theory of evolution. V. On the reconstruction of stature of prehistoric races. *Phil. Trans. R. Soc. London Ser. A* **192**, 169–244.

Persson, T. and Rootzen, H. (1977). Simple and highly efficient estimators for a Type I censored normal sample. *Biometrika* **63**, 123–128.

Rao, C. R. (1947). The problem of classification and distance between two populations. *Nature* **159**, 30–31.

Rao, C. R. (1948a). A statistical criterion to determine the group to which an individual belongs. *Nature* **160**, 835–836.

Rao, C. R. (1948b). The utilization of multiple measurements in problems of biological classification (with discussion). *J. R. Stat. Soc. Ser. B*, **10**, 159–203.

Risley, H. H. (1891). *Tribes and Castes of Bengal, Ethnographic Glossary*, Calcutta, India.

Roy, S. C. (1915). *The Oraons of Chota-Nagpur*, Calcutta, India.

Sinha, B. K. and Giri, N. (1975). On the distribution of a random matrix. *Commun. Stat.* **4**, 1057–1063.

Sitgreaves, R. (1952). On the distribution of two random matrices used in classification procedure. *Ann. Math. Stat.* **23**, 263–270.

Smith, C. A. B. (1947). Some examples of discrimination. *Annals of Eugenics* **13**, 272–282.

Srivastava, M. S. and Khatri, C. G. (1979). *An Introduction to Multivariate Statistics*, North Holland, Amsterdam.

Stoller, D. C. (1954). Univariate two population distribution-free discrimination. *J. Am. Stat. Assoc.* **49**, 770–775.

Subrahmaniam, K. and Ching'anda, E. F. (1978). Robustness of the linear discriminant function to non-normality: Edgeworth series distribution. *J. Stat. Plann. Inf.* **2**, 79–91.

Tan, W. Y. (1985). On Tiku's robust procedure – a Bayesian insight. *J. Stat. Plann. Inf.* **11**, 329–340.

Tan, W. Y. and Balakrishnan, N. (1986). Bayesian insight into Tiku's robust procedures based on asymmetric censored samples. *J. Stat. Comput. Simul.* **24**, 17–31.

Tan, W. Y. and Balakrishnan, N. (1988). Bayesian approach to robust comparison of two means based on asymmetric Type-II censored samples. *J. Stat. Comput. Simul.* **30**, 81–102.

Tiku, M. L. (1967). Estimating the mean and standard deviation from a censored normal sample. *Biometrika* **54**, 155–165.

Tiku, M. L. (1968). Estimating the parameters of lognormal distribution from censored samples. *J. Am. Stat. Assoc.* **63**, 134–140.

Tiku, M. L. (1970). Monte Carlo study of some simple estimators in censored normal samples. *Biometrika* **57**, 207–211.

Tiku, M. L. (1980). Robustness of MML estimators based on censored samples and robust test statistics. *J. Stat. Plann. Inf.* **4**, 123–143.

Tiku, M. L. (1982). Testing linear contrasts of means in experimental design without assuming normality and homogeneity of variances. *Biometrical Journal* **24**, 613–627.

Tiku, M. L. (1988). Modified maximum likelihood estimation for the bivariate normal. *Commun. Stat. - Theor. Meth.* **17**, 893–910.

Tiku, M. L. and Balakrishnan, N. (1984). Robust multivariate classification procedures based on the MML estimators. *Commun. Stat. - Theor. Meth.* **13**, 967–986.

Tiku, M. L. and Balakrishnan, N. (1985). Robust classification procedures based on the MML estimators. In: S. Das Gupta and J. K. Ghosh, eds., *Advances in Multivariate Statistical Analysis*, Indian Statistical Institute, Calcutta, 513–536.

Tiku, M. L. and Balakrishnan, N. (1986). A robust test for testing the correlation coefficient. *Commun. Stat. - Simul. Comput.* **15**, 945–971.

Tiku, M. L. and Balakrishnan, N. (1988a). Robust Hotelling-type T^2-statistics. *Biometrical Journal* **30**, 283–293.

Tiku, M. L. and Balakrishnan, N. (1988b). Robust Hotelling-type T^2-statistics based on the MML estimators. *Commun. Stat. – Theor. Meth.* **17**, 1789–1810.

Tiku, M. L. and Balakrishnan, N. (1988c). Generalization of the robust bivariate T^2-statistic to multivariate populations. *Commun. Stat. – Theor. Meth.* **17**, 3899–3911.

Tiku, M. L. and Balakrishnan, N. (1989). Robust classification procedures based on the MML estimators. *Commun. Stat. – Theor. Meth.* **18**, 1047–1066.

Tiku, M. L., Balakrishnan, N. and Ambagaspitiya, R. S. (1989). Error rates of a robust classification procedure based on dichotomous and continuous random variables. *Commun. Stat. – Simul. Comput.* **18**, 571–588.

Tiku, M. L., Gill, P. S. and Balakrishnan, N. (1989). A robust procedure for testing the equality of mean vectors of two bivariate populations with unequal covariance matrices. *Commun. Stat. – Theor. Meth.* **18**, 3249–3265.

Tiku, M. L., Tan, W. Y. and Balakrishnan, N. (1986). *Robust Inference*, Marcel Dekker, New York.

Tukey, J. W. and McLaughlin, D. H. (1963). Less vulnerable confidence and significance procedures for location based on a single sample: trimming/Winsorization 1. *Sankhya, Ser. A* **25**, 331–352.

Wald, A. (1944). On a statistical problem arising in the classification of an individual into one of two groups. *Ann. Math. Stat.* **15**, 145–162.

Welch, B. L. (1939). Note on discriminant functions. *Biometrika* **31**, 218–220.

C. R. Rao and R. Chakraborty, eds., *Handbook of Statistics, Vol. 8*
© Elsevier Science Publishers B.V. (1991) 203–254

7

Analysis of Population Structure: A Comparative Study of Different Estimators of Wright's Fixation Indices

Ranajit Chakraborty and Heidi Danker-Hopfe

1. Introduction

Computations of Wright's fixation indices (F_{IT}, F_{ST}, and F_{IS}) are pivotal for studying the genetic differentiation of populations. It is well known that these indices can be conceptually defined in terms of correlations between uniting gametes (Wright, 1943, 1951); as functions of heterozygosities and their Hardy–Weinberg expectations (Nei, 1973, 1977), or as functions of variance components from a nested analysis of variance (Cockerham, 1969, 1973; Weir and Cockerham, 1984; Long, 1986). Nei (1977) and Nei and Chesser (1983) considered the question of estimating the fixation indices through a decomposition of gene diversity in the total population, while Cockerham (1969, 1973) and Weir and Cockerham (1984) provided estimation procedures by a variance component analysis. Long (1986) extended the variance component approach of estimation to the case of multiple (greater than two) allelic loci, which gives numerically different results from the Weir–Cockerham estimates (see equation (10) of Weir and Cockerham, 1984 vs. equations (9a), (10a) and (11a) of Long, 1986). Although there are several studies drawing comparisons of these different estimates in simulated data (Van Den Bussche et al., 1986; Chakraborty and Leimar, 1987; Slatkin and Barton, 1989), it is not generally known how these different estimates differ in real data in practice. Furthermore, there is no comprehensive computer algorithm which computes all of these estimates simultaneously.

The purpose of this chapter is twofold:

(1) to review the different conceptualizations of Wright's fixation indices using an uniform set of notations, and to examine the question of estimation of parameters and hypothesis testing in the context of an analysis of categorical data; and

(2) to document a computational algorithm for deriving the different estimators (with their standard errors, and test criteria) developed here (called WRIGHT, with three components: NEI, CLARK, and LONG) that can be used for any given data for population structure analysis.

In doing so, we provide the description of the parameters and express the estimators as functions of the observed data statistics, since there is a misconception that some of the formulations are in terms of the data statistics, and not the underlying parameters. The estimation equations are given encompassing the situations where the genotype or the allele frequencies are available. Note that when allele frequencies are used as observed data characteristics (which is usually the case for loci involving dominance relationships among alleles at a locus or in analysis of data collected from the literature), because of the lack of information on observed heterozygosities within populations, the two fixation indices F_{IT} and F_{IS} cannot be estimated, hence the only parameter that needs estimation is F_{ST}.

Empirical comparisons of these different estimators are provided with a gene diversity analysis of the populations of Sikkim, India published by Bhasin et al. (1986). Finally, we discuss the relative merits of these different estimators in terms of their complexity of computation, and generality in various practical situations. While there are several recent reviews of the difficulties of the estimators of F_{IS}, F_{ST}, and F_{IT} in the literature (see, e.g., Curie-Cohen, 1982; Robertson and Hill, 1984; Weir and Cockerham, 1984), they do not encompass all of the different estimators as fully as presented here. Consequently, these reviews do not explicitly demonstrate why the different methods of estimation produce numerically different results, or how different they can be in practice. Therefore, this review, together with the documentation of a single computer program (available from the authors upon request) should serve as an up-to-date description of the applicability of the estimators of Wright's fixation indices to the analysis of any combination of immunological (blood groups, immunoglobulin-Gm, HLA), biochemical (red-cell isozymes and serum proteins), and DNA polymorphism (Restriction Fragment Length Polymorphism, RFLP's) data in the study of the genetic structure of a subdivided population.

2. Parameters of population structure

2.1. Wright's fixation indices and Nei's gene diversity

When F_{IT} and F_{IS} are defined as correlations between two uniting gametes to produce the individuals relative to the total population and relative to the subpopulations, respectively, the correlation between two gametes drawn at random for each subpopulation (F_{ST}) is known to satisfy the identity (Wright, 1943, 1951)

$$1 - F_{IT} = (1 - F_{IS})(1 - F_{ST}).   \tag{2.1}$$

Consider a population which is subdivided into s subpopulations in each of which Hardy–Weinberg equilibrium (HWE) does not necessarily hold (i.e., $F_{IS} \neq 0$). For a locus with r alleles (denoted as $A_1, A_2, \ldots, A_r$), deviation from HWE can be fully specified by $\frac{1}{2}r(r-1)$ F_{IS} parameters (Rao et al. 1973).

However, if only the homozygotes are considered, r F_{IS} parameters are enough to specify deviations from HWE.

In the latter event, the frequency of the homozygotes for the k-th allele $(A_k A_k)$ in the i-th subpopulation may be written as

$$P_{ik} = p_{ik}^2 + F_{ISik} p_{ik}(1 - p_{ik}), \qquad (2.2)$$

where p_{ik} is the frequency of the A_k allele in the i-th subpopulation for $i = 1, 2, \ldots, s$; $k = 1, 2, \ldots, r$. Therefore the allele-specific F_{IS} in the i-th subpopulation can be written as

$$F_{ISik} = (P_{ik} - p_{ik}^2)/[p_{ik}(1 - p_{ik})]. \qquad (2.3)$$

The deviation from HWE in the total population, with reference to the same homozygote frequency, can be parameterized in the same fashion, giving

$$P_{\cdot k} = \bar{p}_{\cdot k}^2 + F_{ITk}\bar{p}_{\cdot k}(1 - \bar{p}_{\cdot k}), \qquad (2.4)$$

where

$P_{\cdot k} = \sum_{i=1}^{s} w_i P_{ik}$ is the proportion of $A_k A_k$ genotypes in the total population,
$\bar{p}_{\cdot k} = \sum_{i=1}^{s} w_i p_{ik}$ is the frequency of the A_k allele in the total population, and
w_i = weight of the i-th subpopulation relative to the total population size, which yields

$$F_{ITk} = (P_{\cdot k} - \bar{p}_{\cdot k}^2)/[\bar{p}_{\cdot k}(1 - \bar{p}_{\cdot k})]. \qquad (2.5)$$

With these notations the average F_{IS} (within population deviation from HWE) over all subpopulations for the k-th allele, takes the form

$$F_{ISk} = \sum_{i=1}^{s} w_i(P_{ik} - p_{ik}^2) \Bigg/ \left[\sum_{i=1}^{s} w_i p_{ik}(1 - p_{ik}) \right]$$
$$= (P_{\cdot k} - \overline{p_{\cdot k}^2})/(\bar{p}_{\cdot k} - \overline{p_{\cdot k}^2}), \qquad (2.6)$$

where

$$\overline{p_{\cdot k}^2} = \sum_{i=1}^{s} w_i p_{ik}^2.$$

From equation (2.1), we therefore have

$$F_{STk} = (\overline{p_{\cdot k}^2} - \bar{p}_{\cdot k}^2)/(\bar{p}_{\cdot k} - \bar{p}_{\cdot k}^2). \qquad (2.7)$$

Note that, in this formulation, the definitions of allele-specific F_{IS}, F_{IT}, and F_{ST} values are indeed parameters defined in terms of allele frequencies in the population.

Furthermore, to obtain the locus specific values of these fixation indices, we can sum the numerators and denominators over all alleles at a locus $(k = 1, 2, \ldots, r)$ to get the following formulae. From equation (2.6), we have

$$
\begin{aligned}
F_{\text{IS}} &= \left[\sum_{k=1}^{r} \sum_{i=1}^{s} w_i (P_{ik} - p_{ik}^2) \right] \bigg/ \left[\sum_{k=1}^{r} \sum_{i=1}^{s} w_i p_{ik} (1 - p_{ik}) \right] \\
&= \left[\sum_{i=1}^{s} w_i \sum_{k=1}^{r} p_{ik} (1 - p_{ik}) - \sum_{i=1}^{s} w_i \left(1 - \sum_{k=1}^{r} P_{ik} \right) \right] \\
&\quad \times \left[\sum_{i=1}^{s} w_i \sum_{k=1}^{r} p_{ik} (1 - p_{ik}) \right]^{-1} \\
&= (H_{\text{S}} - H_0)/H_{\text{S}} , \qquad\qquad\qquad\qquad\qquad\qquad\qquad (2.8)
\end{aligned}
$$

where

$$
H_{\text{S}} = \sum_{i=1}^{s} w_i \sum_{k=1}^{r} p_{ik} (1 - p_{ik}) = \sum_{i=1}^{s} w_i H_{\text{S}i}
$$

is the average within population heterozygosity expected under HWE, and

$$
H_0 = \sum_{i=1}^{s} w_i \left(1 - \sum_{k=1}^{r} P_{ik} \right) = 1 - \sum_{i=1}^{s} \sum_{k=1}^{r} w_i P_{ik}
$$

is the actual proportion of heterozygotes in the total population. Similarly, from equation (2.5),

$$
\begin{aligned}
F_{\text{IT}} &= \left[\sum_{k=1}^{r} (P_{\cdot k} - \bar{p}_{\cdot k}^2) \right] \bigg/ \left[\sum_{k=1}^{r} \bar{p}_{\cdot k} (1 - \bar{p}_{\cdot k}) \right] \\
&= \left[\sum_{k=1}^{r} \bar{p}_{\cdot k} (1 - \bar{p}_{\cdot k}) - \left(1 - \sum_{k=1}^{r} P_{\cdot k} \right) \right] \bigg/ \left[\sum_{k=1}^{r} \bar{p}_{\cdot k} (1 - \bar{p}_{\cdot k}) \right] \\
&= (H_{\text{T}} - H_0)/H_{\text{T}} , \qquad\qquad\qquad\qquad\qquad\qquad\qquad (2.9)
\end{aligned}
$$

where

$$
H_{\text{T}} = \sum_{k=1}^{r} \bar{p}_{\cdot k} (1 - \bar{p}_{\cdot k}) = 1 - \sum_{k=1}^{r} \bar{p}_{\cdot k}^2
$$

is the heterozygosity in the total population (expected under HWE).

Lastly, from equation (2.7), we have

$$
F_{\text{ST}} = (H_{\text{T}} - H_{\text{S}})/H_{\text{T}} . \qquad\qquad\qquad\qquad\qquad\qquad\qquad (2.10)
$$

Therefore, estimation of the fixation indices are equivalent to estimation of the parameters H_{T}, H_{S}, and H_0 for a locus (Nei, 1973, 1977). Note that these parametric relationships also hold when the fixation indices are defined by pooling over several loci. In this case, H_{S}, H_{T}, and H_0 are the respective heterozygosities averaged over all loci. The criticism that the relationship between fixation indices with heterozygosities is true for data statistics (and not for parameters) is not valid. Weir and Cockerham's (1984) comments are perhaps due to the misconception that under the mutation-drift model, the expectation of F_{ST} in a population with a finite number of subpopulations (expectation under the evolutionary process) is a function of the number of subpopulations as well. Two comments are worth noting at this point.

First, the above parameterization does not depend upon the evolutionary model of genetic differentiation among subpopulations, and hence the relationships (2.8)–(2.10) hold for any general mating system, irrespective of the selective differentials that may exist among the alleles. Second, even though Nei (1977) defined the pooled F_{IS} in terms of a weighted average of the subpopulation-specific F_{IS_i} values (equation (4) of Nei, 1977), with Wright's (1965) and Kirby's (1975) weight functions, such weights are not needed if we first define F_{IS}, F_{IT}, and F_{ST} as allele-specific parameters and obtain the locus-specific parameters by summing numerators and denominators over all alleles at a locus. It is also clear from equations (2.8)–(2.10) that while estimation of F_{IS} and F_{IT} would require sampling of genotypes from all subpopulations (as they are functions of the actual proportion of heterozygotes in the subpopulations, H_0), F_{ST} can be estimated with allele frequency data alone without making any assumption regarding F_{IS}. Furthermore, estimates of F_{ST} from genotype or allele frequency data should be identical, as long as the allele frequencies are obtained by the gene counting method. These issues will be detailed in the estimation section to follow.

2.2. Fixation indices and Cockerham's variance component representation

Cockerham (1969, 1973) redefined the fixation indices in terms of intra-class correlation derived from an analysis of variance of allele frequencies. In this formulation, indicator variables are defined for both alleles of a random individual sampled, which are in turn expressed as a linear model of additive effects of between-subpopulations (a), between-individuals within a subpopulation (b), and within-individual (c or w) variations. Following the classical analysis of variance model of random effects, where the subpopulations are treated as replicates of each other (Weir and Cockerham, 1984), Cockerham (1969, 1973) showed that the component of variance ascribed to the above factors (a, b and c) yield a parametric relationship with the fixation indices. In particular, Cockerham's results for a specific allele can be written in terms of our notation as

$$F_{ITk} = (a_k + b_k)/(a_k + b_k + c_k),$$ (2.11)

$$F_{ISk} = b_k/(b_k + c_k),$$ (2.12)

$$F_{STk} = a_k/(a_k + b_k + c_k),$$ (2.13)

where a_k, b_k, and c_k are the variance components associated with the above factors, in which the genotype frequencies in the population are tabulated with regard to a specific allele, A_k (and thus only the frequencies of the three genotypes $A_k A_k$, $A_k \overline{A}_k$, and $\overline{A}_k \overline{A}_k$ enter into the analysis, $\overline{A}_k$ being a combination of all alleles of type other than A_k).

Before reviewing the estimation equations for these variance components, it is worthwhile to examine how these variance component parameters translate into the gene frequency parameters in a subdivided population.

It is easy to note that

$$a_k + b_k + c_k = \overline{p}_{\cdot k}(1 - \overline{p}_{\cdot k}), \tag{2.14}$$

where $\overline{p}_{\cdot k} = \sum_{i=1}^{s} w_i p_{ik}$, as defined in equation (2.4). Invoking equation (2.13) into equation (2.2), we also have

$$a_k = \overline{p^2_{\cdot k}} - \overline{p}^2_{\cdot k} = \sum_{i=1}^{s} w_i (p_{ik} - \overline{p}_{\cdot k})^2, \tag{2.15}$$

and similarly, from equations (2.4) and (2.12), we have

$$b_k = P_{\cdot k} - \overline{p^2_{\cdot k}} = \sum_{i=1}^{s} w_i (P_{ik} - p_{ik}^2)$$

$$= \sum_{i=1}^{s} w_i F_{IS\,ik} p_{ik}(1 - p_{ik}), \tag{2.16}$$

since $P_{ik} = p_{ik}^2 + F_{IS\,ik} p_{ik}(1 - p_{ik})$, according to our equation (2.2).

Putting equations (2.15) and (2.16) in equation (2.14), we get

$$c_k = \overline{p}_{\cdot k} - P_{\cdot k} = \sum_{i=1}^{s} w_i (p_{ik} - P_{ik}). \tag{2.17}$$

Since $-p_{ik}/(1 - p_{ik}) \leqslant F_{IS\,ik} \leqslant 1$ for all $i = 1, 2, \ldots, s$ and all k, it is easy to see that $c_k \geqslant 0$. Because of equation (2.15), it is also ensured that $a_k \geqslant 0$. However, there is no guarantee that b_k is non-negative. It is, therefore, peculiar that even a parametric value of the variance component due to between individual variation can assume negative values in this formulation. Cockerham (1969) acknowledged this feature, and ascribed this to either a mating system where mates are less related than the average within a subpopulation, or to certain types of selection (Cockerham, 1969, p. 74). Since this arises for $F_{ST} > F_{IT}$ ($\theta > F$, in Cockerham's 1969 notation), this occurs whenever F_{IS} takes negative values (see equation (2.1)).

The above translation of parameters reveals that the negative value of b may not necessarily arise only in estimation; it is an inherent feature of the proposed linear model itself (Cockerham, 1969, 1973). It is particularly uncomfortable, since the linear model is not supposed to produce negative variance components.

In Cockerham's formulation, the locus-specific parameters are defined by summing a_k, b_k, and c_k values over all alleles, and expressing the fixation indices as respective ratios of sums, analogous to equations (2.11)–(2.13). The same pooling algorithm is suggested for definition of parameters pooled over all loci studied (see Weir and Cockerham, 1984, equation (10)).

2.3. Long's extension of Cockerham's model

Long (1986) and Smouse and Long (1988) provided a multivariate extension of the Cockerham model, where a pair of $(r - 1)$-dimensional indicator vectors is defined for a r-allelic genotypic system. This yields a multivariate decomposition of the total dispersion matrix; Σ_a, Σ_b, Σ_c in the analogy of a, b, and c of a bi-allelic locus. With $\Sigma = \Sigma_a + \Sigma_b + \Sigma_c$, the locus-specific fixation indices take the form of

$$F_{\text{IT}} = (r - 1)^{-1} \, \text{tr}[\Sigma^{-1/2}(\Sigma_a + \Sigma_b)\Sigma^{-1/2}], \tag{2.18}$$

$$F_{\text{ST}} = (r - 1)^{-1} \, \text{tr}[\Sigma^{-1/2} \Sigma_a \Sigma^{-1/2}], \tag{2.19}$$

$$F_{\text{IS}} = (r - 1)^{-1} \, \text{tr}[(\Sigma_b + \Sigma_c)^{-1/2} \Sigma_b (\Sigma_b + \Sigma_c)^{-1/2}], \tag{2.20}$$

where tr denotes the trace of a matrix. In this formulation, again, while Σ_a and Σ_c are positive semi-definite matrices, the parametric form of Σ_b can be negative-definite, introducing peculiarities in the interpreting of the decomposition of dispersion matrices.

Note that for $r = 2$, equations (2.18)–(2.20) are mathematically identical to Cockerham's definition of parameters; but for $r > 2$, since equations (2.18)–(2.20) involve covariances of allele or genotype frequencies within and between subpopulations (off-diagonal elements of the Σ-matrices), the locus-specific fixation indices, according to Long's approach, are parametrically different from Weir–Cockerham's parameters. A multi-locus extension of Long's formulation is also available, where the respective Σ matrices for a group of loci are written as block-diagonal locus-specific Σ matrices (see Long, 1986, equation (8)).

In summary, the above parameterization of the genetic structure of a subdivided population indicates that Wright's fixation indices can be expressed in terms of the actual proportion of heterozygotes (H_0) and its expectation (under HWE) in the total population (H_T) and within subpopulations (H_S), without invoking any specific model of the mating system or gene differentiation between or within subpopulations. This mathematical equivalence is shown in the form of parameters, and they are consistent with Wright's identity (equation (2.1)), while the variance–component parameterization is more complex in nature, and could yield possible inconsistencies (e.g., $b < 0$, whenever F_{IS} is negative) for certain evolutionary factors (selection) or social structure of subdivision. Having defined the parameters, let us now turn to estimation and hypothesis testing issues.

3. Estimation of fixation indices

The above discussion indicates that while the heterozygosities or variance components are quadratic functions of allele and/or genotype frequencies within each subpopulation, and their weighted (by relative subpopulation size) averages, the fixation indices are ratios of functions of parameters. While estimation of a ratio of parametric functions is an unpleasant statistical problem for categorical data, to the extent that we may approximate the expectation of a ratio by the ratio of expectations, reasonable estimators of fixation indices may be obtained. Weir and Cockerham (1984) called such estimators (ratio of unbiased estimators of a numerator and a denominator) 'unbiased', while in the strict statistical sense, such estimators are at best consistent (i.e., approach the true value in terms of probability in large samples). A further problem arises, because of the categorical nature of the observations (allele frequencies, or genotype frequencies). The properties of ratio estimators are generally studied in the statistical literature for continuous traits which have Gaussian probability distributions. Even those who are concerned with distinctions between parameters and statistics have been rather cavalier about this aspect of the problem.

In this section we consider some estimators and present estimating equations in terms of the observed frequencies, which in turn indicate how much bias might arise in using these estimating equations. We might also mention that the definition of sample size has been quite elusive in the literature; because it is not always explicit whether it refers to the number of genes sampled, or that of individuals (see, e.g., Weir and Cockerham, 1984).

3.1. Estimation from genotype data

Let us first consider the case where all genotypes are recognizable, so that unequivocally all different alleles can be counted in a sample. As noted before, the F_{IT}, F_{IS}, and F_{ST} parameters depend on the sizes of subpopulations, relative to the total population size. In practice these are unknown, and furthermore, subpopulation sizes generally fluctuate over an evolutionary time period. The temporal change in population sizes has a substantial effect on the coefficients of gene-differentiation as well as heterozygosity (see, e.g., Nei et al., 1975; Chakraborty and Nei, 1977). Therefore, we shall assume all subpopulations to have equal size. This assumption is also explicitly made in Weir and Cockerham (1984, p. 1359). This, however, does not imply that the numbers of individuals sampled from the subpopulations are all equal.

3.2. Estimators of fixation indices by Nei's approach

As before consider a r-allelic locus, and define N_{ikl} to be the number of individuals of genotype $A_k A_l$ in the i-th subpopulation ($k = 1, 2, \ldots, r$; $l = k, \ldots, r$; $i = 1, 2, \ldots, s$). Let N_i be the total number of individuals (sample size) sampled from the i-th subpopulation. The total sample size (number of individuals) sampled from the entire subdivided population is

$$N = \sum_{i=1}^{s} N_i, \quad \text{where } N_i = \sum_{k \geqslant l=1}^{r} \sum N_{ikl}.$$

When $(N_{ikl}; \ k = 1, 2, \ldots, r; \ l = k, \ldots, r)$ is a genotype-specific categorized subdivision of a random sample of N_i individuals from the i-th subpopulation, it is easy to note that

$$X_{ik} = N_{ikk}/N_i \tag{3.1}$$

and

$$x_{ik} = \left(2N_{ikk} + \sum_{l > k = 1}^{r} N_{ikl} \right) \Big/ 2N_i, \tag{3.2}$$

are unbiased estimates of P_{ik} and p_{ik}, the proportion of $A_k A_k$ homozygotes, and the allele frequency of A_k in the i-th subpopulation.

An unbiased estimator for p_{ik}^2 can be obtained as

$$\hat{p_{ik}^2} = x_{ik}(2N_i x_{ik} - 1)/(2N_i - 1), \tag{3.3}$$

which in turn, provides an unbiased estimator of $p_{ik}(1 - p_{ik})$, namely,

$$\frac{2N_i}{2N_i - 1} x_{ik}(1 - x_{ik}). \tag{3.4}$$

Note that if $x_{ik}(1 - x_{ik})$ is used as an estimator for $p_{ik}(1 - p_{ik})$, the extent of bias is

$$b = [(2N_i - 1)/(2N_i) - 1]p_{ik}(1 - p_{ik})$$
$$= -p_{ik}(1 - p_{ik})/(2N_i), \tag{3.4a}$$

i.e., $x_{ik}(1 - x_{ik})$ is an under-estimator of $p_{ik}(1 - p_{ik})$, with proportional bias being $1/2N_i$.

With equations (3.2)–(3.4), the estimator of F_{ISik} (given by equation (2.3)) is

$$\hat{F}_{ISik} = \frac{X_{ik} - x_{ik}(2N_i x_{ik} - 1)/(2N_i - 1)}{2N_i x_{ik}(1 - x_{ik})/(2N_i - 1)}, \tag{3.5}$$

which is a consistent estimator to the extent that the numerator and denominator of the ratio are estimated by their respective unbiased estimators.

Note that if x_{ik}^2 is used as an estimator for p_{ik}^2 (with a negative bias of the order $1/2N_i$), the estimator for F_{ISik} becomes

$$\hat{F}'_{ISik'} = (X_{ik} - x_{ik}^2)/[x_{ik}(1 - x_{ik})], \tag{3.5a}$$

which is identical to Curie-Cohen's (1982) estimator $\hat{f}_1 = 1 - (y/x)$, where y is the observed heterozygosity for the A_k allele in the sample, and $x = 2N_i x_{ik}(1 - x_{ik})$, an estimator of its expectation under HWE.

It might be further noted that Nei's unbiased estimator (equation (3.5)) takes the form

$$\hat{F}_{\mathrm{IS}\,ik} = 1 - [1 - 1/(2N_i)]\, y/x\,, \tag{3.5b}$$

which will be useful in deriving its standard error, shown in the next section.

Curie-Cohen (1982) showed that these equations have a natural multiple-allele extension, when y and x are interpreted as the total observed (H_0) and expected $(H_{\mathrm{E}}$, under HWE) heterozygosity for all alleles at a locus. He, however, did not note the equivalence of his f_1 estimator with Nei's estimate, written in terms of H_0 and H_{S}, at a locus (Nei, 1977).

Let us now consider the joint analysis of data from several subpopulations. Since $\bar{p}_{\cdot\,k} = \sum_{i=1}^{s} w_i p_{ik}$, we have

$$\bar{p}_{\cdot\,k}^2 = \sum_{i=1}^{s} w_i^2 p_{ik}^2 + \sum_{i \neq i' = 1}^{s} w_i w_{i'} p_{ik} p_{i'}k\,,$$

and hence, when the samples from the subpopulations are drawn independently of each other (as usually is the case), we obtain an unbiased estimator of $\bar{p}_{\cdot\,k}^2$, given by

$$\widehat{\bar{p}_{\cdot\,k}^2} = \sum_{i=1}^{s} w_i^2 \frac{x_{ik}(2N_i x_{ik} - 1)}{2N_i - 1} + \sum_{i \neq i' = 1}^{s} w_i w_{i'} x_{ik} x_{i'k}$$

$$= \left(\sum_{i=1}^{s} w_i x_{ik} \right)^2 - \sum_{i=1}^{s} w_i^2 \frac{x_{ik}(1 - x_{ik})}{2N_i - 1}\,.$$

Therefore, estimating the numerators and denominators in an unbiased fashion, we obtain the following estimators of the allele-specific fixation indices at a particular locus:

$$\hat{F}_{\mathrm{IS}k} = \frac{\sum_{i=1}^{s} w_i (X_{ik} - x_{ik}^2) + \sum_{i=1}^{s} w_i x_{ik}(1 - x_{ik})/(2N_i - 1)}{\sum_{i=1}^{s} w_i 2N_i x_{ik}(1 - x_{ik})/(2N_i - 1)}\,, \tag{3.6}$$

$$\hat{F}_{\mathrm{IT}k} = \frac{\sum_{i=1}^{s} w_i X_{ik} - (\sum_{i=1}^{s} w_i x_{ik})^2 + \sum_{i=1}^{s} w_i^2 x_{ik}(1 - x_{ik})/(2N_i - 1)}{\sum_{i=1}^{s} w_i x_{ik} - (\sum_{i=1}^{s} w_i x_{ik})^2 + \sum_{i=1}^{s} w_i^2 x_{ik}(1 - x_{ik})/(2N_i - 1)}\,, \tag{3.7}$$

$$\hat{F}_{\mathrm{ST}k} = \frac{\sum_{i=1}^{s} w_i x_{ik}^2 - (\sum_{i=1}^{s} w_i x_{ik})^2 - \sum_{i=1}^{s} w_i(1 - w_i)x_{ik}(1 - x_{ik})/(2N_i - 1)}{\sum_{i=1}^{s} w_i x_{ik} - (\sum_{i=1}^{s} w_i x_{ik})^2 + \sum_{i=1}^{s} w_i^2 x_{ik}(1 - x_{ik})/(2N_i - 1)}\,. \tag{3.8}$$

Note that while all of these estimators are consistent, to the extent that the numerators and denominators of the parameters defined in equations (2.4)–(2.6) are estimated with their respective unbiased estimators, in applying these equations we need the relative sizes (w_i's) for all subpopulations. These are, however not known in practice; nor can they always be reliably substituted by relative sample sizes. Nei (1977) and Nei and Chesser (1983), therefore assumed that the w_i's are all equal, $w_i = 1/s$ for all i. In that event, equations (3.6)–(3.8) take the form

$$\hat{F}'_{1Sk} = \frac{\sum_{i=1}^{s} [(X_{ik} - x_{ik}^2) + x_{ik}(1 - x_{ik})/(2N_i - 1)]}{\sum_{i=1}^{s} 2N_i x_{ik}(1 - x_{ik})/(2N_i - 1)}, \tag{3.6a}$$

$$\hat{F}'_{1Tk} = \frac{\sum_{i=1}^{s} X_{ik} - (1/s)(\sum_{i=1}^{s} x_{ik})^2 + (1/s)\sum_{i=1}^{s} x_{ik}(1 - x_{ik})/(2N_i - 1)}{\sum_{i=1}^{s} x_{ik} - (1/s)(\sum_{i=1}^{s} x_{ik})^2 + (1/s)\sum_{i=1}^{s} x_{ik}(1 - x_{ik})/(2N_i - 1)}, \tag{3.7a}$$

and

$$\hat{F}'_{STk} = \frac{\sum_{i=1}^{s} x_{ik}^2 - (1/s)(\sum_{i=1}^{s} x_{ik})^2 - (1 - 1/s)\sum_{i=1}^{s} x_{ik}(1 - x_{ik})/(2N_i - 1)}{\sum_{i=1}^{s} x_{ik} - (1/s)(\sum_{i=1}^{s} x_{ik})^2 + (1/s)\sum_{i=1}^{s} x_{ik}(1 - x_{ik})/(2N_i - 1)}. \tag{3.8a}$$

When the sample sizes are large enough, so that $2N_i \approx 2N_i - 1$ and $\sum_{i=1}^{s} x_{ik}(1 - x_{ik})/(2N_i - 1)$ is negligible, these equations take a much simpler form:

$$\hat{F}'_{1Sk} \approx \sum_{i=1}^{s} (X_{ik} - x_{ik})^2 \Big/ \sum_{i=1}^{s} x_{ik}(1 - x_{ik}), \tag{3.6b}$$

$$\hat{F}'_{1Tk} \approx \left[\sum_{i=1}^{s} X_{ik} - \frac{1}{s}\left(\sum_{i=1}^{s} x_{ik}\right)^2\right] \Big/ \left[\sum_{i=1}^{s} x_{ik} - \frac{1}{s}\left(\sum_{i=1}^{s} x_{ik}\right)^2\right], \tag{3.7b}$$

$$\hat{F}'_{STk} \approx \left[\sum_{i=1}^{s} x_{ik}^2 - \frac{1}{s}\left(\sum_{i=1}^{s} x_{ik}\right)^2\right] \Big/ \left[\sum_{i=1}^{s} x_{ik} - \frac{1}{s}\left(\sum_{i=1}^{s} x_{ik}\right)^2\right]. \tag{3.8b}$$

Note that equation (3.8b) takes the well-known form

$$\hat{F}'_{STk} \approx s_k^2/x_{.k}(1 - x_{.k}), \tag{3.8c}$$

where s_k^2 is the variance of the A_k-allele frequency over all subpopulations, $s_k^2 = \sum_{i=1}^{s}(x_{ik} - \bar{x}_{.k})^2/s$, with $\bar{x}_{.k}$ representing the average frequency of the A_k-allele over all subpopulations, $\bar{x}_{.k} = \sum_{i=1}^{s} x_{ik}/s$.

When the investigators have sufficient reason to believe that the sampling from each subpopulation has been conducted in such a manner that the relative sample sizes (N_i/N) reflect their respective relative sizes (population values), one might replace the w_i's in equations (3.6)–(3.8) by their respective sample size weights, $\hat{w}_i = N_i/N$, and obtain the allele-specific estimates of the fixation indices. However, note that while this weighting may serve the purpose of taking into account the relative contribution of each subpopulation in the total population in the current generation, they are not evolutionary stable, as N_i's can fluctuate drastically over time.

Pooling over all alleles at a locus, the locus-specific estimates of F_{IS}, F_{IT} and F_{ST} values can be obtained easily, since the respective parameter values have been defined by summing the numerators and denominators over all alleles at a locus (see equations (2.8)–(2.10). Since these equations are represented in terms of heterozygosities in the population, it may be worthwhile to express the unbiased estimators of H_S, H_0, and H_T explicitly. Nei and Chesser (1983) obtained such estimators, with the assumption that all w_i's are equal ($= 1/s$).

In our terminology, with any general weight (w_i's unequal), we may use the above mentioned unbiased estimators of p_{ik}, P_{ik}^2, and $p_{\cdot k}^2$ to obtain

$$\hat{H}_0 = 1 - \sum_{i=1}^{s} \sum_{k=1}^{r} w_i \hat{P}_{ik} = 1 - \sum_{i=1}^{s} \sum_{k=1}^{r} w_i X_{ik}, \tag{3.9}$$

$$\hat{H}_S = 1 - \sum_{i=1}^{s} \sum_{k=1}^{r} w_i \widehat{p_{ik}^2}$$

$$= 1 - \sum_{i=1}^{s} \sum_{k=1}^{r} \frac{w_i}{2N_i - 1} \left[2N_i \sum_{k=1}^{r} x_{ik}^2 - 1 \right], \tag{3.10}$$

and

$$\hat{H}_T = \sum_{k=1}^{r} \left[\sum_{i=1}^{s} w_i x_{ik} \left(1 - \sum_{i=1}^{s} w_i x_{ik} \right) + \sum_{i=1}^{s} w_i^2 x_{ik}(1 - x_{ik}) \bigg/ (2N_i - 1) \right] \tag{3.11}$$

as respective unbiased estimators of H_0, H_S, and H_T. Substitution of these estimators in equations (2.8)–(2.10) provide consistent estimators of the locus specific fixation indices.

When the w_i's are all equal, equations (3.9)–(3.11) reduce to

$$\hat{H}_0' = 1 - \frac{1}{s} \sum_{i=1}^{s} \sum_{k=1}^{r} X_{ik}, \tag{3.9a}$$

$$\hat{H}_S' = \frac{1}{s} \sum_{i=1}^{s} 2N_i \hat{H}_{Si} \bigg/ (2N_i - 1), \tag{3.10a}$$

$$\hat{H}_T' = \sum_{k=1}^{r} \left[\bar{x}_{\cdot k}(1 - \bar{x}_{\cdot k}) + \frac{1}{s^2} \sum_{i=1}^{s} x_{ik}(1 - x_{ik}) \bigg/ (2N_i - 1) \right], \tag{3.11a}$$

where

$$\hat{H}_{Si} = 1 - \sum_{k=1}^{r} x_{ik}^2 \quad \text{and} \quad \bar{x}_{\cdot k} = \sum_{i=1}^{s} x_{ik}/s.$$

Note that these estimators are exact unbiased estimators of H_0, H_S, and H_T when all subpopulations are of equal size (but N_i's need not be equal), whereas the estimators given by Nei and Chesser (1983) involve some approximations (see their equation (8) in particular). As before, when the N_i's are large, we may equate $2N_i/(2N_i - 1)$ to unity, and neglect the last term of $\hat{H}'_T$, to get

$$\hat{H}'_S \approx \frac{1}{s} \sum_{i=1}^{s} \hat{H}_{Si} \tag{3.10b}$$

and

$$\hat{H}'_T \approx 1 - \sum_{k=1}^{r} \bar{x}^2_{\cdot k}. \tag{3.11b}$$

Thus, when all genotypes are recognizable, unbiased estimators of H_0, H_S, and H_T can be obtained simply by enumerating all allele frequencies in each subpopulation (by gene counting) and evaluating the sum total of all homozygotes (X_{ik}'s). The resulting estimators

$$\hat{F}_{IS} = 1 - \hat{H}_0/\hat{H}_S, \tag{3.12}$$

and

$$\hat{F}_{IT} = 1 - \hat{H}_0/\hat{H}_T, \tag{3.13}$$

$$\hat{F}_{ST} = 1 - \hat{H}_S/\hat{H}_T, \tag{3.14}$$

are again consistent, to the extent that in these the numerators and denominators are estimated by their respective unbiased statistics.

Estimation of parameters pooled over several loci can be achieved exactly in the same manner, by defining the heterozygosities (H_0, H_S, H_T) as averages over all loci.

3.3. Estimators by Cockerham's approach

As shown in equations (2.11)–(2.13), Cockerham (1973) derived the allele-specific fixation indices in terms of components of variance in a nested analysis of variance. In this approach, the estimation of fixation indices reduces to the problem of estimating the components a_k, b_k, and c_k. Weir and Cockerham (1984) gave the explicit forms of these estimators, they are

$$\hat{a}_k = \frac{\overline{N}}{N_c} \left[s_k^2(\hat{w}) - \frac{1}{\overline{N} - 1} \left[\bar{x}_{\cdot k}(\hat{w})(1 - \bar{x}_{\cdot k}(\hat{w})) - \frac{s-1}{s} s_k^2(\hat{w}) - \tfrac{1}{4}\bar{h}(\hat{w}) \right] \right], \tag{3.15}$$

$$\hat{b}_k = \frac{\overline{N}}{\overline{N} - 1} \bar{x}_{\cdot k}(\hat{w})[1 - \bar{x}_{\cdot k}(\hat{w})] - \frac{s-1}{s} s_k^2(\hat{w}) - \frac{2\overline{N} - 1}{4\overline{N}} \bar{h}(\hat{w}), \tag{3.16}$$

and

$$\hat{c}_k = \tfrac{1}{2}\bar{h}(\hat{w}), \tag{3.17}$$

where

$\bar{N} = \sum_{i=1}^{s} N_i/s$ is the average number of individuals sampled per subpopulation,

$N_c = [s\bar{N} - \sum_{i=1}^{s} N_i^2/s\bar{N}]/(s - 1) = \bar{N}(1 - C^2/s)$, where C is the coefficient of variation of sample sizes (N_i's),

$\bar{x}_{\cdot k}(\hat{w}) = \sum_{i=1}^{s} N_i x_{ik}/s\bar{N}$, the weighted average allele frequency of A_k per subpopulation,

$s_k^2(\hat{w}) = \sum_{i=1}^{s} N_i(x_{ik} - \bar{x}_{\cdot k}(\hat{w}))^2/(s - 1)\bar{N}$, is the variance of A_k allele frequencies over subpopulations,

$\bar{h}(\hat{w}) = \sum_{i=1}^{s} N_i h_i(\hat{w})/s\bar{N}$, the average observed heterozygote frequency for allele A_k.

In parallel to equations (2.11)–(2.13), the estimators F_{ITk}, F_{ISk}, and F_{STk} become

$$\hat{F}_{ITk} = (\hat{a}_k + \hat{b}_k)/(\hat{a}_k + \hat{b}_k + \hat{c}_k), \tag{3.15a}$$

and
$$\hat{F}_{ISk} = \hat{b}_k/(\hat{b}_k + \hat{c}_k), \tag{3.16a}$$

$$\hat{F}_{STk} = \hat{a}_k/(\hat{a}_k + \hat{b}_k + \hat{c}_k). \tag{3.17a}$$

Note that these expressions are defined in terms of weighted variance components, where sample sizes from the subpopulations are taken as weights, irrespective of their true relative population sizes (i.e., $\hat{w}_i = N_i/N$). These explicit forms are obtained by algebraic manipulations of the estimated mean square errors in Table 3 (Cockerham, 1973). It should be noted that Cockerham's Table 3 (Cockerham, 1973, p. 688) has an inadvertent error, where the expressions S_a and S_a' should have an additional coefficient 2, which is missing.

Cockerham (1973) also gave an explicit estimator for F_{ISik}, the F_{IS} estimator for a specific allele (A_k) in the i-th subpopulation, which has the form

$$\hat{F}_{ISik} = 1 - \frac{4(N_i - 1)[N_i x_{ik} - N_{ikk}]}{4N_i^2 x_{ik}(1 - x_{ik}) - 2(N_i x_{ik} - N_{ikk})}, \tag{3.18}$$

that can be computed from the respective subpopulation-specific genotype data. A pooled estimator of F_{IS}, pooled over all alleles at a locus, can be obtained by summing the numerator and the denominator of equation (3.18), as done in the other cases.

In particular, when N_i is large, the pooled estimator over all alleles at a locus takes the form

$$\hat{F}_{ISi} = 1 - \frac{1}{2N_i}\left[\sum_{i=1}^{r} h_{ik}\right]\Big/\left[1 - \sum_{i=1}^{r} x_{ik}^2\right], \tag{3.18a}$$

where h_{ik} is the observed number of heterozygotes carrying the A_k allele in the i-th subpopulation.

While equation (3.18) can be derived even without invoking the variance components (see Cockerham, 1969, pp. 689–690), this is different from Nei's estimator (our equation (3.5)), which estimates $F_{\text{IS}ik}$ as a ratio estimator, based on equation (2.3). Both estimators are asymptotically unbiased (since each of them estimates the numerator and denominator by their respective unbiased statistics).

Setting up the equivalence of Cockerham's (1973) and Curie-Cohen's (1982) notations, it may be shown that the above estimator takes the form

$$\hat{F}_{\text{IS}ik} = [2N_i(x - y) + y]/(2N_ix - y),\qquad(3.18b)$$

where $x = 2N_i x_{ik}(1 - x_{ik})$, and $y\ (= \sum'_{l>k} N_{ikl})$ is the observed heterozygosity for the A_k-allele in a particular subpopulation. This equivalence will also be useful in deriving the standard error of this estimator (discussed in the next section).

At this stage, since we have three alternative estimators of $F_{\text{IS}ik}$: Nei's unbiased (equation (3.5)), biased (equation (3.5a)), and Cockerham's (equation (3.18)), it might be worthwhile to study how they behave for a given sample.

It can be shown that

$$\hat{F}_{\text{IS}ik} - \hat{F}'_{\text{IS}ik} = (x_{ik} - X_{ik})/2N_i x_{ik}(1 - x_{ik}),\qquad(3.19)$$

where $\hat{F}_{\text{IS}ik}$ is from equation (3.5) and $\hat{F}'_{\text{IS}ik}$ is from equation (3.5a).

Since $x_{ik} \geqslant X_{ik}$ in any given sample, we have the inequality

$$\text{Nei's unbiased estimator} \geqslant \text{Nei's biased estimator},\qquad(3.20)$$

over the entire sample space.

Furthermore, the expected difference of these two estimators,

$$\text{E}[\hat{F}_{\text{IS}ik} - \hat{F}'_{\text{IS}ik}] \approx (1 - F_{\text{IS}ik})/(2N_i - 1),\qquad(3.21)$$

which is usually very small, of the order $(2N_i - 1)^{-1}$.

Similarly, we can show that

$$\text{Cockerham's estimator (equation (3.18))} > \text{Nei's biased estimator}$$
$$\text{(equation (3.5a))},\qquad(3.22)$$

over the entire sample space.

The relationship between Nei's unbiased and Cockerham's estimator is a little bit more involved. For simplicity, using Curie-Cohen's notation [y = observed number of heterozygotes and x = an estimator of the expected number of heterozygotes, for a specific allele = $2N_i x_{ik}(1 - x_{ik})$], we get

$$\text{Cockerham's estimator (equation (3.18)} - \text{Nei's unbiased estimator}$$
$$\text{(equation 3.5)} = y/(2N_i x)\cdot\text{Cockerham's estimator (equation (3.18))}.$$
$$(3.23)$$

Hence, when Cockerham's estimator is negative we have the string of inequalities

$$\text{equation } (3.5) \geqslant \text{equation } (3.18) \geqslant \text{equation } (3.5a),\qquad(3.24)$$

i.e., Cockerham's estimator is bounded by Nei's biased and unbiased estimators.
   However, when Cockerham's estimator is positive, from equation (3.23) we
have

$$\text{equation } (3.18) \geqslant \text{equation } (3.5) \geqslant \text{equation } (3.5a),\qquad(3.25)$$

i.e., Nei's unbiased estimator is bounded by his biased estimator and that of
Cockerham.
   These inequalities also hold for locus-specific estimators, irrespective of the
number of alleles and allele frequencies. To our knowledge, this mathematical
relationship among these three estimators has not been demonstrated before.
Since the expected differences are of the order of inverse of the number of genes
sampled $(2N_i)$ in a subpopulation, they are generally much smaller than their
standard errors, which will be shown later.
   It is worthwhile to note that while Nei's (1977) or Nei and Chesser's (1983)
estimate of F_{STk} (see equation (3.8) or (3.8a)) is only a function of allele frequencies in all subpopulations, Weir and Cockerham's (1984) estimator of F_{STk}
also depends on the frequencies of observed heterozygosity for the A_k allele in the
sample.
   Weir and Cockerham (1984) also gave explicit expressions for approximations
for these general estimators under several special cases. In particular, they note
that when the N_i's are large, the above estimators take the form

$$\hat{F}'_{ITk} = 1 - \frac{[1 - C^2/s]\bar{h}(\hat{w})}{2[1 - C^2/s]\bar{x}_{.k}(\hat{w})\{1 - \bar{x}_{.k}(\hat{w})\} + 2[1 + (s - 1)/s \cdot C^2]s_k^2(\hat{w})/s},$$
$$(3.15b)$$

$$\hat{F}'_{ISk} = 1 - \frac{\bar{h}(\hat{w})}{2\bar{x}_{.k}(\hat{w})\{1 - \bar{x}_{.k}(\hat{w})\} - 2(s - 1)s_k^2(\hat{w})/s},\qquad(3.16b)$$

and

$$\hat{F}'_{STk} = \frac{s_k^2(\hat{w})}{[1 - C^2/s]\bar{x}_{.k}(\hat{w})\{1 - \bar{x}_{.k}(\hat{w})\} + [1 + (s - 1)/s \cdot C^2]s_k^2(\hat{w})/s},$$
$$(3.17b)$$

in which $\hat{F}'_{STk}$ can be calculated only from allele frequency data. In addition to
the N_i's being large, if s (the number of subpopulations) is also large, Weir–
Cockerham's estimate of F_{STk} takes the well known form of

$$\hat{F}''_{STk} = s_k^2/\bar{x}_{.k}(1 - \bar{x}_{.k}).$$

Note that, while the general estimator of F_{STk} in Cockerham's approach depends upon the genotype frequencies (equation (3.17a)), its large sample approximation (equation (3.17b)) is only dependent on allele frequencies.

Weir and Cockerham (1984) suggested that locus-specific estimators for F_{IS}, F_{IT}, and F_{ST} can be derived by summing $\hat{a}_k$, $\hat{b}_k$, and $\hat{c}_k$ over all alleles, so that

$$\hat{F}_{IS} = \sum_{k=1}^{r} \hat{b}_k \bigg/ \sum_{k=1}^{r} (\hat{b}_k + \hat{c}_k), \tag{3.26}$$

$$\hat{F}_{IT} = \sum_{k=1}^{r} (\hat{a}_k + \hat{b}_k) \bigg/ \sum_{k=1}^{r} (\hat{a}_k + \hat{b}_k + \hat{c}_k), \tag{3.27}$$

and

$$\hat{F}_{ST} = \sum_{k=1}^{r} \hat{a}_k \bigg/ \sum_{k=1}^{r} (\hat{a}_k + \hat{b}_k + \hat{c}_k). \tag{3.28}$$

Although other methods of pooling data of multiple alleles exist (e.g., Wright, 1965; Kirby, 1975; Robertson and Hill, 1984), Weir and Cockerham (1984) advocate that the method presented above (equations (3.26)–(3.28)) is more appropriate for ratio estimators (see also Reynolds et al., 1983).

Note that since the parametric value of b_k can be negative (see equation (2.16)), it is quite possible that in this approach $\hat{F}_{ST}$ can often exceed $\hat{F}_{IT}$. Van Den Bussche et al. (1986) also noted that negative estimates of F_{STk} (or F_{ST}) can arise in Weir–Cockerham's approach when the following inequality holds:

$$s_k^2(\hat{w}) < \frac{1}{N-1} \left[\bar{x}_{\cdot k}(\hat{w}) \{1 - \bar{x}_{\cdot k}(\hat{w})\} - \frac{s-1}{s} s_k^2(\hat{w}) - \tfrac{1}{4}\bar{h}(\hat{w}) \right]. \tag{3.29}$$

While it is possible that Nei and Chesser's (1983) estimator of F_{ST} can also be negative (where $\hat{H}_S > \hat{H}_T$ occur), several simulation studies show that the negative estimates of F_{ST} are more common in the variance component approach (Chakraborty and Leimar, 1987; Van Den Bussche et al., 1986; Slatkin and Barton, 1989).

Finally, equations (3.26)–(3.28) can be extended to obtain pooled estimators of all indices, summing the numerators and denominators over all alleles over several loci (see equation (10) of Weir and Cockerham, 1984, p. 1364).

3.4. Long's estimators for multiple alleles and multiple loci

Long (1986) provided an interesting extension of Cockerham's approach for multiple alleles. He noted that when multiple alleles ($r > 2$) are involved at a locus, summation of a_k, b_k, and c_k over alleles (as suggested by Weir and Cockerham, 1984) ignores the correlation of allele and genotype frequencies (that is inherent in a multinomial sampling of genotypes) within subpopulations. Although this idea is imbedded in Weir–Cockerham's work (see their Appendix, termed as matrix estimation method), the formulation is explicitly stated in Long (1986) in terms of the decomposition of multivariate dispersion matrices. The parameters, as defined by equations (2.18)–(2.20), can be estimated substituting the estimators for the Σ_a, Σ_b, and Σ_c matrices. Long (1986) provided computational formulae for such estimators (see Appendix of Long, 1986) which involve the genotype and allele counts within each subpopulation and their totals over all subpopulations.

Since there are several misprints in the formulae in Long's (1986) paper (see pp. 646–647), we present the general estimation procedure for a r-allelic codominant locus. This has two purposes: first, this exposition clearly indicates how Weir and Cockerham's (1984) expressions have their natural multivariate extensions and, second, it will indicate why Long's algorithm gives numerical results different from those of Weir and Cockerham for a multiallelic locus ($r > 2$). Furthermore, we derive here the explicit closed expressions for the Σ_a, Σ_b, and Σ_c matrices, that are not available in Long (1986). For a single subpopulation, closed expressions for $\Sigma_b + \Sigma_c$ matrix are also shown through this exposition.

For a specific subpopulation, when an estimator for F_{IS} is sought (in parallel to F_{ISik} estimator, as done for Nei's and Cockerham's method earlier—only difference being in Long's procedure we need a different pooling algorithm over all alleles), a multivariate variance–covariance decomposition can be done in analogy of Table 3 of Cockerham (1973). The within-individual mean-square cross-product matrix (MSCP) S_c (equivalent to S_{wk} of Cockerham) for the i-th subpopulation takes the form, whose k-th diagonal element,

$$h_{ik}/2N_i, \quad \text{where } h_{ik} = \sum_{l > k = 1}^{r} N_{ikl},$$

is the observed number of heterozygotes with reference to the A_k-allele in the i-th subpopulation, and the (k, l)-th off-diagonal element of the S_c matrix is $-h_{ikl}/2N_i$, where $h_{ikl} = N_{ikl}$, the observed number of $A_k A_l$ heterozygotes in the i-th subpopulation.

Algebraic manipulation of the MSCP matrix for between individual source of variation, S_b matrix has:

$$k\text{-th diagonal element} = \frac{4N_i x_{ik}(1 - x_{ik}) - h_{ik}}{2(N_i - 1)} \tag{3.30a}$$

and

$$(k, l)\text{-th off-diagonal element} = \frac{h_{ikl} - 4N_i x_{ik} x_{il}}{2(N_i - 1)}, \tag{3.30b}$$

where the x_{ik}'s are as defined in equation (3.2).

These matrices are square matrices of dimension $r - 1$, since the linear constraint of allele frequencies (summation of all allele frequencies at a particular locus being one) has to be used in order to make such matrices non-singular (a requirement needed for the computations done in the sequel).

Estimator of Σ_b matrix (variance–covariance component due to between-individual source of variation) is obtained as

$$\hat{\Sigma}_b = \tfrac{1}{2}[\text{MSCP}(b) - \text{MSCP}(c)] = \tfrac{1}{2}[S_b - S_c], \tag{3.31}$$

since

$$E[\text{MSCP}(b)] = \Sigma_c + 2\Sigma_b \quad \text{and} \quad E[\text{MSCP}(c)] = \Sigma_c$$

(see Cockerham (1973, p. 688).
    Therefore, $\hat{\Sigma}_b$ matrix has the form, whose

$$k\text{-th diagonal element} = \frac{4N_i^2 x_{ik}(1 - x_{ik}) - (2N_i - 1)h_{ik}}{4N_i(N_i - 1)} \tag{3.32a}$$

and

$$(k, l)\text{-th element} = \frac{h_{ikl}(2N_i - 1) - 4N_i^2 x_{ik} x_{il}}{4N_i(N_i - 1)}, \tag{3.32b}$$

for $k, l = 1, 2, \ldots, r - 1$.
    In order to estimate $F_{\text{IS}i}$, we need the matrix $\hat{\Sigma}_b + \hat{\Sigma}_c$, whose

$$k\text{-th diagonal element} = \frac{4N_i^2 x_{ik}(1 - x_{ik}) - h_{ik}}{4N_i(N_i - 1)} \tag{3.33a}$$

and

$$(k, l)\text{-th element} = \frac{h_{ikl} - 4N_i^2 x_{ik} x_{il}}{4N_i(N_i - 1)}, \tag{3.33b}$$

for $k, l = 1, 2, \ldots, r - 1$.
    With these computations, the estimator for $F_{\text{IS}i}$ is

$$\hat{F}_{\text{IS}i} = \frac{1}{r - 1} \operatorname{tr}[(\hat{\Sigma}_b + \hat{\Sigma}_c)^{-1/2} \hat{\Sigma}_b (\hat{\Sigma}_b + \hat{\Sigma}_c)^{-1/2}]. \tag{3.34}$$

Although no closed explicit expression for $\hat{F}_{\text{SI}i}$ can be given in general (for $r > 2$), the explicit expressions for the elements of $\hat{\Sigma}_b + \hat{\Sigma}_c$ and $\hat{\Sigma}_c$ matrices are instructive to understand why the numerical values of Long's estimators are different from Weir–Cockerham's estimators. For example, even if all off-diagonal elements are neglected, equation (3.34) would yield

$$\hat{F}'_{\text{IS}i} = \frac{1}{r - 1} \sum_{k=1}^{r-1} \left[\frac{4N_i^2 x_{ik}(1 - x_{ik}) - (2N_i - 1)h_{ik}}{4N_i^2 x_{ik}(1 - x_{ik}) - h_{ik}} \right]$$

$$= 1 - \frac{1}{r - 1} \sum_{k=1}^{r-1} \left[\frac{2(N_i - 1)h_{ik}}{4N_i^2 x_{ik}(1 - x_{ik}) - h_{ik}} \right], \tag{3.34a}$$

whereas Weir and Cockerham's (1984) algorithm would suggest the computation of

$$\hat{F}''_{ISi} = 1 - \left[\sum_{k=1}^{r} 2(N_i - 1)h_{ik} \right] \bigg/ \left[\sum_{k=1}^{r} [4N_i^2 x_{ik}(1 - x_{ik}) - h_{ik}] \right].$$

$$\tag{3.34b}$$

While for a bi-allelic locus ($r = 2$), equations (3.34), (3.34a), and (3.34b) are identical, there are a number of practical limitations of equation (3.34) which are worth noting. For instance, suppose that there are multiple alleles ($r > 2$) in the total population, but in each subpopulation one or several are not present (either in the sample, or in the subpopulation as a whole), and the missing alleles vary across subpopulations. In such an event, for each subpopulation the S_b and S_c matrices will be of different dimension, and would refer to different sets of alleles. Therefore, in the strict sense F_{ISi} values computed from equation (3.34) cannot be contrasted across subpopulations, since they are based on different sets of alleles even when they belong to the same locus.

Nevertheless, the large sample estimator for F_{ISi}, following the matrix method has a closed form, not noted by Weir and Cockerham (1984) or Long (1986). Note that when the N_i's are large, ignoring terms of the order N_i^{-2}, we have

$$(\hat{\Sigma}_b + \hat{\Sigma}_c)_{kl} = \begin{cases} x_{ik}(1 - x_{ik}) & \text{for } k = l, \\ - x_{ik} x_{il} & \text{for } k \neq l, \end{cases} \qquad \begin{array}{l} (3.33c) \\ (3.33d) \end{array}$$

for $k, l = 1, 2, \ldots, r - 1$ at a locus.

The (k, l)-th element of the $(\hat{\Sigma}_b + \hat{\Sigma}_c)^{-1}$ matrix has the form

$$(\hat{\Sigma}_b + \hat{\Sigma}_c)_{kl}^{-1} = \begin{cases} 1/x_{ik} + 1/x_{ir} & \text{for } k = l, \\ 1/x_{ir} & \text{for } k \neq l, \end{cases}$$

for $k, l = 1, 2, \ldots, r - 1$.

Therefore, if we estimate F_{ISi} by

$$\hat{F}'_{ISi} = (r - 1)^{-1} \text{tr}[(\hat{\Sigma}_b + \hat{\Sigma}_c)^{-1} \hat{\Sigma}_b],$$

it has a closed form

$$\hat{F}'_{ISi} = 1 - [2N_i(r - 1)]^{-1} \sum_{k=1}^{r} h_{ik}/x_{ik}$$

$$= (r - 1)^{-1} \left[\sum_{k=1}^{r} (x_{ik}/p_{ik}) - 1 \right], \tag{3.35}$$

while Cockerham's estimator, pooled over alleles has a large sample form given in equation (3.18a).

Note that equation (3.35) is identical to the estimator $\hat{f}_2$ used by Curie-Cohen (1982), although he arrived at this estimator by a different logic.

When several subpopulations are analysed together, nested multivariate variance–covariance analysis was performed by Long (1986), to obtain the estimators for three variance–covariance component matrices (VCCM's) as

$$S_c = \text{MSCP}(c), \tag{3.36}$$

$$S_b = \tfrac{1}{2}[\text{MSCP}(b) - \text{MSCP}(c)], \tag{3.37}$$

$$S_a = (1/2N_c)\,[\text{MSCP}(a) - \text{MSCP}(b)], \tag{3.38}$$

where N_c is as defined in equations (3.15)–(3.17). Here again, each of these matrices are square matrices of dimension $(r - 1)$. As in the univariate case (equations (3.15)–(3.17)), explicit closed forms of these three matrices can be written which are not given in Long (1986). Long's equation for the MSCP(c) matrix (called MSCP(W) in Long, 1986) for a three allelic locus has a misprint (see his equation on top of p. 647) which fails to show how such a matrix can be computed for a multi-allelic locus. If we write the (k, l)-th element of S_a, S_b, and S_c as a_{kl}, b_{kl}, and c_{kl}, respectively, algebraic manipulation yields

$$\hat{a}_{kk} = \frac{\overline{N}}{N_C}\left\{s_k^2(\hat{w}) - \frac{1}{\overline{N}-1}\left[\overline{x}._k(\hat{w})\{1 - \overline{x}._k(\hat{w})\}\right.\right.$$
$$\left.\left. - [(s-1)/s]s_k^2(\hat{w}) - \tfrac{1}{4}\overline{h}_k(\hat{w})\right]\right\}, \tag{3.37a}$$

$$\hat{a}_{kl} = \frac{\overline{N}}{N_C}\left\{s_{kl}(\hat{w}) - \frac{1}{\overline{N}-1}\left[\overline{x}._k(\hat{w})\overline{x}._l(\hat{w})\right.\right.$$
$$\left.\left. + [(s-1)/s]s_{kl}(\hat{w}) - \tfrac{1}{4}\overline{h}_{kl}(\hat{w})\right]\right\}, \tag{3.37b}$$

$$\hat{b}_{kk} = \frac{\overline{N}}{\overline{N}-1}\left\{\overline{x}._k(\hat{w})\{1 - \overline{x}._k(\hat{w})\} - [(s-1)/s]s_k^2(\hat{w})\right.$$
$$\left. - [(2\overline{N}-1)/4\overline{N}]\overline{h}_k(\hat{w})\right\}, \tag{3.38a}$$

$$\hat{b}_{kl} = \frac{\overline{N}}{\overline{N}-1}\left\{[(2\overline{N}-1)/4\overline{N}]\overline{h}_{kl}(\hat{w}) - \overline{x}._k(\hat{w})\overline{x}._l(\hat{w})\right.$$
$$\left. - [(s-1)/s]s_{kl}(\hat{w})\right\}, \tag{3.38b}$$

$$\hat{c}_{kk} = \tfrac{1}{2}\overline{h}_k(\hat{w}), \tag{3.39a}$$

$$\hat{c}_{kl} = \tfrac{1}{2}\overline{h}_{kl}(\hat{w}), \tag{3.39b}$$

for $k, l = 1, 2, \ldots, r - 1$, where $\overline{x}._k(\hat{w})$ and $s_k^2(\hat{w})$ are as defined in the context of equations (3.15)–(3.17), and

$$s_{kl}(\hat{w}) = \left[\sum_{i=1}^{s} N_i x_{ik} x_{il} - s\overline{N}\overline{x}._k(\hat{w})\overline{x}._l(\hat{w})\right]\Big/\overline{N}(s - 1)$$

is the covariance of the allele frequencies of A_k and A_l over all subpopulations; $\bar{h}_k(\hat{w})$, the observed heterozygote frequency of the A_k allele over subpopulations ($= s\bar{N} \sum_{i=1}^{s} h_{ik}$), and $\bar{h}_{kl}(\hat{w}) = \sum_{i=1}^{s} h_{ikl}/s\bar{N}$ is the average observed frequency of a specific heterozygote $A_k A_l$ over all subpopulations.

Note that equations (3.37a), (3.38a), and (3.39a) are identical to the A_k-allele specific variance components described by Weir and Cockerham (1984), while equations (3.37b), (3.38b), and (3.39b) are direct extensions of these with multinomial sampling of genotypes.

With these explicit general closed form expressions of the elements of S_a, S_b, and S_c matrices one can compute the F_{IS}, F_{IT}, and F_{ST} estimators:

$$\hat{F}_{IS} = \frac{1}{r-1} \operatorname{tr}[(S_b + S_c)^{-1/2} S_b (S_b + S_c)^{-1/2}], \tag{3.40}$$

$$\hat{F}_{ST} = \frac{1}{r-1} \operatorname{tr}[(S_a + S_b + S_c)^{-1/2} S_a (S_a + S_b + S_c)^{-1/2}], \tag{3.41}$$

and

$$\hat{F}_{IT} = \frac{1}{r-1} \operatorname{tr}[(S_a + S_b + S_c)^{-1/2} (S_a + S_b)$$
$$\times (S_a + S_b + S_c)^{-1/2}], \tag{3.42}$$

with far more ease than following Long's (1986) suggestion. Note that like the one-subpopulation situation, even if the off-diagonal elements ($\hat{a}_{kl}$, $\hat{b}_{kl}$, $\hat{c}_{kl}$) are neglected, instead of Weir–Cockerham's estimates (equations (3.26)–(3.28)), equations (3.40)–(3.42) take the respective forms

$$\hat{F}'_{IS} = \frac{1}{r-1} \sum_{k=1}^{r-1} \frac{\hat{b}_{kk}}{(\hat{b}_{kk} + \hat{c}_{kk})}, \tag{3.40a}$$

$$\hat{F}'_{ST} = \frac{1}{r-1} \sum_{k=1}^{r-1} \frac{\hat{a}_{kk}}{(\hat{a}_{kk} + \hat{b}_{kk} + \hat{c}_{kk})}, \tag{3.41a}$$

$$\hat{F}'_{IT} = \frac{1}{r-1} \sum_{k=1}^{r-1} \frac{\hat{a}_{kk} + \hat{b}_{kk}}{(\hat{a}_{kk} + \hat{b}_{kk} + \hat{c}_{kk})}, \tag{3.42a}$$

which perform worse than the estimators (3.26)–(3.28) in Weir and Cockerham's (1984) simulation experiments. Furthermore, the F_{IS} estimator obtained from equation (3.40) is not a weighted average of the subpopulation-specific $\hat{F}_{ISi}$ values obtained from equation (3.34) since for each specific subpopulation the matrices can have different dimensions for reasons stated earlier.

At this point it is worthwhile to mention that this multivariate extension has not been presented explicitly before. Although Weir and Cockerham (1984) found that the estimators by such matrix method have the smallest standard errors in comparison with various other estimators they examined, their computations of the matrix estimators are somewhat different from those of Long (1986). Instead of $\Sigma^{-1/2}\Sigma_a\Sigma^{-1/2}$, Weir and Cockerham used $\Sigma^{-1}\Sigma_a$. Since the Σ matrices, as well as their estimators, are always symmetric square matrices, it is not clear why Long's procedure of pre- and post-multiplication with $-\frac{1}{2}$ power of the $S_a + S_b + S_c$ or $S_b + S_c$ matrices is needed. In fact, since such estimators can be computed only for non-singular $S_a + S_b + S_c$ and $S_b + S_c$ matrices, if we define the fixation indices by

$$F_{IT} = (r-1)^{-1}\,\mathrm{tr}[\Sigma^{-1}(\Sigma_a + \Sigma_b)]\,, \tag{2.18a}$$

$$F_{ST} = (r-1)^{-1}\,\mathrm{tr}[\Sigma^{-1}\Sigma_a]\,, \tag{2.19a}$$

$$F_{IS} = (r-1)^{-1}\,\mathrm{tr}[(\Sigma_b + \Sigma_c)^{-1}\Sigma_b]\,, \tag{2.20a}$$

instead of equations (2.18)–(2.20), only matrix-inversion routines are needed as opposed to the evaluation of eigen values and eigen vectors and inverse computations of the eigen vector matrices that are required in Long's algorithm.

Like the Weir and Cockerham estimator of F_{ST} (equation (3.28)), the estimator given by equation (3.41) also depends on the observed frequency of heterozygotes (see equations (3.37a) and (3.37b)) in addition to allele frequency data, which makes these estimators qualitatively different from that in Nei's approach (equation (3.8a)). Since in most practical situations the off-diagonal elements (a_{kl}, b_{kl}, c_{kl} for $k \neq l$) are small, because the subpopulations are sampled independently; the complexity of computations can be greatly reduced when Weir–Cockerham estimators are computed (according to equations (3.26)–(3.28)) for multi-allelic loci in the variance–component approach to estimation.

3.5. Estimation where genotype data are not available

Sometimes population structure analyses may have to be done in the absence of genotype data. Such is the case where the population structure is to be inferred from the allele frequency data reported in the literature, or the allele frequencies are estimated from phenotypic data at loci where complex dominance relationships exist among various alleles or haplotypes (e.g., ABO, Rh, and HLA system in man). Obviously, since such data do not provide any direct information regarding the observed number (or proportion) of homozygotes or heterozygotes, a somewhat different estimation procedure must be adopted.

In this case, Nei's approach can be easily adopted for estimating F_{ST}, since H_S and H_T parameters can be obtained simply from the estimated allele frequencies (with the assumption that the x_{ik}'s are multinomial proportions from a sample of $2N_i$ genes sampled from the i-th subpopulation). Equation (3.8) or its variant, equation (3.8a) with $w_i = 1/s$, is the estimator of preference here. Since F_{IS} is defined in terms of the deviation of genotype frequencies from their HWE expectations, no direct estimation of this quantity is possible. However, some approximate theory of estimation may be suggested.

Note that in the case of genotype data, the goodness-of-fit χ^2 statistic (of testing for deviation from HWE expectations) for a r-allelic locus is $\chi^2 = N_i(r - 1)F_{\text{IS}i}$ (Li, 1955), and hence an approximate absolute value of F_{IS} can be obtained from $\sqrt{\chi^2/N_i(r - 1)}$ where N_i is the number of individuals sampled from the i-th subpopulation. However, this suggested estimator is quite approximate, since for the loci in a dominance system, the goodness-of-fit statistic has a more complex parametric form (see Rao and Chakraborty, 1974). Furthermore, the sign of F_{IS} cannot be directly inferred from the χ^2 statistics. We advocate that for such data, only F_{ST} estimation is legitimate.

If one prefers the analysis of variance approach even the exact estimation of F_{ST} is not possible, unless large sample approximations are made. This is so because Weir and Cockerham's (1984) estimator of F_{ST} requires estimation of the observed heterozygosity for each allele (see equation (3.15)) and so is the case with Long's approach (see equations (3.37a), (3.37b) and (3.41)). Under the assumption $F_{\text{IS}} = 0$ (random union of gametes within subpopulations), since $F_{\text{IT}} = F_{\text{ST}}$, Weir and Cockerham (1984, 1963) obtained the estimator

$$\hat{F}_{\text{ST}k} = \frac{s_k^2(\hat{w}) - \left[\overline{x}_{\cdot k}(\hat{w})(1 - \overline{x}_{\cdot k}(\hat{w})) - \dfrac{s-1}{s} s_k^2(\hat{w})\right]\Big/[2\overline{N} - 1]}{\left\{1 - \dfrac{2\overline{N}C^2}{(2\overline{N} - 1)s}\right\}\overline{x}_{\cdot k}(\hat{w})\{1 - \overline{x}_{\cdot k}(\hat{w})\} + \left\{1 + \dfrac{2\overline{N}(s - 1)C^2}{(2\overline{N} - 1)s}\right\}\dfrac{s_k^2(\hat{w})}{s}},$$

(3.43)

where $\overline{x}_{\cdot k}(\hat{w})$ and $s_k^2(\hat{w})$ are the weighted mean and variance of the A_k-allele frequency over all subpopulations (defined in equations (3.15)–(3.17)), and C^2 is the coefficient of variation of N_i's over all subpopulations (note that $1 - C^2/s = N_c$, where N_c is as defined in equations (3.15)–(3.17)). Clearly this estimator depends only on allele frequency data. Therefore, the analysis of variance approach, when applied to allele frequency data, also yields a consistent estimator for F_{ST} under the assumption that $F_{\text{IS}} = 0$. For large sample sizes, this assumption is, however, not needed (see equation (3.17b)).

When all subpopulations have the same sample size (i.e., $N_i = N$), equation (3.43) takes the form

$$\hat{F}_{\text{ST}k} = \frac{s_k^2 - \{\overline{x}_{\cdot k}(1 - \overline{x}_{\cdot k}) - [(s - 1)/s]s_k^2\}/(2\overline{N} - 1)}{\overline{x}_{\cdot k}(1 - \overline{x}_{\cdot k}) + s_k^2/s},$$

(3.43a)

which reduces to the well known formula $s_k^2/\overline{x}_{\cdot k}(1 - \overline{x}_{\cdot k})$ when $\overline{N}$ and s are large.

4. Standard errors and hypothesis testing

The discussions in the earlier sections clearly indicate that the problem of estimation of the fixation indices arises because these are defined as ratios of
functions of allele and genotype frequencies in the subpopulations, and hence,
strictly speaking none of the estimators suggested above can be claimed most
efficient. We arrived at consistent estimators by estimating the numerators and
denominators by their respective unbiased statistics. Although several expressions
for the standard errors of these estimators are suggested, and the question of
hypothesis testing has been addressed in a variety of ways, we agree with
Cockerham (1973) that such procedures are on much less sound grounds than
estimation. Nevertheless, since all estimators derived above are of the general
form $\hat{\theta} = t_1/t_2$, where t_1 and t_2 are the estimators of the numerators and the
denominators of the respective fixation index parameters, using Taylor's expansion (Kendall and Stuart, 1977, p. 247), an approximate formula for the
variance of $\hat{\theta}$ can be written as

$$V(\hat{\theta}) \approx \left[\frac{E(t_1)}{E(t_2)}\right]^2 \left[\frac{V(t_1)}{E^2(t_1)} + \frac{V(t_2)}{E^2(t_2)} - \frac{2\,\text{Cov}(t_1, t_2)}{E(t_1) \cdot E(t_2)}\right], \qquad (4.1)$$

where $E(\cdot)$, $V(\cdot)$, and $\text{Cov}(\cdot,\cdot)$ represent the expectation, variance, and covariance of the respective statistics.

For the analysis of data from a single subpopulation, where only F_{IS} is to be
estimated, Curie-Cohen (1982) derived the sampling variance of such estimators.
As shown earlier, the estimators by Nei's and Cockerham's approach (equations
(3.5) and (3.18)) are related to Curie-Cohen's (1982) estimator $\hat{f}_1 = 1 - (y/x)$,
for which he derived a general expression for $\text{Var}(\hat{f}_1)$ at a multi-alleleic
codominant locus. His expression (equation (5); Curie-Cohen, 1982, p. 345) can
be further reduced to

$$V(\hat{f}_1) = \frac{(1 - F_{IS})\,[(1 - \mu_2) + (1 - F_{IS})\,(1 - \mu_2)^2 - (1 - F_{IS})^2\,(\mu_3 - \mu_2^2)]}{n(1 - \mu_2)^2}, \qquad (4.2)$$

where $\mu_2 = \sum_{k=1}^{r} p_{ik}^2$ and $\mu_3 = \sum_{k=1}^{r} p_{ik}^3$, are parameters that depend upon the
true allele frequencies at a locus. In practice the estimates of F_{IS}, μ_2, and μ_3 based
on sample statistics can be used to estimate $V(\hat{f}_1)$. Using our equation (3.5b),
we may immediately note that Nei's unbiased estimator of F_{ISik} has a sampling
variance

$$[1 - 1/(2N_i)]^2\,V(\hat{f}_1) \qquad (4.2a)$$

while, Cockerham's estimator (equation (3.18)) has the variance

$$\frac{(1 - F_{IS})\,[(\mu_2 - 2\mu_3 + \mu_2^2) + F_{IS}(1 - 2\mu_2 + 4\mu_3 - 3\mu_2^2) - 2F_{IS}^2(\mu_3 - \mu_2^2)]}{N_i(1 - \mu_2)^2}, \qquad (4.2b)$$

in which terms of the order $(2N_i)^{-2}$ or less are neglected.

As mentioned earlier, for large samples $F_{\mathrm{IS}i}$ at a locus, estimated by Long's procedure, is identical to the estimator $\hat{f}_2$, used by Curie-Cohen (1982). Since he derived its sampling variance (equation (7); Curie-Cohen, 1982, p. 346), in our notation for Long's estimator we get

$$V(\hat{F}_{\mathrm{IS}}) = \frac{1 - F_{\mathrm{IS}}}{2N_i(r - 1)^2} \left[2(r - 1) - 2(2r - 1)F_{\mathrm{IS}} + r^2 F_{\mathrm{IS}}^2 \right.$$

$$\left. + F_{\mathrm{IS}}(2 - F_{\mathrm{IS}}) \sum_{k=1}^{r} 1/p_{ik} \right]. \qquad (4.2c)$$

Equations (4.2), (4.2a), (4.2b), and (4.2c), therefore provide the approximate sampling variance of Nei's biased, Nei's unbiased, Cockerham's and Long's estimator for F_{IS} for a specific subpopulation, for any general multi-allelic codominant locus. When estimators of a specific allele are sought, the equations (4.2), (4.2a), and (4.2b) can be used taking $r = 2$, as shown for a specific case by Curie-Cohen (1982).

Although for a given sample, these sampling variances are to be evaluated with sample estimates of $F_{\mathrm{IS}ik}$, μ_2, and μ_3; it is possible to compare the relative efficiencies of Nei's unbiased (equation (3.5)), Nei's biased (equation (3.5a)), Cockerham's estimates (equation (3.18)), and its multivariate extension (equation (3.35)) by contrasting their sampling variances for known parametric values of F_{IS}, μ_2, and μ_3.

Equation (4.2a) suggests that when these parameters are fixed, Nei's unbiased estimator has a smaller sampling variance than the biased estimator. Of course, in reality, when estimates are used in variance evaluation this might not occur in a given set of data (since F_{IS} estimates would differ for these two estimators).

Note that for a bi-allelic locus (with allele frequencies p and q), equations (4.2), (4.2b), and (4.2c) all take the common form

$$N_i V(\hat{F}_{\mathrm{IS}}) \approx \frac{1 - F_{\mathrm{IS}}}{2pq} [2pq + 2(1 - 3pq)F_{\mathrm{IS}} - (p - q)^2 F_{\mathrm{IS}}^2], \qquad (4.3)$$

suggesting that the large-sample standard errors of Nei's biased estimator Cockerham's estimator, and Long's estimator are all identical to the extent that the terms of the order $(1/2N_i)^{-1}$ or less are neglected. Equation (4.3) is also identical to equation (3) of Curie-Cohen (1982).

To our knowledge, this equivalence of the standard errors of the different F_{IS} estimators has not been demonstrated before. In view of this mathematical equivalence, one might wonder why the empirical values of the standard errors of the different estimators vary in the simulation experiments of Weir and Cockerham (1984), Van Den Busche et al. (1986), and Chakraborty and Leimar (1987). Note that the standard error of F_{IS} is dependent on the true value of F_{IS} and the allele frequencies at a locus (equation (4.3)). Hence, in the computation of the empirical values of the standard errors it is customary to replace the true values of the parameters by their respective estimates (i.e., $\hat{F}_{IS}$ is substituted for F_{IS}). Since we have shown earlier that the estimates differ depending upon the method of estimation satisfying the inequalities (3.24) and (3.25), it is obvious that the same analytical formula for variance (evaluated by equation (4.3)) will give different values when F_{IS} is replaced by its different estimates.

In order to study the empirical differences in the standard errors, it is therefore important to see how expression (4.3) varies as a function of F_{IS}. Curie-Cohen (1982) examined this in his Figure 1 (for a two allelic locus) and Figures 5 and 7 (for two different three allelic loci). His Figure 1 is somewhat confusing, since expression (4.3) does not decrease to zero as F_{IS} approaches its lower limit ($-p/q$ for $q > p$). Substituting $F_{IS} = -p/q$, it reduces to $p(q - p)/2q^4$, which is zero only if $p = q$. In Figure 1, we therefore plotted $\{N_i V(\hat{F}_{IS})\}^{1/2}$ as a function of F_{IS} for four values of p ($p = 0.01$, 0.1, 0.25, and 0.5). It is clear that for $F_{IS} = 0$, $V(\hat{F}_{IS}) = 1/N_i$, irrespective of the allele frequencies at a bi-allelic locus. In general, $V(\hat{F}_{IS})$ is a cubic function of F_{IS}, which attains its maximum at a value of F_{IS} depending upon the allele frequencies at the locus. When the allele frequencies are very skewed (p close to zero or one), the curve rises very fast for negative values of F_{IS}, and similarly drops fast when F_{IS} approaches one. Since Cockerham's estimator (equation (3.18)) is always larger than Nei's biased estimator (equation (3.5a)), unless the true value of F_{IS} is large, substitution of the respective estimates will yield smaller standard error for Nei's biased estimator as compared to that of Cockerham's estimator. The nature of the curves in Figure 1 indicate that such is the case for negative values of F_{IS}, irrespective of the allele frequencies at the locus. In theory, the situation can be reversed for large positive F_{IS}. But, since large positive estimates of F_{IS} are rare in natural populations (unless the organism is highly inbred), this theoretical possibility is not commonly seen. For skewed allele frequencies, the difference in the empirical values of the standard errors can be substantial, because of the sharp rise of the curve. We therefore claim that the observed discrepancies in the standard errors of the various estimators of F_{IS} are the artifacts of substituting the estimates in the variance formula (equation (4.3)). Indeed, there is no inherent difference in the standard errors, as seen in the analytical formulae established here.

Another comment regarding the standard error evaluation of Long's large-sample estimator of F_{IS} (or f_2 of Curie-Cohen, 1982) is worth mentioning at this point. Note that for a multi-allelic locus, this estimator is defined by contrasting the observed proportion of the homozygosity of each allele with the respective allele frequency (equation (3.35)). However, when equation (4.2c) is used to evaluate its standard error $\{V(\hat{F}_{IS})\}^{1/2}$, substituting $\hat{p}_{ik}$ for p_{ik} and $\hat{F}_{IS}$ for F_{IS}, one might encounter negative variance estimators, particularly when one (or more) allele is rare in the population, and $\hat{F}_{IS}$ is negative. In the application section to follow, we have several situations when it occurred. There does not appear to be any simple solution to circumvent this problem of a negative variance estimate. We simply note that the substitution of estimates for parameters (e.g., $\hat{F}_{IS}$ for F_{IS}

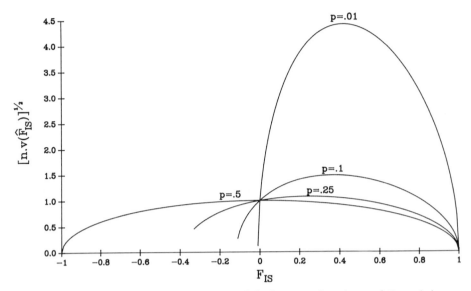

Fig. 1. Relationship between the sampling error of the large-sample estimate of F_{IS} and the true value of parameter (F_{IS}), as studied by plotting $\{n \, \mathrm{Var}(\hat{F}_{IS})\}^{1/2}$ versus F_{IS} for a bi-allelic codominant locus with allele frequencies p and q ($= 1 - p$).

and $\hat{p}_{ik}$ for p_{ik}) in equation (4.2c) yields a poor estimate of $V(\hat{F}_{IS})$ because of the inverse function of $\hat{p}_{ik}$'s (last term of equation (4.2c)).

One solution to this problem, admittedly an ad-hoc one, is to note that when some alleles are rare, since they generally appear in a sample only as heterozygotes, they do not contribute to the estimate of F_{IS} (equation (3.35)). They can be deleted in the variance computation, which is equivalent to computing the $\Sigma \, (1/\hat{p}_{ik})$ term only for alleles that contribute to the estimate of F_{IS}. This avoids the occurrence of a negative variance estimate, as seen in our empirical study. Obviously, more work is needed to provide a justifiable estimator for the standard error of $\hat{F}_{IS}$ in such situations.

The mathematical equivalence of the standard errors shown here apply only for bi-allelic loci. For a general multi-allelic locus such comparisons are more difficult, since the variances also depend on the sum of squares and cubes of allele frequencies (see equations (4.2), (4.2b), and (4.2c)). Nevertheless, for a r-allelic locus with equal gene frequencies (i.e., $p_{ik} = 1/r$ for all k), we have

$$n \, V(\hat{F}_{IS}) = \frac{(1 - F_{IS}) \, [1 + (r - 1)F_{IS}]}{r - 1} \tag{4.4}$$

which holds for all of these estimators.

When data from several subpopulations are jointly used for parameter estimation, equation (4.1) can again be used to obtain approximate variances of these estimators, in which case the variances and covariances reflect the inter-locus variation and covariation of the observed statistics. Chakraborty (1974) was the first to use this idea to evaluate the sampling variance of $\hat{F}_{ST}$, which he represented by

$$V(\hat{F}_{ST}) \approx F_{ST}^2 \left[\frac{V(\hat{H}_S)}{H_S^2} + \frac{V(\hat{H}_T)}{H_T^2} - \frac{2 \, \text{Cov}(\hat{H}_S, \hat{H}_T)}{H_S H_T} \right], \tag{4.5}$$

where the variances and covariances of $\hat{H}_S$ and $\hat{H}_T$ [$V(\hat{H}_S)$, $V(\hat{H}_T)$, and $\text{Cov}(\hat{H}_S, \hat{H}_T)$] are obtained from inter-locus variations of these statistics. While Nei and Chakravarti (1977) demonstrated that the equation (4.5) is approximately adequate, Weir and Cockerham (1984) advocated a jackknife procedure in this context (Miller, 1974; Efron, 1982). In principle, if $\hat{\theta}$ represents an estimator of a parameter θ (not to be confused with Cockerham's notation), based on n observations, then the jackknife variance of $\hat{\theta}$ can be expressed as

$$V(\hat{\theta}) \approx \frac{n-1}{n} \sum_{i=1}^{n} \left[\hat{\theta}(i) - \frac{1}{n} \sum_{i=1}^{n} \hat{\theta}(i) \right]^2, \tag{4.6}$$

where $\hat{\theta}(i)$ is the estimator based on $(n-1)$ observations, omitting the i-th observation. If $\hat{\theta}$ involves some bias in estimating θ (as is the case with ratio estimators), a less biased estimator of θ is

$$\hat{\theta}^* = n\hat{\theta} - [(n-1)/n] \sum_{i=1}^{n} \hat{\theta}(i). \tag{4.7}$$

This technique is adopted in estimating the standard errors of the F_{IS}, F_{IT}, F_{ST} estimators of the variance–component approach by Weir and Cockerham (1984), where jackknifing was done over loci (i.e., estimators of a, b, and c components were computed omitting one locus at a time). In particular, when the L-th locus data is omitted, the respective estimators for F_{IS}, F_{IT}, and F_{ST} used are

$$\hat{F}_{IT}(L) = \left[\sum_{l \neq L} \sum_{k} (\hat{a}_{lk} + \hat{b}_{lk}) \right] \Big/ \left[\sum_{l \neq L} \sum_{k} (\hat{a}_{lk} + \hat{b}_{lk} + \hat{c}_{lk}) \right], \tag{4.8}$$

$$\hat{F}_{IS}(L) = \left[\sum_{l \neq L} \sum_{k} \hat{b}_{lk} \right] \Big/ \left[\sum_{l \neq L} \sum_{k} (\hat{b}_{lk} + \hat{c}_{lk}) \right], \tag{4.9}$$

and

$$\hat{F}_{ST}(L) = \left[\sum_{l \neq L} \sum_{k} \hat{a}_{lk} \right] \Big/ \left[\sum_{l \neq L} \sum_{k} (\hat{a}_{lk} + \hat{b}_{lk} + \hat{c}_{lk}) \right]. \tag{4.10}$$

Note that the same approach can be adopted for Nei's estimation procedure as well, where $\hat{H}_S(L)$, $\hat{H}_T(L)$, and $\hat{H}_0(L)$ values are to be evaluated omitting the L-th locus data.

While the jackknifing over loci may provide standard errors of the estimator, pooled over loci, there has been no explicit formulation for evaluating the sampling errors of individual allele-specific estimators. There is no simple formula for the standard errors of the variance–component estimators for a particular allele, although the sampling theory of categorical analysis of variance (CATANOVA) developed by Light and Margolin (1971), or analysis of diversity (ANODIV) of Rao (1982), indicated in Nayak (1983) may be adopted in this context. Further work is needed to provide computational formulae in this regard.

In principle, under multinomial sampling of genotypes, sampling variances of estimators $\hat{F}_{ITk}$, $\hat{F}_{ISk}$, and $\hat{F}_{STk}$ (equations (3.6)–(3.8)) can be derived, following Nei and Roychoudhury (1974) and Nei (1978) which refer to the sampling variance computations of heterozygosities and genetic distance. No explicit form of the intra-locus standard errors of the fixation indices are yet available.

Although the utility of the estimators is greatly increased when such standard error evaluation is available, this does not immediately resolve hypothesis testing problems, because with categorical data such ratio estimators do not have simple sampling distributions. Nayak (1983) showed that while exact sampling distributions of the variance components (or sum of squares) are not available, in large samples (N_i's large), the mean square error terms can be represented by linear combinations of χ^2 variables. However, the coefficients of such linear combinations are again not estimable, and hence such theory is difficult to apply in practice.

Cockerham (1973) suggested some heuristic test criteria for specific hypotheses. His test criteria require notations somewhat different from the rest of this paper. In order to avoid confusion, let us introduce for each allele (A_k) at a locus, three genotypes $A_k A_k$, $A_k \overline{A}_k$, and $\overline{A}_k \overline{A}_k$, where $\overline{A}_k$ is the combination of alleles except the A_k allele. Let M_{ik2}, M_{ik1}, and M_{ik0} be the observed frequencies of these three genotypes in a sample of N_i individuals from the i-th subpopulation. Note that M_{ikl} represents the number of individuals with l copies ($l = 0, 1, 2$) of the A_k allele in the i-th subpopulation, and $M_{ik0} + M_{ik1} + M_{ik2} = N_i$ for $i = 1, 2, \ldots, s$. As before let $N = N_1 + N_2 + \cdots + N_s$, the total number of individuals in the entire survey.

Furthermore, let x_{ik} represent the estimated allele frequency of A_k in the i-th subpopulation, given by our equation (3.2), which is equivalent to

$$x_{ik} = (2M_{ik2} + M_{ik1})/2N_i, \tag{3.1a}$$

Under the hypothesis that $F_{ST} = 0$ and $F_{ISik} = 0$ for all i and k, the expectations of M_{ikl}'s are given by Cockerham (1973) as:

$$\overline{\eta}_{ik1} = E(M_{ik1}) = 2N_i[2N/(2N - 1)]\overline{x}_{.k}(1 - \overline{x}_{.k}), \tag{4.11}$$

where $\overline{x}_{.k} = \sum_{i=1}^{s} N_i x_{ik}/N$, the weighted average frequency of the A_k allele over all subpopulations,

$$\overline{\eta}_{ik2} = E(M_{ik2}) = N\overline{x}_{.k} - \tfrac{1}{2}\overline{\eta}_{ik1} \tag{4.12}$$

and

$$\bar{\eta}_{ik0} = E(M_{ik0}) = N(1 - \bar{x}._k) - \tfrac{1}{2}\bar{\eta}_{ik1}, \tag{4.13}$$

so that the deviation from $F_{\text{ST}} = 0$ and $F_{\text{IS}ik} = 0$ can be measured by the goodness-of-fit statistic

$$\chi_1^2 = \sum_{i=1}^{s} \sum_{l=0}^{2} (M_{ikl} - \bar{\eta}_{ikl})^2/\bar{\eta}_{ikl}, \tag{4.14}$$

which has a χ^2 distribution with d.f. $2s - 1$ ($2s$ independent genotypes, and one parameter, $\bar{p}._k$ being estimated).

The test-statistic for $F_{\text{IS}ik} = 0$ for all i and k, given by Cockerham (1973) is the sum-total of $s\chi^2$ values measuring deviations from HWE within individual sub-populations. However, since the unbiased estimator of the $A_k\bar{A}_k$ heterozygote proportions in the i-th subpopulation, is $2N_i x_{ik}(1 - x_{ik})/(2N_i - 1)$, under this hypothesis the expectations of M_{ikl}'s are given by

$$\hat{\eta}_{ik1} = E(M_{ik1}) = 4N_i^2 x_{ik}(1 - x_{ik})/(2N_i - 1), \tag{4.15}$$

$$\hat{\eta}_{ik2} = E(M_{ik2}) = N_i x_{ik} - \tfrac{1}{2}\hat{\eta}_{ik1}, \tag{4.16}$$

$$\hat{\eta}_{ik0} = E(M_{ik0}) = N_i(1 - x_{ik}) - \tfrac{1}{2}\hat{\eta}_{ik1}. \tag{4.17}$$

Departure from this hypothesis can be tested by the χ^2 statistic

$$\chi_2^2 = \sum_{i=1}^{s} \sum_{l=0}^{2} (M_{ikl} - \hat{\eta}_{ikl})^2 \bigg/ \hat{\eta}_{ikl}, \tag{4.18}$$

with s d.f.

Cockerham (1973) suggested $\chi_1^2 - \chi_2^2$ as the test criterion with d.f. $s - 1$ for testing the hypothesis $F_{\text{ST}} = 0$. While this may approximately hold for large samples, when the A_k-allele is rare in one or more subpopulations, because of small values of M_{ikl}, or $\bar{\eta}_{ikl}$, or $\hat{\eta}_{ikl}$ this approximation may not be accurate. Workman and Niswander (1970) suggested a more direct test for $F_{\text{ST}} = 0$ by the usual χ^2 test of heterogeneity (Rao, 1965, p. 323) which is commonly employed in most anthropogenetic studies (see, e.g., Chakraborty et al., 1977).

When $F_{\text{IS}ik}$ is assumed to be equal in all subpopulations, the test for $F_{\text{IS}} = 0$ (common value over all subpopulations) can also be tested with a χ^2 statistic. In this case, the expectations of M_{ikl}'s are computed as

$$\hat{\eta}_1 = E\left(\sum_{i=1}^{s} M_{ik1} \right) = 4N \sum_{i=1}^{s} x_{ik}(1 - x_{ik})/(2N - s), \tag{4.19}$$

$$\hat{\eta}_2 = E\left(\sum_{i=1}^{s} M_{ik2} \right) = N\bar{x}._k - \tfrac{1}{2}\hat{\eta}_1, \tag{4.20}$$

$$\hat{\eta}_0 = E\left(\sum_{i=1}^{s} M_{ik0} \right) = N(1 - \bar{x}._k) - \tfrac{1}{2}\hat{\eta}_1, \tag{4.21}$$

which yields

$$\chi_5^2 = \sum_{l=0}^{2} \left[\sum_{i=1}^{s} M_{ikl} - \hat{\eta}_l \right]^2 \bigg/ \hat{\eta}_l,$$

(4.22)

which also has a χ^2 distribution with one d.f.

In a similar vein, Cockerham (1973) suggested a test criterion for $F_{STk} = 0$ from allele frequency data, which takes the form

$$\chi_k^2 = \left[\sum_{i=1}^{s} 2N_i [x_{ik} - \bar{x}_{\cdot k}(\hat{w})]^2 \right] \bigg/ \{\bar{x}_{\cdot k}(\hat{w}) [1 - \bar{x}_{\cdot k}(\hat{w})]\},$$

(4.23)

with $(s - 1)$ d.f., for each specific allele A_k. Although for genotypic data, several alternative test criteria for F_{ST} exist, there is no definitive theory that suggests which should be the preferred one. We might note that expression (4.23) is the most commonly employed test criterion for F_{STk} in empirical studies of population structure (see also Workman and Niswander, 1970).

Although the test criteria (4.13), (4.18), (4.22) and (4.23) have their own intuitive appeal, Rao (1982) and Nayak (1983) showed that when the N_i's are not equal, these χ^2 statistics do not quite reflect an orthogonal decomposition of the total sum of squares in terms of a categorical analysis of variance. Further investigation is needed to address the question of most powerful test criteria in the analysis of such data. Furthermore, since these statistics refer to a single allele (A_k), a combined analysis for multiple allelic loci is not provided by these test criteria.

Long (1986) approached this problem while providing locus-specific estimates of the fixation indices. As shown earlier, Long's (1986) estimators are derived in terms of the three MSCP matrices: MSCP(a), MSCP(b), and MSCP(c), respectively (in Long's notation MSCP(c) = MSCP(W)). He suggested that the significance of F_{IS}, F_{IT} can be tested by

$$\Lambda_1^* = \det[\text{MSCP}(c)]/\det[\text{MSCP}(b)] \approx \Lambda(G, N - s, N),$$

(4.24)

$$\Lambda_2^* = \det[\text{MSCP}(b)]/\det[\text{MSCP}(a)] \approx \Lambda(G, s - 1, N - s),$$

(4.25)

and

$$\Lambda_3^* = \det[(N - 2) \text{MSCP}(b) + 2\text{MSCP}(a)]/\det[N \text{MSCP}(c)]$$

$$\approx \Lambda(G, N - 1, N),$$

(4.26)

respectively, where $\det(Z)$ is the determinant of a matrix Z, G is the dimension of S-matrices (number of independent alleles); and $\Lambda(\text{df}_1, \text{df}_2, \text{df}_3)$ is a Wilk's Λ variate with d.f. df_1, df_2, and df_3 (Anderson, 1984, p. 299).

Although the rationale of these test criteria results from the convergence of the multinomial to the multivariate normal distribution for fairly large sample sizes, there are several problems with these test statistics. First, for unequal sample sizes S_a, S_b, and S_c are not independently distributed (and so too are their respective

MSCP matrices). Nayak (1983) showed that their correlations can be quite substantial, and hence, the criteria Λ_1^*, Λ_2^*, and Λ_3^* do not satisfy the conditions under which Wilk's Λ distribution is valid (see equation (3) of Anderson, 1984, p. 299). Second, the assumption that a MSCP matrix follows a Wishart distribution is true for multivariate normal variates. A multinomial sampling of genotypes where one or more alleles are rare in the population and consequently may be absent in one or more subpopulations, will not approach multivariate normality unless the sample sizes are very large. Third, Wilk's Λ distribution approximation will also require a large number of subpopulations in addition to large N_i values. Since, in the earlier sections we showed that a great deal of work is needed to reduce bias due to small N_i and s values in estimating the fixation indices, the attempt to sweep out all these troubles by using such approximations cannot be generally advocated. Fourth and lastly, as indicated earlier, the variance–component approach may yield a negative-definite MSCP(B) matrix (see Cockerham, 1969, p. 74 for the univariate result), which also makes the Λ-distribution approximation invalid.

In summary, we argue that a rigorous test procedure for studying the significance of the fixation indices is not yet available. All suggested test criteria are only approximate, and caution must be exercised in interpreting their results.

5. An application

Bhasin et al. (1986) studied the genetic structure of the people of Sikkim of North India in order to determine the extent of genetic differentiation among the various subdivisions of their social units. They recognized 13 social groups in this population: North Sikkim, Sherpas, Tamangs, Gurungs, Rais, Limboos, Pradhans, Brahmins, Chhetris, and Scheduled Castes who are ethnohistorically as well as socially isolated to a certain extent. They studied 17 polymorphic blood groups and protein loci in each of these subpopulations. Of these loci, 11 are codominant. Haptoglobin (Hp), Group-Specific Component (Gc), Transferrin (Tf), Acid phosphatase (aP), Phosphoglucomutase-1 (PGM$_1$), 6-phosphogluconate dehydrogenase (6-PGD), Esterase D (EsD), Adenylate kinase (Ak), Hemoglobin (Hb), Duffy (Fy), and Kidd (Ik), at which the number of detected alleles vary from 2 to 5 (Gc and aP have 3 alleles each, Tf has 5 alleles, all of the remaining having 2 alleles each). The remaining six loci—AB0, MNSs, Rh, Kell and Immunoglobulin Gm and Km—have variable degrees of complex dominant relationships among their alleles/haplotypes, so that at each of these loci not all genotypes are distinguishable. The genotype/phenotype data at these loci for each subpopulation are presented in Bhasin et al. (1986).

We consider this survey for illustrating empirically the differences in the various estimators for several reasons. First, as part of a larger study of the extent of genetic differentiation among the populations of Sikkim, the estimates of the fixation indices provide the basis of our further studies, and hence it is important to determine their stability over different estimation methods employed. Second, as the sample sizes of this study drastically differ over subpopulations as well as over loci, this example should also provide insight regarding the stability of parameter estimates with or without invoking the large sample approximations discussed in our theoretical exposition. Finally, the availability of loci with and without dominance relationships in this study will help us to examine some features of the statistical properties of the parameter estimates based on genotype vs. allele frequency data, not commonly found in all applications of this nature. Notwithstanding these issues, we should note that since this review deals with a comparative study of the various estimation procedures, and *not* the population structure of the Sikkimese people, only the results pertaining to the comparative analyses are reported here.

5.1. Comparison of the estimates of F_{IS} in a single subpopulation

We have seen earlier that estimation of F_{IS} is possible from genotype data by several methods. Since only codominant loci can be used for this purpose, Table 1 contrasts the estimates of F_{IS} in the Lepchas of North Sikkim for 11 loci, as computed using equations (3.5) (Nei's unbiased estimator), (3.5a) (Nei's biased estimator) and (3.18) (Cockerham's estimator). Although several other estimators of F_{IS} are available in such situations (Li and Horvitz, 1953; Curie-Cohen, 1982; Robertson and Hill, 1984), we contend that the data presented in Table 1 are sufficient to contrast most of the theoretically justifiable estimators. Note that Curie-Cohen's (1982) estimator $\hat{f}_1$ is identical to Nei's biased estimator, and his $\hat{f}_2$ is exactly the same as the multivariate large sample estimate at a locus, while his $\hat{f}_3$ estimator can be computed from the χ^2 values presented in this table. We computed two different χ^2 values, one corresponding to Nei's unbiased estimator (evaluated by equation (4.18) with $s = 1$), and the other corresponding to the biased estimator (where the expected genotype frequencies are computed by $N_i x_{ik}^2$ for the genotype $A_k A_k$ and $2N_i x_{ik} x_{il}$ for the genotype $A_k A_l$). The χ^2 values for Cockerham's estimator are identical to those of Nei's unbiased estimator, and hence they are not repeated in the table. The allele-specific χ^2's have a single d.f. in every case, while the locus-specific χ^2's have d.f. $r(r - 1)/2$, where r is the number of segregating alleles at a locus in the specific subpopulation. Since for the bi-allelic loci the estimators and χ^2 values are identical for both alleles, and hence their locus-specific values are exactly the same as those based on any specific allele, only allele-specific estimates are given for such loci (Hp, PGM$_1$, PGD, EsD, Ak, Hb, Duffy, and Kidd in our example). Further note that although the transferrin locus has 5 segregating alleles in the total Sikkim population, in Lepchas of North Sikkim only 3 segregating alleles were found (Bhasin et al., 1986), and hence this was treated as a 3-allelic locus for this computation.

We chose not to present the estimator of F_{IS} based on χ^2 values for two reasons. First, since χ^2 values represent deviation from HWE for the two-sided alternative $F_{IS} \neq 0$, the sign of F_{IS} cannot be inferred from the value of χ^2 (Li and Horvitz, 1953; Curie-Cohen, 1982), and second, the χ^2 values can be greatly

Table 1
Allele-specific estimates of F_{ISik} for the Lepchas sampled in North Sikkim

Locus	Allele	N	Frequency	Nei's unbiased estimate F_{ISik} ± s.e.	χ^2	d.f.	Nei's biased estimate F_{ISik} ± s.e.	χ^2	d.f.	Cockerham's estimate F_{ISik} ± s.e.[a]	Long's estimate Weighted[b]	Large sample[c]
Hp	Hp¹	65	0.1154	0.177 ± 0.162	2.18	1	0.171 ± 0.163	1.90	1	0.179 ± 0.164	0.179 ± 0.127	0.171 ± 0.163
Gc	Gc¹ᶠ		0.6129	0.393 ± 0.119	9.75[f]	1	0.388 ± 0.120	9.34[f]	1	0.395 ± 0.120		
	Gc¹ˢ		0.2581	0.415 ± 0.130	10.97[g]	1	0.410 ± 0.131	10.44[g]	1	0.417 ± 0.131		
	Gc²		0.1290	0.004 ± 0.127	0.00	1	−0.005 ± 0.126	0.00	1	0.004 ± 0.128		
	Pooled	33		0.320 ± 0.106	13.70[f]	3	0.314 ± 0.107	13.44[f]	3	0.322 ± 0.107	0.233	0.225 ± 0.108
Tf	Tf^C1		0.7097	−0.087 ± 0.120	0.48	1	−0.096 ± 0.120	0.57	1	−0.088 ± 0.121		
	Tf^C2		0.2823	−0.066 ± 0.121	0.28	1	−0.075 ± 0.122	0.35	1	−0.067 ± 0.122		
	Tf^D		0.0081	0.000 ± 0.126	0.00	1	−0.008 ± 0.008	0.00	1	0.000 ± 0.127		
	Pooled	62		−0.075 ± 0.117	0.74	3	−0.084 ± 0.117	0.82	3	−0.076 ± 0.118	−0.037	−0.047 ± 0.124[d]
AP	P^a		0.1638	−0.061 ± 0.115	0.22	1	−0.070 ± 0.113	0.28	1	−0.061 ± 0.116		
	P^b		0.8190	0.020 ± 0.134	0.02	1	0.012 ± 0.133	0.01	1	0.020 ± 0.135		
	P^c		0.0172	−0.009 ± 0.093	0.01	1	−0.018 ± 0.012	0.02	1	−0.009 ± 0.094		
	Pooled	58		−0.018 ± 0.116	1.79	3	−0.027 ± 0.115	1.86	3	−0.018 ± 0.117	−0.028	−0.037 ± 0.122[d]
PGM₁	PGM₁¹	50	0.6600	−0.324 ± 0.115	5.37[e]	1	−0.337 ± 0.114	5.68[e]	1	−0.328 ± 0.115	−0.328 ± 0.115	−0.337 ± 0.114
PGD	PGD^A	53	0.8679	0.022 ± 0.143	0.03	1	0.012 ± 0.141	0.01	1	0.022 ± 0.144	0.022 ± 0.144	0.012 ± 0.141
EsD	EsD¹	50	0.7400	0.485 ± 0.139	12.17[g]	1	0.480 ± 0.141	11.53[g]	1	0.488 ± 0.141	0.488 ± 0.141	0.480 ± 0.141
AK	AK¹	58	0.9914	0.000 ± 0.130	0.00	1	−0.009 ± 0.009	0.00	1	0.000 ± 0.131	0.000 ± 0.131	−0.009 ± 0.009
Hb	Hb^A	61	0.9754	−0.017 ± 0.075	0.03	1	−0.025 ± 0.015	0.04	1	−0.017 ± 0.075	−0.017 ± 0.075	−0.025 ± 0.015
Duffy	Fy^a	66	0.8485	0.181 ± 0.149	2.27	1	0.175 ± 0.149	2.02	1	0.182 ± 0.150	0.182 ± 0.150	0.175 ± 0.149
Kidd	Ik^a	47	0.4681	−0.268 ± 0.139	3.45	1	−0.282 ± 0.139	3.73	1	−0.272 ± 0.140	−0.272 ± 0.140	−0.282 ± 0.139

a The χ^2 values for Cockerham's estimate of F_{ISik} are exactly the same as that for Nei's unbiased estimates.
b Long's weighted estimates are locus-specific estimates, which are identical to Cockerham's estimator for two-allelic loci.
c Long's large sample estimator is identical to that of Curie-Cohen's (1982) estimator f_2, and hence their standard errors are also the same (see text for details).
d These s.e. values are computed deleting the alleles that do not contribute to the estimate (see text for details).
e $p < 0.05$.
f $p < 0.01$.
g $p < 0.001$.

affected by rare genotypes and their expected values, giving unstable estimates in specific situations, an example of which will be discussed later in this section.

The standard errors of the three estimators are evaluated by equation (4.2) (for Nei's biased estimator), (4.2a) (for Nei's unbiased estimator) and (4.2b) (for Cockerham's estimator), where $r = 2$ is used for allele specific values, and the entire locus data used for locus-specific standard errors (represented by s.e. in the table).

In terms of the values of the estimates, it is clear that Nei's unbiased, biased, and Cockerham's estimates of F_{IS} are quite close to each other, with biased estimates always being the smallest. The differences of these estimates (the allele-specific ones as well as their pooled values over all alleles at a locus) are always encompassed by their respective standard errors (see Table 1).

The standard errors of Nei's unbiased and Cockerham's estimates are also very similar, while for negative $\hat{F}_{ISik}$ (or $\hat{F}_{ISi}$) values Cockerham's estimators have slightly larger standard errors, the situation reverses when the estimates are positive. The differences in the standard errors are however very small, and in no case change the qualitative results of hypothesis testing ($F_{IS} = 0$) either by the χ^2 value shown in the table, or by a crude test of the normal deviate $[\hat{F}_{ISik}/\text{s.e.}(\hat{F}_{ISik})]$–the latter test not explictly shown in this table. As noted earlier, for the bi-allelic loci the differences in the standard errors are produced only because of substituting the respective estimates of F_{IS} in equation (4.3). Curie-Cohen (1982) also showed that in multiallelic loci the standard errors of the various estimators are only slightly different (see Figure 5 of Curie–Cohen, 1982, p. 352).

A comment regarding the standard errors of Nei's biased estimators is worth noting. While these are quite close to those of Nei's unbiased and Cockerham's estimators, where the allele frequency is close to 0 or 1 (e.g., Tf^D, p^C, Ak^1, and Hb^A) the s.e.'s of the biased estimators are substantially smaller than those of Nei's unbiased and Cockerham's estimators. This feature may not be intuitively clear. Nevertheless, Figure 1 indicates that for skewed allele frequencies (p, small), the s.e. of Nei's biased estimate sharply rises from a very small value in the range of negative F_{IS} values. Since we evaluated the s.e. of each estimate by substituting the obtained F_{IS} estimates of the same method, these computations are indeed a comparison of different trajectories. For example, in the case of the Tf^D allele at the Transferrin locus, the standard error of Nei's unbiased estimator is evaluated at $F_{IS} = -0.008$, while that of Nei's unbiased and Cockerham's estimates is evaluated for $F_{IS} = 0.0$. The frequency of this allele in the Lepcha subpopulation is 0.008. For this allele frequency, even for the biased estimator of Nei, the s.e. rises from 0.008 to 0.127 as F_{IS} is changed from -0.008 to 0.0. Therefore, the differences in the standard errors noticed in Table 1 are largely due to the fact that the estimates are somewhat different in these three methods. When allele frequencies are at an intermediate range, small differences between parameter estimates do not substantially change the standard errors, but for skewed allele frequencies even minute changes in the estimates can induce a large difference in standard errors, particularly when the F_{IS} estimate is negative. In spite of such differences, there is no change in the conclusion regarding hypothesis testing either from the χ^2 values or from normal deviates. Even though we do not present similar analyses for the other 12 subpopulations studied by Bhasin et al. (1986), this statement is valid in general.

Table 1 further shows that of the 17 allele-specific tests performed, significant deviation from $F_{IS} = 0$ is found in 4 cases, due to 3 loci (Gc, PGM_1, and EsD). One of these significant deviations is due to a negative F_{IS} (at the PGM_1 locus). This finding is consistent for all three methods employed in this analysis. We also have evidence of negative significant F_{IS} values for the Tf^{C1} allele in Tamangs and Scheduled Castes, Tf^{C2} allele in Rais, Gurungs and Scheduled Castes and for the Kidd locus (either allele) in Rais and Gurungs.

Table 2 presents a summary of the significant (at 5% level) positive and negative $\hat{F}_{ISik}$ (or $\hat{F}_{ISi}$) values in 158 independent allele-specific and 127 locus-specific tests in the total data on 11 loci in the 13 subpopulations mentioned earlier. For allele-specific tests five test procedures are considered in this table: 2 χ^2 tests (one based on biased estimates of genotype frequencies, and the other based on unbiased estimates), and 3 normal deviates (based on Nei's biased and unbiased estimators, and that of Cockerham). For locus-specific tests, in addition to the above five test procedures, normal deviates based on Long's large-sample estimates of F_{ISi} (which is identical to the estimator $\hat{f}_2$ of Curie-Cohen, 1982) are also used, since the standard error of such estimators is known (see equation (4.2c)).

The total number of significant deviations from $F_{IS} = 0$ is almost the same for each χ^2 statistic. The normal deviates based on Nei's unbiased and Cockerham's estimators also behave identically, as do the normal deviates based on Nei's biased and Long's large-sample estimates in the case of locus-specific tests. Furthermore, the numbers of positive and negative significant F_{IS} values according to the χ^2 statistics are not equal; there are far more positive significant values than negative ones.

Table 2
Number of significant ($p < 0.05$) F_{IS} values in the Sikkim survey as detected by various estimators

Test criterion	Allele-specific tests with F_{ISik} value[a]		Locus-specific tests with F_{ISi} value[b]	
	Positive	Negative	Poitive	Negative
χ^2: Unbiased	22	7	12	5
Biased	20	10	10	7
Normal Deviate based on				
Nei's unbiased estimate	16	17	8	14
Nei's biased estimate	17	29	18	27
Cockerham's estimate	16	18	9	13
Long's large sample est.	–	–	18	25

[a] Total number of independent allele-specific tests = 158.
[b] Total number of independent locus-specific tests = 127.

These features are not unique to this data alone, and can be explained on the basis of the theory we presented before. First, note that χ^2 statistics only detect deviation in either direction, and since the range of negative F_{IS} is narrower ($F_{ISik} \geqslant -p_{ik}/(1 - p_{ik})$ for every allele A_k) than the range of positive F_{IS} ($F_{IS} \leqslant 1$), it is expected that more significant positive F_{IS} values will be encountered based on χ^2 goodness-of-fit. Second, since Nei's biased estimator has empirically smaller sampling variance than that of Cockerham's estimator for negative F_{IS} (Figure 1) for all allele frequencies (unless the alleles are equi-frequent), it is expected that this will pick up more significant negative $\hat{F}_{IS}$ values than the normal deviates based on Cockerham's estimator. This is also predicted from Figure 1, which shows that the sampling variance sharply drops off even if the $\hat{F}_{ISik}$ values are slightly decreased, particularly when $\hat{F}_{ISik}$ is negative. Since in all cases Nei's biased estimate is smaller than all other estimates, a normal deviate based on this estimator would necessarily pick up more significant negative F_{IS} values as compared to any other test criteria.

The estimate of F_{IS} for a single subpopulation, combining all alleles at a locus shows exactly the same picture. We have not explicitly shown the behavior of the test criteria based on Long's weighted estimator for the reason that its sampling variance is not yet available. However, its large-sample variance can be computed based on (4.2c), which is used in the computations shown in this table.

5.2. Comparison of F_{IT}, F_{IS}, and F_{ST} estimates over all subpopulations

In Tables 3, 4, and 5 we provide a comparative study of the estimators of the three fixation indices over all 13 subpopulations of Sikkim. Nei's weighted, unweighted, and large sample estimates are computed by equations (3.6)–(3.8), (3.6a)–(3.8a), and (3.6b)–(3.8b), respectively. While the standard errors of these estimators for allele- and locus-specific cases cannot be evaluated, approximate tests for allele-specific F_{IS} and F_{ST} values may be conducted by χ^2 statistics, according to equations (4.22), and (4.23), respectively. These results are shown in Table 3. In Table 4 computations of Cockerham's estimators are shown for the same data. In addition to the weighted (equations (3.15a), (3.16a), and (3.17a)) and large sample estimates (equations (3.15b), (3.16b), and (3.17b)), Cockerham's estimators of F_{ST} are also obtained under the assumption of $F_{IS} = 0$ (equation (3.43)) whose values and χ^2 test criteria are shown in this table. It should be noted that the χ^2 test criteria for F_{IS} and F_{ST} for Cockerham's general estimates are exactly the same as those of Nei's weighted estimators (shown in Table 3), and hence they are not repeated in Table 4. Multivariate estimators, according to the generalized formulation of variance component analysis (equations (2.18a)–(2.20a)), are presented in Table 5. Only locus-specific estimates are needed here, and for two-allelic loci these estimates are identical to those of Cockerham's weighted analysis, as shown earlier.

Several interesting findings emerge from these computations. First, the weighted estimators of each fixation index are nearly the same for Cockerham's and Nei's method. Second, while the F_{ST} estimates of Nei's weighted and unweighted

Table 3
Nei's allele- and locus-specific F_{IT}, F_{IS}, and F_{ST} estimates

Locus	Allele	Weighted							Unweighted			Large N		
		F_{IT}	F_{IS}	χ^2	d.f.	F_{ST}	χ^2	d.f.	F_{IT}	F_{IS}	F_{ST}	F_{IT}	F_{IS}	F_{ST}
Hp	Hp[1]	-0.007	-0.021	0.25	1	0.013	23.96[a]	12	-0.013	-0.029	0.016	-0.014	-0.042	0.027
Gc	Gc[1F]	0.278	0.214	23.48[c]	1	0.081	133.39[c]	12	0.261	0.190	0.087	0.260	0.181	0.097
	Gc[1S]	0.238	0.192	19.76[c]	1	0.057	92.08[c]	12	0.207	0.161	0.055	0.206	0.151	0.066
	Gc[2]	0.188	0.162	14.08[c]	1	0.031	43.86[c]	12	0.193	0.166	0.032	0.192	0.156	0.043
	Pooled	0.240	0.192			0.059			0.224	0.173	0.061	0.223	0.163	0.072
Tf	Tf[C1]	-0.151	-0.163	14.62[c]	1	0.010	26.20[a]	12	-0.162	-0.177	0.013	-0.163	-0.193	0.025
	Tf[C2]	-0.166	-0.176	17.20[c]	1	0.009	22.72[a]	12	-0.179	-0.192	0.011	-0.180	-0.209	0.024
	Tf[C3]	-0.009	-0.017	0.11	1	0.008	24.78[a]	12	-0.011	-0.018	0.006	-0.013	-0.034	0.021
	Tf[C12]	-0.003	-0.009	0.03	1	0.006	23.39[a]	12	-0.004	-0.010	0.006	-0.005	-0.027	0.022
	Tf[D]	-0.003	-0.003	0.00	1	-0.000	7.66	12	-0.002	-0.002	-0.000	-0.002	-0.011	0.008
	Pooled	-0.152	-0.163			0.009			-0.163	-0.177	0.012	-0.164	-0.193	0.024
aP	P[A]	0.080	0.067	2.28	1	0.013	29.55[b]	12	0.055	0.039	0.016	0.054	0.026	0.029
	P[B]	0.088	0.077	2.98	1	0.012	28.81[b]	12	0.065	0.051	0.014	0.064	0.038	0.027
	P[C]	0.193	0.187	9.40[b]	1	0.008	458.95[c]	12	0.253	0.246	0.010	0.252	0.234	0.024
	Pooled	0.087	0.075			0.013			0.066	0.052	0.015	0.065	0.039	0.028
PGM₁	PGM₁[1]	0.025	0.025	0.27	1	-0.000	11.98	12	0.070	0.074	-0.005	0.068	0.057	0.013
PGD	PGD[A]	0.150	0.127	6.09[a]	1	0.26	30.95[b]	12	0.176	0.147	0.034	0.175	0.133	0.048
EsD	EsD[1]	0.145	0.143	11.19[c]	1	0.003	13.78	12	0.134	0.132	0.003	0.134	0.121	0.015
Ak	Ak[1]	0.097	0.067	1.24	1	0.032	188.82[c]	12	0.126	0.100	0.029	0.125	0.088	0.041
Hb	Hb[A]	-0.007	-0.016	0.08	1	0.008	22.21[a]	12	-0.007	-0.019	0.011	-0.008	-0.031	0.022
Duffy	Fy[a]	0.152	0.122	3.52	1	0.034	46.81[c]	12	0.180	0.142	0.045	0.179	0.114	0.072
Kidd	Ik[a]	-0.204	-0.210	11.30[c]	1	0.005	12.67	12	-0.210	-0.218	0.007	-0.213	-0.259	0.037

[a] $p < 0.05$.
[b] $p < 0.01$.
[c] $p < 0.001$.

Table 4
Cockerham's allele- and locus-specific F_{IT}, F_{IS}, and F_{ST} estimates

Locus	Allele	Weighted[a]			Under $F_{IS} = 0$			Large N		
		F_{IT}	F_{IS}	F_{ST}	F_{ST}	χ^2	d.f.	F_{IT}	F_{IS}	F_{ST}
Hp	Hp^1	−0.006	−0.021	0.015	0.014	27.72[c]	12	−0.006	−0.032	0.025
Gc	Gc^{1F}	0.284	0.216	0.086	0.088	108.84[d]	12	0.284	0.206	0.098
	Gc^{1S}	0.242	0.193	0.061	0.063	80.96[d]	12	0.242	0.183	0.073
	Gc^2	0.190	0.163	0.032	0.034	49.15[d]	12	0.191	0.153	0.044
	Pooled	0.245	0.194	0.063	0.065			0.245	0.184	0.075
Tf	Tf^{C1}	−0.150	−0.165	0.014	0.012	23.71[b]	12	−0.149	−0.177	0.023
	Tf^{C2}	−0.165	−0.179	0.012	0.010	22.18[b]	12	−0.164	−0.190	0.022
	Tf^{C3}	−0.008	−0.021	0.012	0.012	24.37[b]	12	−0.008	−0.032	0.024
	Tf^{C12}	−0.002	−0.013	0.011	0.011	23.26[b]	12	−0.001	−0.025	0.023
	Tf^D	−0.003	0.001	−0.004	−0.004	7.62	12	−0.003	−0.010	0.007
	Pooled	−0.151	−0.166	0.013	0.011			−0.151	−0.177	0.022
aP	P^A	0.081	0.068	0.014	0.015	26.23[c]	12	0.081	0.056	0.027
	p^B	0.089	0.077	0.013	0.014	25.08[b]	12	0.089	0.065	0.026
	p^C	0.194	0.187	0.009	0.011	22.41[b]	12	0.194	0.175	0.023
	Pooled	0.088	0.076	0.013	0.014			0.088	0.064	0.026
PGM_1	PGM_1^1	0.025	0.025	−0.000	0.000	12.30	12	0.025	0.010	0.015
PGD	PGD^A	0.152	0.131	0.025	0.027	32.19[c]	12	0.152	0.115	0.042
EsD	EsD^1	0.145	0.144	0.002	0.003	15.38	12	0.146	0.132	0.015
Ak	Ak^1	0.100	0.067	0.036	0.036	51.88[d]	12	0.101	0.056	0.047
Hb	Hb^A	−0.007	−0.016	0.009	0.009	21.91[b]	12	−0.006	−0.027	0.020
Duffy	Fy^a	0.156	0.126	0.035	0.038	29.76[c]	12	0.157	0.101	0.062
Kidd	Ik^a	−0.203	−0.216	0.011	0.005	14.41	12	−0.202	−0.241	0.031

[a] The χ^2 for Cockerham's weighted estimates of F_{IS} and F_{ST} are exactly the same as those for Nei's weighted estimates (see Table 2).
[b] $p < 0.05$.
[c] $p < 0.01$.
[d] $p < 0.01$.

analyses are almost identical, there are some differences in the corresponding F_{IT} and F_{IS} estimates. Third, the large sample F_{ST} estimates of Nei and Cockerham are almost identical, even though these two methods yield somewhat different F_{IT} and F_{IS} values in large samples. Fourth, even when the estimate of allele- and locus-specific F_{IS} is significantly different from zero (tested by the χ^2 values), Cockerham's special case estimate of F_{ST} (equation (3.43) under $F_{ST} = 0$) is almost identical to that of his weighted analysis (Table 4). Fifth, while the large sample values of F_{IT} are very similar to those based on weighted analysis (in Nei's as well as Cockerham's approaches), the F_{ST} values do not behave similarly. Indeed, F_{ST} values are generally larger when large sample approximations are

Table 5
Locus-specific estimates of F_{IT}, F_{IS}, and F_{ST} by the multivariate technique of Long (1986)

Locus	Weighted estimators			Large sample estimators		
	F_{IT}	F_{IS}	F_{ST}	F_{IT}	F_{IS}	F_{ST}
Hp	− 0.006	− 0.021	0.015	− 0.006	0.032	0.025
Gc	0.238	0.185	0.065	0.233	0.164	0.083
Tf	− 0.156	− 0.182	0.022	− 0.167	− 0.204	0.031
AP	0.081	0.063	0.019	0.073	0.045	0.029
PGM$_1$	0.025	0.025	0.000	0.025	0.010	0.015
PGD	0.152	0.131	0.025	0.152	0.115	0.042
EsD	0.145	0.144	0.002	0.146	0.132	0.015
AK	0.100	0.067	0.036	0.101	0.056	0.047
Hb	− 0.007	− 0.016	0.009	− 0.006	− 0.027	0.020
Duffy	0.156	0.126	0.035	0.157	0.101	0.062
Kidd	− 0.203	− 0.216	0.011	− 0.202	− 0.241	0.031

made. Because of equation (2.1), the F_{IS} should be under-estimated in large sample approximations (since F_{IT} does not change substantially). This is the case for every comparison of Cockerham's estimators, while there are some minor discrepancies in Nei's approach. These differences are due to changes in F_{IT} values in large sample vs. weighted analysis. Sixth, the multivariate estimators are the most deviant ones. There is no general trend of these estimators as compared to Nei's and Cockerham's estimators. This is also theoretically justifiable, since the wieghting scheme in the multivariate approach is quite different (equations (2.18a)–(2.20a)).

5.3. Comparison of the estimates pooled over loci

Table 6 presents the estimates and their standard errors pooled over all co-dominant loci. As mentioned before, pooling over loci can be done in two ways for every method of estimation:
(1) by taking the ratio of sums of locus-specific estimates, and
(2) by the technique of jackknifing.
Since each fixation index is described as a function of ratios of parameters (population allele frequencies and their inter-locus variances across subpopulations), Weir and Cockerham (1984) advocated the jackknifing procedure suggesting that this might reduce the bias of estimation and in turn make the standard errors more reliable (Miller, 1974; Efron, 1982). We, however, do not see any substantial change in the estimates as well as in their standard errors through jackknifing. In fact, there is a tendency for the jackknife estimators to have somewhat larger s.e.'s for each fixation index. This table also shows that while Cockerham's and Nei's estimators are virtually identical (weighted as well as large sample), the large sample approximations involve over-estimation of F_{ST} and under-estimation of F_{IS}, F_{IT} remaining very similar. The small difference of

Table 6
Estimates of F_{IT}, F_{IS}, and F_{ST} pooled over loci and their standard errors by the three different methods

	Ratio of sums			Jackknife		
	F_{IT}	F_{IS}	F_{ST}	F_{IT}	F_{IS}	F_{ST}
Nei's estimates						
Weighted	0.045 ± 0.060	0.025 ± 0.055	0.020 ± 0.009	0.045 ± 0.064	0.025 ± 0.058	0.020 ± 0.009
Unweighted	0.044 ± 0.060	0.023 ± 0.055	0.022 ± 0.009	0.044 ± 0.063	0.023 ± 0.058	0.022 ± 0.010
Large N	0.043 ± 0.060	0.005 ± 0.058	0.038 ± 0.008	0.043 ± 0.063	0.005 ± 0.061	0.038 ± 0.009
Cockerham's estimates						
Weighted	0.047 ± 0.060	0.025 ± 0.056	0.022 ± 0.009	0.047 ± 0.064	0.025 ± 0.059	0.022 ± 0.010
Under $F_{IS} = 0$			0.022 ± 0.009			0.022 ± 0.010
Large N	0.047 ± 0.060	0.011 ± 0.057	0.037 ± 0.009	0.048 ± 0.064	0.011 ± 0.060	0.037 ± 0.009
Long's estimates						
Weighted	0.048 ± 0.063	0.028 ± 0.059	0.021 ± 0.011	0.048 ± 0.064	0.029 ± 0.060	0.023 ± 0.011
Large N	0.046 ± 0.062	0.011 ± 0.058	0.036 ± 0.010	0.047 ± 0.063	0.010 ± 0.059	0.038 ± 0.009

Long's estimators as compared with others is mainly produced by the difference of the pooling algorithm in his procedure, as noted earlier. However, unless a survey has a large number of multi-allelic loci, this method is likely to produce an almost identical qualitative conclusion about the genetic structure of the population, as seen in this example.

5.4. Comparison of the estimates of F_{ST} from allele frequency data

As mentioned earlier, analysis of population structure is sometimes necessary from allele frequency data alone. This occurs when either the loci involves dominance relationships among their alleles, or the allele frequency data are collected from the literature for comparative studies. In such cases, the only estimable parameter is F_{ST}. It is shown earlier, that in Nei's gene diversity approach, the estimators (weighted or unweighted) remain the same even if allele frequencies are used in estimation instead of genotype data (see equations (3.8) and (3.8a)). Cockerham's estimator of F_{ST} takes the form of equation (3.43), whose multivariate extension (Long's approach) is obvious from equation (3.41). The variance–component approach (univariate or multivariate) of estimation of F_{ST} from allele frequency data is therefore mathematically equivalent to the estimation of the same parameter from genotype data with the additional assumption that $F_{IS} = 0$. In order to compare the empirical values of these estimators from allele frequency data, we computed Nei's weighted unbiased, Cockerham's, and Long's estimates of F_{ST} for all 17 loci studied by Bhasin et al. (1986). The allele frequencies used in these

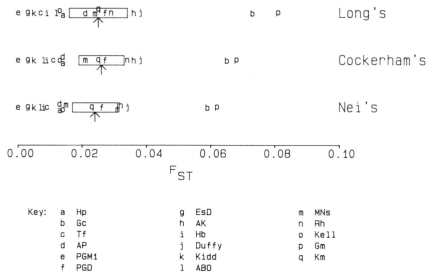

Fig. 2. A comparison of three locus-specific estimators of F_{ST} from allele frequency data on 17 loci in 13 subpopulations of Sikkim, India (Bhasin et al., 1986). The loci are indexed alphabatically (see Key). The averages over loci are indicated by arrow, and the boxes around these means represent $\pm$ s.e. range of the estimates (see text for the explanation of the estimators).

computations are the same as the ones reported in Bhasin et al. (1986). Figure 2 shows a diagramatical comparison of these locus-specific estimates, where the loci are indexed as a to q (see Key of Figure 2). The pooled estimates of F_{ST} over loci are indicated by an arrow, the box around which indicates the range with $\pm$ s.e.

It is clear that the estimates are again empirically very similar. Long's estimates are identical to the Cockerham estimates for all bi-allelic loci, although for multiallelic loci (Gc, Tf, AP, AB0, MNSs, Rh, and Gm) some discrepancies are noticable due to the different pooling (over alleles) algorithm employed in this method, as noted earlier. Nevertheless, the pooled estimates over loci are strikingly similar. Finally, we note that while the computation of the standard errors shown in this figure are based on equation (4.5), the jackknife estimates (equation (4.6)) of these standard errors are almost identical to the ones shown here. Hence, as in the case of genotype data, estimates of F_{ST} from allele frequency data also have similar empirical properties.

6. Discussion

As mentioned in Section 1, the purpose of this paper is to make a comprehensive comparative analysis of estimation of fixation indices by Nei's gene diversity approach with that of the variance component approach developed by Cockerham, or its multivariate extension. Keeping a distinction of parameters and sample statistics, throughout our presentation we have shown that these methods yield empirically very similar results. Even though these approaches have been described in a number of publications (see, e.g., Cockerham, 1969, 1973; Cockerham and Weir, 1986, 1988; Weir and Cockerham, 1984; Nei, 1973, 1977; Nei and Chesser, 1983; Chakraborty, 1974; Chakraborty and Leimar, 187; Long, 1986; Smouse and Long, 1988) and several other related statistics have been developed by others (Haldane, 1954; Li and Horvitz, 1953; Smith, 1970, 1977; Curie-Cohen, 1982; Robertson and Hill, 1984), to our knowledge, the analytical relationships between the two major approaches have not been studied explicitly before. In this discussion, first we re-iterate the new results presented here; and then we provide some arguments regarding the method we would suggest to practioners. Nevertheless, since during the conduct of this study, we developed a comprehensive computer-program for analyzing data on population structure, every estimator discussed in this paper can be computed by our computer algorithm. Interested readers can obtain a copy of the FORTRAN source codes of these programs by writing to the authors (compatible for IBM-AT type computers with a numerical co-procesor).

Our new results are as follows. First, the string of inequalities for the F_{IS} estimators in a single subpopulation shows that the expected differences among the estimators are of the order $1/2N$, N being the number of individuals sampled. While Nei's biased estimator of F_{IS} is always the smallest for any allele, Cockerham's variance–component estimator can be larger (when positive) or smaller (when negative) than Nei's unbiased estimator. Second, even though Long (1986) and Smouse and Long (1988) generalized Cockerham's approach for a multivariate case (three or more alleles at a locus), they failed to note that their method yields F_{IS} estimators mathematically identical to Curie-Cohen's $\hat{f}_2$ (based on the ratio of observed and expected homozygote frequencies), in large samples. Third, for a single subpopulation, Nei's unbiased, biased (identical to $\hat{f}_1$ of Curie-Cohen, 1982–although not stated by him) and Cockerham's estimators of F_{IS} have closed form expressions of standard errors, for specific alleles as well as for the locus as a whole, which are also documented here for the first time. Much of the ground work for these derivations was, however, done by Curie-Cohen (1982) and Robertson and Hill (1984).

These new findings allow more rigorous comparative analyses of the different estimators, than the ones done before. Our empirical data analysis shows the closeness of the different estimators, which are based on somewhat different premises. There have been a number of misconceptions about the gene diversity approach, which should be clarified in this context. Note that the gene diversity approach does not need the correlation interpretation of the fixation indices. The total heterozygosity in subdivided populations is decomposed here on the basis of the number of extant subpopulations. No assumption of the replicative nature of subpopulations is needed. While Cockerham's linear model (of random effects) makes the assumption that the subpopulations studied are replicates from the universe of all subpopulations that exist within the total population, a situation that might apply to experimental populations, in the context of evolutionary significance, it is not clear if this assumption is realistic. In the specific example considered here, Sikkimese people are indeed subdivided into the present 13 subpopulations which during their history have assembled in this geographic region by following different migration routes (Bhasin et al., 1986). They are not replicates of each other, and indeed there may not be any further subpopulation among the people of Sikkim. If a statistical framework forms the basis of the variance–component analysis, the question is: should we treat the underlying linear model (Cockerham, 1969, 1973) as a random effects model in such a situation? Our answer to this question would be no as this subdivided structure represents a fixed-effect model. On the contrary, our exposition clearly indicates that Nei's gene diversity approach has a formal statistical basis, since all components of the decomposition of heterozygosity can be represented in terms of the underlying parameters, and they can be related with Wright's fixation indices without invoking their interpretation through correlations.

At this point it is worthwhile to note that for a single subpopulation the probabilistic interpretation of F_{IS} has been used by Haldane and Moshinsky (1939), Cotterman (1940), and Malécot (1948), where F_{IS} is interpreted as the probability that the two genes at a locus in an individual are identical. This probabilistic interpretation implicitly implies that the F_{IS} can take only non-negative values in the unit interval. Similar probabilistic interpretations of F_{IT} and F_{ST} are also used by Crow and Kimura (1970, pp. 105–106) to prove the Wright's identity (equation (2.1)). They, however, note that since F_{IT} and F_{IS} can be negative, correlational interpretations of these fixation indices also yield the Wright's identity (Crow and Kimura, 1970, pp. 107–108). It is apparently implicit in their derivation that the subpopulations do not exchange migrants during the process of gene differentiation, so that the allele frequency variations across subpopulations do not depend upon the F_{IS} values within the subpopulations. In contrast, in Nei's formulation of gene diversity analysis the Wright's identity is established simply by the notion that F_{IS} and F_{IT} represent summary measures of deviations from the Hardy–Weinberg expectations in the subpopulations and in the total population, respectively, and F_{ST} represents the extent of genetic differentiation (standardized variance of allele frequencies across subpopulations). No assumption regarding migration and selection is needed in such derivation (Nei, 1973, 1977). The Wright's identity (equation (2.1)) simply becomes a mathematical consequence of the parametric definitions of F_{IT} (equation (2.5)), F_{IS} (equation (2.6)), and F_{ST} (equation (2.7)).

When the parameters are so defined, our equations (2.8), (2.9), and (2.10) suggest that all fixation indices have their natural bounds, namely F_{ST} lies between 0 and 1, while F_{IS} and F_{IT} can take positive as well as negative values, depending on H_S being smaller or larger than H_0 for F_{IS} (equation (2.8)) and H_T being smaller or larger than H_0 for F_{IT} (equation (2.9)). In such formulations no assumption is needed regarding the evolutionary mechanism that determines the process of genetic differentiation within and between subpopulations.

Since the variance–component approach can yield a negative value for the variance component b (equation (2.16)), in order to interpret the linear model (equation (16) of Cockerham, 1969) one must assume that $\sum_{k=1}^{r} w_i F_{ISik} p_{ik}(1 - p_{ik})$ must be positive. Cockerham (1969, 1973) recognized this feature of his model, and yet justified it on the ground that evolutionary factors that generally produce negative F_{IS} values are not usually strong enough to produce large negative F_{IS} (or F_{IT}) values. Our data analysis provides evidence contrary to this argument. We indeed found several negative estimates of F_{IS} (Tables 1 and 2). Even if their significance is discounted, because the normal deviates or the χ^2 statistics may not attain their large sample distribution in samples of the size analyzed here, it is unpleasant to deal with a linear model with negative variance components (not only the estimates, but also in parametric form).

Nei (1986) addressed some of these issues along with other evidences where the implicit assumptions of the variance component formulations are unrealistic for natural populations. He also noted that his original definition of F_{ST} ($= D_{ST}/H_T$, called G_{ST} by Nei, 1973) has one deficiency, since it is dependent on the number of subpopulations (s). He suggested one modification, defining $D'_{ST} = sD_{ST}/(s - 1)$, to take into account this deficiency (Nei, 1986). According to this suggestion, H'_T, the gene diversity in the total population is defined as $D'_{ST} + H_S$, yielding the three fixation indices $F_{IS} = H_0/H_S$ (unchanged from the previous definition), $F'_{IT} = H_0/H'_T$, and $F'_{ST} = D'_{ST}/H'_T$, for which the estimation technique presented here works with only minor modifications (see also Nei and Chesser, 1983). When s is large (say, 10 or more), these re-defined fixation indices change only slightly, and hence they are not computed in our application (since for the present example $s = 13$). However, when s is small, it is preferable to calculate these modified indices with the above modifications. Also note that the re-defined F'_{ST} is identical to the parameter β defined by Cockerham and Weir (1988), not recognized by these authors. Therefore an estimator of β can also be obtained by estimating F'_{ST} in the gene diversity approach without the intraclass correlation interpretation. Nevertheless, we must reiterate the point that adjustment for the number of subpopulations does not necessarily help in comparing the coefficient of gene differentiation estimates in different data sets from different natural populations. An extrapolation of such estimates from one set of populations to another can be misleading, since their evolutionary histories are usually different. Cockerham's approach is more ideally suited for experimental populations, where the number of subpopulations represent the replicate of populations designed with a given experimental situation, and hence extrapolation from one experiment to another must need adjustments for variations in number of replicate subpopulations within each experiment.

Notwithstanding these philosophical differences, given the empirical similarity of the various estimators, a recommendation regarding the choice of estimators should be of interest to investigators who deal with real data. On the basis of statistical principles, unfortunately, there is no general recommendation. We claim this for several reasons. First, in every formulation, we have shown that consistent estimators can be derived. The study of large sample variances either by theoretical variances evaluated with intra-locus data, or by empirical evaluation of inter-locus variation shows that all estimators are subjected to similar sampling fluctuations. Second, even though with the aid of computer-algorithms the numerical task of computation can be left to computers, the choice is simply a matter of taste.

Since the gene diversity approach relates F_{ST} to the average genetic distances among subpopulations (Nei's minimum distance; Nei, 1972) a genetic distance interpretation of the coefficient of gene differentiation is also possible. Note that this interpretation does not assume, again, any evolutionary mechanism, and hence this interpretation should hold with or without mutation and selection. While Cockerham's F_{ST} parameters, and its multivariate extension have been shown to have a genetic distance interpretation as well (Reynolds et al., 1983; Long et al., 1987) in order for the measures of co-ancestry to be interpreted as genetic distances one must assume that genetic differentiation occurs without the aid of mutation and selection (Reynolds et al., 1983; Weir and Cockerham, 1984). Furthermore, in this latter paper they also assume that the same population size is maintained for all subpopulations and for all generations. While these assumptions are not needed in formulating Nei's genetic distances, thus far the evolutionary expectation and drift variances of genetic distances have been worked out under the neutral model of evolution without constant population size (Li and Nei, 1976; Nei and Chakravarti, 1977).

We advocate the use of the gene diversity approach for its simplicity and generality for natural populations. No loss of statistical rigor is attendant to this recommendation, as explicitly shown here—because we did not make use of any evolutionary model in this presentation, and as a method of estimation, what we used can be called the method of moments in the terminology of statistical inference. This is the only appropriate estimation technique that yields analytically closed form estimators. We might add here that Curie-Cohen (1982) and Robertson and Hill (1984) investigated the properties of the maximum likelihood estimators of F_{IS} based on multinomial sampling of genotypes, which behave worse than Nei's biased estimator in most practical situations (Curie-Cohen, 1982).

Although we presented analytically closed form expressions of intra-locus variances of F_{IS} estimators, these are applicable only for single-locus data. Generally, large sample sizes are needed to apply these formulae, since the estimators are rather unstable (the drift variance is quite large; as shown by Li and Nei, 1976—for heterozygosities, Nei and Chakravarti, 1978—for $G_{ST} \approx F_{ST}$), and the power of detection of significant deviations of these indices is generally low (Brown, 1970; Ward and Sing, 1970; Chakraborty and Rao, 1972; Haber, 1980; Emigh, 1980). Evolutionary interpretation of the coefficients of gene differentiation or deviations from $F_{IS} = 0$, $F_{IT} = 0$ should be based on data on multiple loci. We have shown that multi-allelic and/or multiple-loci can be analyzed easily without the aid of Long's (1986) multivariate extensions. Indeed Nei's formulation of the decomposition of gene diversity is philosophically based on samples of genomes drawn from the population. He defined gene diversity as the complement of the probability that the two genomes are identical at each locus. Therefore, he computed gene diversity based on a sample of loci (polymorphic and monomorphic, see Nei, 1975, 1987). Even though the parameter F_{ST} (in Nei's terminology, G_{ST}) or its estimate does not change even if the monomorphic loci are excluded, the absolute value of H_T, H_S, and H_0 (averaged over all loci in a genome) changes. Even with a limited number of loci, we have shown that the variances of these quantities can be examined by studying their inter-locus variation, which yields inter-locus variances of the fixation indices as well. Since the inter-locus variance is the major component of the total sampling variance (Nei and Roychoudhury, 1974; Nei, 1978), jackknifing helps a little to provide a more reliable estimate of the extent of sampling variance. This finding is in disagreement with Mueller's study (Mueller, 1979) on genetic distance, but it is consistent with our own simulation results published before (Chakraborty, 1985). Weir and Cockerham (1984) claimed that jackknifing worked 'satisfactorily' in the two-population situation in their simulations (Reynolds et al., 1983), while we find that jackknifing does not add any particular advantage in terms of parameter estimates or their standard errors (Table 6).

Finally, we should return to the issue of hypothesis testing in the context of population structure data analysis. Considerable labor is needed to provide estimators adjusting for the effect of limited sample sizes. It is seen that when sample sizes are small (of the order of 100 or less individuals per locus per subpopulation, in the specific example given here), the use of large sample approximations yield over-estimates of F_{ST} and under-estimates of F_{IS} (F_{IT} remaining almost identical), irrespective of the method of analysis (Nei's vs. Cockerham's). Since such estimates are invoked in evaluating standard errors or for computing test criteria, the question is: are these test criteria reliable, and can we justify the large sample properties of these test criteria? Our answer, although we cannot prove it analytically, is a probable no. We say so, for the reason that if the normal deviates are to be regarded as reliable, we must evaluate the standard errors accurately. We have seen that in some region of the parametric space, the standard errors can be drastically affected, even by a minute change in the parameter estimates. For the χ^2 tests, on the other hand, we must regard the variance components as independently distributed. This assumption, we might note, is also needed in Long's (1986) Wilk's Λ-test criteria. Nayak (1983) has shown that when the genotype data from several subpopulations are represented in the form of an analysis of variance of categorical data (Light and Margolin, 1971), the mean square errors of the different sources of variation are not independently distributed. The correlations between them can often be substantial. Furthermore, for every source of variation, the large sample distribution of the mean square errors is of the form of composite χ^2's, where the coefficients are also functions of unknown parameters. They cannot be simply equated to a χ^2 statistics as is done commonly invoking large sample theory of continuously distributed random variables. Therefore, we argue that the test statistics generally suggested for population structure analysis have much poorer statistical justifiability than the parameter estimates. Cockerham (1973) arrived at this general conclusion, although the sampling theory of weighted categorical data analysis was not available at that time.

7. Summary

A comprehensive comparative study of the various estimators of the fixation indices (F_{IT}, F_{IS}, and F_{ST}) shows that the properties of the estimators based on Nei's gene diversity and Cockerham's variance component analysis are very similar, in spite of their philosophical differences. In the analysis of genotypic data from a single population, a string of inequalities of the different estimators of F_{IS} is mathematically established, with regard to which the discrepancies in the sampling precision of these estimators can be reconciled. The analytical expression for the large sample variance of these estimators suggests that the parametric value of their sampling variance is identical. Empirical evaluation of the bias and standard errors of the three fixation indices from a genetic survey of 17 loci from 13 subpopulations of Sikkim, India suggests that for these ratio estimators the Jackknife method and Taylor's series approximation yield almost identical bias and standard error. These conclusions also hold for the estimation of F_{ST} from allele frequency data alone. A comprehensive computer program for obtaining all estimators has been developed, and is available from the authors upon request.

Acknowledgements

This paper is dedicated to the memory of Sewall Wright, the pioneer of population structure analysis, who passed away during the progress of this work. Dr Masatoshi Nei and Dr William J. Schull are to be acknowledged for their help in critically reviewing earlier versions of this review. Thanks are also due to Mr R. Schwartz for his help in computation and graphic works. This work was supported by grants from the National Institutes of Health and National Science Foundation. HDH was supported by the German research fellowship program of DAAD during the conduct of this work.

References

Anderson, T. W. (1984). *An Introduction to Multivariate Statistical Analysis* (2nd edition). John Wiley, New York.

Bhasin, M. K., Walter, H., Chahal, S. M. S., Bhardwaj, V., Sudhakar, K., Danker-Hopfe, H., Dannewitz, A., Singh, I. P., Bhasin, V., Shil, A. P., Sharma, M. B. and Wadhavan, D. (1986). Biology of the people of Sikkim, India. 1. Studies on the variability of genetic markers. *Z. Morph. Anthrop.* **77**, 49–86.

Brown, A. H. D. (1970). The estimation of Wright's fixation index from genotype frequencies. *Genetica* **41**, 399–406.

Chakraborty, R. (1974). A note on Nei's measure of gene diversity in a substructured population. *Humangenetik* **21**, 85–88.

Chakraborty, R. (1985). Genetic distance and gene diversity: Some statistical considerations. In: *Multivariate Analysis – VI*, P. R. Krishnaiah (ed.). Elsevier, Amsterdam, 77–96.

Chakraborty, R., Chakravarti, A. and Malhotra, K. C. (1977). Variation of allele frequencies among caste groups of the Dhangars of Maharashtra, India: An analysis with Wright's *F*-statistics. *Ann. Hum. Biol.* **4**, 275–280.

Chakraborty, R. and Nei, M. (1977). Bottleneck effect with stepwise mutation model of electrophoretically detectable alleles. *Evolution* **31**, 347–356.

Chakraborty, R. and Leimar, O. (1987). Genetic variation within a subdivided population. In: *Population Genetics and Fishery Management*, N. Ryman and F. Utter (eds.). Sea Grant Program, University of Washington Press, Seattle, WA, 89–120.

Chakraborty, R. and Rao, D. C. (1972). On the detection of *F* from ABO blood group data. *Am. J. Hum. Genet.* **24**, 352–353.

Cockerham, C. C. (1969). Variance of gene frequencies. *Evolution* **23**, 72–84.

Cockerham, C. C. (1973). Variance of gene frequencies. *Evolution* **27**, 679–700.

Cockerham, C. C. and Weir, B. S. (1986). Estimation of inbreeding parameters in stratified populations. *Ann. Hum. Genet.* **50**, 271–281.

Cockerham, C. C. and Weir, B. S. (1987). Correlations, descent measures: Drift with migration and mutation. *Proc. Natl. Acad. Sci. USA* **84**, 8512–8514.

Cotterman, C. W. (1940). A calculas for statistic-genetics. Ph.D. dissertation, Ohio University, Columbus, OH.

Crow, J. F. and Kimura, M. (1970). *An Introduction to Population Genetics Theory*. Harper & Row, New York.

Curie-Cohen, M. (1982). Estimates of inbreeding in a natural population: A comparison of sampling properties. *Genetics* **100**, 339–358.

Efron, B. (1982). *The Jackknife, the Bootstrap and other Resampling Plans*. Society for Industrial and Applied Mathematics, Philadelphia.

Haber, M. (1980). Detection of inbreeding effects by the chisquare test on genotypic and phenotypic frequencies. *Am. J. Hum. Genet.* **32**, 754–760.

Haldane, J. B. S. (1954). An exact test for randomness of mating. *J. Genet.* **52**, 631–635.

Haldane, J. B. S. and Moshinsky, P. (1939). Inbreeding in Mendelian populations with special reference to human cousin marriage. *Ann. Eugen.* **9**, 321–340.

Kendall, M. and Stuart, A. (1977). *The Advanced Theory of Statistica. Vol. 1* (4th edition). MacMillan, New York.

Kirby, G. C. (1975). Heterozygote frequencies in small subpopulations. *Theoretical Population Biology* **8**, 31–48.

Li, C. C. (1955). *Population Genetics*. University of Chicago Press, Chicago, IL.

Li, C. C. and Horvitz, D. G. (1953). Some methods of estimating the inbreeding coefficient. *Am. J. Hum. Genet.* **5**, 107–117.

Li, W.-H. and Nei, M. (1975). Drift variances of heterozygosity and genetic distance in transient states. *Genet. Res.* **25**, 229–248.

Light, R. J. and Margolin, B. H. (1971). An analysis of variance for categorical data. *J. Am. Stat. Assoc.* **66**, 534–544.

Long, J. C. (1986). The allelic correlation structure of Gainj- and Kaam-speaking people. I. The estimation and interpretation of Wright's *F*-statistics. *Genetics* **112**, 629–647.

Malécot, G. (1948). *Les Mathématiques de l'Hérédité*. Masson et Cie, Paris.

Miller, R. G. (1974). The jackknife – A review. *Biometrika* **61**, 1–15.

Mueller, L. D. (1979). A comparison of two methods for making statistical inferences on Nei's measure of genetic distance. *Biometrics* **35**, 757–763.

Nayak, T. K. (1983). Applications of entropy functions in measurement and analysis of diversity. Ph.D. Dissertation, University of Pittsburgh, Pittsburgh, PA.

Nei, M. (1972). Genetic distance between populations. *Am. Nat.* **106**, 283–292.

Nei, M. (1973). Analysis of gene diversity in subdivided populations. *Proc. Natl. Acad. Sci. USA* **70**, 3321–3323.

Nei, M. (1975). *Molecular Population Genetics and Evolution*. North-Holland, Amsterdam.

Nei, M. (1977). *F*-statistics and analysis of gene diversity in subdivided populations. *Ann. Hum. Genet.* **41**, 225–233.

Nei, M. (1978). Estimation of average hterozygosity and genetic distance from a small number of individuals. *Genetics* **89**, 583–590.

Nei, M. (1986). Definition and estimation of fixation indices. *Evolution* **40**, 643–645.

Nei, M. (1987). *Molecular Evolutionary Genetics*. Columbia University Press, New York.

Nei, M. and Chakravarti, A. (1977). Drift variances of F_{ST} and G_{ST} statistics obtained from a finite number of isolated populations. *Theor. Pop. Biol.* **11**, 307–325.

Nei, M. and Chesser, R. K. (1983). Estimation of fixation indices and gene diversities. *Ann. Hum. Genet.* **47**, 253–259.

Nei, M., Maruyama, T. and Chakraborty, R. (1975). The bottleneck effect and genetic variability in populations. *Evolution* **29**, 1–10.

Nei, M. and Roychoudhury, A. K. (1974). Sampling variance of heterozygosity and genetic distance. *Genetics* **76**, 379–390.

Rao, C. R. (1965). *Linear Statistical Inference and Its Applications*. Wiley, New York.

Rao, C. R. (1982). Diversity: Its measurement, decomposition, apportionment and analysis. *Sankhya A* **44**, 1–21.

Rao, C. R., Rao, D. C. and Chakraborty, R. (1973). The generalized Wright's model. In: *Genetic Structure of Populations*, N. E. Morton, (ed.). University of Hawaii Press, Honolulu, 55–59.

Rao, D. C. and Chakraborty, R. (1974). The generalized Wright's model and population structure. *Am. J. Hum. Genet.* **26**, 444–453.

Reynolds, J., Weir, B. S. and Cockerham, C. C. (1983). Estimation of the coancestry coefficient: Basis for a short-term genetic distance. *Genetics* **105**, 767–779.

Robertson, A. and Hill, W. G. (1984). Deviation from Hardy–Weinberg proportions: Sampling variances and use in estimation of inbreeding coefficients. *Genetics* **107**, 703–718.

Slatkin, M. and Barton, N. H. (1989). A comparison of three methods for estimating average level of gene flow. *Evolution* **43**, 1349–1368.

Smith, C. A. B. (1970). A note on testing the Hardy–Weinberg law. *Ann. Hum. Genet.* **33**, 377–383.

Smith, C. A. B. (1977). A note on genetic distance. *Ann. Hum. Genet.* **40**, 463–479.

Smouse, P. E. and Long, J. C. (1988). A comparative F-statistics analysis of the genetic structure of human populations from the Lowland South America and Highland New Guinea. In: *Quantitative Genetics*, B. S. Weir, E. J. Eison, M. M. Goodman and G. Namkoong (eds.). Sinaur Association Inc., Sunderland, 32–46.

Van Den Bussche, R. A., Hamilton, M. J. and Chesser, R. K. (1986). Problems of estimating gene diversity among populations. *The Texas Journal of Science* **38**, 281–287.

Weir B. S. and Cockerham, C. C. (1984). Estimating F-statistics for the analysis of population structure. *Evolution* **38**, 1358–1370.

Workman, P. L. and Niswander, J. D. (1970). Population studies on Southwestern Indian tribes. II. Local genetic differentiation in the Papago. *Am. J. Hum. Genet.* **22**, 24–49.

Wright, S. (1943). Isolation by distance. *Genetics* **28**, 114–138.

Wright, S. (1951). The genetical structure of populations. *Ann. Eugenics* **15**, 323–354.

Wright, S. (1965). The interpretation of population structure by F-statistics with special regard to systems of mating. *Evolution* **19**, 395–420.

C. R. Rao and R. Chakraborty, eds., *Handbook of Statistics, Vol. 8*
© Elsevier Science Publishers B.V. (1991) 255–269

8

Estimation of Relationships from Genetic Data

Elizabeth A. Thompson

1. Introduction: Genetic similarity and genealogical relationships

The basis of heredity in all living organisms is through the replication of DNA, and thence the fact that the DNA of an offspring individual will be a copy of some part of the DNA of the parent. Thus, since individuals carry copies of the DNA of their ancestors, the more closely related a pair of individuals the higher the probable degree of genetic similarity between them. Conversely, under an appropriate model for the inheritance of genetic material, it is possible to infer likely relationships from the degree of genetic similarity.

More precisely, genetic data for diploid organisms (such as humans) usually relate to traits determined by *autosomal Mendelian loci*. That is, to paraphrase Mendel (1866), every individual carries two genes controlling the trait of interest, one a copy of a gene in his mother, the other a copy of a gene in his father, and every individual copies independently to each offspring a random one of the two genes that he carries. The *phenotype* of an individual is his observable type for a trait. His *genotype* is the unordered pair of types of the two genes that he carries. There are many possible genotype–phenotype correspondence patterns.

Relationships may be

(1) genealogical on a generation-to-generation time scale;

(2) relationships among small populations between which related individuals migrate;

(3) longer-term relationships among populations without recent contact, whose similarity reflects the residue of common ancestry, and, at the extreme of this scenario;

(4) relationships between species.

Each time scale has its own appropriate data, measures of similarity, and models underlying these statistics. In this chapter, we shall proceed from the smallest to the largest time scale. In Section 2, we consider relationships between individuals: this will be the most detailed section, in which likelihood analysis of relationship estimation will be addressed. In Section 3, we consider relationships among small populations under models of admixture and migration. In Section 4, we consider long-term effects of population ancestry under models of population fission and divergence. For reasons of space, it is not possible to cover all three topics in the same detail: in Section 2, details and examples will be given, but for Sections 3 and 4, we can provide only an overview, with references to further reading.

2. Mendelian segregation and the estimation of familial relationships

The focus of this section will be upon the smallest time scale: the estimation of relationships between individuals from genetic data. Although, in biological and medical practice, estimation of relationship is dominated by paternity testing, this is a very small and specific part of a much wider statistical problem.

2.1. Gene identity by descent. The parameter space

Genes in related individuals are said to be identical by descent (IBD) if they are copies of the same ancestral genes received via repeated segregations from parent to offspring. Such gene identity is defined only relative to a given pedigree. Genes that are IBD are of the same allelic type (barring mutation, which is unimportant on the time scale considered here). Non-IBD genes may or may not be. Therefore similarities and differences in genotype or phenotype for traits determined by Mendelian loci whose population-genetic characteristics are known can be used to estimate the extent to which individuals carry IBD genes.

For simplicity we restrict attention to sexual diploid organisms, and consider independently inherited (*unlinked*) autosomal Mendelian loci. More complex types of data will be reviewed briefly later. We consider first the estimation of relationship between two individuals. This special case will suffice for development of the statistical approach. At each locus, an individual carries a gene IBD to one of the two genes of his mother, each of the mother's two genes having an equal 50% chance of being the one copied to this offspring. The other gene in the individual is analogously IBD to one of the two genes of his father. We shall also suppose, for simplicity, that the mother and the father of each individual are unrelated to each other; that is, the individuals are not *inbred* relative to the pedigree of interest. This also is a realistic assumption, or at least a close approximation, if, as is usually the case, we are interested in estimating only the proximal relationship between individuals, and this proximal relationship is close. In this case the parameter space of relationships is straightforward. The genealogy determines probabilities with which, independently for each locus, the two individuals share 2, 1 or 0 genes IBD. Thus the space of estimable parameters is a triangle

$$K = \{(k_2, k_1, k_0); k_2 + k_1 + k_0 = 1\}$$

(see Figure 1), where k_i denotes the probability that the two individuals share i genes IBD. This parametrisation of relationships was first introduced by Cotterman (1940), and the probabilities k_i are analogous to the Cotterman k-coef-

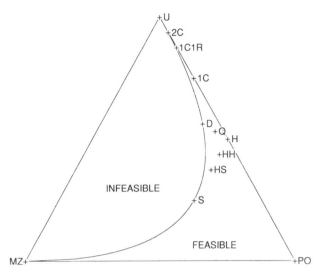

Fig. 1. The parameter space for the estimation of relationship between two non-inbred individuals. The relationships shown are those tabulated in Table 1.

ficients, although we use k_1 in place of the $2k_1$ of Cotterman (1940). Rather than direct estimation of genealogy, estimation is of (k_2, k_1, k_0). Although each genealogical relationship determines values (k_2, k_1, k_0) in the triangle K, the function is unfortunately neither injective nor surjective. More precisely, for all relationships, $k_1^2 \geqslant 4k_0 k_2$, and all dyadic rationals satisfying this contraint are obtainable from many genealogies. However, the (k_2, k_1, k_0) values for standard relationships (sibs, parent–offspring, first cousins, etc.) are distinct, with the notable exception of those for half sibs, uncle–niece, and grandparent–grandchild. Some examples are given in Table 1.

2.2. Unlinked loci. Likelihood form and likelihood estimation

Leaving aside the relationship between genealogy and (k_2, k_1, k_0), consider the estimation of the latter. For any trait determined by a simple Medelian locus, the probability of an ordered genotype or phenotype pair for two individuals, conditional on their having 2, 1 or 0 genes IBD at the trait locus, can be written down. For genotypic data, for which the types of the two genes carried at a given locus by each individual are observable, the form of the probabilities are shown in Table 2. Derivations are straightforward. For example, if two individuals B_1 and B_2 have ordered genotypes $a_i a_j$ and $a_j a_k$ and have one gene IBD, then the shared gene must be that of type a_j. Thus, there are three distinct genes in these two individuals, at this locus, one each of types a_i, a_j and a_k and the genes are distinguishable in the sense that the types of the gene carried only by B_1, the shared gene, and the gene carried only by B_2, are all uniquely identified. The relevant probability is thus $p_i p_j p_k$, the probability that three ordered genes chosen

E. Thompson

Table 1
The (k_0, k_1, k_2) values for some standard relationships

Relationship	(Figure 1)	k_0	k_1	k_2	$\frac{1}{4}(k_1 + 2k_2)$
Self (monozygous twin)	MZ	0	0	1	0.5
Parent–offspring	PO	0	1	0	0.25
Full sib	S	$\frac{1}{4}$	$\frac{1}{2}$	$\frac{1}{4}$	0.25
Half sib, uncle–niece	H	$\frac{1}{2}$	$\frac{1}{2}$	0	0.125
grandparent–grandchild					
Half sibs whose unshared parents are:					
sibs	HS	$\frac{3}{8}$	$\frac{1}{2}$	$\frac{1}{8}$	0.1875
half sibs	HH	$\frac{7}{16}$	$\frac{1}{2}$	$\frac{1}{16}$	0.1563
First cousin	1C	$\frac{3}{4}$	$\frac{1}{4}$	0	0.0625
Double first cousin	D	$\frac{9}{16}$	$\frac{3}{8}$	$\frac{1}{16}$	0.125
Quadruple half first cousin	Q	$\frac{17}{32}$	$\frac{7}{16}$	$\frac{1}{32}$	0.125
First cousin once removed	1C1R	$\frac{7}{8}$	$\frac{1}{8}$	0	0.0312
Second cousin	2C	$\frac{15}{16}$	$\frac{1}{16}$	0	0.0156
Unrelated	U	1	0	0	0.0

Table 2
Probabilities of pairwise genotypes given the number of IBD genes

Genotypes for		Probabilities given m genes IBD		
Individual [a] B_1	Individual [a] B_2	$m = 0$ P_0	$m = 1$ P_1	$m = 2$ P_2
$a_i a_i$	$a_i a_i$	p_i^4	p_i^3	p_i^2
$a_i a_j$	$a_i a_i$	$2p_i^3 p_j$	$p_i^2 p_j$	0
$a_i a_i$	$a_j a_j$	$p_i^2 p_j^2$	0	0
$a_i a_j$	$a_i a_j$	$4p_i^2 p_j^2$	$p_i p_j (p_i + p_j)$	$2p_i p_j$
$a_i a_j$	$a_j a_l$	$4p_i p_j^2 p_l$	$p_i p_j p_l$	0
$a_i a_i$	$a_j a_l$	$2p_i p_j p_l^2$	0	0
$a_i a_j$	$a_k a_l$	$4p_i p_j p_k p_l$	0	0

[a] The two individuals are ordered: the event that the genotypes of (B_1, B_2) are $(a_i a_j, a_i a_i)$ is distinguished from the event that they are $(a_i a_i, a_i a_j)$. However, since all probabilities are symmetric between the two genotypes, these two events have the same probabilities: only a necessary minimum of cases is tabulated above. The two alleles within a genotype are, by convention, unordered: $a_i a_j$ is the same genotype as $a_j a_i$. Allele a_i has population frequency p_i.

at random from the population gene pool would have these three types. For brevity, we shall not go into further cases. More details are given, for example, by Thompson (1986).

For the types of loci normally used to make estimates of relationship, there is a deterministic genotype–phenotype correspondence: a known set of genotypes gives rise to each phenotype. Thus for phenotypic data, such as that shown for some loci in Table 3, probabilities of phenotype pairs are obtained by summing over the relevant genotype pairs. An important point is that all these probabilities depend on the population genetic characteristics of the trait – the population allele frequencies, and (in general) the probabilities of phenotype given genotype. These are assumed known in the estimation of relationship. Fortunately, estimation is quite robust to small variation in assumed population allele frequencies.

Table 3
Estimation of pairwise relationship: an example

Locus	Allele frequencies				Phenoltypes		Likelihood contributions		
	a_1	a_2	a_3	a_4	B_1	B_2	P_0	P_2	P_3
1	0.4	0.6	0	0	a_1a_2	a_1a_2	0.2304	0.2400	0.4800
2	0.3	0.2	0.5	0	a_1-	a_3a_3	0.0975	0.0750	0.0000
3	0.4	0.6	0	0	a_1a_2	a_2a_2	0.1728	0.1440	0.0000
4	0.3	0.4	0.3	0	a_1a_2	a_1-	0.0648	0.0720	0.0000
5	0	0.2	0.4	0.4	a_2-	a_3-	0.1728	0.0960	0.0000
6	0	0.5	0.5	0	a_3a_3	a_3a_3	0.0625	0.1250	0.2500
7	0.3	0.7	0	0	a_1a_1	a_2a_2	0.0441	0.0000	0.0000
8	0.1	0.1	0.3	0.5	a_1-	a_4a_4	0.0425	0.0250	0.0000
9	0.2	0.2	0.6	0	a_3a_3	a_1a_2	0.0288	0.0000	0.0000
10	0.0	0.3	0.7	0	a_3a_3	a_2-	0.2499	0.1470	0.0000
11	0	0.1	0.9	0	a_2-	a_2-	0.0361	0.1090	0.1900
12	0	0.2	0.1	0.7	a_3-	a_3-	0.0225	0.0710	0.1500
13	0.8	0.2	0	0	a_2a_2	a_2a_2	0.0016	0.0080	0.0400
14	0.1	0.9	0	0	a_1a_1	a_1a_1	0.0001	0.0010	0.0100
15	0	0.2	0.8	0	a_2-	a_2-	0.1296	0.2320	0.3600
16	0	0.3	0.7	0	a_3a_3	a_3a_3	0.2401	0.3430	0.4900
17	0.2	0.8	0	0	a_1a_1	a_1a_2	0.0128	0.0320	0.0000
18	0.2	0.6	0.2	0	a_1-	a_1-	0.0144	0.0400	0.1200
19	0.1	0.9	0	0	a_1a_1	a_1a_2	0.0018	0.0090	0.0000
20	0.1	0.9	0	0	a_1a_2	a_1a_1	0.0018	0.0090	0.0000

Let the probability of the observed data at locus j, conditional on the two individuals having i genes IBD at that locus be $P_i^{(j)}$; then the probability of the observed locus-j data is

$$k_2 P_2^{(j)} + k_1 P_1^{(j)} + k_0 P_0^{(j)},$$ (2.1)

and the log-likelihood for parameters (k_2, k_1, k_0) for data from s independently inherited loci is the concave function

$$\sum_{j=1}^{s} \log_e (k_2 P_2^{(j)} + k_1 P_1^{(j)} + k_0 P_0^{(j)}) .$$

An important aspect of the estimation problem is that the probabilities k_i are the proportions of loci at which the pair of individuals share i genes IBD. Given only data at a single locus, however informative, the contours of log-likelihood are linear, and the maximum likelihood estimate is at a vertex of the relationship triangle. For the estimation of relationship, it is thus more important to have data for a large number of independently segregating loci than that each locus should provide a clear locus-specific indication of the number of IBD genes.

2.3. An example

We give here a numerical example that exemplifies and extends the above discussion. We assume two individuals, with observed phenotypes at 20 loci as given in Table 3. At every locus, a_1 and a_2 are *codominant*, and are *dominant* to a_3, where all three alleles exist. All three alleles are *dominant* to a_4. There are thus five phenotypes:

a_1– comprising genotypes $\{a_1 a_1, a_1 a_3, a_1 a_4\}$,
a_2– comprising genotypes $\{a_2 a_2, a_2 a_3, a_2 a_4\}$,
a_3– comprising genotypes $\{a_3 a_3, a_3 a_4\}$, and
$a_4 a_4$ and $a_1 a_2$ each comprising a single genotype only.

(For some loci, these phenotypes can be simplified. For example, where there is no a_4, phenotype a_3– becomes genotype $a_3 a_3$.) The allele frequencies are also given in Table 3, as are the coefficients of the gene-identity-by descent (GIBD) probabilities for each contribution to the likelihood surface. The net log-likelihood is shown in Figure 2. The expected concavity of the surface is indeed obtained, and we have a unique maximum likelihood estimate, and contours of the likelihood surface providing likelihood-based confidence regions are as shown. The clear inference is that the individuals are sibs, but the general method of conversion from estimates of GIBD probabilities to an estimate of genealogical relationship is an open question. There are no relationships that provide GIBD probabilities in the 'infeasible' range (Figure 1); maximum likelihood estimation should be restricted to the 'feasible' space. Conversely, there are an infinite array of genealogies providing GIBD probabilities arbitrarily close to any point in the feasible region. In practice, comparison of likelihoods of standard simple relationships may be a more useful procedure than maximum likelihood estimation. For our example, we may give the log-likelihood difference between the sib hypothesis and other standard alternatives, such as those of Table 1. Note, however, that even among simple relationships there is not a 1–1 correspondence between GIBD probabilities and relationship; half-sibs, grandparent–grandchild and uncle–nephew all correspond to the same point in the gene identity space, and are thus indistinguishable on the basis of the type of data discussed here.

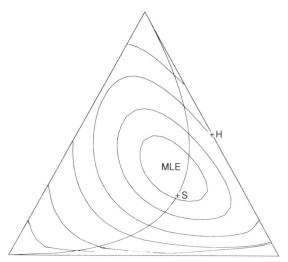

Fig. 2. Contours of log-likelihood for the IBD parameters of relationships between two non-inbred individuals, for the example data of Table 3. The log-likelihood differences between the maximum and each contour are 0.25, 1.75, 2.75, 4.25, 6.75 and 13.25, respectively.

2.4. Linked loci in the estimation of relationship

The simplest cases discussed so far refer to data at unlinked Medelian loci. However, as detailed data on individual genomes become more readily available, the presence of linkage among the data loci becomes more usual. In principle, the problem remains the same. Relationships imply probabilities of identity by descent, although these are now two-locus (or multilocus) identity probabilities, and, given the genetic characteristics of the data loci, these parameters are estimable from phenotypic data at the linked loci. However, there is an intrinsic difference now in that these gene identity probabilities depend not only upon the genealogical structure, but upon the recombination fraction (or degree of linkage) between the loci in question. Thus, the IBD parameters of relationship now also involve the genetic parameters of the data loci.

The details of the analysis will not be pursued here, but it is of interest that linkage plays a paradoxical role in the estimation of relationship. On the one hand the positive correlation in IBD among the data loci in general means that there is less information to discriminate among relationships of different degree than for equally informative independent (unlinked) loci. While the mean numbers at which the two relatives carry, 2, 1, or 0 IBD genes remain unchanged, in any sample of loci, the variance of these numbers is inflated by the positive correlation. On the other hand, data at linked loci can discriminate between relationships which are indistinguishable (non-identifiable) on the basis of data at unlinked loci. The relationships half-sib, grandparent–grandchild and uncle–nephew discussed above have identical single-locus probabilities of GIBD patterns, but except for complete linkage or free recombination, have distinct two-locus GIBD probabilities (Thompson, 1986).

At the limit of linked loci, are data, such as that of DNA sequence, which provide almost infinite density of information along a continuous genome. In principle, the lengths of sections of IBD genome can also provide information to resolve genealogical relationships (Fisher, 1949; Donnelly, 1983), but neither data nor appropriate statistical methods are yet available.

2.5. Bayesian estimates of relationship. Paternity testing

The estimation of relationship is perhaps most practicaly relevant in the practice of paternity testing, on which there is an enormous literature which we shall not even attempt to summarise. Some statistical issues are addressed by Chakraborty and Ferrell (1982). It is often proposed that prior probabilities of paternity be incorporated into the likelihood-based inferences, providing a Bayesian posterior probability of paternity. Several methods of estimating an appropriate prior probability of paternity have been proposed (Baur et al., 1981).

Whether Bayesian or non-Bayesian, most methods are in effect a test of the hypothesis of a (child, mother, father) genealogical trio against the alternative of (child, mother, unrelated); that is, only the likelihoods or posterior probabilities of these two alternatives are considered. However, there seems to be some advantage to regarding the problem in the framework of estimation of relationship rather than of testing, although this is not the standard legal/medical practice. An estimation approach was given by Goldgar and Thompson (1988). Here, a (child, mother, unknown) trio is assumed, and the unknown is parametrised via the relationship of that individual to the father of the child. (This father is assumed unrelated to the mother.) This unknown relationship may then be estimated by either a likelihood or a Bayesian approach, the Bayesian prior distribution now being over the kinship coefficients between the putative father and the true one.

2.6. Use of DNA data in the estimation of relationship

There are three types of DNA-based data currently available (or potentially available in the near future) for use in estimation of relationships; none of these is sequence data which is far more expensive and difficult to obtain. All three are determined through the use of restriction enzymes which cut the sample DNA at points where specific short sequences (usually 4 or 6 bases) of DNA occur. Differences in the DNA of individuals are reflected in the different lengths of fragments of DNA thereby obtained. The first restriction fragment length polymorphisms (RFLPs) (Botstein et al., 1980) detected differences arising from mutations or a small insertion or deletion in the DNA. Use of RFLP data for the estimation of biological relationships differs little from the use of classical markers, although normally one expects that these DNA markers, being more highly variable in a population, will convey more information.

The second DNA data type are provided by Jeffreys' probes (Jeffreys et al.,

1985). Here the differences in DNA fragment lengths arise from variable numbers of repeats of a short DNA section. Moreover, since the probes will match (is homologous to) multiple regions of DNA in each individual, the data are the joint outcome over multiple loci. The genotypes provided by individual loci cannot be identified, but the degree of similarity of the set of fragment lengths shown by relatives reflects the degree of closeness of relationship between them. Although there are interesting statistical questions in the estimation of genealogical relationships from this type of data, these may not be of long-term practical importance, since, for the human species at least, Jeffreys probe information is becoming superceded rapidly by data on Variable-Number-of-Tandem-Repeat loci (Nakamura et al., 1987). The principle is similar to that of Jeffreys' probe data, but since the VNTR probe is homologous to only a single region of DNA the resulting information in similar to that of 'classical' RFLPs in reflecting varying fragment lengths at a single genetic locus. The statistical situation is thus again that of Section 2.2, but for loci that are even more highly informative than RFLPs. These VNTR loci provide large amounts of information both for the identification of individuals, and estimation of relationships.

2.7. *Genealogy reconstruction versus estimation of relationship*

In theory, the estimation of relationship between pairs of individuals extends also to the estimation of joint relationships between numerous individuals, and thence to the reconstruction of genealogies from genetic data. Any joint genealogical relationship provides a probability distribution over a set of GIBD patterns for the genes of the individuals involved, and likelihood estimation of this distribution from joint phenotype data on these individuals is possible. However, the number of possible GIBD patterns increases very rapidly with the number of individuals considered (Thompson, 1974). Further, the number of GIBD patterns of positive probability for a given genealogy also increases rapidly, and the number of alternative hypotheses of genealogical relationship between a set of individuals is immense. Note also that, as for the simple case of pairwise relationships between non-inbred individuals, the GIBD probabilities that are to be estimated are the expected proportion of loci exhibiting that GIBD pattern. Thus, unless there are data for many more loci than there are GIBD patterns of positive probability, the estimate will have undesirable properties.

In practice, the problem of reconstructing a genealogy must be partitioned into smaller components. This raises several statistical questions, most notably the following: although likelihood estimation of relationship between a given set of individuals is (subject to identifiability) consistent, estimation of the individuals who should be placed in a given relationship in the reconstructed genealogy is not. Whereas the true relationship between a set of individuals has maximum expected log-likelihood, the individuals giving maximum expected log-likelihood for a given relationship are not necessarily those for whom this is the true relationship. More specifically, an individual's 'most likely parent' can be *in expectation* a full sib.

Thompson (1976) details this problem further. There are many open problems in the methodology for the reconstruction of genealogies from genetic data.

One may ask the purpose of a large-scale detailed genealogical construction as opposed to more specific estimates of relationship. While such is a fascinating possibility, it seems unlikely that any reconstructed genealogy of a large natural population will ever have sufficient accuracy to provide direct measures of population biological parameters, such as, for example, the variance of male reproductive success (Meagher and Thompson, 1987). More direct estimation of such parameters (e.g., Roeder et al., 1989) under a population model which incorporates the genotype arrays of parent and offspring generations may be a more useful approach.

3. Relationships between populations

Over time, Mendelian segregation, acting within a population on the frequencies of alleles at polymorphic loci, results in the process known as *random genetic drift*. Allele frequencies in populations change simply due to the random demographic processes and the random Mendelian segregation of genes from parent to offspring within the population pedigree. Populations which are closely related, in the sense of containing individuals who have higher probabilities of sharing genes IBD, will also tend to show similar allele frequencies. These similarities, which are a residue of common ancestry and past gene flow, between populations can be used to infer relationships between populations.

There are two approaches to the analysis of relationships between populations from genetic data. One focuses upon measures of current genetic similarity or difference, while the other focuses more on the parameters of the demographic and genetic processes (admixture, migration, drift, mutation and selection) that lead to current observations. In this paper we shall focus upon the latter, not because it is necessarily to be preferred, but because this chapter relates to the estimation of parameters of relationship, while the chapter of Rao and Chakraborty (1991) covers the topic of genetic distance measures.

3.1. Population kinship and random genetic drift

The idea that GIBD between populations could be estimated from genetic similarities was formalised by Morton et al. (1971), who proposed that the level of gene identify between genes randomly chosen from two populations could be best estimated from a measure of allelic identity between such genes, the latter being a function of the population allele frequencies. However, the variance of the processes involved is high, and genetic similarity may not reflect gene identity with high precision.

An alternative approach is to consider directly the probability distributions of allele frequencies in populations subject to random genetic drift, and to model observed frequencies. Additionally, actual data consists of sample allele frequencies, and the sampling process provides additional levels of genetic divergence between observed sample frequencies. This multinomial sampling is akin to the multinomial sampling from generation to generation of the drift process. A useful approximation is that a set of $k - 1$ functions of the k allele frequencies at a single locus perform approximately independent Brownian motions under the drift process, with rate parameter $1/(8N)$, where N is the population size (Edwards, 1971), and that the multinomial sampling in current populations provide additional independent random increments, with mean 0 and variance $1/(8n)$, where n is the sample size. Thus these transformations of sample frequencies have approximately multivariate Normal distributions over populations, which simplifies analysis. Independently inherited loci contribute independent dimensions to drift process. The population size N affects the process at all loci equally, although, of course, sample sizes n may vary between loci. On the basis of a demographic framework detailing population origins, admixture, or migration patterns, and under the model of genetic change implies by random genetic drift, and under the sampling of current populations, it is possible, in some simple cases, to derive a probability for current data as a function of unknown demographic parameters, and thence to estimate these parameters from this likelihood function.

3.2. Population relationships resulting from admixture

Where a population is a mixture of two ancestral populations, it is possible to estimate this mixture proportion, provided data are available both on the allele frequencies in the unmixed gene pools, and in the mixture. An early example was given by Glass and Li (1953), but the stochastic process of random genetic drift in the ancestral populations was not incorporated into the model. Thompson (1973) provides an example of likelihood estimation of a mixture proportion on the basis of a model which incorporates random genetic drift, not just in the mixed population, but also in the two founder gene pools, and also models explicitly the sampling variation in all three current gene pools. The population of Iceland, which was used as an example in this analysis, probably comes closer than most real examples to fulfilling the many assumptions of the model, but estimates must be viewed with caution. The power of genetic data to resolve details of history on this (40-generation) time scale is not great.

3.3. Population relationships resulting from interpopulation migration

Where a population has a well defined discrete subpopulation structure and a time homogeneous pattern of intersubpopulation migration, this pattern and level of migration will be reflected in current levels of interregional variation. The theoretical models of differentiation have been well developed (Weiss and Kimura, 1965; Maruyama, 1972). The statistical aspects of estimation of migration relationships between populations from the realised genetic data have been less considered, although Bodmer and Cavalli-Sforza (1968) address some data issues, and analyses of genetic microdifferentiation have a long history (e.g., Friedlaender et al., 1969).

For populations i, $i = 1, \ldots, k$, the migration pattern is most conveniently summarised by a backwards migration matrix

$$M_{ij} = P(\text{Random parent of individual born in } i \text{ was born in } j).$$

Superimposing random genetic drift and sampling, Jacquard (1974) gives recursive formulae for the covariance in allele frequencies between populations at generation t, which lead to simpler approximate formulae for the covariances, $V_{ij}^{(t)}$ between the transformed allele frequencies:

$$V_{ij}^{(t+1)} = \left[\sum_l \sum_{l'} M_{il} M_{jl'} V_{ll'}^{(t)} \right] (1 - \delta_{ij}/2N_i) + \delta_{ij}/2N_i,$$

and hence, in principle, likelihoods for migration parameters are available. However, except in a few special cases, the complexity of demographic history is unlkely to be adequately summarised by a migration matrix model with few parameters. Wood et al. (1985) consider some such models, and the testing of migration structure hypotheses from actual migration counts. Rogers and Harpending (1983) reduce the pattern of genetic differentiation to a single function, and relate the expectation of this measure to a function of the eigenvalues of the migration matrix. In effect, therefore, the level of migration is summarised by a single parameter, which can be estimated from equilibrium genetic differentiation. Thompson (1983) uses a migration model with two migration parameters, one reflecting external migration (α) from a population '0' and one reflecting internal migration (K), with migration matrix

$$M_{ij} = (1 - \alpha)\left[(1 - K)\delta_{ij} + KN_j \middle/ \sum_{l=1}^{k} N_l \right] + \alpha\delta_{j0}, \quad i = 1, \ldots, k, \quad j = 0, 1, \ldots, k,$$

to analyse genetic differentiation among the regions of the Faroe Islands. The parameter K is estimated by likelihood from the degree of genetic differentiation in polymorphic allele frequencies, on the basis of a model of sampling and random genetic drift. It is possible to make likelihood-based estimates of the parameters from current genetic data, but the precision of estimation is not high, even assuming the simplified model approximates the true migration structure.

Although the model does not assume equilibrium, one important general point that emerges is that the parameter estimates will imply a time scale over which equilibrium levels of divergence can be expected to be established. Interregional migration patterns predating this time period will no longer be reflected in patterns or levels of genetic similarity, and are non-identifiable on the basis of current genetic data.

3.4. Isolation by distance

Often populations are not well modelled as discrete subunits, but nonetheless the population has a migration-induced relationship structure, which is reflected in increasing genetic divergence at larger geographic distances. Wright (1943) first developed the theory of the decay of expected correlations in allele frequency with distance, under models of migration within a population continuum. Malecot (1948) approached the same problem in a different way, deriving the form of the correlation, as a function of geographic distance. By fitting observed correlations to this form, it is possible to investigate not only the degree and scale of migration, but also its dimensionality.

4. Phyletic relationships

As the complete genealogy of individuals is viewed from greater and greater distance, the details of individuals and individual relationships become blurred, leaving eventually a genealogy of populations (Felsenstein, 1983). Where populations fission, but do not undergo fusion, relationships between them are *phyletic*, and can be represented as a tree. The basic assumptions underlying models of genetic change in such populations, is that distinct populations evolve independently. Thus genetic changes, leading to patterns of genetic similarities and differences, consist of a process of independent increments between the fissions of populations, the nodes of the evolutionary tree.

4.1. Intraspecific relationships

Random genetic drift can also provide a model for polymorphic intraspecific variation on a larger time scale, between populations that have ot exchanged genes in millenia. For populations that show the same alleles differing only in frequency, this process of drift leading to population divergence is an appropriate model. Edwards (1970) developed a branching diffusion model; functions of the allele frequencies change within each population under a diffusion process, and populations fission as a branching tree. Felsenstein (1973) and Thompson (1975) present likelihood approaches to this estimation problem (see also Felsenstein, 1983). Although the likelihood function from given data can easily be written down for a given set of parameter values and for a given form of tree, likelihood estimation of the phyletic relationships between the populations raises many interesting statistical questions. Whereas estimates of divergence times within a given tree topology are fairly readily obtained, these are often viewed mainly as nuisance parameters. The primary question is to estimate the tree topology, but the number of labelled tree topologies increases rapidly with the number of current populations (the 'tips' of the tree), and each constitutes a separate discrete hypothesis.

4.2. Interspecific relationships

On a still greater evolutionary time scale, there are not only changing frequencies of polymorphic alleles, but a more general *substitution* process in which alleles are lost from populations, or new mutations arise and some even become fixed in some phyletic groups. The models required are then those that describe this substitution process. This gives rise to distance measures which, in expectation, increase linearly with time of divergence under this model (Nei, 1973) and can thus be used to estimate relative divergence times. Alternatively, models for substitution of alleles, or DNA bases, over periods in which evolution proceeds independently in different populations can be developed directly, and a direct likelihood estimation approach taken to the analysis of data (Felsenstein, 1983).

5. Conclusion

Relationships between individuals, and thence between the populations containing individuals, are reflected in genetic data, and can thus be estimated from genetic data, provided an appropriate model is developed. For relationships between individuals, the model for the data is well defined. Although there are computational problems inherent in the reconstruction of large genealogies from genetic data, the likelihood for any specific genealogical hypothesis is readily obtainable.

For relationships between populations, there are many alternative models of migration history, none of which are likely to be close to a long-term reality. The depth of history which is identifiable from current genetic data is limited to the period over which history affects current data. Long-past migration, and distant speciation events will no longer be reflected in current divergence patterns. On the other hand, some past migrations of considerable antiquity have left their imprint on current genetic similarities, and at least qualitative inferences can then be made. Ammerman and Cavalli-Sforza (1984) provide an example in their analysis of the Neolithic diffusions of Indo-European populations.

Acknowledgement

I am grateful to Ellen Wijsman for many helpful comments on an earlier draft, and to Amy Huisken for producing the diagrams.

References

Ammerman, A. J. and Cavalli-Sforza, L. L. (1984). *The Neolithic Transition and the Genetics of Populations in Europe.* Princeton University Press, Princeton, NJ.
Baur, M. P., Rittner, C. and Wehner, H. D. (1981). The prior parameter in paternity testing: Its relevance and estimation by maximum likelihood. In: *Lectures of the Ninth International Congress of the Society for Forensic Hemogenetics,* Bern, 389–392.
Bodmer, W. F. and Cavalli-Sforza, L. L. (1968). A matrix migration model for the study of random genetic drift. *Genetics* **59**, 565–592.
Botstein, D., White, R. L., Skolnick, M. H. and Davis, R. W. (1980). Construction of a genetic linkage map in man using restriction fragment length polymorphisms. *Am. J. Hum. Genet.* **32**, 314–331.
Chakraborty, R. and Ferrell, R. E. (1982). Correlation of paternity index with probability of exclusion and efficiency criteria of genetic markers for paternity testing. *Forensic Sci. Int.* **19**, 113–124.
Chakraborty, R. and Rao, C. R. (1991). Measurement of genetic studies. In: *Handbook of Statistics* **8**, C. R. Rao and R. Chakraborty (eds.). North-Holland, Amsterdam, 271–316.

Cotterman, C. W. (1940). A calculus for statistico-genetics. Ph.D. Thesis, Ohio State University. In: *Genetics and Social Structure*. P. Ballonoff (ed.), Benchmark papers in Genetics, Dowden, Hutchinson and Ross, 1975, 157–272.

Donnelly, K. P. (1983). The probability that related individuals share some section of the genome identical by descent. *Theor. Pop. Biol.* **23**, 34–63.

Edwards, A. W. F. (1970). Estimation of the branch points of a branching diffusion process. *J. Roy. Statist. Soc. (B)* **32**, 154–174.

Edwards, A. W. F. (1971). Distances between populations on the basis of gene frequencies. *Biometrics* **27**, 873–881.

Felsenstein, J. (1973). Maximum likelihood estimation of evolutionary trees from continuous characters. *Am. J. Hum. Genet.* **25**, 471–492.

Felsenstein, J. (1983). Statistical inference of phylogenies. *J. R. Stat. Soc. Ser. A* **146**, 246–272.

Fisher, R. A. (1949). *The Theory of Inbreeding*. Academic Press, New York.

Friedlaender, J. S., Sgaramella-Zonta, L. A., Kidd, K. K., Lai, L. Y. C., Clark, P. and Walsh, R. J. (1971). Biological divergences in South-Central Bougainville: An analysis of blood polymorphisms and anthropometric measurements utilising tree models, and a comparison of these variables with linguistic, geogrpahic and migrational 'distances'. *Am. J. Hum. Genet.* **23**, 253–270.

Glass, B. and Li, C. C. (1953). The dynamics of racial intermixture – An analysis based upon the American Negro. *Am. J. Hum. Genet.* **5**, 1–20.

Goldgar, D. E. and Thompson, E. A. (1988). Bayesian interval estimation of genetic relationships: Application to paternity testing. *Am. J. Hum. Genet.* **42**, 135–142.

Jacquard, A. (1974). *The Genetic Structure of Populations*. Springer-Verlag, New York.

Jeffreys, A. J., Wilson, V. and Thein, S. L. (1985). Individual-specific 'fingerprints' of human DNA. *Nature* **316**, 76–79.

Malecot, G. (1948). *Les Mathématiques de l'Hérédité*. Masson, Paris.

Maruyama, T. (1972). The rate of decay of genetic variability in a geographically structure finite population. *Math. Biosci.* **14**, 325–335.

Meagher, T. R. and Thompson, E. A. (1987). Analysis of parentage for naturally established seedlings within a population of *Chamaelirium luteum* (Liliaceae). *Ecology* **68**, 803–812.

Mendel, G. (1866). Experiments in plant hybridisation. In: English translation, with a commentary by R. A. Fisher, Oliver and Boyd, Edinburgh, 1965.

Morton, N. E., Yee, S., Harris, D. E. and Lew, R. (1971). Bioassay of kinship. *Theor. Pop. Biol.* **2**, 507–524.

Nakamura, Y., Leppert, M., O'Connell, P., Wolff, R., Hom, T., Culver, M., Martin, C., Fujimoto, E., Hoff, M., Kumlin, E. and White, R. (1987). Variable number of tandem repeat (VNTR) markers for human gene mapping. *Science* **235**, 1616–1622.

Nei, M. (1973). The theory and estimation of genetic distance. In: *Genetic Structure of Populations*, N. E. Morton (ed.). Honolulu University Press, Honolulu.

Roeder, K., Devlin, B. and Lindsay, B. G. (1989). Application of maximum likelihood methods to population genetic date for the estimation of individual fertilities. *Biometrics* **45**, 363–379.

Rogers, A. R. and Harpending, H. C. (1983). Population structure and quantitative characters. *Genetics* **105**, 985–1002.

Thompson, E. A. (1973). The Icelandic admixture problem. *Ann. Hung. Genet.* **37**, 69–80.

Thompson, E. A. (1974). Gene identities and multiple relationships. *Biometrics* **30**, 667–680.

Thompson, E. A. (1975). *Human Evolutionary Trees*. Cambridge University Press, Cambridge.

Thompson, E. A. (1976). Inference of genealogical structure. *Soc. Sci. Inform.* **15**, 477–526.

Thompson, E. A. (1983). Inferring migration history from genetic data: Application to the Faroe Islands. *SSHB Symposium* **23**, R. W. Hiorns (ed.). Taylor and Francis, London, 123–142.

Thompson, E. A. (1986). *Pedigree Analysis in Human Genetics*. Johns Hopkins University Press, Baltimore, MD.

Weiss, G. H. and Kimura, M. (1965). A mathematical analysis of the stepping stone model of genetic correlation. *J. Appl. Probab.* **2**, 129–149.

Wood, J. W., Smouse, P. E. and Long, J. C. (1985). Sex-specific dispersal patterns in two human populations of highland New Guinea. *Am. Nat.* **125**, 747–768.

Wright, S. (1943). Isolation by distance. *Genetics* **28**, 114–138.

C. R. Rao and R. Chakraborty, eds., *Handbook of Statistics, Vol. 8*
© Elsevier Science Publishers B.V. (1991) 271–316

9

Measurement of Genetic Variation for Evolutionary Studies

R. Chakraborty and C. R. Rao

1. Introduction

One of the major objectives of employing population genetic principles in evolutionary studies is to describe the amount of genetic variation among individuals within as well as between populations. The rationale behind such studies is to examine the pattern of genetic variation within and between populations from which inferences regarding the mechanism of maintenance of genetic variation can be drawn. Measures of genetic variation employed, therefore, should reflect temporal as well as spatial changes. Furthermore, the utility of indices of variation are enhanced when the mathematical properties of such measures are such that the relative roles of the different evolutionary factors, such as mutation, genetic drift and selection, are distinguishable, so that their relative importance in governing the mode of evolution can be understood from the magnitude and pattern of such measures of variation.

In this chapter our objective is to review the existing measures of genetic variation from the above perspective particularly emphasizing the concepts of indices of genetic variation that have well-defined biological interpretations. For these measures we shall also describe their mathematical properties to examine their utility in the context of evolutionary studies. Also statistical issues of estimation and hypothesis testing will be discussed with particular attention to examine whether or not such measures can be used to understand the underlying evolutionary processes that led to the existing amount of variation in contemporary populations. Two classes of measures will be considered, those describing: (i) genetic variability within populations, and (ii) genetic variability between populations. This distinction is important, but it is worthwhile to note that the conceptualization of measures of variation within and between populations should be under a unified framework. For each of these we have to deal with their pattern of changes over time under different evolutionary forces so that the genetic affinities between populations can be studied with the same principles that dictate genetic variations among individuals within a population.

Before proceeding to a detailed discussion of the issues raised above, we must note that before the discovery of serological methods for detecting genetic variation, morphological (qualitative as well as quantitative) variation was the most popular means of examining variability in almost all organisms. In fact, the classic definition of species, genera, and evolutionary kingdoms are based primarily on morphological variation and reproductive isolation of organisms. Even though morphological variation is the most convenient means of identifying differences between individuals, populations or organisms, the main problem of measuring genetic variation at such a level is that we are not certain to what extent such variation is truly genetic. This is so because environmental factors have a large role on morphological variation, and it is difficult to quantify the effect of environmental factors on morphological variation particularly with population-based studies. In a similar vein, also the genetics of reproductive isolation is not a well-known phenomenon. Therefore, quantification of this aspect for studying genetic variation within and between populations is problematic.

With the advent of biochemical methods for detecting genetic variation, and more recently with the discovery of recombinant DNA technology, these difficulties are circumvented to a large extent, because genetic variation can now be detected at a molecular level so that the uncertainty of the extent of the environmental role can almost be totally neglected. Therefore, our major discussions will be restricted to genetic variation only, although at the end we shall make some remarks to suggest some obvious connections of morphological variation with the genetic distance measures.

2. Genetic variation within populations

Almost all discussions in population genetics start with genetic characteristics of populations, without even describing what the term 'population' means. Simply, a population can be defined as a group of interbreeding individuals that exist together in time and space (Hedrick, 1983). Elsewhere (Chakraborty and Danker-Hopfe, this volume) there is a detailed discussion of the genetics of a substructured population. This is the case where the population is fragmented because of the fact that interbreeding is not uniform (or random) across the various segments of the population. For the purpose of this chapter, we consider that within a population the interbreeding is random, so that every member of the population has an equal chance of mating with another, irrespective of their genetic make-up. This is a conceptualized ideal population, often called a random-mating (panmictic) population. Although not explicitly stated, Fisher's (1953) assertion – "it is often convenient to consider a natural population not so much as an aggregate of living individuals but as an aggregate of gene ratios" – in fact is applicable only to a random-mating population.

Given this definition of a population, we might start with several biologically meaningful measures of genetic variation within populations.

2.1. Proportion of polymorphic loci

It is estimated that the genome of higher organisms probably contains somewhere between four to fifty thousand structural loci. Complete genetic characterizations of individuals would need determining arrangements of nucleotides at all these loci. Even though this is the ultimate goal of molecular genetics, at present this is clearly impossible. With electrophoretic or restriction fragment length polymorphism (RFLP) techniques, however, it is possible to examine several of these loci to examine if individuals are of similar or dissimilar types at such loci. If all individuals are identical, and in particular homozygous for the same allele at a locus, the locus is called monomorphic (or fixed). Sometimes, however, rare variation is found. Therefore, a more generous definition of polymorphism is that the frequency of the most common allele is equal to or less than an arbitrarily chosen large fractional number ($q < 1$). Generally, $q = 0.99$ or 0.995 is taken as a criterion of polymorphism. The proportion of loci satisfying this criterion is called the proportion of polymorphic loci (P), and it provides a crude measure of genetic variation in a population. This index is simple, but not without caveats. First, the value of q is not truly independent of sample size. For example, in a sample of n diploid individuals any allele with frequency p ($0 < p < 1$) has a probability $(1 - p)^{2n}$ of not being represented in a sample, and hence in spite of the fact that there may be multiple alleles segregating at a locus, when n is small there is a high likelihood of only one allele being observed in the sample. Second, when the number of loci examined (r) is small, the proportion of polymorphic loci is subject to a large sampling error and often this measure becomes useless. In spite of these difficulties, this measure had been advocated in many ecological studies (Magurran, 1988). In the context of evolutionary studies, Kimura and Ohta (1971) showed the utility of this index in supporting the neutral mutation theory of molecular evolution. Nevertheless, we conclude that unless the number of individuals surveyed per locus is large and a large number of loci are sampled from the genome, this measure of genetic variation is too crude for any detailed investigation of evolutionary processes.

2.2. Gene diversity

An alternative measure of genetic variation that does not depend upon the arbitrary definition of polymorphism is called gene diversity. In the genetic literature gene diversity was popularized in the early seventies (Nei, 1972; Lewontin, 1972) where it is defined as the probability that two genes randomly chosen from the population are dissimilar in their genetic types. If p_1, p_2, ..., p_k represent the true frequencies of k segregating alleles at a locus in a population, the gene-diversity at the locus is defined as

$$h = 1 - \sum_{i=1}^{k} p_i^2 . \tag{2.1}$$

In a random-mating population this is equivalent to the proportion of heterozygous individuals in the population. Therefore, the terms gene diversity and heterozygosity are often interchangeably used in population genetics literature. It is important to note that for organisms (and/or populations) whose mating behavior is not random, the actual heterozygosity in the population can be much different from the gene diversity predicted by equation (2.1).

In the ecological literature this measure of diversity has a much older history. Originally, it was suggested as a measure of ecological diversity by Gini (1912) and reinvented by Simpson (1949) (see also Rao, 1982a).

Irrespective of the origin of this measure, its advantages in evolutionary studies are several. First, gene-diversity defined this way can easily be extended over several loci, simply by averaging it over loci to define average gene diversity over the genome of any organism. Second, this measure is also applicable to haploid or polyploid organisms. For example, in the case of the mitochondrial genome (which generally behaves like a single-locus haploid genome, generally maternally transmitted) if the different mitochondrial morphs (detected either by RFLP or sequence studies) are treated as different alleles, h can be used without any modification to depict genetic variation at the mitochondrial genome in any population. Third, this measure can be used to study genetic variation irrespective of the mating system prevalent in the population. Nei (1972) defined the complement of h ($J = \sum p_i^2$) as gene identity, which in certain occasions has interesting biological implications. For example, for a random-mating population this is equivalent to homozygosity in a population and therefore can be used to study the effect of inbreeding in a population. The reciprocal of gene identity is also called the effective number of alleles (Kimura and Crow, 1964).

Note that the quantity h, defined by equation (2.1), is always nonnegative. Its minimum value (zero) is reached when there is perfect monomorphism ($p_i = 1$ for a single i, and other p_i's are all equal to zero) in the population. The maximum value, $(k - 1)/k$, is however dictated by the number of segregating alleles in the population. The maximum gene diversity is reached when all segregating alleles have equal probability of occurrence in the population. The maximum value approaches one as k increases ($k \rightarrow \infty$). Therefore, under the infinite allele model (Wright, 1948; Kimura and Crow, 1964) the measure of gene diversity (equation (2.1)) has a well-defined range between 0 and 1.

2.3. Shannon-information index

In information science the concept of entropy functions leads to another measure of diversity

$$h_s = - \sum_{i=1}^{k} p_i \log_e p_i, \tag{2.2}$$

which also has been used in the context of evolutionary and ecological studies (see, e.g., Lewontin, 1972; Magurran, 1988; Rao, 1982b,c). Even though entropy functions have useful applications in information sciences and theoretical physics, in the context of evolutionary studies this measure has several limitations. First, it does not have a clear cut biological meaning. Second, although its minimum value (zero) is attained for a monomorphic locus, it can easily take very large values. Its maximum value is $\log_e k$, again obtained when each of the k segregating alleles are equally frequent in the population. It should be noted, however, that for independently segregating loci, h_s measured over all loci can be averaged to evaluate the average per locus diversity within populations, because of the additive property of the logarithms of products.

2.4. Number of alleles

Although gene diversity and Shannon-information index have received maximum attention for measuring genetic variation within populations, one disturbing feature of these measures is that their magnitude is insensitive to the number of alleles that are present in very low frequencies in a population. Therefore, even if allelic diversity is very large in a population, if one or a few of them are predominant and the others are rare, gene diversity on Shannon-index for such a locus will be small. Consequently, an alternative measure of genetic variation can simply be the number of different allelic types at the locus, irrespective of their frequencies. In ecological literature this is well known as species richness (see, e.g., Pielou, 1975). It is a simple measure and, as will be discussed later, has a number of unique statistical properties for certain evolutionary models of maintenance of variation. Nevertheless, its principal drawback stem from the fact that this measure disregards the frequencies with which the various alleles occur in a population. Furthermore the expected number of different alleles observed in a sample is critically dependent on the sample size, while the expectation of the observed gene diversity approaches its asymptotic value comparatively faster. Consequently, gene diversity (or Shannon-information index) is a relatively robust measure subject to smaller sampling fluctations. Lastly, and probably more importantly, for some methods of genetic assay there is an artificial limit on the number of different alleles whose gene products can be distinguished. For example, for the serological tests based on the antigen–antibody reaction, the number of alleles that can be distinguished is simply restricted by the number of different types of antibodies available for the assay. For RFLP's where we can detect the presence or absence of a specific restriction enzyme recognition sequence, the number of allelic types is only two. Of course, the recent discovery of Variable Number of Tandem Repeat (VNTR) polymorphisms make this measure an useful and informative summary statistics of genetic variation within a population.

2.5. Unified measures of diversity

Although each of the above measures of genetic variation within populations had been advocated by their respective proponents at length in the context of ecological and evolutionary studies, attempts to unify these concepts under a single mathematical framework has started only recently. Most notable work in this direction is a characterization theorem, first postulated by Rao (1982a), which can be stated as follows.

For a single attribute (locus) with k different categories (alleles) occurring with relative frequencies $p_1, p_2, \ldots, p_k$ in a population, let $h(p) = h(p_1, p_2, \ldots, p_k)$ denote a measure of diversity. Assume that $h(p)$ satisfies the postulates:

  (i) $h(p)$ is symmetric with respect to the components of the p-vector and attains its maximum for $e = (1/k, 1/k, \ldots, 1/k)$; i.e. when all of the k categories are equally frequent,

  (ii) $h(p)$ admits partial derivatives up to the second order for the $k - 1$ independent components of the p-vector and the $h''(p)$ matrix of second partial derivatives, $h''(p) = ((h''_{ij}(p)))$ for $i, j = 1, 2, \ldots, k - 1$ with

$$h''_{ij}(p) = \partial^2 h(p)/\partial p_i \partial p_j,$$

is continuous and not null at $p = e$ and

  (iii) $\qquad h\{(p + e)/2\} = \frac{1}{2}\{h(p) + h(e)\} = c\{h(e) - h(p)\},$

where c is a constant.

Then $h(p)$ must be of the form

$$h(p) = a\left[1 - \sum_{i=1}^{k} p_i^2\right] + b, \qquad (2.3)$$

where $a > 0$ and b are constants. In other words, postulates (i) through (iii) essentially characterizes the Gini–Simpson measure of gene diversity.

Rao (1982b) and Rao and Boudreau (1984) also indicated that along with the diversity measures given by equations (2.1) and (2.2), the following four other indices,

  (a) *α-order entropy of Havrda and Charavát*, defined by:

$$h_\alpha(p) = [1 - \sum p_i^\alpha]/[2^{\alpha-1} - 1] \quad \text{for } \alpha > 0 \quad \text{and} \quad \alpha \neq 1, \qquad (2.4)$$

  (b) *paired Shannon entropy*, defined as:

$$h_p(p) = -\sum p_i \log_e p_i - \sum (1 - p_i) \log_e(1 - p_i), \qquad (2.5)$$

  (c) *α-degree entropy of Renyi*, defined as:

$$h_R(p) = (1 - \alpha)^{-1} \log_e (\sum p_i^\alpha) \quad \text{for } 0 < \alpha < 1, \qquad (2.6)$$

and
  (d) *γ-entropy function*, defined as:

$$h_\gamma(p) = [1 - (\sum p_i^{1/\gamma})^\gamma]/[1 - 2^{\gamma-1}] \quad \text{for } \gamma > 0, \ \gamma \neq 1, \qquad (2.7)$$

all satisfy the two conditions

C_0: $h(p) = 0$ if and only if all components of the p-vector are zero except one (i.e., $p_i = 1$ for one i and the remaining p_i's are all zero), and

C_1: $h\{\lambda p + (1 - \lambda)q\} \geqslant \lambda h(p) + (1 - \lambda)h(q)$, with equality if and only if $p = q$ (the concavity property).

Therefore, all of these could be used as diversity measures of genetic variation. Even though these quantities do not have any easy biological interpretation, it may be noted that the Gini–Simpson measure of diversity is a special case of equation (2.4) for $\alpha = 2$ and equation (2.7) for $\gamma = \frac{1}{2}$.

In the context of ecological studies, Hill (1973) proposed another method of unification of the various diversity measures by defining an index

$$h_{\mathrm{H}}(p) = \left[\sum_{i=1}^{k} p_i^{\alpha} \right]^{1/(1-\alpha)}, \tag{2.8}$$

which has a number of interesting properties. For example, it reduces to the number of alleles for $\alpha = 0$, reciprocal of gene identity for $\alpha = 2$, and exponential Shannon-information index for $\alpha = 1$.

We thus conclude that apart from the concept of proportion of polymorphic loci, all other measures of genetic variation within populations (gene diversity or heterozygosity, number of alleles, or Shannon-information index) belong to a general class of diversity measures. Therefore, the biologically interpretable concepts of the number of segregating alleles and gene diversity also have a sound mathematical basis.

2.6. Estimation of diversity indices

In the above formulation note that nothing is assumed regarding the production and maintenance of genetic variability in defining any of the indices of genetic variation within populations. The only assumption made is that the allelic distinctions are discrete, so that in a population, alleles can be unequivocally categorized in different classes, and their relative frequencies can be determined without error. The statistical formulation of the estimation problem, therefore, can be based on the multinomial distribution theory. In some population genetics texts there is a misconception that Nei's measure of gene identity (complement of equation (2.1)) assumes selective neutrality of alleles (see, e.g., Futuyama, 1979), which is obviously incorrect.

In a survey of n genes ($\frac{1}{2}n$ diploid individuals) randomly selected from a population, let n_1, n_2, ..., n_k be the frequencies of k observed alleles in the sample ($n = \Sigma n_i$). It is well known that $x_i = n_i/n$ is an unbiased estimate of the true population frequency of the ith allele ($i = 1, 2, \ldots, k$). The most popular approach of estimating any of the above diversity indices is to substitute x_i for p_i in the above definitions to get the respective estimators.

From the simple logic of multinomial sampling, Nei and Roychoudhury (1974) and Nei (1978a) showed that

$$\hat{h}_1 = 1 - \sum x_i^2 , \qquad (2.9)$$

is not an unbiased estimator of h (equation (2.1)), since

$$E(\hat{h}_1) = (1 - n^{-1}) \cdot (1 - \sum p_i^2) . \qquad (2.10)$$

Therefore, these authors suggested the unbiased estimator

$$\hat{h} = n \cdot (1 - \sum x_i^2)/(n - 1) , \qquad (2.11)$$

although the bias of $\hat{h}_1$ is generally quite small when 50 or more individuals are surveyed for genetic variation ($n > 100$) within populations. It should be noted, however, that the proportional bias of estimator (2.9) is also dependent on the true gene diversity in the population. We recommend the use of equation (2.11) for all sample sizes, since it is not difficult to compute, and its sampling variance can also be computed by

$$V(\hat{h}) = [2(n - 2)\{\sum x_i^3 - (\sum x_i^2)^2\} + \sum x_i^2 - (\sum x_i^2)^2]/[n(n - 1)] . \qquad (2.12)$$

Similar logic also applies to the Shannon-information index (equation (2.2)), where substitutions of p_i by x_i leads to the estimator

$$\hat{h}_s = - \sum x_i \log_e x_i , \qquad (2.13)$$

whose expectation is approximately given by

$$E(\hat{h}_s) \approx - \sum p_i \log_e p_i - \frac{k - 1}{n} + \frac{1 - \sum p_i^{-1}}{12n^2} + \frac{\sum \{p_i^{-1} - p_i^{-2}\}}{12n^3} , \qquad (2.14)$$

shown by Hutcheson (1970) and Bowman et al. (1971). Because of the extra terms on the right-hand side of this equation, it is difficult to suggest any simple unbiased estimator of h_s. Nevertheless, Peet (1974) showed that in practice the contributions of the last two terms of equation (2.14) are not substantial. Therefore,

$$\hat{h}_s' = - \sum x_i \log_e x_i + (k - 1)/n \qquad (2.15)$$

corrects for a substantial part of the bias of the estimator (2.13), when the sample size is not very small (say, $n > 100$). The sampling variance of $\hat{h}_s$ (equation (2.13)) is approximately given by

$$V(\hat{h}_s) \approx \frac{\sum p_i(\log_e p_i)^2 - (\sum p_i \log_e p_i)^2}{n} + \frac{k - 1}{2n^2} , \qquad (2.16)$$

which can be used for testing whether or not different loci have substantially different extent of variation (see Whittaker, 1972, and Magurran, 1988, for some ecological applications of such test procedures).

In the above formulation it is assumed that the number of alleles (k) at a locus is known, and it is identical to the one observed in any sample. Particularly with molecular techniques of detection of genetic variation this is not always the case. The number of different possible allelic types may be unknown, particularly if we assume that every mutation (at a molecular level) yields a different allele, not previously seen in the population (the classic infinite allele model – Wright, 1948; Kimura and Crow, 1964), and hence, we may treat k as an unknown parameter as well. In this situation, the observed number of alleles in a sample (k_0) can be treated as a random variable. Based on multinomial property of sample frequencies of the different alleles, Chakraborty et al. (1988) and Chakraborty (1991) have shown that k_0 has the expectation and variance given by

$$E(k_0) = k - \sum_{i=1}^{k} (1 - p_i)^n \approx k - \sum_{i=1}^{k} e^{-np_i}, \tag{2.17}$$

and

$$V(k_0) = \sum_{i=1}^{k} (1 - p_i)^n \cdot \left[1 - \sum_{i=1}^{k} (1 - p_i)^n \right] + \sum_{i=1}^{k} \sum_{\substack{j=1 \\ i \neq j}}^{k} (1 - p_i - p_j)^n$$

$$\approx \sum_{i=1}^{k} e^{-np_i} \cdot \left[1 - \sum_{i=1}^{k} e^{-np_i} \right] + \sum_{i=1}^{k} \sum_{\substack{j=1 \\ i \neq j}}^{k} e^{-n(p_i + p_j)}. \tag{2.18}$$

Using an approach similar to the classical occupancy problem (Feller, 1957), Chakraborty (1991) also evaluated the sampling distribution of k_0 for known values of the k and the p-vector. Furthermore, the correlation between k_0 and $\hat{h}_1$ (equation (2.9)) can also be numerically evaluated (see Chakraborty and Fuerst, 1979; Chakraborty and Griffiths, 1982), showing that the gene diversity index and the observed number of alleles are generally highly correlated. Contrasts of equations (2.10) and (2.17) demonstrate another distinctive feature of the sampling distributions of gene diversity and the number of alleles. While the expectation of $\hat{h}_1$ very quickly approaches its asymptotic value (true value of gene diversity in the population, suggesting that this biased estimator is consistent in the strict statistical sense), the convergence of the expected k_0 to its true value (k) is comparatively slower. The coefficient of variation of k_0 is also generally larger than that of the gene diversity estimator.

From these considerations, we might argue that the gene diversity measure of genetic variation has the advantages: (i) it is easily interpreted in biological terms (probability of non-identity of alleles, or heterozygosity in a random-mating population), (ii) it is estimable with an unbiased estimator (equation (2.11)), whose sampling variance is also available, and (iii) it is less affected by sample size, particularly when sample size is large.

2.7. Sampling distributions and hypotheses testing issues

The estimator (2.9) of gene diversity has also received considerably more attention from statistical consideration compared with others. For example, Lyons and Hutcheson (1979) derived exact expressions for the first four moments of this estimator. Using standard asymptotic techniques (see Rao, 1973, Chapter 6, or Bishop et al., 1975, Chapter 15), it is also possible to derive an asymptotic distribution of such a statistic. For example, under the sampling frame postulated above we know that

$$\sqrt{n}[(x_1 - p_1), (x_2 - p_2), \ldots, (x_{k-1} - p_{k-1})] \to N_{k-1}(\mathbf{0}_n, \Sigma) \qquad (2.19)$$

in probability as $n \to \infty$, where Σ is a $(k-1) \times (k-1)$ symmetric square matrix with elements

$$\Sigma_{ij} = \begin{cases} p_i(1 - p_i) & \text{for } i = j, \\ -p_i p_j & \text{for } i \neq j. \end{cases}$$

Using (2.19) it is easy to show that for any general function

$$h_\alpha = h_\alpha(\mathbf{p}) = \sum_i p_i^\alpha \quad \text{for } \alpha > 0, \qquad (2.20)$$

its estimator $\hat{h} = \sum x_i^\alpha$ will have an asymptotic distribution given by

$$\sqrt{n}(\hat{h}_\alpha - h_\alpha) \to N(0, \sigma^2) \quad \text{as } n \to \infty, \qquad (2.21)$$

where

$$\sigma^2 = \alpha^2 \left[\sum_i p_i^{2\alpha - 1} - \left(\sum_i p_i^\alpha \right)^2 \right].$$

Nayak (1983) used this result or simple modifications of this to derive asymptotic variances and distributions of estimators of other generalized diversity measures, such as the ones noted in equations (2.4)–(2.7). Furthermore, note that the asymptotic property (2.21) can be further extended for any regular function of h_α as well. For example, defining

$$f = \phi[h_\alpha(\mathbf{p})] \qquad (2.22)$$

as a new estimable parameter, where ϕ is a continuous function admitting first derivative, we immediately get from equation (2.20) the following result

$$\sqrt{n}[\hat{f} - \phi(h_\alpha)] \to N(0, \tau^2) \quad \text{as } n \to \infty, \qquad (2.23)$$

where

$$\tau^2 = [\phi'(h_\alpha)]^2 \cdot \sigma^2.$$

Clearly, this result can be used to study the large sample distribution of the generalized diversity function $h_H(p)$ (see equation (2.8)) proposed by Hill (1973), of which the Gini–Simpson (gene diversity) and Shannon-information indices are special cases, as noted earlier.

Such large sample properties are also useful to compare diversity values across populations and/or loci, or to postulate various correlates of diversity indices. For example, one objective of population genetics studies could be whether or not variation within populations at various loci (e.g., say protein loci) are proportional to the sub-unit size of the protein molecules, or whether gene diversities at specific loci across populations are proportional to their respective effective population sizes. Rationale of such hypotheses has direct implication with regard to the mechanism of maintenance of genetic variation that will be discussed in the next section.

For this purpose, first note that the various diversity measures introduced earlier can be written in the form

$$D = D(p) = \sum_{i=1}^{k} \phi(p_i),$$
(2.24)

where ϕ is a continuous function admitting derivatives up to the second order.

Note that consistent (in the strict statistical sense, see Rao, 1973) estimators of D can be obtained by substituting x_i for p_i. Defining such as estimator as $\hat{D}$, application of the asymptotic result (2.22) yields

$$\sqrt{n}(\hat{D} - D_0) \to N(0, \tau^2),$$
(2.25)

where D_0 is the hypothesized value of D and τ is defined in (2.22) with appropriate modification of the ϕ function.

Multiple loci (populations) extensions of this result is also possible. For example, if D_1 and D_2 refer to diversity values for two loci (or two populations), and n_1 and n_2 refer to their respective sample sizes (both assumed to be large enough for appealing to asymptotic results), the test-statistic

$$U = \frac{\hat{D}_1 - \hat{D}_2}{\{(\hat{\tau}_1/n_1) + (\hat{\tau}_2/n_2)\}^{1/2}},$$
(2.26)

has an asymptotic standard normal distribution.

Note that Nei and Roychoudhury (1974) and Nei (1987) made use of such asymptotic distributions in the context of applying the gene diversity concept of genetic variation, even though such results are not formally rationalized in these works. In further extensions of such asymptotic distributions see Agresti and Agresti (1978), Rao (1982a,b,c) and Nayak (1983).

2.8. Evolutionary dynamics of genetic variation within a population

As mentioned before, while the definitions of the above measures of variation within populations hold for any mechanism of production and maintenance of genetic variation, it is important to examine their pattern under different evolutionary models. Since natural populations of almost all organisms are of finite size, stochastic models of evolutionary dynamics of genetic variation are generally more realistic for such purposes (Kimura, 1955; Kimura and Crow, 1964; Nei, 1975, 1987; Ewens, 1979).

Under the assumption that the population size is finite and constant over generations, every mutation leads to an allele not seen before in the population, and all alleles are selectively neutral; following Kimura (1964) it can be derived (Kimura and Crow, 1964) that the expected number of alleles whose frequency lies in the interval $(p, p + \mathrm{d}p)$ is given by $\Phi(p)\,\mathrm{d}p$, where $\Phi(p)$ is of the form

$$\Phi(p) = \theta p^{-1}(1 - p)^{\theta - 1}, \tag{2.27}$$

in a population which is at steady-state under the mutation-drift balance. In equation (2.27), θ is a composite parameter, $\theta = 4N_e v$, where N_e is the effective population size and v is the mutation rate per locus per generation. The function $\Phi(p)$ is also called the steady-state allele frequency spectrum. The functional form of $\Phi(p)$ plays a crucial role in determining the evolutionary expection of any function of allele frequencies in a population. For example, for the gene identity index, $J = \sum p_i^2$, the evolutionary expectation in a steady-state population under the assumption of the neutral mutation hypothesis is given by (Kimura and Crow, 1964)

$$J = 1/(\theta + 1), \tag{2.28}$$

so that the gene diversity in an equilibrium population becomes $\theta/(\theta + 1)$. Watterson (1974) and Stewart (1976) further derived that the variance of gene diversity due to evolutionary fluctuations is given by

$$V(h) = 2\theta/[(1 + \theta)^2(2 + \theta)(3 + \theta)], \tag{2.29}$$

so that the relationship between mean gene diversity and its variance across loci can be used to examine whether or not the pattern of gene diversity seen in any survey of genetic variation in a natural population conforms to the predictions of the neutral mutation hypothesis (Nei et al., 1976b; Fuerst et al., 1977).

The allele frequency spectrum function (equation 2.27) can also be used to determine the evolutionary expectation of other measures of genetic variation. For example, the expected number of alleles that exist in an equilibrium population is given by

$$n_a = \int_{1/2N}^{1} \Phi(x)\,\mathrm{d}x = \theta \cdot \sum_{r=0}^{2N-1} (\theta + r)^{-1}, \tag{2.30}$$

whereas the effective number of alleles has the evolutionary expectation

$$n_{\mathrm{e}} = 1/E\left(\sum_i p_i^2\right) = (\theta + 1).$$ (2.31)

Furthermore, the expected proportion of polymorphic loci (P) defined earlier in terms of the arbitrary criteria of polymorphism ($q < 1$) is related with the parameter θ by the functional relationship

$$P = 1 - q^\theta,$$ (2.32)

shown by Kimura (1971).

The seminal work of Ewens (1972) laid the foundation of using these features of a steady-state allele frequency spectrum to derive the sampling theory of selectively neutral alleles. Chakraborty and Fuerst (1979) and Chakraborty and Griffiths (1982) used this theory to show that for a random sample of n genes ($\frac{1}{2}n$ individuals) drawn from a steady-state population, the evolutionary expectation and variance of gene-identity are given by

$$E(J) = n^{-1} + (n-1)/[n(1+\theta)],$$ (2.33)

and

$$V(h) = \frac{n-1}{2n^3}\left[\frac{5n(n-1)+2}{1+\theta} - \frac{8(n-1)(n-2)}{2+\theta} + \frac{3(n-2)(n-3)}{3+\theta}\right.$$

$$\left. - \frac{n-1}{n^3}\left(2 + \frac{n-1}{1+\theta}\right)\right]\Big/(1+\theta).$$ (2.34)

Analogous expressions for the expectation and variance of the number of distinct alleles in a sample of n genes are given by Ewens (1972) which require the computation of series

$$E(n_{\mathrm{a}}) = \theta \sum_{r=0}^{n-1} (\theta + r)^{-1}$$ (2.35)

and

$$V(n_{\mathrm{a}}) = \theta \sum_{r=0}^{n-1} r/(\theta + r)^2.$$ (2.36)

These analytical results have found several applications in the evolutionary literature. For example, Nei et al. (1976b) and Fuerst et al. (1977) have used the relationship between the mean and variance of gene diversity in different organisms to conclude that the pattern of genetic variation in natural populations are to a large extent consistent with the neutral mutation hypothesis. Hedrick and Thomson (1983) used the relationship between gene diversity and number of alleles to show that the HLA polymorphism in humans is probably maintained by a form of balancing selection. Chakraborty et al. (1988) and Chakraborty (1990a) showed that the relationship between gene diversity and number of alleles can also be used to detect population substructuring within populations. Chakraborty (1990b) and Chakraborty and Schwartz (1990) have shown that this relationship (h versus n_a) can also be used to study population structure even if these statistics are obtained from single locus data, provided that the locus is highly polymorphic, as is the case with the mitochondrial genome, or the recently discovered hypervariable VNTR loci.

For these applications, it is important to note that the composite parameter $\theta = 4N_e v$ has to be estimated from the available data. The observed level of gene diversity or the number of alleles can be used to estimate this parameter. Details of such estimation procedures are discussed in Nei (1987), Zouros (1979), Chakraborty and Schwartz (1990) and Chakraborty and Weiss (1991).

We avoid the recapitulations of these results, but note that while the number of alleles provide a maximum likelihood estimator (MLE) of θ, its standard error is often larger than the moment estimator obtained from gene diversity. None of these estimators are, however, unbiased. The bias correction is easily possible for the moment estimator based on gene diversity, while an appropriate bias correction is not possible for the MLE based on the number of alleles. A further problem of the maximum likelihood estimator is that it is very sensitive to departure from the assumption that the sample is drawn from a homogeneous population. In the presence of population subdivision, since this statistic can easily be substantially inflated, the resultant estimator of θ can often be an overestimate, while the gene diversity is relatively less disturbed by population subdivision, particularly when the genetic difference among the subpopulations is not large (Chakraborty et al., 1988).

Another application of the sampling theory of selectively neutral alleles is the estimation of the components of the parameter θ. Nei (1977) used Ewens' (1972) sampling theory to derive the expected number of rare alleles in a population, from which the parameter v could be estimated. Chakraborty (1981) revised Nei's estimator taking into account the sampling property of the number of rare alleles observed in a sample.

These methods, as well as others based on the same features of the steady-state allele frequency spectrum (see Nei, 1987 and Chakraborty and Neel, 1989 for references) have some common drawback, namely, the effective population size must be known for the population, and second, data from several loci must be pooled to derive an estimator for the average mutation rate for the specific group of loci. Efforts to estimate the two parameters, N_e and v, separately started from the work of Zouros (1979) who derived relative values of N_e for several populations, and relative values of v for several loci, utilizing the relationship between number of alleles and gene diversity for several population-loci data combinations. Chakraborty and Neel (1989) used the same data structure (data on allele frequencies collected for a set of common loci surveyed in different populations) to estimate the locus-specific mutation rates and population-specific effective sizes simultaneously. Their results suggest that even if the steady-state assumption is correct, there is a substantial difference between mutation rates for different protein loci, so that data from different loci cannot be generally pooled to apply the predicted equation (equation (2.27)) for the allele frequency spectrum.

Variations of mutation rate over loci have been incorporated in the evolutionary studies to derive the equilibrium properties of gene diversity and the number of alleles by Nei et al. (1976a) and Chakraborty and Fuerst (1979).

All of the above results are applicable to the case where it is assumed that mutations always lead to a new allele, so that the alleles that are detected to be similar by a specific genetic assay are assumed to be identical at the molecular level. This is clearly not the case with electrophoretic or restriction fragment studies, since both of these approaches have certain degree of hidden variation. The Stepwise Mutation Model was proposed by Ohta and Kimura (1973) to take this hidden variation into account. The sampling properties of allele frequency distribution as well as equilibrium values of gene diversity and number of alleles under this model are also well studied (see, e.g., Ohta and Kimura, 1973; Kimura and Ohta, 1978; Chakraborty et al., 1978, 1980; Chakraborty and Nei, 1976; Chakraborty and Fuerst, 1979).

Further extension of the equilibrium models of genetic diversity within a population has been made by relaxing the assumption of selective neutrality of alleles. Although in practice it is difficult to mathematically model any selection mechanism, and the selective pressure is generally quite unstable over evolutionary time, various types of selection have been incorporated in such studies. For example, overdominant selection has been studied by Kimura and Crow (1964), Ewens (1964), Watterson (1977), Li (1978), Yokoyama and Nei (1979), Maruyama and Nei (1981), among others. The equilibrium properties of a population under the force of slightly disadvantageous selection have been examined by Li (1977, 1978), Ohta (1976), Kimura (1979) and Ewens and Li (1980), and the effect of fluctuating selection intensity has been studied by Nei and Yokoyama (1976), Takahata and Kimura (1979), Takahata (1981) and Gillespie (1983, 1985). These studies generally indicate that when the different features of genetic variation within a population are simultaneously analyzed in conjunction with each other, selection coefficients of magnitude larger than the inverse of population size can be detected, although the statistical power of tests based on sampling properties of any single measure of genetic variation is generally low.

The most troublesome assumption of evolutionary inference based on measures of genetic variation within a population is the constancy of population size over its evolutionary history. Although ecologists as well as geneticists were aware of the limitation of this assumption, and some postulates have been suggested with regard to the effect of severe fluctuations of population size (Wright, 1931; Mayr, 1963), systematic studies of the effect of fluctuation of population size on genetic variation started from the work of Nei et al. (1975), Chakraborty (1977) and Chakraborty and Nei (1977). In order to understand the effect of fluctuations of population size, first note that even if the population size remains constant, the gene identity (J_t) in a population at the tth generation is related with that of the previous generation by the relationship

$$J_t = (1 - v)^2 [(2N)^{-1} + \{1 - (2N)^{-1}\}J_{t-1}],$$ (2.37)

first derived by Malécot (1948), where N represents the effective population size. When the fluctuations of effective population size over generations can be mathematically modelled, the above equation still holds with the change that N should be replaced by N_{t-1}, the effective size at the $(t - 1)$th generation. The above difference equation can then be approximated by the differential equation

$$dJ_t/dt = (2N_t)^{-1} + (4N_t v + 1)J_t/(2N_t),$$ (2.38)

which yields the solution

$$J_t = \int_0^t \frac{e^{-2vy}}{2N_y} \exp\left[-\int_0^y (2N_x)^{-1}\,dx \right] dy$$
$$+ J_0 \exp\left[-2vt - \int_0^t (2N_x)^{-1}\,dx \right],$$ (2.39)

that can be evaluated for any analytical relationship of N_t with t (Nei et al., 1975). It is clear from equation (2.39) that when population size fluctuates over generations, the changes of gene diversity $1 - J_t$ over any evolutionary time period depends on the harmonic mean of population size, so that the equilibrium value of J, J_∞, may be approximated by $(4\tilde{N}v + 1)^{-1}$, where $\tilde{N}$ is the harmonic mean of effective population size over time. Since the harmonic mean is dictated more by the smallest population size ever reached, equation (2.39) implies that when a population goes through a severe bottleneck, the gene diversity can be drastically reduced. It takes a fairly long time before any substantial recovery of the lost variation even when the population size quickly returns to its carrying capacity.

Equation (2.39) can also be used to examine the changes of gene identity over time for a population of constant size ($N_t = N$ for all t). In this case, it takes the simple form

$$J_t = J_\infty + (J_0 - J_\infty) \exp[-(1 + \theta)t/(2N)],$$ (2.40)

in which $\theta = 4Nv$ and $J_\infty = (1 + \theta)^{-1}$. Note that the equation (2.40) is identical to Malécot's (1948) formula.

Similar analysis of the evolutionary dynamics of the number of alleles has also been conducted by Nei et al. (1975), Maruyama and Fuerst (1984) and Watterson (1984). Although these analytical results are comparatively more tedious than the one indicated above for gene identity, the general conclusion is that the number of alleles after a bottleneck is also drastically reduced, but its recovery process is faster than gene diversity, when the population size starts to grow. As a result, the relationship between gene diversity and number of alleles departs from the predictions of the constant population size model. If the parameter $\theta = 4Nv$ is estimated from gene diversity, in the presence of bottleneck, there is an apparent excess of the observed number of alleles, and this excess is primarily due to the excess of rare alleles.

It should be noted, however, that other mechanisms may also produce a similar effect. For example, if there is substructuring in the population, and the extent of genetic differences between the subpopulations is large, ignoring substructuring may indicate an excess number of alleles in comparison with gene diversity (see, e.g., Chakraborty et al., 1988; Chakraborty, 1990a,b). Therefore, excess number of alleles in comparison with the neutral expectations for an equilibrium population by itself is not firm evidence of the effects of a past bottleneck in the population.

In summary, the above discussion indicates that at least the two measures of genetic diversity within populations (Gini–Simpson index and number of alleles) can be used to study the evolutionary changes of such variation within a population from which the mechanism of maintenance of polymorphism can be inferred. Although there are some concerns regarding the statistical power of such test procedures, applications of these two measures to empirical data on protein polymorphism in natural populations indicate that the effect of selection on genetic variation can be distinguished from that of the simple mutation-drift pressure when the selection coefficient is large. Furthermore, the presence of substructuring within a population and past fluctuations of population sizes can also be detected with these measures of genetic variation. Therefore, the above two measures not only have sound mathematical basis, their utility in evolutionary studies is also well established.

3. Genetic variation between populations

3.1. Concept of distance and brief review of genetic distance measures

The formulation of the previous section can be directly extended to define the genetic variation (distance) between populations. In this case the two individuals should be chosen at random, one from each of the two populations. A point, however, must be noted. Since the concept of distance between populations, in a sense refers to the magnitude of allelic diversity between them, it is necessary to define central positioning of the populations in addition to which corrections for within population variation among individuals are to be incorporated in the definition of the distance measure.

This rationale stems from the topological interpretation of the distance concept, where a distance function should satisfy the property of a metric. Stated mathematically, distance between populations i and j, D_{ij}, should satisfy the conditions of (i) non-negative (i.e., $D_{ij} \geqslant 0$, the equality holding if and only if $i = j$), and (ii) triangular inequality (i.e., for any three populations i, j and k we should have $D_{ij} \leqslant D_{ik} + D_{jk}$). Note that the condition (i) immediately implies that the definition of D_{ij} must take into account within population variations in both populations. The two conditions, jointly imply that the populations can be conceptualized as points in an Euclidean space and their distance measured as the geometric distance between the corresponding points in that space. In classical numerical taxonomy this conceptualization is given critical importance because cluster analytic methods of classifying populations based on principal component decomposition of distance matrices for a set of populations require both of these conditions.

As mentioned before, the origin of genetic distance studies predates the discovery of genetic markers and examination of allele frequency differences. The above rationale is well justified for a set of quantitative traits, because the dimension of the reference sampling universe is unambiguously determined for such traits by the number of traits with respect to which variability is being measured. For evolutionary studies with genetic markers and allele frequencies, this is not necessarily the case, because with reference to a multiallelic locus, the dimension of the Euclidian space is dictated by the number of segregating alleles in a population, and this is variable over time (due to genetic drift and mutations). Further problems of geometric interpretation of genetic distance in the context of evolutionary studies are also discussed in Nei (1973, 1987).

In the context of genetic studies the number of different formulations of genetic distance measures far exceed the ones proposed for studying within-population variability. We avoid an exhaustive review of them, because there are several excellent reviews on this subject (e.g., Smith, 1977; Nei, 1978b; Jorde, 1980 and Lalouel, 1980) that are still up-to-date. Historically, even though Sanghvi (1953) was the first to use the concept of genetic distance for evolutionary studies, similar concepts for measuring population differences were proposed by Czekanowski (1909) and Pearson (1926), which were later modified for statistical reasons by Fisher (1936) and Mahalanobis (1936). Cavalli-Sforza and Edwards (1964, 1967) employed an angular transformation of gene frequencies to arrive at a genetic distance measure that they used to draw an evolutionary tree of human racial groups in terms of gene frequencies. This spawned a flurry of interest to human geneticists and anthropologists.

For a topical review of the existing indices of genetic distances, they can be classified into two broad classes: (i) ones that are intended for population classification, and (ii) others that are intended for the study of evolution.

Nei (1987) classified Czekanowski's (1909) mean differences and its variation ('Manhattan metric', see Sneath and Sokal, 1973), Pearson's coefficient of racial likeness (Pearson, 1926), Rogers' distance (Rogers, 1972), Mahalanobis' distance (Mahalanobis, 1936), Sanghvi's (1953) distance and its variant (Balakrishnan and Sanghvi, 1968), Kurczynski's (1970) D^2, Bhattacharyya (1946) and Cavalli-Sforza and Edwards' (1967) distance into the first category. He also grouped the distance indices based on Wright's F_{ST} index (e.g., Latter's (1973) distance, Cavalli-Sforza's (1969) distance), Morton's kinship indices (Morton, 1975) and his own distance indices (Nei, 1972) into the latter category. The rationale of this categorization is based on the premise that when the evolutionary dynamics of gene frequencies are formulated in terms of a defined evolutionary mechanism (such as mutation, drift and migration) the first category of distance indices generally do not satisfy a well defined pattern or trend. For example, several of them do not exhibit an increasing trend with the evolutionary time of isolation of populations. On the other hand, the distance indices that Nei called useful for evolutionary studies can be subjected to evolutionary models to provide their predicted value as functions of evolutionary time, extent of gene migration between populations and their effective sizes. Even though this rationale may not be universally accepted, because it ties the definition of genetic distance with the evolutionary models, if a genetic distance measure is to be used for the study of evolution this type of categorization is important.

3.2. *Relationships between some specific distance indices*

Even though the purpose of using genetic distance must be well specified before defining an appropriate index, it is worthwhile to examine whether or not the various distance indices have any analytical relationship so that we can understand what aspect of gene frequency differences between populations are emphasized to what degree in these various formulations. Such attempts might help to resolve some of the controversies surrounding the choice of distance functions. In this direction some work has been done (see, e.g., Rao 1982a–c; Chakraborty, 1980; Nei, 1987), which may be worthwhile to mention in the process of providing unified measures of genetic dissimilarity between populations.

For studying the analytical relationships between various distance indices it should first be noted that for any population α, the allele frequencies $(p_{\alpha 1}, p_{\alpha 2}, \ldots, p_{\alpha k})$ at a k-allelic locus define the population on a $(k - 1)$-dimensional space. If V denotes a symmetric square matrix of rank $k - 1$ such that its (i, j)th element, v_{ij}, is defined by

$$v_{ij} = \begin{cases} p_{\alpha i}(1 - p_{\alpha i}) & \text{for } i = j = 1, 2, \ldots, k - 1, \\ -p_{\alpha i} p_{\alpha j} & \text{for } i \neq j = 1, 2, \ldots, k - 1, \end{cases} \tag{3.1}$$

then the inverse of this matrix $V^{-1} = ((v^{ij}))$ is given by

$$v^{ij} = \begin{cases} 1/p_{\alpha k} & \text{for } i \neq j = 1, 2, \ldots, k - 1, \\ 1/p_{\alpha i} + 1/p_{\alpha k} & \text{for } i = j = 1, 2, \ldots, k - 1. \end{cases} \tag{3.2}$$

Now for two populations α and β, if we define their genetic distance in terms of Mahalanobis' distance, computed on the basis of the first $k - 1$ components of their allele frequency vectors, we have

$$D^2 = \sum_{i=1}^{k-1} \sum_{j=1}^{k-1} (p_{\alpha i} - p_{\beta i}) S^{-1} (p_{\alpha j} - p_{\beta j}), \tag{3.3}$$

when S is the variance–covariance matrix based on their pooled allele frequencies,

$\frac{1}{2}(p_{\alpha i} + p_{\beta i})$ for $i = 1, 2, \ldots, k$. Algebraic simplification of equation (3.3) using the result (3.2) yields

$$D^2 = \sum_{i=1}^{k} (p_{\alpha i} - p_{\beta i})^2 / [2(p_{\alpha i} + p_{\beta i})], \tag{3.4}$$

which is equivalent to Sanghvi's (1953) distance. Note that although Sanghvi (1953) and Balakrishnan and Sanghvi (1968) emphasized that for computing their distance function one allele should be deleted from each locus (because of the linear constraint of allele frequencies), the above derivation shows that if the equivalence with Mahalanobis' distance is to be established, then in computing equation (3.4) there is no need to throw out any allele for defining an appropriate index of genetic distance.

When the two populations, α and β, are not very dissimilar in their allele frequencies, the above index, D^2 (equation (3.4)), is also equivalent to Bhattacharyya's (1946) distance, θ^2, where

$$\cos \theta = \sum_{i=1}^{k} \sqrt{p_{\alpha i} p_{\beta i}}. \tag{3.5}$$

This is so, because

$$\cos \theta = \sum_{i=1}^{k} \sqrt{p_{\alpha i} p_{\beta i}} = \frac{1}{2} \sum [(p_{\alpha i} + p_{\beta i})^2 - (p_{\alpha i} - p_{\beta i})^2]^{1/2}$$

$$= \frac{1}{2} \sum_{i=1}^{k} (p_{\alpha i} + p_{\beta i}) \left[1 - \frac{(p_{\alpha i} - p_{\beta i})^2}{(p_{\alpha i} + p_{\beta i})^2} \right]^{1/2}$$

$$\approx 1 - \frac{1}{4} \sum_{i=1}^{k} \frac{(p_{\alpha i} - p_{\beta i})^2}{(p_{\alpha i} + p_{\beta i})}, \tag{3.6}$$

obtained by series expansion of the square-root term under the summation sign, and neglecting fourth and higher powers of allele frequency differences between the populations.

Furthermore, since for small θ, $\cos \theta \approx 1 - \frac{1}{2}\theta^2$, we obtain from equation (3.6) an approximate formula

$$\theta^2 \approx \frac{1}{2} \sum_{i=1}^{k} (p_{\alpha i} - p_{\beta i})^2 / (p_{\alpha i} + p_{\beta i}), \tag{3.7}$$

which is identical to equation (3.4). Therefore, when two populations are genetically close to each other, Sanghvi's (1953), Mahalanobis' (1936) and Bhattacharyya's (1946) distances are all equal.

Note that in equation (3.4) or (3.7), the allele frequency difference between populations are weighted by the inverse of their average frequencies in the two populations. In a genetic analysis of multiple (more than two) populations, this variance–covariance matrix (S, see equation (3.3)) has to be defined over all populations (based on the assumption that the within-population multinomial sampling variances are homogeneous across populations). This raises a problem, since the definition of each pairwise distance between populations then depend upon how many and which other populations are included in the study. In one of Nei's genetic distance measures, therefore, he gave equal weight to all allele frequency differences between populations to define

$$D_{\mathrm{m}} = \tfrac{1}{2} \sum_{i=1}^{k} (p_{\alpha i} - p_{\beta i})^2 , \tag{3.8}$$

as the minimum genetic distance between populations α and β.

Rao (1982a) provided a geometric interpretation of Nei's minimum distance (equation (3.8)), by noting that D_{m} can be written as

$$D_{\mathrm{m}} = (\sqrt{J_{\alpha\alpha}} - \sqrt{J_{\beta\beta}})^2 + 2\sqrt{J_{\alpha\alpha}J_{\beta\beta}}(1 - \cos\theta_{\alpha\beta}) , \tag{3.9}$$

where $J_{\alpha\alpha} = \sum_{i=1}^{k} p_{\alpha i}^2$ and $J_{\beta\beta} = \sum_{i=1}^{k} p_{\beta i}^2$ are the gene identies within populations α and β, and $\cos\theta_{\alpha\beta} = \sum_{i=1}^{k} p_{\alpha i} p_{\beta i}/\sqrt{J_{\alpha\alpha}J_{\beta\beta}}$ depends on the angle ($\theta_{\alpha\beta}$) between the allele frequency vectors of the two populations. Nei's minimum genetic distance, therefore, has two components; the first measuring the differences in lengths of the gametic vectors (or difference in homozygosity or diversity within populations) and the second depending on the angle between the vectors (difference in orientation). Thus, Nei's minimum distance measures the overall in orientation). Thus, Nei's minimum distance measures the overall difference both in length and orientation of the gametic vectors. Rao (1982a) advocated the use of square-root of D_{m} for cluster analysis of populations, because this transformation of D_{m} for cluster analysis of populations, because this transformation of D_{m} makes the index a proper metric, satisfying the non-negativity as well as triangular inequality conditions, mentioned before. However, this transformation will distort the relationship between Nei's minimum distance and evolutionary time of divergence between the populations, as will be shown later.

With the above notations, Nei's standard genetic distance can be defined as,

$$D_{\mathrm{s}} = -\log_{\mathrm{e}}(J_{\alpha\beta}/\sqrt{J_{\alpha\alpha}J_{\beta\beta}}) , \tag{3.10}$$

where $J_{\alpha\beta} = \sum_{i=1}^{k} p_{\alpha i} p_{\beta i}$, is the probability of gene identity when two genes are chosen one each from populations α and β, and $J_{\alpha\alpha}$, $J_{\beta\beta}$ are the gene identities within populations α and β. Rao (1982a) further showed that D_{s} (equation (3.10)) is only a function of the orientation difference of the allele frequency vectors in populations α and β. This is so, because in terms of the definition of $\theta_{\alpha\beta}$ in equation (3.9), equation (3.10) translates to

$$D_{\mathrm{s}} = -\log_{\mathrm{e}} \cos\theta_{\alpha\beta} . \tag{3.11}$$

It should again be noted that D_s also is not a proper metric, even though it has an interesting evolutionary property which will be discussed later. Because of equation (3.11), Rao (1982a) advocates that

$$\theta_{\alpha\beta} = \cos^{-1}[e^{-D_s}] \tag{3.12}$$

can be used for cluster analysis of populations that rely on the metric properties of distance indices (for other relevant results, see Rao, 1984).

Equations (3.9) and (3.12) furthermore indicate that when the gene diversities in two populations are equal (i.e., $J_{\alpha\alpha} = J_{\beta\beta}$),

$$
\begin{aligned}
D_m &= 2J_{\alpha\alpha}(1 - e^{-D_s}) \\
&\approx 2J_{\alpha\alpha}D_s, \quad \text{when } D_s \text{ is much smaller than } 1,
\end{aligned}
\tag{3.13}
$$

so that these two distance indices are proportional to each other when the two populations are at steady state with respect to their internal population structures.

Finally, writing the allele frequencies at the m loci as a large vector of dimension $k = k_1 + k_2 + \cdots + k_m$, where k_r is the number of alleles at the rth locus, Rao (1982a) showed that Nei's standard distance (equation (3.10) or (3.11)) derived from the combined multinomial allele frequency (of dimension k) is the same as Nei's maximum distance, defined by

$$D_{\max} = -\log_e\left[\prod_{r=1}^{m} \cos\theta_{\alpha\beta}^{(r)}\right], \tag{3.14}$$

where $\cos\theta_{\alpha\beta}^{(r)} = \sum_{i=1}^{k_r} p_{\alpha i}^{(r)} p_{\beta i}^{(r)} / \sqrt{J_{\alpha\alpha}^{(r)} J_{\beta\beta}^{(r)}}$ is the orientation function defined based on allele frequencies at the rth locus alone. Because of this result, again an appropriate metric transform of $D_{\max}$ is

$$\theta_{\alpha\beta}^{(\max)} = \cos^{-1}(e^{-D_{\max}}), \tag{3.15}$$

that can be used for population cluster analysis.

These relationships between the various distance indices indicate that: (i) the criticism that Nei's genetic distance indices are not proper metric functions can be circumvented by constructing their appropriate transforms, although such transformed functions do not have any clear-cut biological interpretation, (ii) Nei's three indices have geometric interpretations as well, and (iii) distance indices that are based on generalized distance of Mahalanobis can also be written in terms of orientation functions of allele frequency vectors. Chakraborty (1980) studied similar analytic relationships between distance indices based on Wright's F_{ST} and multivariate multinomial vectors. In addition, empirical studies also demonstrate that at least for closely related populations or organisms, many measures of genetic distance are highly correlated (Richmond, 1972; Rogers, 1972; Hedrick, 1975; Chakraborty and Tateno, 1976; Nei, 1976) even though they are often based on different biological and mathematical premises.

3.3. Estimation of genetic distance

Before reviewing the estimation procedures of genetic distances some general comments are in order, since there are several apparent misconceptions in this regard in the population genetic literature. First, note that all of the above mentioned distance functions are formulated in terms of population parameters such as allele frequencies for all loci in the genome. Second, since the main purpose of studying genetic distance is to measure the extent of genetic differentiation among a given set of populations (two or more), the sampling of populations is not a relevant issue for estimation. However, two features of sampling must be incorporated in estimating genetic distance appropriately. One aspect of sampling results from the selection of individuals from the populations. Most genetic surveys involve assaying genetic variation in a small fraction of individuals drawn from the populations. We shall assume that these individuals have been randomly drawn, so that the allele counts at a specific locus can be considered an unbiased estimate of the true allele frequencies from the populations at such a locus. The second aspect of sampling is that of the selection of loci from the genome. At present we can only examine a small fraction of the genome for genetic studies. Here as well, we shall assume that the loci are selected at random from the genome. This second assumption implies that the monomorphic loci should also be incorporated in the estimation of genetic distance, even though they do not contribute to variation per se. The number of monomorphic loci surveyed allows the determination of an appropriate denominator when we want to scale the distance measure on the per locus basis, and for evaluating the inter-locus variation of estimates of genetic distances between populations.

Estimation of genetic distance is usually a troublesome issue, since the above formulations indicate that the distance functions defined above generally involved functions of the population parameters (allele frequencies at all loci in the populations), and hence the sampling distributions of such complicated functions of the sampled allele frequencies are unknown. Therefore, efficient estimation procedures, such as the maximum likelihood method cannot be generally followed. However, the method of moments can be employed, and procedures of bias correction can be adopted to suggest (asymptotically) unbiased estimators. Since the determination of the precision of estimators is also important in judging the suitability of a given estimator, the sampling variance of the suggested estimators should also be evaluated. Nei and Roychoudhury (1974) and Nei (1978a) made some important contributions in this regard, particularly dealing with the estimation issues of Nei's distance measures. Chakraborty (1985) evaluated some empirical sampling properties of such estimators based on extensive computer simulations. First we briefly review the evaluation of bias and bias correction methods for the moment estimators of genetic distances, and then outline the sampling variance computation procedures. For two reasons only Nei's distance functions are considered. These are the ones most well-studied in this regard and second, since these distance functions involve quadratic functions of allele frequencies and transformations such as exponentiation or logarithms, Taylor-series approximation can be adopted to approximate their large sample variances and biases. It should be noted that similar studies should be done on other distance functions as well for an appropriate comparative analysis.

From equation (3.8) it is clear that if x_i and y_i represent the sampled allele frequencies of the ith allele at a locus in populations α and β when n_α and n_β individuals have been randomly drawn from these populations, respectively, then the moment estimator of Nei's minimum distance is given by

$$\hat{D}_m^* = (1 - j_{\alpha\beta}^*) - \tfrac{1}{2}[(1 - j_\alpha^*) + (1 - j_\beta^*)], \tag{3.16}$$

where j_α^*, j_β^*, and $j_{\alpha\beta}^*$ are $\Sigma\, x_i^2$, $\Sigma\, y_i^2$, and $\Sigma\, x_i y_i$, respectively. In a random-mating population, the expectations of these three sample statistics are known under multinomial sampling of alleles (Johnson and Kotz, 1969; Nei and Roychoudhury, 1974). Specifically, the statistics j_α^* and j_β^* are biased estimators of their respective population values ($j_{\alpha\alpha}$ and $j_{\beta\beta}$) while $j_{\alpha\beta}^*$ is an unbiased estimator of $j_{\alpha\beta}$ [e.g., $E(j_\alpha^*) = j_{\alpha\alpha} + (1 - j_{\alpha\alpha})/2n_\alpha$]. Therefore, the bias of the estimator $\hat{D}_m^*$ can be evaluated exactly from the relationship

$$E(\hat{D}_m^*) = D_m + \tfrac{1}{2}[(1 - j_{\alpha\alpha})/2n_\alpha + (1 - j_{\beta\beta})/2n_\beta], \tag{3.17}$$

which may be corrected easily by suitably altering the estimators j_α^* and j_β^* in (3.16). Nei (1978a) suggested estimators $\hat{j}_\alpha = (2n_\alpha \Sigma\, x_i^2 - 1)/(2n_\alpha - 1)$, $\hat{j}_\beta = (2n_\beta \Sigma\, y_i^2 - 1)/(2n_\beta - 1)$ for $j_{\alpha\alpha}$ and $j_{\beta\beta}$ and $\hat{j}_{\alpha\beta} = \Sigma\, x_i y_i$ for $j_{\alpha\beta}$ to derive an unbiased estimator of D_m given by

$$\hat{D}_m = (\hat{j}_{\alpha\alpha} + \hat{j}_{\beta\beta})/2 - \hat{j}_{\alpha\beta}. \tag{3.18}$$

Equation (3.17) shows that the bias of the usual moment estimator of D_m (equation (3.16)) depends on the sample sizes from each population and their within-population variation (gene diversity).

Since all of these equations can be arranged over all loci surveyed to obtain an estimate of the per-locus minimum genetic distance, this bias term (last part of equation (3.16)) should be carefully examined. For example, in the context of electrophoretic variation at the protein loci, since the average gene diversity within the major human populations is generally 10% or less (see Nei and Livshits, 1989), one might argue that when 100 or more individuals are surveyed this bias is generally too small compared with the estimate of the distance, and hence, the bias correction (equation (3.18)) will not substantially alter the estimate. This is not true for loci that have a much higher variation. For example, gene diversity for the Variable Number of Tandem Repeat (VNTR) loci in many human populations sometimes approaches 90% or higher. In such cases, the bias correction is extremely critical, since $\hat{D}_m^*$ is an overestimate of the true minimum genetic distance and without bias correction one might erroneously over-emphasize the extent of genetic differentiation between populations.

While the bias-correction method suggested above for the estimation of Nei's minimum genetic distance is simple to employ, this is not so direct for the estimation of his standard genetic distance index. This is so because the standard genetic distance function (equation (3.10)) involves a logarithmic transformation of the $j_{\alpha\alpha}$, $j_{\beta\beta}$, and $j_{\alpha\beta}$ values. Taylor-series approximation yields that when j_α^*, j_β^*, and $j_{\alpha\beta}^*$ are used in equation (3.10) to obtain an usual moment estimator $(\hat{D}_s^*)$ of the standard genetic distance, its expectation becomes

$$E(\hat{D}_s^*) \approx D_s + \tfrac{1}{4}[(1 - j_\alpha)/n_\alpha j_\alpha + (1 - j_\beta)/n_\beta j_\beta]\,, \tag{3.19}$$

so that again the usual moment estimator is biased upwards. Nei (1978a) suggests the bias correction by altering the j_α^*, j_β^*, and $j_{\alpha\beta}^*$ values by $\hat{j}_\alpha$, $\hat{j}_\beta$, and $\hat{j}_{\alpha\beta}$, as done in the context of estimating the minimum genetic distance. This yields a modified moment estimator of D_s given by

$$\hat{D}_s = -\log_e[\hat{j}_{\alpha\beta}/(\hat{j}_{\alpha\alpha}\hat{j}_{\beta\beta})^{1/2}]\,, \tag{3.20}$$

whose expectation is nearly equal to the true value of D_s, to the order of n_α^{-2} and n_β^{-2} terms of the corresponding Taylor-series expansion. Nei (1978a) called this estimator an unbiased standard genetic distance, although strictly speaking it is only asymptotically unbiased. For small sample sizes, the bias of $\hat{D}_s$ is of course much smaller than that of $\hat{D}_s^*$. Like the biased estimator of the minimum genetic distance, $\hat{D}_s^*$ might also substantially overemphasize the degree of genetic differentiation between populations when the within-population variation is large (as in the case of the hypervariable VNTR loci).

While the above bias-correction methods substantially circumvents the possibility of overemphasizing the degree of genetic differentiation between populations when the sample sizes are small or the extent of within-population variation is large, the disturbing feature of these estimates is their possible negative values that might occur when the true genetic distance is very small. Even though in principle negative estimates can be avoided by increasing the sample size, such spurious negative values are generally taken as indicators of small genetic differentiation between populations (Nei, 1978a).

The above method of bias correction can also be used for any other distance functions as long as they involve up to the second moment of the allele frequencies. For example, an unbiased estimator of Latter's ϕ^* (Latter, 1973) can be obtained from

$$\hat{\phi}^* = [\tfrac{1}{2}(j_{\alpha\alpha} + j_{\beta\beta}) - j_{\alpha\beta}]/(1 - j_{\alpha\beta})\,. \tag{3.21}$$

Reynolds et al. (1983) suggested similar bias-correction methods for several other distance functions. However, when a distance function involves absolute values of allele frequency differences among populations, the above method of bias-correction cannot be employed.

A final comment on combining data over several loci is important to remember. While for the minimum genetic distance averaging over the different loci can either be done in computing the $\hat{j}_{\alpha\alpha}$, $\hat{j}_{\beta\beta}$, and $\hat{j}_{\alpha\beta}$ values or in computing the final estimate of the distance, for the standard distance computation averaging of these gene-identity values must be done to arrive at the estimate. This is so, because the minimum genetic distance based on r loci can also be written as

$$D_m = \sum_{k=1}^{r} d_{mk}/r \,, \tag{3.22}$$

where d_{mk} is the minimum genetic distance based on allele frequency data at the kth locus. Such a statement cannot be made for the standard genetic distance, since when data on all r loci are to be considered to compute the standard genetic distance, it involves average values of the gene identities (see equation (3.10) before employing the logarithmic transformation.

This difference introduces substantial changes in the computation of sampling variance of the estimators of minimum and standard distances. In the case of the minimum genetic distance, because of the independent segregation of alleles at different loci equation (3.22) immediately gives

$$V(\hat{D}_m) = \sum_{i=1}^{r} V(\hat{d}_{mk})/r \,, \tag{3.23}$$

where

$$V(\hat{d}_{mk}) = \sum_{i=1}^{r} (\hat{d}_{mk} - \hat{D}_m)^2/(r-1) \,.$$

No approximation is involved in the above equations. On the other hand, the sampling variance of the estimator of the standard genetic distance can only be evaluated approximately through the equation

$$V(\hat{D}_s) \approx \left[\frac{V(j_{\alpha\alpha k})}{4j_{\alpha\alpha}^2} + \frac{V(j_{\alpha\alpha k})}{4j_{\alpha\alpha}^2} + \frac{V(j_{\alpha\beta k})}{j_{\alpha\beta}^2} + \frac{\mathrm{cov}(j_{\alpha\alpha k}, j_{\beta\beta k})}{j_{\alpha\alpha}j_{\beta\beta}} \right.$$
$$\left. - \frac{\mathrm{cov}(j_{\alpha\alpha k}, j_{\alpha\beta k})}{j_{\alpha\alpha}j_{\alpha\beta}} - \frac{\mathrm{cov}(j_{\beta\beta k}, j_{\alpha\beta k})}{j_{\beta\beta}j_{\alpha\beta}} \right] \Big/ r \,, \tag{3.24}$$

in which $j_{\alpha\alpha k}$, etc. are the gene-identity values for the kth locus and $V(j_{\alpha\alpha k})$, $\mathrm{cov}(j_{\alpha\alpha k}, j_{\alpha\beta k})$ terms are the variances and covariances of the specific gene-identity values across all loci samples. For example,

$$\mathrm{cov}(j_{\alpha\alpha k}, j_{\alpha\beta k}) = \sum_{k=1}^{r} (j_{\alpha\alpha k} - \hat{j}_{\alpha\alpha})(j_{\alpha\beta k} - \hat{j}_{\alpha\beta})/(r-1) \,.$$

Equation (3.24) can be computed directly from the sampled allele frequencies at the k loci in populations α and β. Similar variance computation formulae for other statistics related with gene differentiation (e.g., normalized gene identity and

Nei's maximum genetic distance) are given in Nei and Roychoudhury (1974), Nei (1978a) and Nei (1987).

In trying to interpret the magnitude of standard errors of an estimated distance we must remember that there are two types of sampling errors involved in the estimation, as mentioned before. Nei and Roychoudhury (1974) and Li and Nei (1975) have shown that the sampling variance of gene diversity or genetic distance has three components: (i) interlocus interclass variance (due to variation of the $j_{\alpha\alpha k}$, $j_{\alpha\beta k}$, and $j_{\beta\beta k}$ values across loci at the time of evolutionary split of the populations from their common ancestry; a quantity on which no direct observation is available), (ii) interlocus intraclass variance (due to random genetic drift and possibly differential selection, if any, that occurs within populations α and β after their split from their common ancestry; which may be measured in expectation when the same set of random loci are used for both populations and by considering that the pair of populations is a random realization of their evolutionary history), and (iii) intralocus variance (due to a contemporary sampling of genes from the existent populations α and β, which can be evaluated from the theory of multinomial sampling of alleles – see, e.g., equation (2.12). Therefore, even though the formal definition of the genetic distances are not tied with any particular model of the evolution of gene differentiation, the relative contributions of the above three components to the total sampling variance of genetic distance is related to the process of evolution. Theoretical studies (Li and Nei, 1975) indicate that as long as the process of gene differentiation proceeds with a balance between mutation and random genetic drift pressures, the interlocus intraclass variance constitutes a major fraction of the total variance (in the range 78% or higher, increasing as a function of the product of the mutation rate and effective population size). Variance due to contemporary sampling of alleles (the third component), on the contrary, is shown to constitute a much smaller fraction of the total variance (Chakraborty et al., 1978). These results, together, imply that the single most important factor of the sampling variance of genetic distance (and gene-diversity) is the random genetic drift (that contributes to the second, namely interlocus intraclass, component of the total variance).

The above properties of the total sampling variance of the genetic distance estimators are important for determining whether or not the Taylor-series approximation rule (sometimes called the delta method – see Mueller, 1979) is appropriate for bias reduction of the estimators. Mueller (1979) argued that the above methods are not strictly adequate for bias reduction. He came to this conclusion through some simulation studies and suggested bootstrapping as a more appropriate method for bias correction. His simulation replications, however, consist of resampling from the same allele frequencies for a set of loci. Hence, the only source of variation he was controlling through the simulation replications was the intralocus (contemporary sampling of alleles) component, which is perhaps the smallest of the three. Therefore, his results are not conclusive proof of the inadequacy of the delta method, which is also shown in simulation studies of Chakraborty (1985). Table 1 shows the summary of this later study, where a set of 6 populations sequentially diverging from a common ancestry at time points

$(t/2N) = 1, 2, \ldots, 6$ from their common ancestry was simulated. The allele frequencies from these six extant populations were sampled following the algorithm of Griffiths and Li (1983). The observed genetic distances (minimum and standard) in this table are the average values over 20 loci in 500 replicates, using the value of the composite parameter, $4Nv$ (four times the product of effective population size and mutation rate per locus per generation), equal to 0.1. The observed variances are their empirical values computed from the interreplication variation of the average distances over 20 loci. The expectations of the unbiased distances and their variances are computed using equations (3.20) and (3.22)–(3.24); i.e., by using the delta method, as suggested by Nei (1978a). From the numerical values of this table it is clear that there is good agreement of simulated values (observed) and their expectations based on the delta method. However, for shorter time of divergence ($t/2N < 2$), the expected variance, computed by the delta method, seems to be somewhat larger than the ones observed. Nevertheless, since the simulation algorithm (Griffiths and Li, 1983) is also an approximate one (particularly when $t/2N$ is small), we may regard these agreements quite satisfactory, especially since these variances are also subject to large stochastic fluctuations. Contrary to Mueller's (1979) conclusion, these results indicate that the delta method of estimation does not always yield upward bias in the estimates. Note that only the mean standard distance is consistently overestimated in the simulated data.

Table 1
Comparison of the observed (simulated) and expected (Li and Nei, 1975) values of mean and variance of Nei's minimum and standard genetic distances computed over 20 loci. The simulation is conducted over 500 independent replications with 5 time of divergence values following the algorithm of Griffiths and Li (1983).[a]

		Time of divergene (in units of $2N$ generations)				
		1.0	2.0	3.0	4.0	5.0
Minimum distance						
Mean	Obs.	0.091	0.168	0.239	0.298	0.353
	Exp.	0.087	0.165	0.236	0.300	0.358
Variance	Obs.	0.00196	0.00431	0.00616	0.00806	0.00992
	Exp.	0.00271	0.00558	0.00759	0.00892	0.00974
Standard distance						
Mean	Obs.	0.107	0.208	0.311	0.408	0.517
	Exp.	0.100	0.200	0.300	0.400	0.500
Variance	Obs.	0.00311	0.00833	0.01550	0.02386	0.03182
	Exp.	0.00404	0.00907	0.01662	0.02377	0.03162

[a] All computations refer to the case with $4Nv = 0.1$.

3.4. *Sampling distribution and hypothesis testing issues*

The use of genetic distance for evolutionary studies may require testing two types of hypothesis. In the first type of questions, we might ask whether or not a specific pair of populations are substantially differentiated in terms of their allele frequency distributions (i.e., whether or not $D_{XY} = 0$ for any pair of populations X and Y). The second type of questions relate to testing whether or not the distance between any specific population contrasts are equal (e.g., $D_{XY} = D_{XZ}$ or $D_{XY} = D_{ZW}$). In either case, it is necessary to employ the sampling distribution of the relevant linear contrasts of the estimators of the genetic distances (e.g., in the first case D_{XY} itself, or in the second case, $\hat{D}_{XY} - \hat{D}_{XZ}$ or $\hat{D}_{XY} - \hat{D}_{ZW}$). Unfortunately, not many theoretical studies of such nature have been done. Nei (1987) showed that for the first type of hypotheses (i.e., $D_{XY} = 0$), Sanghvi's (1953) distance measure can be used, since it is directly related with the heterogeneity chi-square statistic of allele frequency differences between populations. For example, the statistic

$$\chi^2 = 2n_X n_Y \sum \frac{(x_i - y_i)^2}{n_X x_i + n_Y y_i} \, , \tag{3.25}$$

where the summation sign is over all alleles segregating in either of the two populations has a chi-square distribution when the sample size (number of individuals) from the two populations are large. The degrees of freedom equals the number of segregating alleles in the total population minus one. Even though a significant value of this statistic generally should indicate that any distance statistic between them should be substantial, the trouble with this test statistic is that it is very sensitive in the presence of rare alleles. In applying this statistic it might be necessary to combine all of the rare alleles into a single class, so that the chi-square approximation can be applied with suitable alteration of the degrees of freedom. The test of second type of hypotheses is much more involved. This is so, because even if i, j, k, and l are all different, it does not guarantee that the relevant estimators of the distances are totally independent. This follows from our previous discussion on the components of sampling variance, one of which relates to the evolutionary history of the four populations involved in such a hypothesis. Chakraborty (1985) and Nei (1987) suggested an approximate procedure for testing such hypothesis. Since the estimator of Nei's minimum genetic distance based on a group of loci can be written in terms of the single locus estimates through equation (3.22), for testing H_0: $D_{XY} = D_{ZW}$, we might consider the test statistic

$$t = r^{-1} \sum_{k=1}^{r} \{\hat{d}_{XYk} - \hat{d}_{ZWk}\} / \sqrt{V(\hat{d}_{XYk} - \hat{d}_{ZWk})/r} \, , \tag{3.26}$$

where $\hat{d}_{XYk}$ is the unbiased stimator of the minimum genetic distance between populations X and Y based on allele frequency data at the kth locus and the variance of the specific contrasts of such distance estimators across the r loci. The large sample distribution of this statistic is a t-distribution with degrees of freedom $r - 1$ (Chakraborty, 1985). Since this test is based on the minimum genetic distance, if this shows significant differences between distances of populations X and Y in comparison with that between populations Z and W, we might infer that the evolutionary separation between (X, Y) and (Z, W) are statistically different. It should be noted that Chakraborty's (1985) study is quite empirical based on computer simulations, and a formal proof of this result is highly desirable along the lines indicated earlier in the context of sampling theory of estimators of gene diversity within a population.

Since the standard genetic distance is not equal to the mean of single-locus distances, no such test statistic can be constructed with estimators of standard genetic distances. But a test procedure based on the estimators of the standard distance is desirable, since the minimum genetic distance has the undesirable property that in the presence of polymorphisms in both populations but no shared allele between the two populations, their minimum genetic distance is simply equal to the average gene-identity within them which can take any arbitrary value irrespective of the evolutionary time of differentiation between them.

3.5. Relationship between genetic distance and evolutionary time

Even though the above discussions of genetic distance measures do not explicitly assume any evolutionary model of gene differentiation between populations, in order to establish the relationship between genetic distance and evolutionary time of separation of populations, we must invoke the process of gene differentiation between populations. Two types of models must be distinguished in establishing such theoretical relationships. First, when the populations are genetically close, one can assume that the evolutionary separation between populations is so small that the allele frequency differences between them are strictly due to the pressure of random genetic drift operating within them and no mutation has been accumulated in either of them. In this case, Wright's F_{ST} for two populations has the following approximate expectation

$$E(F_{ST}) \approx 2(1 - e^{-t/2N})/(1 + e^{-t/2N}),\tag{3.27}$$

where t is the time (in generations) of separation between two populations, each of which maintained a constant effective size N throughout their evolution since their split from a common ancestry and did not exchange any migrant between them. When $t/2N$ is much smaller than one, equation (3.27) can be further approximated by

$$E(F_{ST}) \approx t/2N,\tag{3.28}$$

so that F_{ST} becomes proportional to the time of divergence between populations. In the presence of mutation, however, this relationship breaks down (Nei, 1975).

Latter's (1973) $\phi*$ also has a similar property, since in the absence of mutation, the expectation of $\phi*$ is given by

$$E(\phi*) = 1 - e^{-t/2N}, \tag{3.29}$$

while in the presence of mutation and when the initial population is at mutation-drift equilibrium, Nei (1987) derived the relationship

$$E(\phi*) = J_\infty(1 - e^{-2vt})/(1 - J_\infty e^{-2vt}), \tag{3.30}$$

where $J_\infty = (1 + 4Nv)^{-1}$. Therefore, both of these F_{ST}-related genetic distance functions are not always linear increasing functions of evolutionary time, particularly when the populations are evolutionarily distantly related.

Similar analysis with Cavalli-Sforza's (1969) and Rogers' (1972) distances indicate that their proportionality with the time of separation between populations also breaks down as $t/2N$ becomes large (Nei, 1976).

Nei's minimum and standard distances, on the contrary, have better relationships with the time of divergence. When the rate of gene substitution is constant over loci and over generations and the populations are at steady-state under the mutation-drift balance, Nei's minimum genetic distance has the expectation

$$E(D_m) = J_\infty(1 - e^{-2vt}), \tag{3.31}$$

which suggests that D_m increases almost linearly when t is small, but the rate of increase gradually decreases. The minimum distance, of course, does not increase indefinitely as t increases, since the asymptotic value of $E(D_m)$ is equal to $J_\infty = (1 + 4Nv)^{-1}$. On the other hand, under the same set of assumptions the standard genetic distance has the expectation

$$E(D_s) = 2vt, \tag{3.32}$$

for all values of t, so that the time of divergence between populations can be directly obtained from the standard genetic distance when the rate of gene substitution is known (Nei, 1972).

Note that when the minimum or standard distance functions are transformed to make them proper metric functions (see equations (3.9) and (3.12)), the above relationships with t are distorted, so that even though such transformed distance functions become suitable for cluster analysis, they do not necessarily give any simple estimate of the time of divergence between populations.

The above relationships between distance and time of divergence is obviously dependent on two critical assumptions. First, the rate of gene substitution is constant over all loci and second, all mutations lead to distinguishable alleles. Neither of these two assumptions strictly is valid for electrophoretically detectable polymorphisms, since differences in the rate of amino acid substitutions have been noticed across loci (Dayhoff, 1969, 1972), and electrophoretic techniques fail to reveal about $\frac{2}{3}$ to $\frac{3}{4}$ of all new mutations (Nei and Chakraborty, 1973). Theoretical as well as empirical studies indicate that both of these assumptions may be relaxed. Nei et al. (1976a) proposed a model of variable mutation rates, where the variation of mutation rates across loci was assumed to follow a Gamma distribution. Using this model, Nei et al. (1976a) suggested a modified distance function,

$$D_v = (1 - I)/I , \qquad (3.33)$$

where I is the normalized gene identity between two populations. Nei et al. (1976a) and Nei (1987) showed that when the standard deviation of mutation rates across loci is roughly equal to the mean mutation rate (over loci), D_v is again linearly related with the time of divergence. Therefore, instead of taking the negative value of the logarithm of normalized gene identity (as done in computing the standard distance), equation (3.33) can be used to compute the genetic distance that preserves the linearity with t, when the mutation rates are variable across loci. Estimation of D_v and evaluation of standard error of the estimator of D_v can be done exactly in the same fashion as that of D_s (see Nei, 1987, for details).

The problem of hidden variation of electrophoresis can be circumvented by appealing to the stepwise mutation model, originally proposed by Ohta and Kimura (1973). Nei and Chakraborty (1973), Li (1976a) and Chakraborty and Nei (1977) have studied the expected gene identity under the stepwise mutation model. Although the exact relationship is rather involved, the expected relationship between the normalized gene identity and t can be approximated by

$$I_E = e^{-2vt} \sum_{r=0}^{\infty} (vt)^{2r}/(r!)^2 , \qquad (3.34)$$

where v is the mutation rate per generation, and I_E is the normalized gene identity between populations using electrophoretic allele frequencies.

When the mutation rate (v) varies from locus to locus following a Gamma distribution with coefficient of variation 1, the average value of I_E is given by

$$I_{EA} = \frac{1}{1 + 2\bar{v}t}\left[1 + \sum_{r=1}^{\infty} \frac{(2r)!}{(r!)^2}\left(\frac{\bar{v}t}{1 + 2\bar{v}t}\right)^{2r}\right], \qquad (3.35)$$

where $\bar{v}$ is the mean of v over all loci. This relationship can be used to estimate the time of separation between populations from electrophoretic allele frequency data. Admittedly, this is a tedious relationship for estimation purposes. Nei (1987) provides some numerical computations of I_{EA} versus t, and also the relationship between $-\log_e I_{EA}$ versus t, which suggests that even though (3.35) relaxes both of the assumptions involved in the strict linear relationship between D_s and t, up to the time point $t = v^{-1}$, the distortions produced by hidden variation of electrophoresis and variation of mutation rates across loci are not substantial, and hence, even the simple linear relationship between D_s and t is satisfactory for rough estimation of time of divergence between populations.

Perhaps more important is to examine how the relationship between distance and time is distorted when the populations exchange genes over generations. This is worthwhile to investigate, because at the early stage of differentiation between populations, they are not generally totally isolated. For example, the different ethnic populations of humans are not strictly isolated and the history of migration between ethnic groups is quite long in certain cases.

Many different models of migration have been examined, although the qualitative conclusions are quite similar. For example, it is intuitively clear that in the presence of migration, the asymptotic value of normalized gene identity between populations can not be zero (which happens when the two populations are completely isolated). Nei and Feldman (1972) showed that when the rate of migration (m) is much larger than the mutation rate (v), the steady-state value of the normalized gene identity between two populations is given by

$$I_\infty = m/(m + v),$$ (3.36)

but the rate of approach to this equilibrium value is quite slow; the number of generations required to approach I_∞ is almost equal to the inverse of the mutation rate. Li (1976b) studied the transient expectation of gene identity between populations under the joint effects of mutation, migration and random genetic drift, showing that the accumulation of genetic difference between populations can be substantially retarded in the presence of even a small amount of migration.

Finally, the relationship between distance and time of divergence can be distorted by fluctuations of population sizes as well. Since it is impossible for any natural population to maintain a constant size during its entire history of evolution, this aspect received considerable attention from intuitive as well as analytical points of view. Chakraborty and Nei (1977) showed that when one or both of the diverging populations experiences population bottleneck, the genetic distance between them rapidly increases initially and the rate of increase is higher when the bottleneck size is small than when it is large. However, as the population sizes approaches their carrying capacity, the linear relationship between distance and time is restored. If only one of the populations goes through a bottleneck, Chakraborty and Nei (1974) showed that a simple modification of the standard distance preserves the linear relationship with time quite satisfactorily. In this case, we may define

$$D_b = -\log_e(J_{XY}/J_{XX}),$$ (3.37)

where J_{XX} is the gene identity within the population which remained constant in its size.

In the above theoretical relationships the term expectation is used in the sense that we considered evolutionary replicates of pairs of populations evolving under exactly similar evolutionary processes. This is perfectly valid since the observation on a single pair of populations is only a single realization of the stochastic evolutionary process, and it may differ from the theoretical relationship expected under the forces of evolution due to the large stochastic fluctuations of allele frequencies.

3.6. Empirical data on relationship between distance and time of divergence

The development of the theory of genetic distance between populations, species or any taxa spawned a great deal of interest among population biologists to measure the time of divergence between such organisms, particularly when data on fossil records, separation of lands and seas, island formation are not available to calibrate the level of electrophoretic variation between taxa. Since several assumptions are implicit in relating genetic distance with the time of divergence, Nei (1987) collected data on genetic distance between several taxa for which time of divergence information are available from other sources. Earlier, Nei and Roychoudhury (1974) analyzed the genetic variability among the major three human racial groups to suggest that the time of divergence (in years) can be estimated from his standard genetic distance by the relationship

$$t = 5 \times 10^6 D_s, \tag{3.38}$$

which implies that D_s and t should follow a linear regression line passing through the origin.

Figure 1 shows the summary of data collected by Nei (1987) in which D_s is plotted against t (measured in units of 100000 years). Fitting a linear regression through the origin we find that the regression coefficient of D_s on t is 0.0122 ± 0.0015, and the predicted linear regression is statistically adequate (t-statistic with 15 degrees of freedom is 0.63, $P > 0.25$). Translating this into the problem of estimating t from D_s, we find a predicting equation

$$t = (8.1 \pm 2.9) \times 10^6 D_s, \tag{3.39}$$

where the constant of proportionality is somewhat larger (not significantly, however) than the one suggested by Nei and Roychoudhury (1974). Even though the 16 data points of Figure 1 fit a straight line through the origin fairly well, before concluding that these observations are indicative of a molecular clock (constant rate of evolution), there are several issues that must be considered. For example, the proteins used in estimating D_s widely varies in the surveys on these different organisms, and the constant of proportionality is known to be dependent on the protein loci (Dayhoff, 1969, 1972). The hidden variation of electrophoresis is also a problem, and the extent to which different alleles are detectable through electrophoresis also varies from laboratory to laboratory. Further, since the standard error of the estimate of D_s is critically dependent on the number of loci samples (and it varies between 18 and 62 in these data, see Nei, 1987), the measuremental errors of all data points are not truly uniform. Lastly, the time estimates taken

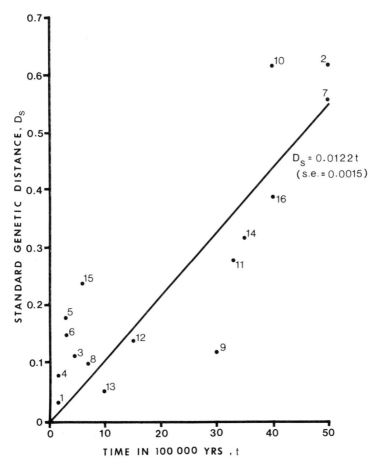

Fig. 1. Empirical relationship between Nei's standard genetic distance and time of divergence (from other sources) for several organisms. The points refer to genetic distances between: (1) Negroid and Mongoloid; (2) Man vs. Chimpanzee; (3) Two macaque species; (4) Pocket gophers; (5) Woodrats; (6) Deer mice species; (7) Ground squirrels species; (8) Ground squirrels subspecies; (9) Galapagos finches; (10) Bipes species; (11) Lizards; (12) Cave and surface fishes; (13) Minnows; (14) Panamanian fishes — set 1; (15) Panamanian fishes — set 2; and (16) Panamanian sea urchins. See Nei (1987, Table 9.4) for exact sources of data.

from other sources are not really as certain as they might look, because the sources are rather varied, and not always related to the specific taxa being contrasted. In spite of these difficulties, the near perfect linear relationship (correlation coefficient = 0.86) between t and D_s in these data is suggestive of the utility of a rough prediction of time of divergence from genetic distances between populations.

4. Practical utility of measures of variation

It is obvious from the above discussions that any measure of variation, within-as well as between-populations, is necessarily only a summary statistic obtained from the multi-dimensional allele frequency data. Although this leads to loss of information, the simplicity of these measures lies in their easy interpretability for the purpose of specific applications. In this section we mention the most frequent applications of genetic diversity measures, principally outlining the adequacies and limitations. Jorde (1985) also addressed some of these issues from a different point of view.

4.1. Applications for studying origin and relatedness

Elsewhere in this volume Thompson (Chapter 8) and Saitou (Chapter 10) discussed the methods for determining the relatedness of individuals and inferring evolutionary relationships among populations (or taxa) from the measures of genetic variation. While under a set of specific assumptions such methods lead to important evolutionary conclusions, one should note that the evolutionary divergence between populations or species may occur through space as well as time. Therefore, other features of dissimilarities must not be overlooked when attributing genetic variation between populations only to their evolutionary history. For example, the migration structure between different geographic populations of a species can be quite complex and sometimes it is not unrelated with human migration. Furthermore, genetic variation among spatially separated populations may also be under differential selection pressures. Phylogenetic analysis, generally, do not take into account these two factors adequately. As a result, estimates of divergence time may be vitiated. Lathrop (1982) provided a partial solution to the migration issue by calculating simultaneous MLE estimates of phylogenies and admixture rates between populations. His solution is critically dependent on the assumption that the history of gene migration is far shorter (and more recent) than the time of separation of populations from their common ancestry. While this assumption may be realistic for studying genetic differentiation between certain geographic populations, in general the applicability of such methods is rather limited. In this sense, Lathrop's treatment of invoking admixture between populations into a phylogenetic analysis is strictly valid for microevolutionary studies.

Construction of evolutionary tree or clustering of populations, which have predominated the interest of most users of gene diversity indices, are primarily motivated by the classical statistical theory of estimation and hypothesis testing. This is so, because in these procedures the underlying evolutionary processes are modeled in terms of a set of parameters and the trees/clusters represent subsets of such parametric spaces which are examined in contrast with some hypothesized values of the parameters obtained from other sources (e.g., biogeography, fossil records, etc.). The robustness of these techniques in the presence of violation of the underlying assumptions (e.g., constancy of the evolutionary rates of substitution in different lineages, absence of selection differential across the ecological niches of habitation of the populations) is not a well-studied area of research. Furthermore, the distributional assumptions needed for the clustering techniques are not validated in most applications. Consequently, some attempts have been made suggesting non-parametric methods based on the principles of exploratory data analysis (Karlin et al., 1979, 1982) and spatial (and temporal) autocorrelation analysis (Sokal and Friedlaender, 1982; Sokal and Menozzi, 1982; Sokal and Wartenberg, 1983). In these methods primary importance is placed on robustness and freedom from the distributional assumptions. However, the genetic basis of postulations of these methods is of limited nature, as a result of which the genetic maps of populations based on autocorrelation or clustering of populations based on the exploratory data analysis are difficult to interpret in terms of an evolutionary model. As in the case of evolutionary tree reconstruction, the inclusion of a long-standing admixture is also a problematic issue in such non-parametric analysis.

Many investigators constructed diversity-related (or gene frequency based) genetic maps of populations by differential shading or coloring, placing them on the specific geographic regions which represent their habitation or location. Such genetic contours were initially suggested in the encyclopedic work of Mourant et al. (1976). Although computer-assisted techniques can now generate such genetic maps of populations based on allele frequency data at any arbitrary number of loci, the multi-dimensional scaling techniques implicit in these methods necessarily rely on major co-ordinates of allele frequency differences among populations. To what extent these major co-ordinates are truly multi-locus characteristics of genetic differentiation has not explicitly been demonstrated. Evolutionary studies of allele frequency changes indicate that the single-locus measures of genetic diversity are subject to large stochastic errors (see e.g., Nei and Roychoudhury, 1974; Li and Nei, 1975; Nei, 1978a). Should the major coordinates reflect genetic variation at one or a few loci alone, the stochastic errors of the resultant genetic maps of populations can also be quite large. This is parallel to the problem of constructing a reliable phylogenetic tree based on data on a small set of loci (Nei et al., 1983 and Saitou, this volume).

4.2. Application for detecting selection differential

As mentioned before, one criterion for judging the appropriateness of an index of genetic variation is that it should take distinctly different values for loci/populations for which the evolutionary forces of maintenance of polymorphisms are different. Since some of the proposed measures of genetic variation have such discriminatory powers of detecting the presence/absence of selection, numerous studies utilized such indices of genetic variation to test the neutral mutation hypothesis. The utility of diversity-related statistics in detecting selection is certainly compromised by the data summarization that are implicit in the definition of such statistics. Therefore, a gene diversity analysis or a genetic distance analysis does not necessarily constitute a formal proof of demonstrating that one or more of the evolutionary forces are not major factors of maintenance of genetic variation. However, when the patterns of genetic variation within and between populations are supplemented by other features of allelic variability (e.g., allele frequency distribution, proportion of shared alleles of various gene frequency class, etc.), congruence of observations with the predictions of certain evolutionary models can provide inference with regard to the major features of evolution. In a series of studies (Fuerst et al., 1977; Chakraborty et al., 1978, 1980; Nei and Graur, 1984) this rationale was used to arrive at a conclusion that in general electrophoretic variation in natural populations is largely maintained by counteracting forces of random genetic drift and mutation. Although the statistical power of the test procedures for detecting differential selection employed in these studies has been contested (Gilpin et al., 1976), such tests jointly imply that even if selection plays an important role in maintaining genetic variation in natural populations, the extent of selection differential must be very small ($Ns \ll 1$).

This does not, of course, indicate that differential selection does not occur at any locus. Recent studies have shown that the genetic variation at the HLA-gene complex in humans and the MHC-gene region in other organisms are perhaps maintained by some form of positive Darwinian selection (Hedrick and Thomson, 1983; Hughes and Nei, 1988).

4.3. Applications of genetic diversity in genetic epidemiology

Studies of genetic diversity also have important implications in understanding the origin and spread of genetic diseases of simple Mendelian inheritance. In addition, the variations of prevalence rates of complex diseases among populations can be examined in conjunction with the genetic variation among populations to predict the role of genetic factors in such diseases.

Classic examples of the utility of genetic variations in understanding the epidemiology of simple Mendelian diseases include the demonstration of the multiple origin of the β-globin mutation causing sickle-cell anemia which is found in several parts of the world. Initially the multiple origin of the sickle-cell mutation was proposed on the basis that this disease was found prevalent in populations that do not show any obvious evolutionary closeness, the latter being an observation based on genetic variability at other loci among these populations (Lehmann and Cutbush, 1952). Recent molecular data confirmed this view, because the single nucleotide substitution at the β-globin chain that causes sickle-cell anemia has been found under widely varied haplotypic backgrounds, which is difficult to explain by a single occurrence of this mutation during the human evolution (Pagnier et al., 1984; Chebloune et al., 1988). The genetic heterogeneity of phenylketonuria (PKU), a disease caused by the deficiency of hepatic phenylalanine hydroxylase, was also first postulated from its varied incidence rates in popuations of diverse origin (Flatz et al., 1984; Kidd, 1987). DNA studies now document that PKU mutation(s) bearing chromosomes have haplotypic patterns that are suggestive of multiple mutations, even though in aggregate they produce a simple autosomal recessive transmission of this disease (Scriver et al., 1988). Cost-benefit analysis of screening for life-threatening genetic diseases are also generally based on the genetic affinity of populations in which such diseases are expected to occur in appreciable frequencies.

A good example of the applications of gene diversity analysis for understanding the genetic basis of complex diseases is given by Chakraborty et al., (1986) and Chakraborty and Weiss (1986). In Northern American populations the adult onset non-insulin dependent diabetes mellitus (NIDDM) is found in varied prevalence rates, and in general there is a correspondence of the age-adjusted NIDDM rates with the degree of Amerindian admixture in the populations. This association can be explained assuming genetic susceptibility to this disease (Chakraborty and Weiss, 1986). Although the mode of inheritance of NIDDM cannot be exactly determined from such studies, their importance for determining the best suited populations for familial studies to map such disease-gene cannot be completely over-ruled (Chakraborty and Weiss, 1988).

4.4. Relationship between molecular and phenotypic evolution

For macroevolutionary studies the most significant application of measures of genetic variation comes from the fact that using such indices it is possible to reconcile the different features of evolution at the molecular and phenotypic level. This issue is discussed at length by Nei (1987), where it is argued that there are two ways for explaining why at a molecular level the variations are selectively equivalent, while at a phenotypic level natural selection plays a dominant role. First, we might assume that each mutation, taken singly, might have a selective differential so small that they are almost neutral and their cumulative effects for many loci are responsible for the adaptive change of organisms. The second explanation is that a small proportion of the mutations are adaptive, while most nondeleterious mutations are neutral. In other words, adaptive evolution can be caused by a comparatively smaller number of mutations.

A considerable body of literature has been accumulated studying which of these two alternative explanations explain the observed pattern of morphological and molecular variation better (see Nei, 1987 for references). Nevertheless, it is clear that a complete resolution of this problem must come from understanding the genetic basis of morphological traits, without which we cannot predict how many mutations are truly needed to make drastic changes at the phenotypic level.

5. Summary and conclusions

The main thrust of our presentation in this chapter has been to demonstrate that it is possible to define indices of genetic variation to measure genetic differences at within- as well as between-populations level which are free of evolutionary assumptions regarding the production and maintenance of genetic variation. The statistical properties of such indices of genetic diversity are well-developed in certain cases, although much work in this direction is needed. We further showed that there is a parallelism of the development of such measures of variation for genetic and ecological studies, even though this parallelism has not been explicitly recognized often in the literature of statistical ecology and population genetics. The ecologists generally emphasized the statistical properties of measures of variation, studied in terms of sampling from the contemporary populations, while the emphasis of the population genetic studies had been to determine the effect of evolutionary factors on such summary measures. Our approach of reviewing the various indices of genetic variation is somewhat different from the existing excellent reviews (cited earlier) in the sense that we showed that it is possible to unify the different concepts by considering some general mathematical properties of measures of variation.

We argue that it is not possible to provide any rule of thumb to decide which index of variation is the best. On the contrary, the suitability of a particular index of variation must be judged based on the purpose of the users. The applicant must decide to what extent the index satisfies the premises of the intended applications. For example, if we are using an index of variation to infer the evolutionary mechanism, we must make an effort to test the validity of the assumptions needed to ensure that the particular gene diversity index satisfies the assumptions required for drawing such inferences. In the past, some reviewers argued that the evolutionary studies involving gene diversity related statistics most likely cannot provide further insight about evolutionary mechanisms because of the untestability of assumptions such as constancy of population size, nor can we quantify the extent of variations of the rates of substitution at different regions of the genome (Harpending, 1974). With the advent of DNA technology, such problems are circumvented to a great extent, because these assumptions are now directly testable.

Another important advance of laboratory technology that raises the optimism of utilizing the measures of genetic variation for micro- as well as macro-evolutionary studies is that it is now possible to conduct population-based studies at numerous regions of the genome that contain extremely large amount of variation, measured in terms of number of allelic variants and the extent of gene diversity (e.g., heterozygosity). The Variable Number of Tandem Repeats (VNTR) loci are classic examples of such hyper-polymorphic loci. Inclusion of such loci in genetic distance studies should allow us to understand evolution during a smaller time period, because of the faster rate of mutations at such loci. The large number of variant alleles at such loci should also aid in understanding the genetic structure of populations, because we can conduct analysis at a single-locus level, without compromising the statistical power of the analysis (Chakraborty, 1990a,b). Therefore, we might conclude by saying that new molecular data as well as new computer-assisted data analysis techniques provide sufficient opportunity to increase the potential of future advances that can be made for studies of genetic diversity.

Acknowledgements

This work was partially funded by US Public Health Service Grant GM 43199 from the National Institutes of Health and Grant 90-IJ-CX-0038 from the US National Institutes of Justice. We thank Drs. M. Nei and W. J. Schull for their comments and suggestions during the preparation of this chapter.

References

Agresti, A. and Agresti, B. F. (1978). Statistical analysis of qualitative variation. In: K. F. Schuessler, ed., *Sociological Methodology*. Josey Bass, San Francisco, CA, 204–237.

Balakrishnan, V. and Sanghvi, L. D. (1968). Distance between populations on the basis of attribute data. *Biometrics* **24**, 859–865.

Bhattacharyya, A. (1946). On a measure of divergence between two multinomial populations. *Sankhya* **7**, 401–406.

Bishop, Y. M. M., Fienberg, S. E. and Holland, P. W. (1975). *Discrete Multivariate Analysis: Theory and Practice*. MIT Press, Cambridge, MA.

Bowman, K. O., Hutcheson, K., Odum, E. P. and Shenton, L. R. (1971). Comments on the distribution of indices of diversity. In: G. P. Patil, E. C. Pielou and W. E. Waters, eds., *Statistical Ecology, Vol. 3*. Pennsylvania State Univ. Press, University Park, PA, 315–366.

Cavalli-Sforza, L. L. (1969). Human diversity. In: *Proc. 12th Intl. Congr. Genet. Vol. 3, Tokyo*, 405–416.

Cavalli-Sforza, L. L. and Edwards, A. W. F. (1964). Analysis of human evolution. In: *Proc. 11th Intl. Congr. Genet.*, 923–933.

Cavalli-Sforza, L. L. and Edwards, A. W. F. (1967). Phylogenetic analysis: Models and estimation procedures. *Amer. J. Hum. Genet.* **19**, 233–257.

Chakraborty, R. (1977). The distribution of nucleotide site differences between two randomly chosen cistrons in a population of variable size. *Theor. Pop. Biol.* **11**, 11–22.

Chakraborty, R. (1980). Relationship between single- and multi-locus measures of gene diversity in a subdivided population. *Ann. Hum. Genet.* **43**, 423–428.

Chakraborty, R. (1981). Expected number of rare alleles per locus in a sample and estimation of mutation rates. *Amer. J. Hum. Genet.* **33**, 481–484.

Chakraborty, R. (1985). Genetic distance and gene diversity: Some statistical considerations. In: P. R. Krishnaiah, ed., *Multivariate Analysis — VI*. Elsevier, Amsterdam, 77–96.

Chakraborty, R. (1990a). Mitochondrial DNA polymorphism reveals hidden heterogeneity within some Asian populations. *Amer. J. Hum. Genet.* **47**, 87–94.

Chakraborty, R. (1990b). Genetic profile of cosmopolitan populations: Effects of hidden subdivision. *Anthrop. Anz.* **48**, 313–331.

Chakraborty, R. (1991). Generalized occupancy problem and its applications in population genetics. In: C. F. Sing and C. L. Hanis, eds., *Impact of Genetic Variation on Individuals, Families and Populations*. Oxford Univ. Press, New York, in press.

Chakraborty, R., Ferrell, R. E., Stern, M. P., Haffner, S. M., Hazuda, . P. and Rosenthal, M. (1986). Relationship of prevalence of non-insulin-dependent diabetes mellitus to Amerindian admixture in the Mexican Americans of San Antonio, Texas. *Genet. Epidemiol.* **3**, 435–454.

Chakraborty, R. and Fuerst, P. A. (1979). Some sampling properties of selectively neutral alleles: Effects of variability of mutation rates. *Genet. Res.* **34**, 253–267.

Chakraborty, R., Fuerst, P. A. and Nei, M. (1978). Statistical studies on protein polymorphism in natural populations, II: Gene differentiation between populations. *Genetics* **88**, 367–390.

Chakraborty, R., Fuerst, P. A. and Nei, M. (1980). Statistical studies on protein polymorphism in natural populations, III: Distribution of allele frequencies and the number of alleles per locus. *Genetics* **94**, 1039–1063.

Chakraborty, R. and Griffiths, R. C. (1982). Correlation of heterozygosity and number of alleles in different frequency classes. *Theor. Pop. Biol.* **21**, 205–218.

Chakraborty, R. and Neel, J. V. (1989). Description and validation of a method for simultaneous estimation of effective population size and mutation rate from human population data. *Proc. Natl. Acad. Sci. USA* **86**, 9407–9411.

Chakraborty, R. and Nei, M. (1974). Dynamics of gene differentiation between incompletely isolated populations of unequal sizes. *Theor. Pop. Biol.* **5**, 460–469.

Chakraborty, R. and Nei, M. (1976). Hidden genetic variability in electromorphs in finite populations. *Genetics* **84**, 385–393.

Chakraborty, R. and Nei, M. (1977). Bottleneck effect with stepwise mutation model of electrophoretically detectable alleles. *Evolution* **31**, 347–356.

Chakraborty, R. and Schwartz, R. J. (1990). Selective neutrality of surname distributions in an immigrant Indian community of Houston, Texas. *Amer. J. Hum. Biol.* **2**, 1–15.

Chakraborty, R., Smouse, P. E. and Neel, J. V. (1988). Population amalgamation and genetic variation: Observations on artificially agglomerated tribal populations of Central and South America. *Amer. J. Hum. Genet.* **43**, 709–725.

Chakraborty, R. and Tateno, Y. (1976). Correlations between some measures of genetic distance. *Evolution* **30**, 851–853.

Chakraborty, R. and Weiss, K. M. (1986). The frequency of complex diseases in hybrid populations. *Amer. J. Phys. Anthrop.* **70**, 489–503.

Chakraborty, R. and Weiss, K. M. (1988). Admixture as a tool for finding linked genes and detecting that difference from allelic association between loci. *Proc. Natl. Acad. Sci. USA* **85**, 9119–9123.

Chakraborty, R. and Weiss, K. M. (1991). Genetic variation of the mitochondrial DNA genome in American Indians is at mutation-drift equilibrium. *Amer. J. Phys. Anthrop.*, in press.

Chebloune, Y., Pagnier, J., Trabuchet, G., Faure, C., Verdier, G., Labie, D. and Nigon, V. (1988). Structural analysis of the 5' flanking region of the β-globin gene in African sickle cell anemia patients: Further evidence for three origins of the sickle cell mutation in Africa. *Proc. Natl. Acad. Sci. USA* **85**, 4431–4435.

Czekanowski, J. (1909). Zur Differentialdiagnose der Neandertalgruppe. Korrespondenzblatt Deutsch. *Ges. Anthropol. Ethnol. Urgesch.* **40**, 44–47.

Dayhoff, M. O. (1969). *Atlas of Protein Sequence and Structure, Vol. 4.* Natl. Biomed. Res. Foundation, Silver Springs, MD.

Dayhoff, M. O. (1972). *Atlas of Protein Sequence and Structure, Vol. 5.* Natl. Biomed. Res. Foundation, Silver Springs, MD.

Ewens, W. J. (1964). The maintenance of alleles by mutation. *Genetics* **50**, 891–898.

Ewens, W. J. (1972). The sampling theory of selectively neutral alleles. *Theor. Pop. Biol.* **3**, 87–112.

Ewens, W. J. (1979). *Mathematical Population Genetics.* Springer, Berlin.

Ewens, W. and Li, W. H. (1980). Frequency spectra of neutral and deleterious alleles in a finite population. *J. Math. Biol.* **10**, 155–166.

Feller, W. (1957). *An Introduction to Probability Theory and Its Applications, Vol. 1.* Wiley, New York.

Fisher, R. A. (1936). The use of multiple measurements in taxonomic problems. *Ann. Eugen.* **7**, 179–188.

Fisher, R. A. (1953). Population genetics. *Proc. Roy. Soc. B.* **141**, 510–523.

Flatz, G., Oelbe, M. and Herrmann, H. (1984). Ethnic distribution of phenylketonuria in the northern German population. *Hum. Genet.* **65**, 396–399.

Fuerst, P. A., Chakraborty, R. and Nei, M. (1977). Statistical studies on protein polymorphism in natural populations, I: Distribution of single locus heterozygosity. *Genetics* **86**, 455–483.

Futuyama, D. J. (1979). *Evolutionary Biology.* Sinauer, Sunderland, MA.

Gilpin, M., Soulé, M., Ondrick, A. and Gilpin, E. A. (1976). Overdominance and U-shaped gene frequency distribution. *Nature* **263**, 497–499.

Gillespie, J. H. (1983). Some properties of finite populations experiencing strong selection and weak mutation. *Amer. Nat.* **121**, 691–708.

Gillespie, J. H. (1985). The interaction of genetic drift and mutation with selection in a fluctuating environment. *Theor. Pop. Biol.* **27**, 222–237.

Gini, C. (1912). Variabilitá e mutabilitá, Studi Economicoaguridici della facolta di Giurisprudenza dell, Universite di Cagliari III, Part II.

Griffiths, R. C. and Li, W. H. (1983). Simulating allele frequencies in a population and the genetic differentiation of populations under mutation pressure. *Theor. Pop. Biol.* **23**, 19–33.

Harpending, H. C. (1974). Genetic structure of small populations. *Ann. Rev. Anthropol.* **3**, 229–243.

Hedrick, P. W. (1975). Genetic similarity and distance: Comments and comparisons. *Evolution* **29**, 362–366.

Hedrick, P. W. (1983). *Genetics of Populations*. Science Book Intl., Portola Valley, CA.

Hedrick, P. W. and Thomson, G. (1983). Evidence for balancing selection at HLA. *Genetics* **104**, 49–456.

Hill, M. O. (1973). Diversity and evenness: A unifying notation and its consequences. *Ecology* **54**, 427–431.

Hughes, A. L. and Nei, M. (1988). Pattern of nucleotide substitution at major histocompatibility complex class I loci reveals overdominant selection. *Nature* **355**, 167–170.

Hutcheson, K. (1970). A test for comparing diversities based on Shannon formula. *J. Theor. Biol.* **29**, 151–154.

Johnson, N. L. and Kotz, S. (1969). *Distributions in Statistics: Discrete Distributions*. Houghton Mifflin, Boston, MA.

Jorde, L. (1980). The genetic structure of subdivided human populations: A Review. In: J. H. Mielke and M. H. Crawford, eds., *Current Developments in Anthropological Genetics: Theory and Methods, Vol. 1*. Plenum, New York, 135–208.

Jorde, L. (1985). Human genetic distance studies: Present status and future prospects. *Ann. Rev. Anthrop.* **14**, 343–373.

Karlin, S., Carmelli, D. and Bonné-Tamir, B. (1982). Analysis of biochemical genetic data on Jewish populations, III: The application of individual haplotype measurements for intra- and inter-population comparisons. *Amer. J. Hum. Genet.* **34**, 50–64.

Karlin, S., Kenett, R. and Bonné-Tamir, B. (1979). Analysis of biochemical genetic data on Jewish populations, II: Results and interpretations of heterogeneity indices and distance measures with respect to standards. *Amer. J. Hum. Genet.* **31**, 341–365.

Kidd, K. K. (1987). Phenylketonuria: Population genetics of a disease. *Nature* **327**, 282–283.

Kimura, M. (1955). Solution of a process of random genetic drift with a continuous model. *Proc. Natl. Acad. Sci. USA* **41**, 144–150.

Kimura, M. (1964). Diffusion models in population genetics. *J. Appl. Prob.* **1**, 177–232.

Kimura, M. (1971). Theoretical foundation of population genetics at the molecular level. *Theor. Pop. Biol.* **2**, 174–208.

Kimura, M. (1979). Model of effectively neutral mutations in which selective constraint is incorporated. *Proc. Natl. Acad. Sci. USA* **76**, 3440–3444.

Kimura, M. and Crow, J. F. (1964). The number of alleles that can be maintained in a finite population. *Genetics* **49**, 725–738.

Kimura, M. and Ohta, T. (1971). Protein polymorphism as a phase of molecular evolution. *Nature* **229**, 467–469.

Kimura, M. and Ohta, T. (1978). Stepwise mutation model and distribution of allele frequencies in a finite population. *Proc. Natl. Acad. Sci. USA* **75**, 2868–2872.

Kurczynski, T. W. (1970). Generalized distance and discrete variables. *Biometrics* **26**, 525–534.

Lalouel, J.-M. (1980). Distance analysis and multidimensional scaling. In: J. H. Mielke and M. H. Crawford, eds., *Current Developments in Anthropological Genetics: Theory and Methods. Vol. 1*. Plenum, New York, 209–250.

Lathrop, G. M. (1982). Evolutionary trees and admixture: Phylogenetic inferences when some populations are hybridized. *Ann. Hum. Genet.* **46**, 245–255.

Latter, B. D. H. (1973). Measures of genetic distance. In: N. E. Morton, ed., *Genetic Structure of Populations*, Press of Hawaii, Honolulu, Hawaii.

Lehmann, H. and Cutbush, M. (1952). Sickle-cell trait in Southern India. *Brit. Med. J.* **1**, 404–405.

Lewontin, R. C. (1972). The apportionment of human diversity. *Evol. Biol.* **6**, 381–398.

Li, W.-H. (1976a). Electrophoretic identity of proteins in a finite population and genetic distance between taxa. *Genet. Res.* **28**, 119–127.

Li, W.-H. (1976b). Effect of migration on genetic distance. *Amer. Nat.* **110**, 841–847.

Li, W.-H. (1977). Maintenance of genetic variability under mutation and selection pressures in a finite population. *Proc. Natl. Acad. Sci. USA* **74**, 2509–2513.

Li, W.-H. (1978). Maintenance of genetic variability under the joint effect of mutation, selection, and random genetic drift. *Genetics* **90**, 349–382.

Li, W.-H. and Nei, M. (1975). Drift variances of heterozygosity and genetic distance in transient states. *Genet. Res.* **25**, 229–248.

Lyons, N. I. and Hutcheson, K. (1979). Distributional properties of Simpson's index of diversity. *Comm. Stat. A* **8**(6), 569–574.

Magurran, A. E. (1988). *Ecological Diversity and Its Measurement*. Princeton Univ. Press, Princeton, NJ.

Mahalanobis, P. C. (1936). On the generalized distance in statistics. *Proc. Natl. Inst. Sci. India* **2**, 49–55.

Malécot, G. (1948). *Les Mathématiques de l'Hérédité*. Masson et Cie, Paris.

Maruyama, T. and Fuerst, P. A. (1984). Population bottlenecks and non-equilibrium models in population genetics, I: Allele numbers when populations evolve from zero variability. *Genetics* **108**, 745–763.

Maruyama, T. and Nei, M. (1981). Genetic variability maintained by mutation and overdominant selection in finite populations. *Genetics* **98**, 441–459.

Mayr, E. (1963). *Animal species and Evolution*. Harvard Univ. Press, Cambridge, MA.

Mueller, L. D. (1979). Comparison of two methods for making statistical inferences on Nei's measure of genetic distance. *Biometrics* **35**, 757–763.

Morton, N. E. (1975). Kinship, information and biological distance. *Theor. Pop. Biol.* **7**, 246–255.

Mourant, A. E., Kopec, A. C. and Domaniewska-Sobczak, K. (1976). *The Distribution of the Human Blood Groups and Other Polymorphisms*. Oxford Univ. Press, Oxford.

Nayak, T. K. (1983). Applications of entropy functions in measurement and analysis of diversity. Ph.D. Dissertation, Department of Mathematics and Statistics, University of Pittsburgh, PA.

Nei, M. (1972). Genetic distance between populations. *Amer. Nat.* **106**, 283–292.

Nei, M. (1973). The theory and estimation of genetic distance. In: N. E. Morton, ed., *Genetic Structure of Populations*. Univ. Press of Hawaii, Honolulu, Hawaii.

Nei, M. (1975). *Molecular Population Genetics and Evolution*. American Elsevier, New York.

Nei, M. (1976). Mathematical models of speciation and genetic distance. In: S. Karlin and E. Nevo, eds., *Population Genetics and Ecology*. Academic Press, New York, 723–765.

Nei, M. (1977). Estimation of mutation rate from rare protein variants. *Amer. J. Hum. Genet.* **29**, 225–232.

Nei, M. (1978a). Estimation of average heterozygosity and genetic distance from a small number of individuals. *Genetics* **89**, 583–590.

Nei, M. (1978b). The theory of genetic distance and evolution of human races. *Jap. J. Hum. Genet.* **23**, 341–369.

Nei, M. (1987). *Molecular Evolutionary Genetics*. Columbia Univ. Press, New York.

Nei, M. and Chakraborty, R. (1973). Genetic distance and electrophoretic identity of proteins between taxa. *J. Mol. Evol.* **2**, 323–328.

Nei, M., Chakraborty, R. and Fuerst, P. A. (1976a). Infinite allele model with varying mutation rate. *Proc. Natl. Acad. Sci. USA* **73**, 4164–4168.

Nei, M. and Feldman, M. W. (1972). Identity of genes by descent within and between populations under mutation and migration pressures. *Theory. Pop. Biol.* **3**, 460–465.

Nei, M., Fuerst, P. A. and Chakraborty, R. (1976b). Testing the neutral mutation hypothesis by distribution of single locus heterozygosity. *Nature* **262**, 491–493.

Nei, M. and Graur, D. (1984). Extent of protein polymorphism and the neutral mutation theory. *Evol. Biol.* **17**, 73–118.

Nei, M. and Livshits, G. (1989). Evolutionary relationships of Europeans, Asians, and Africans at the molecular-level. In: N. Takahata and J. F. Crow, eds., *Population Biology of Genes and Molecules*. Baifukan, Tokyo, 59–68.

Nei, M., Maruyama, T. and Chakraborty, R. (1975). The bottleneck effect and genetic variability in populations. *Evolution* **29**, 1–10.

Nei, M. and Roychoudhury, A. K. (1974). Sampling variance of heterozygosity and genetic distance. *Genetics* **76**, 379–390.

Nei, M., Tajima, F. and Tateno, Y. (1983). Accuracy of estimated phylogenetic trees from molecular data, II: Gene frequency data. *J. Mol. Evol.* **19**, 153–170.

Nei, M. and Yokoyama, S. (1976). Effects of random fluctuations of selection intensity on genetic variability in a finite population. *Jap. J. Genet.* **51**, 355–369.

Ohta, T. (1976). Role of very slightly deleterious mutations in molecular evolution and polymorphism. *Theor. Pop. Biol.* **10**, 254–275.

Ohta, T. and Kimura, M. (1973). A model of mutation appropriate to estimate the number of electrophoretically detectable alleles in a finite population. *Genet. Res.* **22**, 201–204.

Pagnier, J., Mears, J. G., Dunda-Belkhodja, O., Schaefer-Rego, K. E., Beldjord, C., Nagel, R. L. and Labie, D. (1984). Evidence for the muticentre origin of the sickle cell hemoglobin gene in Africa. *Proc. Natl. Acad. Sci. USA* **81**, 1771–1773.

Pearson, K. (1926). On the coefficient of racial likeness. *Biometrika* **18**, 337–343.

Peet, R. K. (1974). The measurement of species diversity. *Ann. Rev. Ecol. System.* **5**, 285–307.

Pielou, E. C. (1975). *Ecological Diversity*. Wiley, New York.

Rao, C. R. (1973). *Linear Statistical Inference and Its Application*. Wiley, New York.

Rao, C. R. (1982a). Gini–Simpson index of diversity: A characterization, generalization and applications. *Utilitas Mathematica* **21**, 273–282.

Rao, C. R. (1982b). Diversity and dissimilarity coefficients: A unified approach. *Theor. Pop. Biol.* **21**, 24–43.

Rao, C. R. (1982c). Diversity: Its measurement, decomposition, apportionment and analysis. *Sankhya A* **44**, 1–21.

Rao, C. R. (1984). Use of diversity and distance measures in the analysis of qualitative data. In: G. N. Van Vark and W. H. Howell, eds., *Multivariate Statistical Methods in Physical Anthropology*. Reidel, Dordrecht, 49–67.

Rao, C. R. and Boudreau, R. (1984). Diversity and cluster analyses of blood group data on some human populations. In: A. Chakravarti, ed., *Human Population Genetics: The Pittsburgh Symposium*. Van Nostrand Reinhold, New York, 331–362.

Reynolds, J., Weir, B. S. and Cockerham, C. C. (1983). Estimation of coancestry coefficient: Basis for short-term genetic distance. *Genetics* **105**, 767–779.

Richmond, R. C. (1972). Genetic similarities and evolutionary relationships among the semispecies of Drosophila paulistorum. *Evolution* **26**, 536–544.

Rogers, J. S. (1972). Measures of genetic similarity and genetic distance. In: *Studies in Genetics VII*. Univ. of Texas, Publication 7213, Austin, TX, 145–153.

Sanghvi, L. D. (1953). Comparison of genetical and morphological methods for a study of biological differences. *Amer. J. Phys. Anthrop.* **11**, 385–404.

Scriver, C. R., Kaufman, S. and Woo, S.L.C. (1988). Mendelian hyperphenylalaninemia. *Ann. Rev. Genet.* **22**, 301–321.

Simpson, E. H. (1949). Measurement of diversity. *Nature* **163**, 688.

Smith, C. A. B. (1977). A note on genetic distance. *Ann. Hum. Genet.* **40**, 463–479.

Sneath, P. H. A. and Sokal, R. R. (1973). *Numerical Taxonomy*. Freeman, San Francisco.

Sokal, R. R. and Friedlaender, J. S. (1982). Spatial autocorrelation analysis of biochemical variation on Bougainville island. In: M. H. Crawford and J. H. Mielke, eds., *Current Developments in Anthropological Genetics, Vol. 2: Ecology and Population Structure*. Plenum, New York, 205–227.

Sokal, R. R. and Menozzi, P. (1982). Spatial autocorrelations of HLA frequencies in Europe support demic diffusion of early farmers. *Amer. Nat.* **119**, 1–17.

Sokal, R. R. and Wartenberg, D. E. (1983). A test of spatial autocorrelation analysis using an isolation-by-distance model. *Genetics* **105**, 219–237.

Stewart, F. M. (1976). Variability in the amount of heterozygosity maintained by neutral mutations. *Theor. Pop. Biol.* **9**, 188–201.

Takahata, N. (1981). Genetic variability and rate of gene substitution in a finite population under mutation and fluctuating selection. *Genetics* **98**, 427–440.

Takahata, N. and Kimura, M. (1979). Genetic variability maintained in a finite population under mutation and autocorrelated random fluctuation of selection intensity. *Proc. Natl. Acad. Sci. USA* **76**, 5813–5817.

Watterson, G. A. (1974). Models for the logarithmic species abundance distributions. *Theor. Pop. Biol.* **6**, 217–250.

Watterson, G. A. (1977). Heterosis or neutrality? *Genetics* **85**, 789–814.

Watterson, G. A. (1984). Allele frequencies after a bottleneck. *Theor. Pop. Biol.* **26**, 387–407.

Whittaker, R. H. (1972). Evolutionand measurement of species diversity. *Taxon* **21**, 213–251.

Wright, S. (1931). Evolution in Mendelian populations. *Genetics* **16**, 97–159.

Wright, S. (1948). Genetics of populations. *Encyclopedia Britanica* **10**, 111, 111A–D, 112.

Yokoyama, S. and Nei, M. (1979). Population dynamics of sex-determining alleles in honey bees and self-incopatibility alleles in plants. *Genetics* **91**, 609–626.

Zouros, E. (1979). Mutation rates, population sizes and amounts of electrophoretic variation of enzyme loci in natural populations. *Genetics* **92**, 623–646.

C. R. Rao and R. Chakraborty, eds., *Handbook of Statistics, Vol. 8*
© Elsevier Science Publishers B.V. (1991) 317–346

10

Statistical Methods for Phylogenetic Tree Reconstruction

Naruya Saitou

1. Introduction

Reconstruction of the phylogeny of organisms is one of the most important problems in evolutionary study. A phylogeny is usually illustrated by a tree-like figure. Thus we call it 'phylogenetic tree' or simply 'tree' in this chapter. It may be interesting to note that Darwin (1859) was the first to show such a tree to explain the pattern of divergence of species through evolution, though his tree was an imaginary one.

Previously phylogenetic trees were reconstructed mostly by using morphological data. With the advent of the study of molecular evolution, however, it is now customary to construct phylogenetic trees from molecular data, especially from nucleotide sequences. In this chapter we will therefore be concerned primarily with nucleotide sequence data. Nevertheless, most of the methods discussed in this chapter can also be applied to other types of data, including non-molecular data. I will first discuss theoretical aspects of phylogenetic trees in the next section. Distance matrix methods and character-state methods are explained in Sections 3 and 4, respectively. Lastly, the efficiency of different methods is discussed.

2. Theoretical aspects of phylogenetic trees

2.1. Rooted trees and unrooted trees

Mathematically, a phylogenetic tree is literally a 'tree' in graph theory. (A graph is composed of node(s) and edge(s). A node represents any object and an edge represents the relationship between nodes.) A tree is a special kind of graph: there should be only one path between any two nodes. Thus there is no loop in a tree (e.g., see Figure 1). In evolutionary study, the term 'branch' is used instead of 'edge' and not only the branching pattern (topological relationship between nodes) but the length of each branch is often important.

A tree can be either directed or undirected. In a directed tree there is a

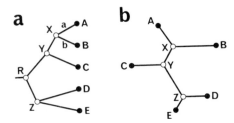

Fig. 1. Examples of a rooted tree (a) and an unrooted tree (b) for five OTUs.

particular node, or root, and there will be a unique path from this node to any other node. Hence a directed tree is also called a rooted tree. Figure 1(a) shows an example of a rooted tree, in which the root is designated as R. In organismal evolution, the direction is of course that of time and the root is the common ancestor. Therefore a phylogenetic tree in an ordinary sense is a rooted tree.

An undirected tree does not have such root, and it is also called an unrooted tree. Although an unrooted tree itself may not be regarded as a phylogenetic tree, it can be converted to a rooted tree if the position of the root is specified. Figure 1(b) is an example of an unrooted tree, and the topological relationship of nodes is identical with that of Figure 1(a) if we ignore the root (R) of Figure 1 and the difference of branch lengths between these two trees. Unrooted trees are sometimes called 'networks', but that has a different meaning in graph theory.

Nodes can be any kind of object, and in evolutionary study, it can be a species, populations, or genes, as will be discussed in Section 2.4. It is useful to distinguish exterior nodes (full circles in Figure 1) and interior nodes (empty circles in Figure 1). Exterior nodes have only one branch but interior nodes have more than one branch. We usually have informations on the exterior nodes only. Exterior nodes are often referred to as operational taxonomic units (OTUs) and interior nodes may be called hypothetical taxonomic units (HTUs).

2.2. *Possible number of trees*

When we consider the phylogenetic relationship of three OTUs, there are three possibilities (Figure 2(a)). The true phylogenetic tree is one of these rooted trees. The number of possible trees rapidly increases with increasing the number of OTUs compared. The general equation for the possible number of bifurcating rooted trees for n ($\geqslant 2$) OTUs is given by

$$(2n - 3)!/(2^{n-2}(n - 2)!) .\tag{2.1}$$

Equation (2.1) was first presented by Cavalli-Sforza and Edwards (1967). The number of bifurcating unrooted trees for n OTUs is given by replacing n by $n - 1$ in equation (2.1). Figure 2(b) shows the three possible unrooted trees for four OTUs. Table 1 gives the possible number of rooted and unrooted bifurcating trees up to 10 OTUs. For the number of multifurcating trees, see Felsenstein (1978a).

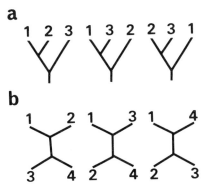

Fig. 2. (a) Three possible rooted trees for three OTUs. (b) Three possible unrooted trees for four OTUs.

Table 1
Possible numbers of rooted and unrooted trees for *i* OTUs

Number of OTUs	Possible number of	
	Rooted trees	Unrooted trees
2	1	1
3	3	1
4	15	3
5	105	15
6	945	105
7	10 395	945
8	135 135	10 395
9	2 027 025	135 135
10	34 459 425	2 027 025

It is clear from Table 1 that the search of the true phylogenetic tree for more than 10 OTUs is as if looking for a needle in a haystack, if we examine trees one by one. Unfortunately, the problem of finding the true tree from all the possible trees belongs to a so-called NP-complete problem, and there is no effective algorithm for this: we have to do an exhaustive search of all possible trees. This is why so many heuristic methods have been proposed for reconstruction of phylogenetic trees.

2.3. Topological differences between trees

The branching pattern of a tree with a given number of OTUs is called a 'topology' in evolutionary study (here too the word has a different meaning in graph theory). Each tree has its own topology, distinguished from those of other trees. However, the amount of topological difference can vary from tree to tree.

Robinson and Foulds (1981) devised a metric to represent the topological difference between trees for the same number of OTUs. Two trees (a and b) are presented in Figure 3. The topological difference between these two trees is defined by the number of operations necessary to transform each other. In this case, we need six operations (three contractions and three decontractions; see Robinson and Foulds, 1981). Tateno et al. (1982) called this metric a 'distortion index'.

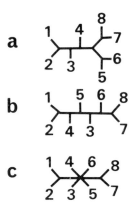

Fig. 3. Two topologically different trees (a and b) for 8 OTUs and their consensus tree (c).

A multifurcating tree c of Figure 3 is sometimes called the 'consensus tree' for trees a and b of Figure 3 (Adams, 1972). When trees a and b are equally likely, one way is to present the consensus tree c. Tree c can be obtained after three contractions of trees a and b.

2.4. Gene trees and species trees

Traditionally a phylogenetic tree automatically means a tree of species. However, genes are usually the units of comparison in molecular evolution, and there are several important differences between the phylogenetic tree of species and that of genes. The former is called 'species tree' and the latter 'gene tree' (Tateno et al., 1982; Nei, 1987). The most prominent difference between these two trees is illustrated in Figure 4. Because a gene duplication occurred before the speciation of species A and B, both species have two homologous genes (1 and 2) in their genomes. In this situation, we should distinguish 'orthology', that is homology of genes reflecting the phylogenetic relationship of species, from 'paralogy', that is homology of genes caused by gene duplication(s) (Fitch, 1970). Thus, genes *A1* and *B1* (or *A2* and *B2*) are 'orthologous', but genes *A1* and *A2*, *B1* and *B2*, *A1* and *B2*, or *A2* and *B2* are 'paralogous'. If one is not aware of the gene duplication event, gene tree for *A1* and *B2* may be misrepresented as the species tree of A and B, and thus a gross overestimation of the divergence time may occur.

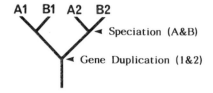

Fig. 4. A gene tree for four genes from two species.

Even when orthologous genes are used, a gene tree may be different from the corresponding species tree. This difference comes from the existence of allelic polymorphism of the ancestral species. A simple example is illustrated in Figure 5. A gene sampled from species A has its direct ancestor in the ancestral species X, and so does a gene sampled from species B. Thus the divergence between two genes sampled from different species always overestimates that of species (see Figure 5(a)). The amount of overestimation is related to the population size of the ancestral species X (see Tajima, 1983, for details). It may be interesting to note that Nei (1972) considered this overestimation and estimated the amount by the average heterozygosity in the present population. Thus Nei's genetic distance is an estimation of the amount of divergence of species or populations, not genes.

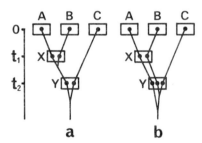

Fig. 5. Two gene trees (a and b) in which the topological relationship of genes is the same as or different from that of the specie trees, respectively.

The difference between species tree and gene tree is not confined to the amount of divergence. If the time $T (= t_2 - t_1$; see Figure 5) is not long, the two ancestral genes sampled from species A and B may coexist in the ancestral species Y. At this stage we also have a gene that is ancestral to a gene sampled from species C. Thus the topological relationship of these three genes is determined by chance alone, and it is possible to have a gene tree in which the topological relationship is different from that of species tree, as shown in Figure 5(b). For a more detailed discussion, readers may refer to Nei (1987).

When we consider phylogenetic relationships of different populations rather than species, the phylogenetic tree of populations may be called 'population tree'. If the effect of migration between populations is ignored, a population tree is essentially the same as a species tree. Thus the above explanation on species trees is applicable to population trees. On the difference between a population tree and a gene tree, readers may refer to Takahata and Nei (1985) and Pamilo and Nei (1988).

If the recombination or gene conversion events are also considered, the problem of constructing gene trees becomes much more complicated. Readers who are interested in this topic may refer to Hudson and Kaplan (1985).

2.5. Expected trees and observed trees

Branch lengths of a phylogenetic tree is ideally proportional to physical time. Thus the branch a and b of Figure 1(a) should have the same length. We call this type of tree 'expected tree'. In real data, however, we may have different lengths for branches a and b. This is because the amount of genetic change is not always the same between these two branches. A tree reconstructed from observed data is called 'observed tree'. Both species tree and gene tree can be either an expected or an observed tree. The distinction between expected and observed trees is similar to that between expected and realized distance trees defined by Nei (1987).

If the rate of evolution is constant among different lineages, or if the molecular clock is assumed, one may think that an observed tree is the same as the expected tree. However, this may not be the case. Even if the rate is constant, an observed tree usually show some variations on the number of nucleotide substitutions for each branch because of stochastic and sampling errors of nucleotides compared. It is thus important to distinguish observed trees from expected trees.

2.6. Classification of methods for tree construction

Many methods have been proposed for finding the phylogenetic tree from observed data. To clarify the nature of each method, it is useful to classify these methods from various aspects. Tree-making methods can be divided into two types in terms of the type of data they use; distance matrix methods and character-state methods. A distance matrix consists of a set of $\frac{1}{2}n(n-1)$ distance values for n OTUs (see Table 2 as an example), whereas an array of character states is used for the character-state methods. Sections 3 and 4 are thus classified.

Another classification is by the strategy of a method to find the best tree. One way is to examine all or a large number of possible trees and choose the best one in terms of a certain criterion. We call this the 'exhaustive search method'. For example, the maximum parsimony method belongs to this category. The other strategy is to examine a local topological relationship of OTUs and find the best tree. This type of method is called the 'stepwise clustering method' (Saitou and Imanishi, 1989). Most of the distance methods are stepwise clustering methods.

Table 2
An example of distance matrix[a]

	C	P	G	H	O
Common chimpanzee	–	0.0117	0.0415	0.0373	0.0895
Pygmy chimpanzee	0.0118 $\pm$ 0.0036	–	0.0405	0.0319	0.0863
Gorilla	0.0427 $\pm$ 0.0069	0.0416 $\pm$ 0.0068	–	0.0362	0.0905
Human	0.0382 $\pm$ 0.0065	0.0327 $\pm$ 0.0060	0.0371 $\pm$ 0.0064	–	0.0873
Orangutan	0.0953 $\pm$ 0.0106	0.0916 $\pm$ 0.0104	0.0965 $\pm$ 0.0107	0.0928 $\pm$ 0.0104	–

[a] Data from Hixson and Brown (1986). Figures above the diagonal indicate the proportion of the nucleotide difference, and those below the diagonal are the estimated number of nucleotide substitution per site (with their SEs).

3. Distance methods

3.1. Distance matrices

In distance methods, a phylogenetic tree is constructed by considering the relationship among the distance values of a distance matrix. There are two kinds of distances; metric and non-metric. The former follows the principle of triangle inequality and the latter does not. The triangle inequality is:

$$D_{ij} \leqslant D_{ik} + D_{jk}, \qquad (3.1)$$

where D_{ij} is the distance between OTUs i and j. In numerical taxonomy, especially in cladistics, use of metric distance is advocated (see Section 4.2 for cladistics). In molecular evolution, however, the estimated number of nucleotide substitutions as an evolutionary distance does not necessarily follow the principle of triangle inequality.

An example of a distance matrix is presented in Table 2. The data are Hixson and Brown's (1986) mitochondrial DNA sequences for human (H), chimpanzee (C), pygmy chimpanzee (P), gorilla (G), and orangutan (O). Gaps in the aligned sequences were excluded, and a total of 939 nucleotides were used for the analysis. Values above the diagonal are the proportions of nucleotide differences, and those below the diagonal are evolutionary distances (numbers of nucleotide substitutions per site).

There are many methods for estimating evolutionary distances (see Nei, 1987, for a review), and we used a simple method of Jukes and Cantor (1969) for this case. Evolutionary distance (d) between two nucleotide sequences is estimated by

$$d = -\tfrac{3}{4}\log[1 - \tfrac{4}{3}p],$$ (3.2)

where p is the proportion of different nucleotides between the two sequences. The standard error of d is estimated by

$$SE(d) = 3/(3 - 4p)[p(1 - p)/L]^{1/2},$$ (3.3)

where L is the number of nucleotides compared (Kimura and Ohta, 1972).

3.2. Phenetic methods

A simple way of classification of organisms is to combine phenotypically similar objects first. This approach is called 'phenetics' in numerical taxonomy. A phylogenetic tree constructed by a phenetic approach is called 'phenogram'. There are many ways to obtain a phenogram from a distance matrix (see Sneath and Sokal, 1973 for a review), and all are step wise clustering methods. In this section, two methods (UPGMA and WPGMA) that are frequently used in molecular evolution will be discussed. Original ideas of UPGMA (Unweighted Pair Group Method by average) and WPGMA (Weighted Pair Group Method by Arithmetic average) were first proposed by Sokal and Michener (1958). UPGMA was independently proposed by Nei (1975) for molecular data.

Let us explain the algorithms of UPGMA and WPGMA using the evolutionary distance matrix of Table 2. We first choose the smallest distance, that is, D_{PC} ($= 0.0118$). Then OTUs P and C are combined and the distances between the combined OTU (PC) and the remaining OTUs are computed as:

$$D_{(PC)i} = \tfrac{1}{2}(D_{Pi} + D_{Ci}),$$ (3.4)

where i represents an OTU other than P or C. Hence there are now only six distance values (see Table 3(a)). At the next step, again the smallest distance $(D_{(PC)H} = 0.0355)$ is chosen from the distance matrix of Table 3(a). Then the OTU

Table 3
An example of the procedure used in UPGMA and WPGMA

(a) First step				(b) Second step[a]			
	PC	G	H		PCH	G	O
G	0.0422			PCH	–	0.0397	0.0932
H	0.0355	0.0371		G	0.0405	–	0.0965
O	0.0935	0.0965	0.0928	O	0.0932	0.0965	–

[a] Figures above and below the diagonal are obtained by WPGMA and UPGMA, respectively.

(PC) and OTU H are further combined into OTU (PCH). When we compute the average distance between OTU (PCH) and other two OTUs (G and O), the difference between UPGMA and WPGMA arises. In the case of UPGMA, the original distances are always used for obtaining the averaged distances, whereas the current distance matrix is used for WPGMA. Thus, for example,

$$D_{(PCH)G} = \tfrac{1}{3}(D_{PG} + D_{CG} + D_{HG}) = 0.0405$$

if we apply UPGMA, and

$$D_{(PCH)G} = \tfrac{1}{2}(D_{(PC)G} + D_{HG}) = 0.0397$$

if we apply WPGMA. If the data strictly follows the constancy of evolutionary rate, these two values are identical. In reality, however, the rate may not be the same. Thus there can be a slight difference between distances obtained by UPGMA and that by WPGMA. Table 3(b) shows the distance matrix after the second step.

After two more steps, all five OTUs are clustered into a single OTU. The final tree thus obtained by UPGMA is shown in Figure 6(a). The tree obtained by WPGMA has the same topological relationship with that tree in the present example. Note that an estimated distance between any pair of OTUs can be obtained by summing all branch lengths connecting these two OTUs.

Boxes of Figure 6(a) represent the ranges of one standard error (SE) of the distance of each branching points from the present time, computed by Nei et al.'s (1985) method. Computation of SEs was done as follows. The SE of distance X

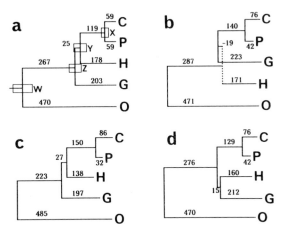

Fig. 6. Phylogenetic trees reconstructed by UPGMA (a), the Fitch–Margoliash method (b), the distance Wagner method (c), and the neighbor-joining method (d). The data are from Table 2. All figures should be multiplied by 10^{-4}. Boxes in tree a denote one SE of the distances between each branching points and the extant species.

$(=D_{XC} = D_{XP})$ of Figure 6(a) is given by $SE(X) = \frac{1}{2}SE(D_{PC})$, where $SE(D_{PC})$ is SE of D_{PC}. Thus $SE(X) = \frac{1}{2} \times 0.0036 = 0.0018$ from Table 2.

Estimation of $SE(Y)$ is more complicated. We first note the relation $SE^2(Y) = V(Y) = \frac{1}{4}V(D_{(PC)H})$, where $V(\cdot)$ denotes a variance. $D_{(PC)H}$ has been estimated by $\frac{1}{2}(D_{PH} + D_{CH})$ applying equation (3.4). Thus

$$V(D_{(PC)H}) = V[\tfrac{1}{2}(D_{PH} + D_{CH})]$$
$$= \tfrac{1}{4}[V(D_{PH}) + V(D_{CH}) + 2\,\text{Cov}(D_{PH}, D_{CH})], \qquad (3.5)$$

where $\text{Cov}(i, j)$ is the covariance between distances i and j. Because the lineages of two chimpanzee species evolved independently after the divergence at point X, $\text{Cov}(D_{PH}, D_{CH}) = V(D_{HX})$, where D_{HX} is estimated by

$$D_{HX} = D_{(PC)H} - \tfrac{1}{2}D_{PC} = \tfrac{1}{2}(D_{PH} + D_{CH}) - \tfrac{1}{2}D_{PC}. \qquad (3.6)$$

Thus,

$$D_{HX} = \tfrac{1}{2}(0.0327 + 0.0382) - \tfrac{1}{2} \times 0.0118 = 0.02955,$$

from Table 2. This value is an estimate of evolutionary distance (d) between nodes X and H of Figure 6(a). However, the proportion (p) of nucleotide difference is used for estimating the variance (or square of SE) of d (see equation 3.3). Therefore we estimate p from d applying equation (3.2) as

$$\hat{p} = \tfrac{3}{4}[1 - e^{-4d/3}]. \qquad (3.7)$$

In the present example, $\hat{p}$ becomes 0.02898. Putting this value into equation (3.3),

$$\text{Cov}(D_{PH}, D_{CH}) = V(L_{HX}) = [SE(L_{HX})]^2 = 0.3242 \times 10^{-4}.$$

Putting this and two other variances (squares of SEs in Table 2) into equation (3.5),

$$V(D_{(PC)H}) = \tfrac{1}{4}(0.0065^2 + 0.0060^2 + 2 \times 0.3242 \times 10^{-4}) = 3.577 \times 10^{-5}.$$

Hence, $SE(Y) = \frac{1}{2}[V(D_{(PC)H})]^{1/2} = 0.0030$. $SE(Z)$ $(= 0.0029)$ and $SE(W)$ $(= 0.0049)$ were computed in a similar way, and these SEs are represented as boxes in Figure 6(a).

Nei et al. (1985) also considered evolutionary distances based on amino acid substitutions, restriction site changes, and Nei's (1972) genetic distance. Although the general principle is the same, equations corresponding to equation (3.7) differ in each distance measure. Recently, R. Chakraborty (personal communication) improved Nei et al.'s method for Nei's genetic distance.

The constancy of the evolutionary rate is implicitly assumed for the phenetic approach of numerical taxonomy. With the discovery of molecular clock, such a phenetic methods, especially UPGMA, has been advocated for reconstructing phylogenetic trees (Nei, 1975). In this connection, it should be noted that UPGMA gives least-squares estimates of branch lengths for the tree obtained (Chakraborty, 1977). That is, UPGMA minimizes the quantity

$$S = \sum (D_{ij} - 2\lambda t_{ij})^2 \, , \tag{3.8}$$

where λ is the evolutionary rate and t_{ij} is the time since divergence between OTUs i and j.

3.3. Fitch and Margoliash's method

Fitch and Margoliash (1967) proposed an exhaustive search method for reconstructing a phylogenetic tree. The first step of this method is to estimate branch lengths of a tree, of which tree topology is expected to be the same or quite similar to that obtained by UPGMA. The principle of the branch length estimation is as follows. Let us designate L_{ij} for the length of branch connecting nodes i and j. Then $D_{ij} = L_{iX} + L_{jX}$ in the tree of Figure 7. This is because the additivity of

Fig. 7. A relationship of three OTUs.

distances is assumed. From this relationship, L_{iX}'s ($i = 1, 2,$ and 3) are estimated by

$$L_{1X} = \tfrac{1}{2}(D_{12} + D_{13} - D_{23}) \, , \tag{3.9a}$$

$$L_{2X} = \tfrac{1}{2}(D_{12} + D_{23} - D_{13}) \, , \tag{3.9b}$$

$$L_{3X} = \tfrac{1}{2}(D_{13} + D_{23} - D_{12}) \, . \tag{3.9c}$$

When we compare n (> 3) OTUs, OTU 3 is a composite OTU, which consists of all the remaining OTUs. Then D_{13} and D_{23} are given by taking averages of D_{1j}'s and D_{2j}'s ($j = 3, 4, \ldots, n$), respectively.

If D_{12} is found to be the smallest, then OTUs 1 and 2 are combined as in the case of UPGMA and the averaged distances between this combined OTU (12) and other OTUs are computed using equation (3.4). L_{1X} and L_{2X} are also computed applying equations (3.9a) and (3.9b) at this step. The same procedure is repeated until all OTUs are clustered to become a single OTU.

At the next step the so-called percent standard deviation (PSD) is used as the criterion. For distance data of n OTUs,

$$\text{PSD} = \left[\frac{2\Sigma \left\{ (D_{ij} - E_{ij})/D_{ij} \right\}^2}{n(n-1)} \right]^{1/2} \times 100, \tag{3.10}$$

where the summation is for all possible pairs of OTUs and E_{ij} is estimated (patristic) distance between OTUs i and j. E_{ij} is obtained by summing estimated lengths of branches connecting OTUs i and j.

Tateno et al. (1982) proposed a criterion (S_0) similar to PSD:

$$S_0 = \left[\frac{2\Sigma (D_{ij} - E_{ij})^2}{(n-1)} \right]^{1/2}. \tag{3.11}$$

We can use either PSD or S_0 as the criterion to find the best tree, and the tree that has the smallest PSD or S_0 is chosen as the best tree through an exhaustive search of all possible trees.

Table 7 shows values of PSD and S_0 for four trees obtained from data of Table 2. Tree 1 has been obtained by UPGMA (see Figure 6(a)). Two chimpanzee species are clustered for trees 2 and 3 as in tree 1, but the branching pattern among human (H), chimpanzees (PC), and gorilla (G) is different each other. Human and pygmy chimpanzee are clustered in tree 4. The Fitch–Margoliash (FM) method chose tree 2 as the best tree among these four trees as did Tateno et al.'s (1982) method (see Table 7). The tree thus obtained is shown in Figure 6(b). Note that there is a negative branch (between H and (PCG) cluster) in this tree.

Because the FM method produces an unrooted tree, orangutan was assumed to be the outgroup species among five species compared in this example, and the root was located on the branch going to orangutan, assuming a constancy of evolutionary rate.

Prager and Wilson (1978) and Sourdis and Krimbas (1987) modified the criterion of the FM method for choosing the best tree. They discarded trees in which negative branch lengths were obtained. Tree 2 is discarded if we use this modified criterion, and instead tree 3 will be chosen. However, it is possible that even the true tree may have a branch with negative distance. A negative value for a branch may appear if there are many backward and parallel substitutions.

Other types of modification of the FM method have been proposed by de Soete (1983) and Elwood et al. (1985). Readers who are interested in their modifications may refer to original papers.

3.4. Distance Wagner method

Farris (1972) proposed a method that can be considered as an application of the principle of maximum parsimony to distance data. Because the technique of reconstructing unrooted trees from character-state data is called 'Wagner network' in cladistics, Farris named this method 'distance Wagner' (DW) method. In this case, however, a distance measure satisfying triangle inequality (a metric) is supposed to be used. Thus in the following, proportion of nucleotide difference of Table 2, that is a metric, is used.

We first connect two OTUs of which distance is the smallest. This is D_{PC} ($= 0.0117$) from Table 2. Then these two OTUs are combined and the distance between this combined OTU (PC) and the remaining OTUs are computed by equation (3.4). Second, the OTU that has the smallest distance from the OTU (PC) is chosen. The appropriate OTU is H. After this choice, L_{PX}, L_{CX}, and L_{HX} are computed by applying equations (3.9a)–(3.9c).

We now proceed to the next step, where one more OTU is added to the unrooted tree for three OTUs. There are three possibilities for each remaining OTU (either G or O) to be connected to the tree. For example, OTU G may be connected to either branch PX, CX, or HX (see Figure 8). Thus lengths of all

Fig. 8. Three possible additions of OTU G to P–C–H tree.

possible branches are computed and the branching pattern that gives the shortest length is chosen to be connected to the three-OTU tree. Branch lengths are computed in a similar way as equation (3.9):

$$L_{G_P Y_P} = \tfrac{1}{2}(D_{GP} + L_{G_P X} - L_{PX}), \tag{3.12a}$$

$$L_{G_C Y_C} = \tfrac{1}{2}(D_{GC} + L_{G_C X} - L_{CX}), \tag{3.12b}$$

$$L_{G_H Y_H} = \tfrac{1}{2}(D_{GH} + L_{G_H X} - L_{HX}), \tag{3.12c}$$

where subscripts of G and Y designate the positions of branch connecting OTU G (see Figure 8). In equation (3.12), L_{PX}, L_{CX}, and L_{HX} have already been computed at the previous step, whereas $L_{G_P X} = L_1$ or L_2, $L_{G_C X} = L_2$ or L_3, and $L_{G_H X} = L_1$ or L_3, where

$$L_1 = D_{GC} - L_{CX}, \quad L_2 = D_{GH} - L_{HX}, \quad L_3 = D_{GP} - L_{PX}. \tag{3.13}$$

Among L_1, L_2, and L_3, the largest value is used for all of the $L_{G_i X}$ ($i = $ P, C, or H) in equation (3.12). Tateno et al. (1982) considered the use of the distance Wagner method for evolutionary distance that are not metric. In this case, a gross overestimation of branch length can happen by this procedure. Thus they used the average of L_1, L_2, and L_3, instead of the largest value. Faith (1985) took a different modification for estimating $L_{G,X}$ of equation (3.12). In this case, equation (3.9) is repeatedly used and the weighted average gives the estimates for these branch lengths.

In the present example, L_3 (= 0.0374) is the largest and putting this and the other values into equations (3.12a)–(3.12c), $L_{G_H Y_H}$ (= 0.0224) turns out to be the smallest. Thus OTU G is connected to the branch HX. The same procedure is repeated until all OTUs are connected. The final tree is presented in Figure 6(c).

As in the case of the FM method, the DW method also produces unrooted trees. Thus we can locate the root at the branch going to orangutan (O). When we have no information on the outgroup species, the location of the root can be estimated as the mid-point of the largest estimated distance, if a rough constancy of evolutionary rate is assumed (Farris, 1972). Estimated (patristic) distance between OTUs C and O (= 0.0970), that is considerably larger than the observed distance (D_{CO} = 0.0895), is the largest in the present example, and the root was placed at the point of which distance is 0.0485 ($= \frac{1}{2} \times 0.0970$) from node O (see Figure 6(c)). Under the assumption of constant evolutionary rate, the root can also be obtained by minimizing the variance of evolutionary rate (Farris, 1972).

Because the proportion of nucleotide difference was used for the DW method, the length (0.0223 + 0.0485 = 0.0708) of the branch going to OTU O of Figure 6(c) is slightly shorter than those of other trees in Figure 6. This is probably because Jukes and Cantor's (1969) evolutionary distances, in which multiple hits were corrected, were used in the latter trees. However, some of the lengths of the other branches of Figure 6(c) are larger than those of the other trees.

3.5. Neighbor-joining method

A pair of OTUs are called 'neighbors' when these are connected through a single interior node in an unrooted, bifurcating tree. For example, OTUs A and B in Figure 1(b) are a pair of neighbors. If we combine these OTUs, this combined OTU (AB) and OTU C become a new pair of neighbors. It is thus possible to define the topology of a tree by successively joining pairs of neighbors and producing new pairs of neighbors. For example, the topology of tree a in Figure 3 can be described by the following pairs of neighbors: [1, 2], [5, 6], [7, 8], [1–2, 3], and [1–2–3, 4]. Note that there is another pair of neighbors, [5–6, 7–8], that is complementary to [1–2–3, 4] in defining the topology. In general, $n - 2$ pairs of neighbors can be produced from a bifurcating tree of n OTUs.

The neighbor-joining (NJ) method of Saitou and Nei (1987) produces a unique final tree by sequentially finding pairs of neighbors. The algorithm of the NJ method starts with a starlike tree, as given in Figure 9(a), which is produced under the assumption that there is no clustering of OTUs. In practice, some pairs of OTUs are more closely related to each other than other pairs are. Consider a tree that is of the form given in Figure 9(b). In this tree the neighboring OTUs [1, 2] are separated from the other OTUs (3, 4, ..., 8) by branch XY. Any pair of OTUs can take the positions of 1 and 2 in the tree, and there are $\frac{1}{2}n(n - 1)$

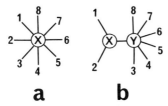

Fig. 9. A star-like tree (a) and a tree (b) that is one step aside from the star-like tree.

ways of choosing them for n OTUs. Among these possible pairs of OTUs, we choose the one that gives the smallest sum of branch lengths. Thus the principle of minimum evolution is used in the NJ method. This pair of OTUs is then regarded as a single OTU, and the next pair of OTUs that gives the smallest sum of branch lengths is again chosen. This procedure is continued until all $n - 2$ neighbors are found.

The sum of the branch lengths is computed as follows. First, the branch length between nodes X and Y in the tree of Figure 9(b) is estimated by

$$L_{XY} = \frac{1}{2(n-2)} \left[\sum_{k=3}^{n} (D_{1k} + D_{2k}) - (n-2)(L_{1X} + L_{2X}) - 2 \sum_{i=3}^{n} L_{iY} \right].$$
(3.14)

Noting the relationships $L_{1X} + L_{2X} = D_{12}$ and $\sum_{i=3}^{n} L_{iY} = [\sum_{3 \leqslant i < j}^{n} D_{ij}]/(n-3)$, we find that the sum (S_{12}) of all branch lengths of the tree in Figure 9(b) becomes

$$S_{12} = L_{1X} + L_{2X} + L_{XY} + \sum L_{iY}$$

$$= \frac{1}{2(n-2)} \left[\sum_{k=3}^{n} (D_{1k} + D_{2k}) \right] + \tfrac{1}{2}D_{12} + \frac{1}{n-2} \sum_{3 \leqslant i < j} D_{ij}.$$
(3.15)

It can be shown that equation (3.15) is the sum of least squares estimates of branch lengths for tree 9b (see Appendix A of Saitou and Nei, 1987). In general, we compute all S_{ij} $(1 \leqslant ij \leqslant n)$ and choose the pair of OTUs i and j that shows the smallest S_{ij} value.

Definition of S_{ij} seems complicated, but it can be computed in a simplified form

$$D_{ij} = -(R_i + R_j)/2(n-2) + \tfrac{1}{2}D_{ij} + Q/(n-2),$$
(3.16)

where $R_i = \sum_{k=1}^{n} D_{ik}$, $R_j = \sum_{k=1}^{n} D_{jk}$, and $Q = \sum_{k<l}^{n} D_{kl}$. Because R_i $(1 \leqslant i \leqslant n)$ and Q can be computed before computation of S_{ij}'s, computation of S_{ij} is actually quite simple (see also Studier and Keppler, 1988). Note that $D_{ij} = D_{ji}$ and $D_{ii} = 0$ are assumed in the computation of R_i's.

Let us apply the NJ method to the evolutionary distance matrix of Table 2. $Q = 0.5803$, and R_i's are presented at the first column of Table 4. From these, S_{ij}'s were computed as shown in Table 4, and we find that S_{PC} $(= 0.1384)$ is the

Table 4
R_i values and S_{ij} matrix for Table 2

	R_i	C	P	G	H
Common chimpanzee	0.1880				
Pygmy chimpanzee	0.1777	0.1384			
Gorilla	0.2179	0.1471	0.1483		
Human	0.2008	0.1477	0.1467	0.1422	
Orangutan	0.3762	0.1470	0.1469	0.1427	0.1437

smallest. Thus OTUs P and C are combined and the distance between the combined OTU (PC) and a remaining OTU i is computed by equation (3.4). The same procedure is repeated for the new distance matrix, and finally tree d of Figure 6 is obtained.

Algorithm of the NJ method is quite similar to that of UPGMA. Instead of choosing the smallest distance, we choose the smallest S_{ij} value at each step, and the distance averaging follows. Therefore the computation is very rapid.

When a distance matrix is strictly additive (any distance is sum of appropriate branch lengths), the NJ method was proved to reconstruct the true tree (Saitou and Nei, 1987; Studier and Keppler, 1988).

3.6. Transformed distance methods

When evolutionary rate varies from lineage to lineage in a phylogenetic tree as in a tree in Figure 10, the following distance transformation may give an improved topology for the average distance method (Farris, 1977)

$$D'_{ij} = \tfrac{1}{2}(D_{ij} - D_{i\mathrm{R}} - D_{j\mathrm{R}}) . \tag{3.17}$$

where R refers to the reference OTU. This property has been independently rediscovered by Klotz et al. (1979) and by Li (1981) (see also Klotz and Blanken, 1981).

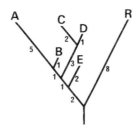

Fig. 10. A phylogenetic tree in which evolutionary rate varies considerably among different lineages. Figures are branch lengths.

The underlying logic of the transformation is as follows. If we change the sign of D'_{ij} of the above equation, $-D'_{ij}$ (a positive value) corresponds to the branch length between the reference OTU R and the interior node connecting OTUs i and j (see equation (3.9)). Thus if we apply UPGMA to the distance matrix, the correct topology should be obtained. The reference OTU can be a composite one that consists of more than one OTU.

Assuming the exact additivity, distances were computed from Figure 10 and they are shown in Table 5(a). If we apply UPGMA to this distance matrix, OTUs C and D are first clustered, but OTUs B and E will be erroneously clustered at the next step. Table 5(b) shows a transformed distance matrix, in which OTU R was treated as a reference. If we apply UPGMA to this matrix, OTUs C and D are first clustered, followed by OTUs A and B and OTUs (CD) and (AB). Thus it is clear that the transformation gives the correct tree topology.

Table 5
Original and transformed distance matrices based on the tree of Figure 8

	(a) Original distances (D)					(b) Transformed distances (D')			
	A	B	C	D	E	A	B	C	D
B	6					− 12			
C	11	7				− 11	− 11		
D	10	6	3			− 11	− 11	− 14	
E	9	5	8	7		− 10	− 10	− 10	− 10
R	17	13	16	15	12				

There is one problem in this method; we usually do not know which OTU is a reference or outgroup. Li (1981) used UPGMA for estimating the root of a tree, and two groups of OTUs separated by this root are alternatively used for transformation of distances of the other group. If the position of the root determined by UPGMA is correct, Li's method is expected to perform efficiently. However, it is possible that UPGMA misdetermines the root.

Let us apply Li's (1981) method to the data of Table 2. UPGMA tree (see Figure 6(a)) is first constructed, and five OTUs are divided into two groups according to the position of the root. Because one group consists of only one OTU (O), transformed distance matrix of Table 6 is computed for OTUs (P, C, H, and G) of the other group. It is clear that tree 3 (see Table 7) will be obtained if we apply UPGMA to this matrix. The tree finally obtained (not shown) is quite similar to tree 6d, which was obtained by the NJ method.

3.7. Other methods

Many other distance methods have been proposed for reconstruction of phylogenetic trees, and we briefly discuss some of them.

Table 6
Transformed distance matrix of Table 2

	C	P	G
Pygmy chimpanzee	− 0.0876		
Gorilla	− 0.0746	− 0.0733	
Human	− 0.0750	− 0.0759	− 0.0761

Let us consider an unrooted tree with four OTUs (see Figure 2(b)), and assume that every distance is the sum of relevant branch lengths or that the strict additivity holds. Then we have the relation $D_{12} + D_{34} < D_{13} + D_{24} = D_{14} + D_{23}$ for the leftmost tree of Figure 2(b). We can use a similar relationship for finding the tree,

$$D_{12} + D_{34} < D_{13} + D_{24} \quad \text{and} \quad D_{12} + D_{34} < D_{14} + D_{23}. \qquad (3.18)$$

This condition is called the four-point metric (Buneman, 1971). The additive condition (Dobson, 1974) or the relaxed additivity condition (Fitch, 1981) is closely related to the four-point metric. It should be noted that the DW, the NJ, and the transformed distance method are all reduced to this condition in the case of four OTUs (Saitou and Nei, 1986, 1987).

Sattath and Tversky (1977) used the four-point metric for inferring tree topology for more than four OTUs. Interestingly, their method has a similarity with the NJ method. Fitch (1981) proposed a method also applying the four-point metric in a somewhat different way. Readers may refer to the original papers.

Edwards and Cavalli-Sforza (1965) proposed a method called 'cluster analysis'. The division of OTUs that gives the largest between-cluster sum of squares (or the smallest within-cluster sum of squares) is sequentially chosen in this method. There are $2^{n-1} - 1$ ways for n OTUs to be divided into two clusters, and all possibilities are examined. The same procedure is applied to each cluster thus found, and finally a rooted tree is obtained after $n - 1$ steps.

Cavalli-Sforza and Edwards (1967) proposed two exhaustive search methods (the additive tree and the minimum evolution methods) and they applied these methods to gene frequency data of human populations. The additive tree method assumes that distances along the tree are additive, and the least square method is used to minimize the errors between observed distances and estimated distances that are obtained by summing estimated branch lengths. This procedure is applied for all possible trees, and the tree that has the smallest sum of squares is chosen. The minimum evolution method for n OTUs is equivalent to the Steiner problem in $n - 1$ dimensions (see Courant and Robbins, 1941, for a review of Steiner problem). Computation of the additive tree and the minimum evolution methods are cumbersome when the number of OTUs is large.

Saitou and Imanishi (1989) proposed a simple method applying the principle of minimum evolution. In this method, branch lengths of a given tree are estimated by the procedure of Fitch and Margoliash (1967), and the tree with the smallest sum of branch lengths (SBL) is chosen as the best tree. It has been shown that the property of this method is similar to that of the NJ method (Saitou and Imanishi, 1989). An example of the minimum evolution (ME) method is shown in Table 7. The method chose tree 3 as the best one, as in the case of the NJ and ML methods. The ME method seems to be closely related to Dayhoff's (1978) method (see Blanken et al., 1982).

Table 7
Results of five exhaustive search methods for data of Table 2 (from Saitou and Imanishi, 1989)[a]

Method	Tree 1: ((PC)H)G	Tree 2: ((PC)G)H	Tree 3: (PC)(HG)	Tree 4: ((PH)C)G
FM	+ 0.60	0	+ 0.47	+ 25.33
TA	+ 0.35	0	+ 0.16	+ 4.45
ME	+ 0.35	+ 1.10	0	+ 6.59
MP	0	+ 1	0	+ 8
ML	− 2.98	− 3.97	0	− 33.86

[a] Values for FM, TA, and ME are PSD, $S_0 \times 1000$, and SBL × 1000, respectively. Values for MP is the required number of nucleotide substitutions, and those for ML is the log-likelihood. Values of the best tree are set to be zero, and the other values represent differences from that of the best tree.

3.8. Statistical tests

There are several methods for testing the statistical significance of a tree obtained. In the case of a UPGMA tree, SEs of the distances of branching points can be computed by Nei et al.'s (1985) method, as has been shown in Section 3.2. Thus by a standard t-test, the difference of the distances of the branching points Y and Z are shown to be not statistically significant, whereas those between X and Y and between Z and W are statistically significant in Figure 6(a).

Hasegawa et al. (1985) applied a generalized least square method for estimating branch lengths for a given tree, under the assumption of a constant evolutionary rate with their own model of nucleotide substitution, and gave equations for computing variances of estimated branch lengths. Later they also applied the bootstrap method (Felsenstein, 1985; see also Section 4.2) for computing variances (Hasegawa et al., 1987). Readers may refer to the original papers.

When we do not have the assumption of the constant rate of evolution, unrooted trees should be considered. In this case the estimation of SEs for each branch length is not easy. However, an approximate SE for each branch can be obtained by applying equation (3.7). That is, estimated branch length (d) is used for estimating the proportion ($\hat{p}$) of nucleotide difference, and this $\hat{p}$ is used to estimate SE of d using equation (3.3). If we apply this simple procedure to tree d of Figure 6, the length (0.0015) of the branch connecting the H–G cluster and the

CP–O cluster is not significantly greater than zero (its SE being 0.0013). Similarly, the length (0.0129) of the branch connecting the C–P cluster and HG–O cluster is significantly greater than zero (its SE being 0.0037). Thus the clustering of chimpanzee (C) and pygmy chimpanzee (P) is supported, whereas that of human (H) and gorilla (G) is not. It should be noted, however, that this method is expected to give a smaller SE than the true value. Thus the test based on this estimation is not conservative.

Templeton (1985) proposed a method (delta Q-test) for a statistical test of different tree topologies. However, Saitou (1986) and Ruvolo and Smith (1986) showed that the delta Q-test is theoretically unjustified. Thus this method is not recommended.

4. Character-state methods

4.1. Character states

Any discrete characters can be used for character-state methods, such as morphological characters, amino acid and nucleotide sequences, and restriction site maps. In principle, each character is considered separately in character-state methods. However, a more essential unit of comparison for the character-state method is 'configuration'. A configuration is a distribution pattern of characters for a given set of OTUs. If there are two characters (ancestral and derived), there are 2^n configurations for n OTUs. For the case of nucleotide sequences, there are four characters (A, G, T, and C) and the number (c) of configuration becomes

$$c = \tfrac{1}{6}(4^{n-1} + 3 \times 2^{n-1} + 2) \tag{4.1}$$

(Saitou and Nei, 1986). For example, there are 5, 15, and 51 configurations for 3, 4, and 5 sequences. Any length of nucleotide sequences for a given set of sequences can be described as an array of configurations, and the distribution pattern of the number of each configuration is essential for the construction of a tree.

The maximum parsimony method and the maximum likelihood method will be discussed in the following.

4.2. Maximum parsimony method

The evolutionary process of morphological characters can be classified into two different aspects. Groups of organisms similar in general levels of organization is called 'grade' and groups of common genetic origin is called 'clade' (Simpson, 1961). Cladistics or cladism is named after clade. However, cladists usually rely on the maximum parsimony method alone for finding the phylogenetic tree. A phylogenetic tree constructed by a cladistic approach is called 'cladogram'.

Camin and Sokal (1965) proposed the principle of maximum parsimony for reconstructing rooted trees from morphological characters. The tree that requires the smallest change of characters is chosen under this principle. When one considers a rooted tree, it is necessary to define the direction of the tree. This is done by determining the state of a character either to a derived one (apomorphy) or to a primitive one (plesiomorphy). A clade is defined by a synapomorphy, or a sharing of a derived state. Whether the state of a character is apomorphous or plesiomorphous depends on the opinion of each researcher. Thus there is a certain level of subjectiveness on the determination of direction of a tree.

On the study of molecular evolution, the direction of a tree is not easy to determine. Thus unrooted trees are usually constructed. An unrooted tree can be converted into a rooted tree by the knowledge of an outgroup OTU or by assuming the constancy of evolutionary rate as discussed earlier. Eck and Dayhoff (1966) proposed the maximum parsimony (MP) method for amino acid sequence data, and Fitch (1971) presented an algorithm for computing the minimum number of nucleotide substitutions for a given tree. A method of estimating the minimum number of nucleotide substitutions from amino acid sequence data was also proposed by Fitch and Farris (1974). Later Fitch (1977) clarified the properties of the MP method for nucleotide sequence data. There are several variations for the maximum parsimony method, and reader may refer to a comprehensive review by Felsenstein (1982).

Let us consider a tree for five nucleotide sequences, and assume that nucleotides A, A, T, G, and G were observed in sequences 1–5 in this order at one nucleotide site (see Figure 11). If tree 11a is considered, interior nodes X and Z should have nucleotides A and G, respectively, from the maximum parsimony principle. However, node Y can have either A, G, or T, because two nucleotide substitutions (denoted by full circles in Figure 11) are required in all three cases.

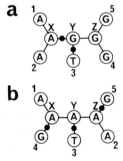

Fig. 11. Two trees (a and b) for five sequences. Nucleotides for each node are shown in circles, and dots denote nucleotide substitutions.

The case of nucleotide G at node Y is shown in tree 11a. If tree 11b is considered, we now need three nucleotide substitutions. Nodes X, Y, and Z can have nucleotides A or G simultaneously. This example shows that the nucleotides of ancestral or interior nodes may not be determined unambiguously, and the estimation of the length of each branch length is often difficult in the MP method.

In the above example, the minimum numbers of required nucleotide substitutions are different in trees 11a and 11b. This kind of nucleotide configuration is informative in choosing the best tree. A nucleotide configuration is 'informative' when there are at least two different kinds of nucleotides, each represented at least two times. Only these informative configurations are used in the MP method and non-informative ones are discarded. Non-informative configurations include invariant sites and singular sites in which only one nucleotides are represented more than one times (Fitch, 1977).

Certain nucleotide configurations can be fitted to a given tree with the minimum number of substitutions (the number of variable nucleotides minus one), whereas the other configurations require more than the minimum. The nucleotide sites with the first group of configurations are called compatible sites, and those with the latter group are called incompatible sites. The nucleotide site considered in the

Table 8
Maximum parsimony analysis of Hixson and Brown's data

	Configuration[a]					Number of observations	Number of substitutions for			
	C	P	G	H	O		Tree 1	Tree 2	Tree 3	Tree 4
1	y	y	x	x	x	8	8	8	8	16
2	x	x	y	x	y	5	5	10	10	5
3	x	x	y	y	x	5	10	10	5	10
4	x	x	x	y	y	4	8	4	8	8
5	x	y	x	y	x	1	2	2	2	1
6	x	y	x	x	y	1	2	2	2	2
7	y	x	y	x	x	1	2	2	2	2
8	y	x	x	x	y	1	2	2	2	2
9	x	x	y	y	z	1	2	2	2	3
	Total					27	41	42	41	49

[a] x, y and z are different nucleotides.

above example is compatible to tree 11a and incompatible to tree 11b. This difference is considered in the compatibility method (LeQuesne, 1969), and the tree that has the largest number of sites that are mutually compatible is chosen. Thus this method has a similarity with the MP method. When the number of OTUs is 4 or 5, these two methods are identical.

Table 8 shows an example of the maximum parsimony analysis using Hixson and Brown's (1986) mitochondrial DNA sequence data. Only informative configurations are listed in the table, and the number of nucleotide sites involved is 27 out of 939 sites. Four trees of Table 7 were examined and trees 1 and 3 are equally parsimonious, though only one additional substitution is required for tree 2. Two chimpanzee species are not clustered in tree 4, and this tree requires much larger number of nucleotide substitutions.

Because the maximum parsimony principle is simple and it is philosophically related to Occam's razor, it has become very popular not only in classical taxonomy but in molecular evolution. When the overall amount of divergence is small, the MP method may be appropriate. However, a gross underestimation of branch lengths occurs when the amount of divergence is large (e.g., Saitou, 1989). Furthermore, the MP method is not appropriate for finding the tree topology in some cases. Felsenstein (1978b) showed a condition in which the MP method and the compatibility method is positively misleading. Thus we should be cautious for the use of the MP method and the compatibility method.

In the standard maximum parsimony method, all changes are equally weighted, since the method was originally applied for morphological data in which the probability of change of each character is rarely known. In nucleotide sequence data, however, we have some knowledge on the probability of nucleotide changes. For example, transitional changes have been known to dominate the substitution process in mitochondrial DNA. In this case, it may be more appropriate to apply the MP method only for transversional changes (e.g., Saitou and Nei, 1986). Noting this kind of property in the real data, Tateno (1990) proposed a general method for giving different weights to each change of nucleotides before applying the principle of maximum parsimony. Readers interested in this method may refer to the original paper.

Recently Lake (1987) proposed a method called 'evolutionary parsimony'. While the standard MP method focuses at signals (compatible configurations for a given tree), Lake's method calls attention to noises (incompatible configurations), and transitional changes and transversional changes are distinguished. Unfortunately, the evolutionary parsimony is applicable only for four OTUs at this stage. Readers who are interested in this method may refer to Lake (1987).

Cavender (1978, 1981) proposed a statistical test for the MP method. However, he considered only four OTUs, and the results seems to be not appropriate for real evolutionary data. Felsenstein (1985) introduced the bootstrap method for tree comparison. This method involves resampling characters from one's own data, with replacement, to create a series of artificial samples of the same size as the original data. The MP method is applied to each of these, and the variation among the resulting trees are taken to indicate the size of the error in the original data. For more detail, readers may refer to the original paper.

Templeton (1983) applied the Wilcoxon signed-rank test to the MP method for restriction site data, and concluded that the clustering of chimpanzee and gorilla are statistically significant by analyzing Ferris et al.'s (1981) data. However, Nei and Tajima (1985, 1987) indicated drawbacks of the MP method through a theoretical study, and a more detailed study of Li (1986) showed that Ferris et al.'s (1981) data were not enough for obtaining statistically significant clustering of chimpanzee and gorilla (see also Smouse and Li, 1988). Thus Templeton's (1983) method is not recommended.

When we compare relatively large number of OTUs and character states, it is necessary to use computers. Platnick (1988) reviewed two computer programs for the MP method: PAUP (version 3.0) by D. L. Swofford and Hennig86 (version 1.5) by J. S. Farris. PHYLIP (version 3.1) by J. Felsenstein also contains several programs for the MP method and its variations.

4.3. Maximum likelihood method

The maximum likelihood (ML) method of tree-making was originally proposed by Cavalli-Sforza and Edwards (1967) for gene frequency data. Later, Kashap and Subas (1974) applied the ML method for three amino acid sequences, assuming the constancy of evolutionary rate. Langley and Fitch (1974) also used the ML method for estimating the branch lengths of a given tree, and compared these estimates with those obtained by the MP method.

Felsenstein (1981) developed the ML method for finding an unrooted tree from nucleotide sequence data. Let us explain the principle of his method. Consider tree *b* of Figure 1 as an example. We first restrict our attention to a specific nucleotide site, and assume that nucleotide N_i was observed at exterior node *i* (*i* = A, B, C, D, or E). On the other hand, nucleotide N_j at interior node *j* (*j* = X, Y, or Z) is unknown, and it can be one of four nucleotides. Then the likelihood (*L*) of this site becomes

$$L = \sum_{N_Y} \left\{ g_Y P_{YC} \left[\sum_{N_X} P_{YX} P_{XA} P_{XB} \right] \left[\sum_{N_Z} P_{YZ} P_{ZD} P_{ZE} \right] \right\}, \qquad (4.2)$$

where g_Y is the probability that node Y has nucleotide N_Y, $P_{ij} \equiv \Pr(N_i, N_j, L_{ij})$ is the probability of observing nucleotide N_i and N_j at nodes *i* and *j*, respectively, with the branch length L_{ij}. Summation is for four possible nucleotides, because N_X, N_Y, and N_Z are variables. To obtain $\Pr(N_i, N_j, L_{ij})$, we must specify the pattern of nucleotide substitution. If we use Jukes and Cantor's (1969) random substitution model,

$$\Pr(N_i, N_j, L_{ij}) = \begin{cases} \frac{1}{4} + \frac{3}{4} \exp(-4L_{ij}/3) & \text{if } N_i = N_j, \qquad (4.3a) \\ \frac{1}{4} - \frac{1}{4} \exp(-4L_{ij}/3) & \text{if } N_i \neq N_j. \qquad (4.3b) \end{cases}$$

It should be noted that the reversibility of time is assumed in the above formulation, a necessary assumption for unrooted trees. When different character-state data such as amino acid sequences or restriction sites are to be used, equation (4.3) should be modified by taking into account the nature of each character state. But the essential nature of the likelihood function of equation (4.2) remains the same.

Likelihood for each nucleotide site defined by equation (4.2) is then multiplied for all sites, and usually log-likelihood is computed for different set of branch lengths for a given tree topology, and the set that shows the highest log-likelihood is numerically searched. Fukami and Tateno (1988) proved that there exists a single ML point in the possible parameter range under the Jukes–Cantor model of nucleotide substitution.

Saitou (1990) showed that the ML estimate of the number of nucleotide substitutions between two nucleotide sequences is identical with that obtained by

Jukes and Cantor's (1969) and by Kimura's (1980) method. In the case of more than two sequences, however, this identity does not hold.

Original formulation of the ML method by Cavalli-Sforza and Edwards (1967) included the probability of tree topology, assuming a Yule process. Felsenstein (1981) took a different procedure, in which the ML value for each tree is compared and the tree with the highest ML value is chosen. Nei (1987) argued that the ML value computed in this way is conditional for each tree. Recently Hasegawa and Kishino (1988) tried to justify Felsenstein's (1981) procedure by applying an information theory. When we consider a gene tree or gene genealogy within a population, however, the probability of observing a specific tree topology should be considered (see Tajima, 1983). Noting this theoretical problem, Saitou (1988) proposed a step-wise tree searching algorithm for the ML method. This is similar to that of the NJ method, in which a star-like tree is first considered. The final tree is nested from a previous tree with a trifurcation, and the difference in the maximum log-likelihood values between the two trees can be used for hypothesis testing. Yet even this procedure has some theoretical problem (see Saitou, 1989b). Thus we should be cautious of conducting a statistical test based on the ML method.

Let us apply the ML method to Hixson and Brown's (1986) data. Program DNAML of Felsenstein's PHYLIP version 3.1 was used to obtain ML values for four trees of Table 7. The transition/transversion ratio was set to be 5.0 and observed frequency of nucleotides were used (Saitou and Imanishi, 1989). Tree 3 showed the highest likelihood value among four trees, and tree 4 was the worst. Interestingly, the rank of these trees in terms of the ML values is identical with that of the minimum evolution method (see Table 7), though the estimated branch lengths (not shown) by the ML method were somewhat different to those of the NJ method (Figure 6(d)).

Felsenstein (1987) developed the 'maximum likelihood' method for DNA–DNA hybridization data. However, he considered several components of experimental errors, and this method is closely related to analysis of variance. Thus it may not be considered as a standard ML method.

5. Relative efficiency of tree-making methods

Many different methods have been proposed for reconstructing phylogenetic trees, as reviewed above. Then which method should we use? It is generally difficult to compare different tree-making methods using actual data, because we rarely know the true phylogenetic tree. Therefore, the relative efficiencies of tree-making methods should be studied through computer simulated data, in which the true tree is known. For example, Peacock and Boulter (1975) simulated amino acid sequence data, Tateno and Nei (1978) simulated nucleotide sequence data, and Nei et al. (1983) simulated gene frequency data. More recently, Fiala and Sokal (1985) simulated morphological data by specifying a transition probability model.

Tateno et al. (1982) did a comprehensive study of tree reconstruction from nucleotide sequences. They considered a phylogenetic tree for eight or more sequences, and a Poisson process was mainly used to simulate nucleotide substitutions for 300 nucleotides. They compared four distance methods (UPGMA, FM, DW, and a modified DW). Their results indicated that the efficiency of each method depended on various conditions. A similar but more extensive studies have been done by Tateno (1985), Tateno and Tajima (1986), and Sourdis and Krimbas (1987). Using a similar scheme of simulation as Tateno et al. (1982) developed, Saitou and Nei (1987) compared six distance methods (UPGMA, DW, a modified DW, Li's (1981) method, Sattath and Tversky's (1977) method, and the NJ method), and they showed that their NJ method and Sattath and Tversky's method were generally better than the other distance methods.

Blanken et al. (1982) considered an addition of one nucleotide sequence to the known phylogenetic tree, thus it is different from ordinary problem of tree construction.

Saitou and Nei (1986) considered trees for relatively small number (up to five) of nucleotide sequences, and derived the expected proportion of each nucleotide configuration for a given tree. Using this information they simulated a multinomial sampling of nucleotides to obtain the simulated sequences. The number of nucleotides required to obtain the correct tree with a probability of 95% has been examined for UPGMA, the FM method, the DW method and the transformed distance method (or the four point metric), and the MP method. Their results for unrooted trees for four sequences show that UPGMA and the FM method are inferior to the other methods. Li (1986) did a similar study for restriction site data.

Hasegawa and Yano (1984) and Saitou (1988) compared the MP and ML methods for the case of four nucleotide sequences, and they showed that the ML method can find the correct tree with an appreciable proportion when the MP method is positively misleading in the sense of Felsenstein (1978b).

Sourdis and Nei (1988) extended Saitou and Nei's (1987) study by including Faith's (1985) modification of the DW method and the MP method for comparison. They showed that the MP method was generally inferior to some distance methods such as the NJ method. For the case of the MP method, they examined trees that are close to the true tree, but this strategy has been shown to be effective by a preliminary study in which all possible trees were examined (Sourdis and Nei, 1988).

Recently, Saitou and Imanishi (1989) compared five exhaustive search methods (MP, ML, FM, FM using S_0, and the minimum evolution (ME) method using SBL) with the NJ method under the model tree for six sequences, and all 105 unrooted trees were examined, except for the ML method in which a limited number of trees were examined. They showed that the NJ, ME, and ML methods performed better than the other three methods. This result was obtained when the evolutionary distance was used for distance methods. When the proportion of nucleotide difference (a metric) was used, all distance methods showed a poor performance. Li et al. (1988) compared the NJ method with Lake's evolutionary parsimony in the case of four OTUs. However, they used only the proportion of nucleotide difference. Therefore, the validity of their conclusion is questionable.

In summary, popular methods such as UPGMA and the FM method have been shown to be generally inferior to the other methods. Considering the computation time when a relatively large number of OTUs is compared, a stepwise clustering method such as the NJ method seems to be the first choice for researchers interested in molecular phylogeny.

Acknowledgments

I thank Dr. Masatoshi Nei for his helpful comments on the manuscript. It should also be mentioned that the overall structure and some details of this review followed chapter 11 of his book (Nei, 1987). This work was partially supported by grants from the Ministry of Education, Science, and Culture of Japan.

References

Adams, E. N. (1972). Consensus techniques and the comparison of taxonomic trees. *Syst. Zool.* **21**, 390–397.

Blanken, R. L., Klotz, L. C. and Hinnebusch, A. G. (1982). Computer comparison of new and existing criteria for constructing evolutionary trees from sequence data. *J. Mol. Evol.* **19**, 9–19.

Buneman, P. (1971). The recovery of trees from measurements of dissimilarity. In: F. R. Hodson, D. G. Kendall and P. Tautu (eds.), *Mathematics in the Archeological and Historical Sciences*, Edinburgh University Press, Edinburgh, 387–395.

Camin, J. H. and Sokal, R. R. (1965). A method for deducing branching sequences in phylogeny. *Evolution* **19**, 311–326.

Cavalli-Sforza, L. L. and Edwards, A. W. F. (1967). Phylogenetic analysis: Models and estimation procedures. *Am. J. Hum. Genet.* **19**, 233–257.

Cavender, J. A. (1978). Taxonomy with confidence. *Math. Biosci.* **40**, 271–280.

Cavender, J. A. (1981). Tests of phylogenetic hypothesis under generalized models. *Math. Biosci.* **54**, 217–229.

Chakraborty, R. (1977). Estimation of time of divergence from phylogenetic studies. *Can. J. Genet. Cytol.* **19**, 217–223.

Courant, R. and Robbins, H. (1941). *What is mathematics?* Oxford University Press, Oxford.

Darwin, C. (1859). *On the origin of species.* John Murray, London.

Dayhoff, M. O. (1978). Survey of new data and computer methods of analysis. In: M. O. Dayhoff (ed.), *Atlas of Protein Sequence and Structure*, National Biomedical Research Foundation, Washington DC, 327.

de Soete, G. (1983). A least squares algorithm for fitting additive trees to proximity data. *Psychometrika* **48**, 621–626.

Dobson, A. J. (1974). Unrooted trees for numerical taxonomy. *J. Appl. Prob.* **11**, 32–42.

Eck, R. and Dayhoff, M. O. (1966). In: M. O. Dayhoff (ed.), *Atlas of Protein Sequence and Structure*, National Biomedical Research Foundation, Silver Springs, MD.

Edwards, A. W. F. and Cavalli-Sforza, L. L. (1965). A method for cluster analysis. *Biometrics* **21**, 362–375.

Elwood, H. J., Olsen, G. J. and Sogin, M. L. (1985). The small-subunit ribosomal RNA gene sequences from the hypotrichous ciliates *Oxytricha nova* and *Stylonychia pustulata. Mol. Biol. Evol.* **2**, 399–410.

Faith, D. P. (1985). Distance methods and the approximation of most-parsimonious trees. *Syst. Zool.* **34**, 312–325.

Farris, J. S. (1972). Estimating phylogenetic trees from distance matrices. *Am. Natur.* **106**, 645–668.

Farris, J. S. (1977). On the phenetic approach to vertebrate classification. In: M. K. Hecht, P. C. Goody and B. M. Hecht (eds.), *Major Patterns in Vertebrate Evolution*, Plenum Press, New York, 823–850.

Felsenstein, J. (1978a). The number of evolutionary trees. *Syst. Zool.* **27**, 27–33.

Felsenstein, J. (1978b). Cases in which parsimony or compatibility methods will be positively misleading. *Syst. Zool.* **27**, 401–410.

Felsenstein, J. (1981). Evolutionary trees from DNA sequences: A maximum likelihood approach. *J. Mol. Evol.* **17**, 368–376.

Felsenstein, J. (1982). Numerical methods for inferring evolutionary trees. *Quart. Rev. Biol.* **57**, 379–404.

Felsenstein, J. (1985). Confidence limits on phylogenies: An approach using the bootstrap. *Evolution* **39**, 783–791.

Felsenstein, J. (1987). Estimation of hominoid phylogeny from a DNA hybridization data set. *J. Mol. Evol.* **26**, 123–131.

Ferris, S. D., Wilson, A. C. and Brown, W. M. (1981). Evolutionary tree for apes and humans based on cleavage maps of mitochondrial DNA. *Proc. Natl. Acad. Sci. USA* **78**, 2432–2436.

Fiala, K. L. and Sokal, R. R. (1985). Factors determining the accuracy of cladogram estimation: Evaluation using computer simulation. *Evolution* **39**, 609–622.

Fitch, W. M. (1970). Distinguishing homologous from analogous proteins. *Syst. Zool.* **19**, 99–113.

Fitch, W. M. (1971). Toward defining the course of evolution: Minimum change for a specific tree topology. *Syst. Zool.* **20**, 406–416.

Fitch, W. M. (1977). On the problem of discovering the most parsimonious tree. *Am. Natur.* **111**, 223–257.

Fitch, W. M. (1981). A non-sequential method for constructing trees and hierarchical classifications. *J. Mol. Evol.* **18**, 30–37.

Fitch, W. M. and Farris, J. S. (1974). Evolutionary tree with minimum nucleotide replacements from amino acid sequences. *J. Mol. Evol.* **3**, 263–278.

Fitch, W. M. and Margoliash, E. (1967). Construction of phylogenetic trees. *Science* **155**, 279–284.

Fukami, K. and Tateno, Y. (1989). On the maximum likelihood method for estimating molecular trees: Uniqueness of the likelihood point. *J. Mol. Evol.* **28**, 460–464.

Hasegawa, M., Kishino, H. and Yano, T. (1985). Dating of the human–ape splitting by a molecular clock of mitochondrial DNA. *J. Mol. Evol.* **22**, 160–174.

Hasegawa, M., Kishino, H. and Yano, T. (1987). Man's place in Hominoidea as inferred from molecular clocks of DNA. *J. Mol. Evol.* **26**, 132–147.

Hasegawa, M. and Yano, T. (1984). Maximum likelihood method of phylogenetic inference from DNA sequence data. *Bull. Biomet. Soc. Jpn.* **5**, 1–7.

Hasegawa, M. and Kishino, T. (1989). Confidence limits on the maximum likelihood estimate of the hominoid tree from mitochondrial-DNA sequences. *Evolution* **43**, 672–677.

Hixson, J. and Brown, W. M. (1986). A comparison of the small ribosomal RNA genes from the mitochondrial DNA of the great apes and humans: sequence, structure, evolution, and phylogenetic implications. *Mol. Biol. Evol.* **3**, 1–18.

Hudson, R. R. and Kaplan, N. L. (1985). Statistical properties of the number of recombination events in the history of a sample of DNA sequences. *Genetics* **111**, 147–164.

Jukes, T. H. and Cantor, C. R. (1969). Evolution of protein molecules. In: H. N. Munro (ed.), *Mammalian Protein Metabolism*, Academic Press, New York, 21–132.

Kashyap, R. L. and Subas, S. (1974). Statistical estimation of parameters in a phylogenetic tree using a dynamic model of the substitutional process. *J. Theor. Biol.* **47**, 75–101.

Kimura, M. and Ohta, T. (1972). On the stochastic model for estimation of mutational distance between homologous proteins. *J. Mol. Evol.* **2**, 87–90.

Klotz, L. C. and Blanken, R. L. (1981). A practical method for calculating evolutionary trees from sequence data. *J. Theor. Biol.* **91**, 261–272.

Klotz, L. C., Komar, N., Blanken, R. L. and Mitchell, R. M. (1979). Calculation of evolutionary trees from sequence data. *Proc. Natl. Acad. Sci. USA* **76**, 4516–4520.

Lake, J. A. (1987). A rate-independent technique for analysis of nucleic acid sequence: Evolutionary parsimony. *Mol. Biol. Evol.* **4**, 167–191.

Langley, C. H. and Fitch, W. M. (1974). An examination of the constancy of the rate of molecular evolution. *J. Mol. Evol.* **3**, 161–177.

LeQuesne, W. J. (1969). A method of selection of characters in numerical taxonomy. *Syst. Zool.* **18**, 201–205.

Li, W. H. (1981). Simple method for constructing phylogenetic trees from distance matrices. *Proc. Natl. Acad. Sci. USA* **78**, 1085–1089.

Li, W. H. (1986). Evolutionary change of restriction cleavage sites and phylogenetic inference. *Genetics* **113**, 187–213.

Li, W. H., Wolfe, K. H., Sourdis, J. and Sharp, P. M. (1989). Reconstruction of phylogenetic trees and estimation of divergence times under nonconstant rates of evolution. Cold Spring Harbor Symp. Quant. Biol. (in press).

Nei, M. (1972). Genetic distance between populations. *Am. Natur.* **106**, 283–292.

Nei, M. (1975). *Molecular Population Genetics and Evolution*, North-Holland, Amsterdam.

Nei, M. (1987). *Molecular Evolutionary Genetics*, Columbia University Press, New York.

Nei, M., Stephens, J. C. and Saitou, N. (1985). Methods for computing the standard errors of branching points in an evolutionary tree and their application to molecular data from humans and apes. *Mol. Biol. Evol.* **2**, 66–85.

Nei, M. and Tajima, F. (1985). Evolutionary change of restriction cleavage sites and phylogenetic inference for man and apes. *Mol. Biol. Evol.* **2**, 189–205.

Nei, M. and Tajima, F. (1987). Problems arising in phylogenetic inference from restriction-site data. *Mol. Biol. Evol.* **4**, 320–323.

Nei, M., Tajima, F. and Tateno, Y. (1983). Accuracy of estimated phylogenetic trees from molecular data. II. Gene frequency data. *J. Mol. Evol.* **19**, 153–170.

Pamilo, P. and Nei, M. (1988). Relationships between gene trees and species trees. *Mol. Biol. Evol.* **5**, 568–583.

Peacock, D. and Boulter, D. (1975). Use of amino acid sequence data in phylogeny and evaluation of methods using computer simulation. *J. Mol. Biol.* **95**, 513–527.

Platnick, N. I. (1988). Programs for quicker relationships. *Nature* **335**, 310.

Prager, E. M. and Wilson, A. C. (1978). Construction of phylogenetic trees for proteins and nucleic acids: empirical evaluation of alternative matrix methods. *J. Mol. Evol.* **11**, 129–142.

Robinson, D. F. and Foulds, L. R. (1981). Comparison of phylogenetic trees. *Mathematical Bioscience* **53**, 131–147.

Ruvolo, M. and Smith, T. F. (1986). Phylogeny and DNA–DNA hybridization. *Mol. Biol. Evol.* **3**, 285–289.

Saitou, N. (1986). On the delta Q-test of Templeton. *Mol. Biol. Evol.* **3**, 282–284.

Saitou, N. (1988). Property and efficiency of the maximum likelihood method for molecular phylogeny. *J. Mol. Evol.* **27**, 261–273.

Saitou, N. (1989). A theoretical study of the underestimation of branch lengths by the maximum parsimony principle. *Syst. Zool.* **38**, 1–6.

Saitou, N. (1990). Maximum likelihood methods. *Methods in Enzymology* **183**, 584–598.

Saitou, N. and Imanishi, T. (1989). Relative efficiencies of the Fitch–Margoliash, maximum-parsimony, maximum likelihood, minimum-evolution, and the neighbor-joining methods of phylogenetic tree construction in obtaining the correct tree. *Mol. Biol. Evol.* **6**, 514–525.

Saitou, N. and Nei, M. (1986). The number of nucleotides required to determine the branching order of three species with special reference to the human–chimpanzee–gorilla divergence. *J. Mol. Evol.* **24**, 189–204.

Saitou, N. and Nei, M. (1987). The neighbor-joining method: A new method for reconstructing phylogenetic trees. *Mol. Biol. Evol.* **4**, 406–425.

Sattath, S. and Tversky, A. (1977). Additive similarity trees. *Psychometrika* **42**, 319–345.

Simpson, G. G. (1961). *Principles of Animal Taxonomy*, Columbia University Press, New York.

Smouse, P. E. and Li, W. H. (1987). Likelihood analysis of mitochondrial restriction-cleavage patterns for the human–chimpanzee–gorilla trichotomy. *Evolution* **41**, 1162–1176.

Sneath, P. H. A. and Sokal, R. R. (1973). *Numerical Taxonomy*, Freeman, San Francisco, CA.

Sokal, R. R. and Michener, C. D. (1958). A statistical method for evaluating systematic relationships. *Univ. Kans. Sci. Bull.* **28**, 1409–1438.

Sourdis, J. and Krimbas, C. (1987). Accuracy of phylogenetic trees estimated from DNA sequence data. *Mol. Biol. Evol.* **4**, 159–166.

Sourdis, J. and Nei, M. (1988). Relative efficiencies of the maximum parsimony and distance-matrix methods in obtaining the correct phylogenetic tree. *Mol. Biol. Evol.* **5**, 298–311.

Studier, J. A. and Keppler, K. J. (1988). A note on the neighbor-joining algorithm of Saitou and Nei. *Mol. Biol. Evol.* **5**, 729–731.

Tajima, F. (1983). Evolutionary relationship of DNA sequences in finite populations. *Genetics* **105**, 437–460.

Takahata, N. and Nei, M. (1985). Gene genealogy and variance of interpopulational nucleotide differences. *Genetics* **110**, 325–344.

Tateno, Y. (1985). Theoretical aspects of molecular tree estimation. In: T. Ohta and K. Aoki (eds.), *Population Genetics and Molecular Evolution*, Japan Scientific Society Press, Tokyo, 293–312.

Tateno, Y. (1990). A method for molecular phylogeny construction by direct use of nucleotide sequence data. *J. Mol. Evol.* **30**, 85–90.

Tateno, Y. and Nei, M. (1978). Goodman et al.'s method for augmenting the number of nucleotide substitutions. *J. Mol. Evol.* **11**, 67–73.

Tateno, Y., Nei, M. and Tajima, F. (1982). Accuracy of estimated phylogenetic trees from molecular data. I. Distantly related species. *J. Mol. Evol.* **18**, 387–404.

Tateno, Y. and Tajima, F. (1986). Statistical properties of molecular tree construction methods under the neutral mutation model. *J. Mol. Evol.* **23**, 354–361.

Templeton, A. R. (1983). Phylogenetic inference from restriction endonuclease cleavage site maps with particular reference to the evolution of humans and apes. *Evolution* **37**, 221–244.

Templeton, A. R. (1985). The phylogeny of the hominoid primates: a statistical analysis of the DNA–DNA hybridization data. *Mol. Biol. Evol.* **2**, 420–433.

C. R. Rao and R. Chakraborty, eds., *Handbook of Statistics, Vol. 8*
© Elsevier Science Publishers B.V. (1991) 347–372

Statistical Models for Sex-Ratio Evolution

Sabin Lessard

1. The notion of reproductive value and optimal sex ratios

1.1. Optimality principles in two-sex populations

The *fundamental theorem of natural selection*, which can be traced back to the period 1920–1930 in early works of S. Wright and R. A. Fisher, states that the mean fitness should increase in one-locus random mating populations if frequency-independent viability selection is acting indifferently on males and females (see, e.g., Ewens, 1979). In multilocus models with recombination of genes, the maximization principle based on the mean fitness happens to be in contradiction with another maximization principle that is related to entropy. As a matter of fact, with recombination but without selection, there is global convergence to Hardy–Weinberg–Robbins equilibria, that is, equilibria such that the gene distributions at different positions are independent. With both recombination and selection, there may exist limit cycles whose occurrence can be intuitively explained by the opposition between two basic tendencies, the increase of entropy and the increase of fitness (see, e.g., Akin, 1979).

With sex-differentiated selection, there is a difficulty of a different kind. In this case, it is not obvious how to compute the mean fitness in the whole population and we have to resort to a notion of reproductive value taking into account that genes are transmitted through males and females. For diplo-diploid random mating populations, Fisher (1930) states: "If we consider the aggregate of an entire generation of (...) offspring, it is clear that the total reproductive value of the males in this group is exactly equal to the total value of all the females, because each sex must supply half the ancestry of all future generations of the species." This statement lead Shaw and Mohler (1953) to define the reproductive value of an individual as

$$\frac{M_0}{2\overline{M}} + \frac{F_0}{2\overline{F}} \quad \text{with mean value 1,} \tag{1.1}$$

Research supported in part by the Natural Sciences and Engineering Research Council of Canada under Grant No. OGP0008833.

where M_0 and F_0 refer to individual fitnesses through male and female functions (or progenies), respectively, with mean values $\overline{M}$ and $\overline{F}$ in the population. In this case, MacArthur (1965) showed that the 'optimal strategy' for an individual should be to maximize $M_0 \times F_0$. By simulations of two-allele one-locus models, Speith (1974) pointed out that it is the product of the mean fitnesses $\overline{M} \times \overline{F}$ (and not the mean of the individual products $M_0 \times F_0$) that "appears to be monotonically nondecreasing after the first few generations. However, one probable exception was observed", without any explanations. Eshel (1975) and Uyenoyama and Bengtsson (1979) described the exact equilibrium structures for particular two-allele one-locus models with linear relationships between male and female fitnesses, which suggested that $\overline{M} \times \overline{F}$ should be maximized at a stable equilibrium. One of the main concerns was the resulting sex ratio in the population. Many authors (see, e.g., Charnov, 1982, and references therein) applied the product maximization principle to treat various aspects of sex allocation (e.g., the evolution of separated sexes versus hermaphroditism) and proposed several extensions for spatially structured populations, non-random mating patterns, haplo-diploid species, age-structured populations, varying environments, etc., but without questioning the validity of the principle or its exact significance in natural populations with genetic constraints.

But the fundamental question remains the following: in what sense and in what conditions is $\overline{M} \times \overline{F}$ maximized in Mendelian populations? Related questions are:

– Does $\overline{M} \times \overline{F}$ increase from one generation to the next one (as in the *fundamental theorem of natural selection*)?

– Does the maximum value of $\overline{M} \times \overline{F}$ correspond to a population state that is uninvadable by any mutant (*evolutionarily stable state* in Maynard Smith's (1982) terminology)?

– Does a mutant allele invade if and only if it renders $\overline{M} \times \overline{F}$ closer to its maximum value (*evolutionary genetic stability property* in Eshel and Feldman's (1982) terminology)?

– Does the product $\overline{M} \times \overline{F}$ increase over successive equilibria as new alleles are introduced (*evolutionary genetic stability property over successive equilibria* or *evolutionary attractiveness property* in Lessard's (1984) terminology)?

– Is a decrease of $\overline{M} \times \overline{F}$ over successive equilibria possible? If yes, in what conditions?

– What happens to the product maximization principle in haplo-diploid populations?

– Is the product maximization principle valid with multifactorial inheritance?

REMARK. ES (*evolutionary stability*) theory owes a great deal of its first developments to Hamilton's (1967) work on 'unbeatable' sex ratios in non-random mating populations, that is, sex ratios that have a 'selective advantage over any other'. But this definition suggests a global selective advantage. In the original game theoretic approach, an ES *strategy* (ESS) was defined as an individual strategy that has greater fitness than any other in randomly pairwise contests once adopted by all members of a population. Then the concept was applied to polymorphic population states (population strategies) and stability properties were proved at least for autogenous populations in continuous time (see Maynard Smith, 1982, and references therein). But in the search of ES states in Mendelian populations, initial increase properties of mutant alleles have become the predominant criteria in use. In this context, an ES *state* corresponds to a population equilibrium that cannot be invaded by any mutant (at least at a geometric rate; see Lessard, 1989). In a more evolutionary perspective, we can look for a population state toward which there is convergence following the invasion of any equilibrium by a mutant allele. If this is the case for a population state at least initially (after the first few generations following invasion) and if this condition is necessary for invasion, then we have the EGS (*evolutionary genetic stability*) property as introduced by Eshel and Feldman (1982) in a study on sex-ratio evolution. In this framework, the long-term behavior over successive equilibria was addressed in Karlin and Lessard (1983, 1986). Global convergence as in Lessard (1984) for two-phenotype frequency-dependent selection models is a challenging problem (see also Nagylaki, 1987, for convergence results under weak selection on fertility and viability and Lande, 1980, for polygenic characters).

1.2. Reproductive values of males and females

The notion of reproductive value has been defined more rigorously since its introduction by Fisher (1930). For males and females in two-sex populations, we introduce the *parent-offspring gene transmission matrix*

$$V = \begin{bmatrix} v_{ff} & v_{fm} \\ v_{mf} & v_{mm} \end{bmatrix}, \tag{1.2}$$

where v_{fm} is the proportion of genes in female offspring transmitted by the male parent and similarly for v_{ff}, v_{mf} and v_{mm}. The matrix V is a stochastic matrix since V is nonnegative – all entries are nonnegative – and $v_{ff} + v_{fm} = v_{mf} + v_{mm} = 1$. If V is irreducible and aperiodic (or equivalently, V^t is positive – all entries are positive – for some integer $t \geq 1$), then there exists a vector $v = (v_f, v_m)$ such that

$$v^T V = v^T, \qquad v_f + v_m = 1, \qquad v_f > 0, \qquad v_m > 0 \tag{1.3}$$

(T for transpose), and

$$V^t \to \begin{bmatrix} v_f & v_m \\ v_f & v_m \end{bmatrix} \quad \text{as } t \to \infty \tag{1.4}$$

(see, e.g., Karlin and Taylor, 1975). The vector v is a positive and normalized leading left eigenvector for V. The components v_f and v_m of v are called the *reproductive values of all females and all males*, respectively. The interpretation of v_f and v_m is that of relative genetic contributions of all females and all males, respectively, to future generations.

Sabin Lessard

In diplo-diploid populations, we have

$$V = \begin{bmatrix} \frac{1}{2} & \frac{1}{2} \\ \frac{1}{2} & \frac{1}{2} \end{bmatrix} \quad \text{and} \quad (v_f, v_m) = (\tfrac{1}{2}, \tfrac{1}{2}) . \tag{1.5}$$

In haplo-diploid populations, we have

$$V = \begin{bmatrix} \frac{1}{2} & \frac{1}{2} \\ 1 & 0 \end{bmatrix} \quad \text{and} \quad (v_f, v_m) = (\tfrac{2}{3}, \tfrac{1}{3}) . \tag{1.6}$$

(See, e.g., Oster et al. (1977) and Benford (1978) for more complex cases with worker-laying of male eggs in social insects.)

One of the main properties of v_f and v_m is that the average of the gene frequencies in female and male offspring weighted by v_f and v_m, respectively, is the same as the corresponding weighted average in female and male parents if there is *no gametic selection*, and in particular *Mendelian segregation* of genes. As a matter of fact, if the proportion of an allele $A^{(k)}$ is $f_v^{(k)}$ in a female of genotype v ($v = 1, \ldots, N_f$ where N_f is the total number of genotypes in females) and $m_\mu^{(k)}$ in a male of genotype μ ($\mu = 1, \ldots, N_m$ where N_m is the total number of genotypes in males), and if the frequencies of genotype v in female parents and female offspring are x_v and $\tilde{x}_v$, respectively, and the frequencies of genotype μ in male parents and male offspring are y_μ and $\tilde{y}_\mu$, respectively, in an infinite population, then we have

$$\begin{bmatrix} \tilde{p}^{(k)} \\ \tilde{q}^{(k)} \end{bmatrix} = \begin{bmatrix} v_{ff} & v_{fm} \\ v_{mf} & v_{mm} \end{bmatrix} \begin{bmatrix} p^{(k)} \\ q^{(k)} \end{bmatrix}, \tag{1.7}$$

where

$$p^{(k)} = \sum_v x_v f_v^{(k)} , \qquad q^{(k)} = \sum_\mu y_\mu m_\mu^{(k)} , \tag{1.8a}$$

$$\tilde{p}^{(k)} = \sum_v \tilde{x}_v f_v^{(k)} , \qquad \tilde{q}^{(k)} = \sum_\mu \tilde{y}_\mu m_\mu^{(k)} , \tag{1.8b}$$

are the frequencies of $A^{(k)}$ in female and male parents and female and male offspring, respectively. Therefore,

$$v_f \tilde{p}^{(k)} + v_m \tilde{q}^{(k)} = v_f p^{(k)} + v_m q^{(k)} . \tag{1.9}$$

Moreover, this quantity remains constant if there is *no zygotic selection* from conception to matury (that is, reproductive age).

The reproductive values v_f and v_m are for all females and all males, respectively. If a female has fitness F_0 from conception to maturity and the mean fitness in the female population is $\overline{F}$, then the reproductive value of that female can be defined as $v_f(F_0/\overline{F})$ whose mean value in the female population is v_f. In the same way, the reproductive value of a male whose fitness is M_0 can be defined as $v_m(M_0/\overline{M})$ where $\overline{M}$ is the mean fitness in the male population such that the mean reproductive value in the male population is v_m. Overall, the mean reproductive value in the population is $v_f + v_m = 1$. For diplo-diploid populations with $v_f = v_m = \frac{1}{2}$, this yields the Shaw–Mohler formula (1.1).

2. Covariance formulas and the evolution of the sex ratio

2.1. Li–Price covariance formula for two-sex populations

Covariance formulas as introduced by Li (1967) and Price (1970) are powerful tools to study the increase or decrease of gene frequencies.

For two-sex populations, let the genotype frequencies over two successive generations be related by the recurrence equations

$$x'_v = \frac{\tilde{x}_v F_v}{\overline{F}}, \quad v = 1, \ldots, N_f, \tag{2.1a}$$

$$y'_\mu = \frac{\tilde{y}_\mu M_\mu}{\overline{M}}, \quad \mu = 1, \ldots, N_m, \tag{2.1b}$$

where F_v and M_μ for $v = 1, \ldots, N_f$ and $\mu = 1, \ldots, N_m$ represent the fitnesses of female and male offspring according to the female and male genotypes, respectively, with mean values in the population

$$\overline{F} = \sum_v \tilde{x}_v F_v \quad \text{and} \quad \overline{M} = \sum_\mu \tilde{y}_\mu M_\mu. \tag{2.2}$$

In general, the male and female fitnesses may depend on the genotype frequencies in the population and then be frequency-dependent.

For the allele frequencies, we get the difference equations

$$\begin{aligned}
\Delta p^{(k)} &= \sum_v x'_v f_v^{(k)} - \sum_v x_v f_v^{(k)} \\
&= \sum_v (\tilde{x}_v F_v f_v^{(k)})/\overline{F} - \sum_v x_v f_v^{(k)} \\
&= \mathrm{Cov}(F, f^{(k)})/\overline{F} + \tilde{p}^{(k)} - p^{(k)},
\end{aligned} \tag{2.3a}$$

and similarly,

$$\Delta q^{(k)} = \mathrm{Cov}(M, m^{(k)})/\overline{M} + \tilde{q}^{(k)} - q^{(k)}, \tag{2.3b}$$

where

$$\text{Cov}(F, f^{(k)}) = \sum_v \tilde{x}_v F_v f_v^{(k)} - \left(\sum_v \tilde{x}_v F_v \right) \left(\sum_v \tilde{x}_v f_v^{(k)} \right), \tag{2.4a}$$

$$\text{Cov}(M, m^{(k)}) = \sum_\mu \tilde{y}_\mu M_\mu m_\mu^{(k)} - \left(\sum_\mu \tilde{y}_\mu M_\mu \right) \left(\sum_\mu \tilde{y}_\mu m_\mu^{(k)} \right). \tag{2.4b}$$

Appealing to (1.9), these equations lead to

$$\Delta(v_f p^{(k)} + v_m q^{(k)}) = v_f \text{Cov}(F, f^{(k)})/\overline{F} + v_m \text{Cov}(M, m^{(k)})/\overline{M} \tag{2.5}$$

(see, e.g., Price, 1970; Uyenoyama et al., 1981; Taylor, 1988).

2.2. Average effects and average excesses of genes

2.2.1. Covariance formula for average effects
Average effects of genes were first defined by Fisher (1918) (see also Fisher, 1930, 1941).

For a quantitative character with genotypic value G defined for every female genotype v at a single autosomal locus as the expected character G_v associated with that genotype ($v = 1, \ldots, N_f$), the additive value (or genic value, or breeding value) at the single autosomal locus with possible alleles $A^{(1)}, \ldots, A^{(k)}, \ldots, A^{(n)}$ is

$$A = \sum_k \pi_f f_v^{(k)} \alpha^{(k)}, \tag{2.6}$$

where π_f is the ploidy of females ($\pi_f = 2$ in diplo-diploid as well as haplo-diploid populations), $f_v^{(k)}$ is the proportion of allele $A^{(k)}$ in genotype v and $\alpha^{(k)}$ for $k = 1, \ldots, n$ are such that

$$S^2 = \sum_v \tilde{x}_v \left[G_v - \overline{G} - \sum_k \pi_f f_v^{(k)} \alpha^{(k)} \right]^2 \tag{2.7}$$

is minimum with the constraint

$$\sum_v \tilde{x}_v \sum_k f_v^{(k)} \alpha^{(k)} = \sum_k \tilde{p}^{(k)} \alpha^{(k)} = 0, \tag{2.8}$$

where $\tilde{x}_v$ is the frequency of genotype v in female offspring ($v = 1, \ldots, N_f$) and

$$\overline{G} = \sum_v \tilde{x}_v G_v. \tag{2.9}$$

The quantity $\alpha^{(k)}$ is called the *average effect* of $A^{(k)}$ on G. The difference

$D = G - \overline{G} - A$ is called the *dominance deviation*. When $D = 0$, we say that there is *no dominance*.

Differentiating S^2 with respect to $\alpha^{(k)}$, equaling to 0 and using the constraint on the average effects yield

$$\text{Cov}(G, f^{(k)}) = \sum_l \pi_f \alpha^{(l)} \, \text{E}(f^{(k)} f^{(l)}) \quad \text{(E for expectation)}$$

$$= \tfrac{1}{2}\{p^{(k)} + q^{(k)}\} \alpha^{(k)} + \sum_l \tfrac{1}{2}\{p^{(k)}q^{(l)} + p^{(l)}q^{(k)}\} \alpha^{(l)}$$

$$\cong \tilde{p}^{(k)} \alpha^{(k)} \tag{2.10}$$

if mating is random and if selection is weak or $p^{(k)} = q^{(k)} = \tilde{p}^{(k)}$ since then

$$\sum_l \tfrac{1}{2}\{p^{(k)}q^{(l)} + p^{(l)}q^{(k)}\} \alpha^{(l)} \cong \tilde{p}^{(k)} \sum_l \tilde{p}^{(l)} \alpha^{(l)} = 0 \tag{2.11}$$

with

$$\tilde{p}^{(k)} = \tfrac{1}{2}(p^{(k)} + q^{(k)}) \quad \text{and} \quad \tilde{p}^{(l)} = \tfrac{1}{2}(p^{(l)} + q^{(l)}). \tag{2.12}$$

The notation $\cong$, which will be used throughout, means an approximation when selection is weak enough such that the differences between $p^{(k)}$, $q^{(k)}$, $\tilde{p}^{(k)}$ and $\tilde{q}^{(k)}$ in the same generation or in consecutive generations can be neglected. Note that we have actually an equality in (2.10) when $p^{(k)} = q^{(k)} = \tilde{p}^{(k)}$.

In general, we have the covariance formula

$$\alpha^{(k)} \cong \text{Cov}(G, f^{(k)})/\tilde{p}^{(k)}. \tag{2.13}$$

An analogous formula is obtained for average effects of genes on a quantitative character in males (diploid or haploid males).

2.2.2. Li–Price covariance formula in terms of average excesses
Following Fisher (1930, 1941), the quantity

$$a^{(k)} = \text{Cov}(G, f^{(k)})/\tilde{p}^{(k)}, \tag{2.14}$$

is called the *average excess* of $A^{(k)}$ in G. (Actually Fisher defined in a different but equivalent way the average excess as well as the average effect of a 'gene substitution'.) The following expression for the average excess ensues:

$$a^{(k)} = G^{(k)} - \overline{G}, \tag{2.15}$$

where

$$G^{(k)} = \sum_v \tilde{x}_v G_v f_v^{(k)} \Big/ \left[\sum_v \tilde{x}_v f_v^{(k)} \right], \tag{2.16}$$

such that $G^{(k)}$ is the *marginal value* of G for $A^{(k)}$.

With respect to average excesses in female and male fitnesses considered as quantitative characters, the Li–Price covariance formula for two-sex populations takes the form

$$\Delta(v_f p^{(k)} + v_m q^{(k)}) = v_f \tilde{p}^{(k)} \left\{ \frac{b^{(k)}}{\overline{F}} \right\} + v_m \tilde{q}^{(k)} \left\{ \frac{d^{(k)}}{\overline{M}} \right\}, \tag{2.17}$$

where

$$b^{(k)} = F^{(k)} - \overline{F} = \sum_v \tilde{x}_v F_v f_v^{(k)} \bigg/ \left[\sum_v \tilde{x}_v f_v^{(k)} \right] - \overline{F}, \tag{2.18a}$$

$$d^{(k)} = M^{(k)} - \overline{M} = \sum_\mu \tilde{y}_\mu M_\mu m_\mu^{(k)} \bigg/ \left[\sum_\mu \tilde{y}_\mu m_\mu^{(k)} \right] - \overline{M}. \tag{2.18b}$$

The quantities $F^{(k)}$ and $M^{(k)}$ are the marginal fitnesses for $A^{(k)}$ in female and male offspring, respectively. Under weak selection, $b^{(k)}$ and $d^{(k)}$ are good approximations of the average effects of $A^{(k)}$ on F and M, respectively, denoted $\beta^{(k)}$ and $\delta^{(k)}$, respectively.

2.3. Sex allocation in random mating populations

2.3.1. Model I: Determination by the offspring's genotype
When sex allocation in hermaphroditic diploid random mating poulations is determined by the offspring's genotype at a single locus such that the female and male fitnesses of an offspring are given by

$$F = G \quad \text{and} \quad M = 1 - G, \tag{2.19}$$

where G is the genotypic value of the offspring for sex allocation ($0 \leqslant G \leqslant 1$), the Li–Price covariance formula for two-sex populations reduces to

$$\Delta(v_f p^{(k)} + v_m q^{(k)}) = v_f \left\{ \frac{\mathrm{Cov}(G, f^{(k)})}{\overline{G}} \right\} + v_m \left\{ \frac{\mathrm{Cov}(1 - G, m^{(k)})}{1 - \overline{G}} \right\}$$

$$= \left\{ \frac{v_f}{\overline{G}} - \frac{v_m}{1 - \overline{G}} \right\} \mathrm{Cov}(G, f^{(k)}), \tag{2.20}$$

where $\overline{G}$ is the mean genotypic value and $\mathrm{Cov}(G, m^{(k)}) = \mathrm{Cov}(G, f^{(k)})$ since the genotype distribution in offspring as males or females is the same. Making $v_f = v_m = \frac{1}{2}$, we get

$$\frac{\Delta(p^{(k)} + q^{(k)})}{p^{(k)} + q^{(k)}} = \left\{ \frac{1 - 2\overline{G}}{2\overline{G}(1 - \overline{G})} \right\} a^{(k)}, \tag{2.21}$$

where $a^{(k)}$ is the average excess of $A^{(k)}$ in G. Therefore at equilibrium,

$$\begin{cases} \text{either (i) } \overline{G} = \tfrac{1}{2} \\ \text{or (ii) } a^{(k)} = 0 \text{ for all } A^{(k)} \text{ represented} . \end{cases} \tag{2.22}$$

In case (i), we have an *even sex ratio* in the population. In case (ii), the equilibrium is said *genotypic*.

2.3.2. Model II: Determination by the mother's genotype

When progeny sex ratio is determined by the mother's genotype at a single autosomal locus in random mating populations with separated sexes (diplo-diploid as well as haplo-diploid populations), the fitnesses of a daughter ($\mathscr{D}$) and a son ($\mathscr{S}$) are given by

$$F_{\mathscr{D}} = G_{\mathscr{M}} \quad \text{and} \quad M_{\mathscr{S}} = 1 - G_{\mathscr{M}}, \tag{2.23}$$

respectively, where $G_{\mathscr{M}}$ is the genotypic value of the mother ($\mathscr{M}$) for progeny sex ratio. In such a case, the Li–Price covariance formula becomes

$$\Delta(v_f p^{(k)} + v_m q^{(k)}) = v_f \left\{ \frac{\text{Cov}(G_{\mathscr{M}}, f_{\mathscr{D}}^{(k)})}{\overline{G}_{\mathscr{M}}} \right\} + v_m \left\{ \frac{\text{Cov}(1 - G_{\mathscr{M}}, f_{\mathscr{S}}^{(k)})}{1 - \overline{G}_{\mathscr{M}}} \right\}$$

$$= \left\{ \frac{v_f}{\overline{G}_{\mathscr{M}}} R_{\mathscr{M} \to \mathscr{D}}^{(k)} - \frac{v_m}{1 - \overline{G}_{\mathscr{M}}} R_{\mathscr{M} \to \mathscr{S}}^{(k)} \right\} \text{Cov}(G_{\mathscr{M}}, f_{\mathscr{M}}^{(k)}), \tag{2.24}$$

where

$$R_{\mathscr{M} \to \mathscr{D}}^{(k)} = \text{Cov}(G_{\mathscr{M}}, f_{\mathscr{D}}^{(k)}) / \text{Cov}(G_{\mathscr{M}}, f_{\mathscr{M}}^{(k)}),$$

$$R_{\mathscr{M} \to \mathscr{S}}^{(k)} = \text{Cov}(G_{\mathscr{M}}, f_{\mathscr{S}}^{(k)}) / \text{Cov}(G_{\mathscr{M}}, f_{\mathscr{M}}^{(k)}), \tag{2.25}$$

and $f_{\mathscr{D}}^{(k)}$, $f_{\mathscr{S}}^{(k)}$ and $f_{\mathscr{M}}^{(k)}$ denote the proportions of $A^{(k)}$ in a daughter, a son and a mother, respectively. Therefore

$$\frac{\Delta(v_f p^{(k)} + v_m q^{(k)})}{v_f p^{(k)} + v_m q^{(k)}} = \left\{ \frac{v_f}{\overline{G}_{\mathscr{M}}} R_{\mathscr{M} \to \mathscr{D}}^{(k)} - \frac{v_m}{1 - \overline{G}_{\mathscr{M}}} R_{\mathscr{M} \to \mathscr{S}}^{(k)} \right\} C_{\mathscr{M}}^{(k)} a_{\mathscr{M}}^{(k)}, \tag{2.26}$$

where $a_{\mathscr{M}}^{(k)}$ is the average excess of $A^{(k)}$ in $G_{\mathscr{M}}$ and

$$C_{\mathscr{M}}^{(k)} = \frac{\tilde{p}_{\mathscr{M}}^{(k)}}{v_f p^{(k)} + v_m q^{(k)}} \simeq 1, \tag{2.27}$$

where $\tilde{p}_{\mathscr{M}}^{(k)}$ denotes the frequency of $A^{(k)}$ at conception in the maternal population.

2.4. Relatedness

Relatedness was first defined in the context of kin selection theory (Hamilton, 1964). In random mating populations, the relatedness coefficient between a donor ($\mathscr{I}$) and a recipient ($\mathscr{I}'$) was defined as the expected fraction of genes identical by descent in $\mathscr{I}'$ to one or more genes in $\mathscr{I}$. Genes are *identical by descent* (i.b.d.) if they are copies of a same ancestor (Malécot, 1948). More generally, the relatedness coefficient of $\mathscr{I}$ to $\mathscr{I}'$ has been defined with respect to an allele $A^{(k)}$ and a genotypic value G as

$$R^{(k)}_{\mathscr{I} \to \mathscr{I}'} = \mathrm{Cov}(G_{\mathscr{I}}, f^{(k)}_{\mathscr{I}'})/\mathrm{Cov}(G_{\mathscr{I}}, f^{(k)}_{\mathscr{I}}) \tag{2.28}$$

(see Michod and Hamilton, 1980).

For any daughter $\mathscr{D}$ in diplo-diploid as well as haplo-diploid populations, we have

$$f^{(k)}_{\mathscr{D}} = \tfrac{1}{2}(\chi^{(k)}_{\mathscr{M}} + \chi^{(k)}_{\mathscr{F}}), \tag{2.29}$$

where $\chi^{(k)}_{\mathscr{M}}$ and $\chi^{(k)}_{\mathscr{F}}$ are the numbers (0 or 1) of gene $A^{(k)}$ at an autosomal locus transmitted to the daughter by her mother $\mathscr{M}$ and her father $\mathscr{F}$, respectively. Assuming random mating, $\chi^{(k)}_{\mathscr{F}}$ is independent of $G_{\mathscr{M}}$ and we have

$$
\begin{aligned}
R^{(k)}_{\mathscr{M} \to \mathscr{D}} &= \mathrm{Cov}(G_{\mathscr{M}}, \tfrac{1}{2}(\chi^{(k)}_{\mathscr{M}} + \chi^{(k)}_{\mathscr{F}}))/\mathrm{Cov}(G_{\mathscr{M}}, f^{(k)}_{\mathscr{M}}) \\
&= \tfrac{1}{2}\mathrm{Cov}(G_{\mathscr{M}}, \chi^{(k)}_{\mathscr{M}})/\mathrm{Cov}(G_{\mathscr{M}}, f^{(k)}_{\mathscr{M}}) \\
&\cong \tfrac{1}{2}
\end{aligned}
\tag{2.30}
$$

under weak selection since then $f^{(k)}_{\mathscr{M}}$ at conception is close to the expected value of $\chi^{(k)}_{\mathscr{M}}$ at maturity for a mother. In the same way, we find

$$R^{(k)}_{\mathscr{M} \to \mathscr{S}} \cong \tfrac{1}{2} \tag{2.31}$$

for a son $\mathscr{S}$ in diplo-diploid populations. In haplo-diploid populations with haploid males, we have $f^{(k)}_{\mathscr{S}} = \chi^{(k)}_{\mathscr{M}}$ and

$$R^{(k)}_{\mathscr{M} \to \mathscr{S}} \cong 1. \tag{2.32}$$

These results are in agreement with Hamilton (1964, 1972). The approximate values of the quantities $R^{(k)}_{\mathscr{M} \to \mathscr{D}}$ and $R^{(k)}_{\mathscr{M} \to \mathscr{S}}$ under weak selection (the right member terms in (2.31) and (2.32)) will be denoted throughout $R_{\mathscr{M} \to \mathscr{D}}$ and $R_{\mathscr{M} \to \mathscr{S}}$, respectively.

2.5. Initial case in frequency of a mutant allele

2.5.1. Initial increase condition in Model I

The Li–Price covariance formula enables us to study the increase in frequency of a mutant allele when rare.

Consider an equilibrium with respect to alleles $A^{(1)}, \ldots, A^{(n)}$ at an autosomal locus and introduce a mutant allele $A^{(n+1)}$. Extending the notation in (2.1), let $\tilde{x}_{N_f+1}, \ldots, \tilde{x}_{N_f+l_f}$ and $\tilde{x}_{N_m+1}, \ldots, \tilde{x}_{N_m+l_m}$ be the frequencies of all genotypes carrying $A^{(n+1)}$ in female and male offspring, respectively. For $\tilde{\mathbf{x}}^{(n+1)} = (\tilde{x}_{N_f+1}, \ldots, \tilde{x}_{N_f+l_f})$ and $\tilde{\mathbf{y}}^{(n+1)} = (\tilde{y}_{N_m+1}, \ldots, \tilde{y}_{N_m+l_m})$, we have over two successive generations near the equilibrium

$$\begin{bmatrix} \tilde{\mathbf{x}}^{(n+1)} \\ \tilde{\mathbf{y}}^{(n+1)} \end{bmatrix}' = L \begin{bmatrix} \tilde{\mathbf{x}}^{(n+1)} \\ \tilde{\mathbf{y}}^{(n+1)} \end{bmatrix} + \text{higher order terms} . \tag{2.33}$$

Iterating t times yields

$$\begin{bmatrix} \tilde{\mathbf{x}}^{(n+1)} \\ \tilde{\mathbf{y}}^{(n+1)} \end{bmatrix}^{(t)} = L^t \begin{bmatrix} \tilde{\mathbf{x}}^{(n+1)} \\ \tilde{\mathbf{y}}^{(n+1)} \end{bmatrix} + \text{higher order terms} . \tag{2.34}$$

If L is nonnegative — all entries are nonnegative — and L^t is positive — all entries are positive — for some integer $t \geqslant 1$, which is often the case and will be assumed throughout, then there exists a quantity $\lambda > 0$ and vectors

$$\tilde{\boldsymbol{\xi}}^{(n+1)} = (\tilde{\xi}_{N_f+1}, \ldots, \tilde{\xi}_{N_f+l_f}),$$

$$\tilde{\xi}_v > 0 \quad \text{for } v = N_f + 1, \ldots, N_f + l_f, \qquad \sum_{v=N_f+1}^{N_f+l_f} \tilde{\xi}_v = 1 , \tag{2.35a}$$

$$\tilde{\boldsymbol{\eta}}^{(n+1)} = (\tilde{\eta}_{N_m+1}, \ldots, \tilde{\eta}_{N_m+l_m}),$$

$$\tilde{\eta}_\mu > 0 \quad \text{for } \mu = N_m + 1, \ldots, N_m + l_m, \qquad \sum_{\mu=N_m+1}^{N_m+l_m} \tilde{\eta}_\mu = 1 , \tag{2.35b}$$

such that

$$L^t \begin{bmatrix} \tilde{\mathbf{x}}^{(n+1)} \\ \tilde{\mathbf{y}}^{(n+1)} \end{bmatrix} \Big/ \lambda^t \to \text{multiple of} \begin{bmatrix} \tilde{\boldsymbol{\xi}}^{(n+1)} \\ \tilde{\boldsymbol{\eta}}^{(n+1)} \end{bmatrix} \quad \text{as } t \to \infty . \tag{2.36}$$

The quantity λ is the leading eigenvalue of L with $(\tilde{\boldsymbol{\xi}}^{(n+1)}, \tilde{\boldsymbol{\eta}}^{(n+1)})$ as a right eigenvector. (See, e.g., Karlin and Taylor, 1975, for a review of the Perron–Frobenius theory for nonnegative matrices.)

The quantity λ is the *growth rate* of the mutant genotype frequencies, and then also the growth rate of the mutant allele frequencies, in female and male offspring near the equilibrium. For Model I, equations (2.15), (2.16), (2.21) and (2.36) yield the expression

$$\lambda - 1 = \left\{ \frac{1 - 2\overline{G}}{2\overline{G}(1 - \overline{G})} \right\} (G^{(n+1)} - \overline{G}), \tag{2.37}$$

where $\overline{G}$ and $G^{(n+1)}$ are evaluated at the equilibrium. In order to evaluate $G^{(n+1)}$ at the equilibrium we use the *stable distribution* $\tilde{\boldsymbol{\xi}}^{(n+1)}$ of genotypes in offspring as females near the equilibrium (which is the same as the stable distribution $\tilde{\boldsymbol{\eta}}^{(n+1)}$ of genotypes in offspring as males near the equilibrium in Model I), that is

$$G^{(n+1)} = \sum_{v = N_f + 1}^{N_f + l_f} \tilde{\xi}_v G_v f_v^{(k)} \bigg/ \left[\sum_{v = N_f + 1}^{N_f + l_f} \tilde{\xi}_v f_v^{(k)} \right], \tag{2.38}$$

at the equilibrium.

There will be *invasion* of $A^{(n+1)}$ at a geometric rate if and only if $\lambda > 1$, or equivalently for Model I,

$$G^{(n+1)} \begin{cases} > \overline{G} & \text{if } \overline{G} < \tfrac{1}{2}, \\ < \overline{G} & \text{if } \overline{G} > \tfrac{1}{2}, \end{cases} \tag{2.39}$$

where $G^{(n+1)}$ and $\overline{G}$ are evaluated at the equilibrium. With the interpretation of $G^{(n+1)}$ as the marginal value of G for $A^{(n+1)}$ at the equilibrium, the interpretation of (2.39) is that a mutant allele invades the equilibrium if and only if it renders the population sex ratio $\overline{G}$ closer to $\tfrac{1}{2}$ at least initially after enough generations (see Lessard, 1989, for a formal proof). In such a case, $\overline{G} = \tfrac{1}{2}$ is said to exhibit the *evolutionary genetic stability* (EGS) property (Eshel and Feldman, 1982).

2.5.2. *Evolutionary genetic stability (EGS) property in Model II*
By analogy with $\overline{G} = \tfrac{1}{2}$ in Model I, the sex ratio $\overline{G}_{\mathcal{M}}$ in Model II that has the EGS property must satisfy

$$\left\{ \frac{v_f}{\overline{G}_{\mathcal{M}}} \right\} R_{\mathcal{M} \to \mathcal{D}} - \left\{ \frac{v_m}{1 - \overline{G}_{\mathcal{M}}} \right\} R_{\mathcal{M} \to \mathcal{S}} = 0 . \tag{2.40}$$

We find

$$\overline{G}_{\mathcal{M}} = \frac{v_f R_{\mathcal{M} \to \mathcal{D}}}{v_f R_{\mathcal{M} \to \mathcal{D}} + v_m R_{\mathcal{M} \to \mathcal{S}}} = \tfrac{1}{2}, \tag{2.41}$$

in diplo-diploid as well as haplo-diploid populations.

3. Exact genetic models for sex-ratio evolution

3.1. Formulation of a general two-sex one-locus model

The following two-sex one-locus model is classical (see, e.g., Karlin, 1978; Ewens, 1979). Consider n alleles $A^{(1)}, \ldots, A^{(n)}$ at a single locus whose frequencies are $p^{(1)}, \ldots, p^{(n)}$ in females and $q^{(1)}, \ldots, q^{(n)}$ in males. An individual with genotype $A^{(k)}A^{(l)}$ is assumed to have constant positive fitness $F_{kl}\ (= F_{lk})$ as a female and $M_{kl}\ (= M_{lk})$ as a male $(k, l = 1, \ldots, n)$. To take into account ecological and/or biological constraints, we suppose further that all possible male and female fitness pairs (M_{kl}, F_{kl}) belong to some fitness set S (i.e., some bounded subset of the positive quadrant).

Assuming an infinite population with discrete non-overlapping generations obtained by random mating in a constant environment, we have the recurrence equations

$$p^{(k)\prime} = \frac{p^{(k)} \sum_l F_{kl} q^{(l)} + q^{(k)} \sum_l F_{kl} p^{(l)}}{2\overline{F}}, \quad k = 1, \ldots, n, \qquad (3.1\text{a})$$

$$q^{(k)\prime} = \frac{p^{(k)} \sum_l M_{kl} q^{(l)} + q^{(k)} \sum_l M_{kl} p^{(l)}}{2\overline{M}}, \quad k = 1, \ldots, n, \qquad (3.1\text{b})$$

where

$$\overline{F} = \sum_{k,l} F_{kl} p^{(k)} q^{(l)} \quad \text{and} \quad \overline{M} = \sum_{k,l} M_{kl} p^{(k)} q^{(l)}. \qquad (3.2)$$

Observe that the mean male and females fitness pair $(\overline{M}, \overline{F})$ belongs to the convex hull of S denoted throughout by $\mathcal{H}(S)$. Model (3.1) is a special case of model (2.1) with constant female and male fitness determined by the offspring genotype in diplo-diploid populations.

3.2. Initial increase condition

Consider an equilibrium of (3.1) with male and female mean fitnesses given by $(\overline{M}, \overline{F})$. Suppose that a mutant allele $A^{(n+1)}$ is then introduced in small frequency into the population. Let $(M^{(n+1)}, F^{(n+1)})$ be the male and female marginal fitnesses of $A^{(n+1)}$ when rare. These marginal fitnesses are computed with respect to the stable distribution of female and male gametes carrying the mutant allele $A^{(n+1)}$ near the equilibrium and are given by

$$F^{(n+1)} = w_{\mathrm{f}}(F\mathbf{q})_{n+1} + w_{\mathrm{m}}(F\mathbf{p})_{n+1}, \qquad (3.3\text{a})$$

$$M^{(n+1)} = w_{\mathrm{f}}(M\mathbf{q})_{n+1} + w_{\mathrm{m}}(M\mathbf{p})_{n+1}, \qquad (3.3\text{b})$$

where

$$(F\boldsymbol{q})_{n+1} = \sum_k F_{n+1,k} q^{(k)}, \qquad (F\boldsymbol{p})_{n+1} = \sum_k F_{n+1,k} p^{(k)}, \qquad (3.4a)$$

$$(M\boldsymbol{q})_{n+1} = \sum_k M_{n+1,k} q^{(k)}, \qquad (M\boldsymbol{p})_{n+1} = \sum_k M_{n+1,k} p^{(k)}, \qquad (3.4b)$$

and (w_f, w_m) is the leading right eigenvector with $w_f > 0$, $w_m > 0$, $w_f + w_m = 1$ for the positive matrix

$$C = \begin{bmatrix} \dfrac{(F\boldsymbol{q})_{n+1}}{2\overline{F}} & \dfrac{(F\boldsymbol{p})_{n+1}}{2\overline{F}} \\[2ex] \dfrac{(M\boldsymbol{q})_{n+1}}{2\overline{M}} & \dfrac{(M\boldsymbol{p})_{n+1}}{2\overline{M}} \end{bmatrix}, \qquad (3.5)$$

with

$$\boldsymbol{p} = (p^{(1)}, \ldots, p^{(n)}) \quad \text{and} \quad \boldsymbol{q} = (q^{(1)}, \ldots, q^{(n)}), \qquad (3.6)$$

evaluated at the equilibrium. The matrix C relates $(p^{(n+1)}, q^{(n+1)})$ over two successive generations in a linear approximation near the equilibrium and there will be invasion of $A^{(n+1)}$ at a geometric rate if and only if

$$\lambda = \frac{F^{(n+1)}}{2\overline{F}} + \frac{M^{(n+1)}}{2\overline{M}} > 1, \qquad (3.7)$$

where λ is the leading eigenvalue of C. (In the case $\lambda = 1$, there is degeneracy in the linear approximations and a further analysis is required to determine the condition for invasion at a lower rate. The formula (3.7) can be extended to multilocus genetic systems and frequency-dependent selection. See Lessard (1989) for details.)

3.3. Evolutionary dynamics

We will now examine what can happen following invasion according to the shape of the fitness set.

CASE 1. $M_{kl} = bF_{kl}$ $(b > 0)$ *for all* k, l.

   This case is equivalent to a standard one-sex model. Therefore the product $\overline{M} \times \overline{F} = b\overline{F}^2$ increases over two successive generations. This is a reformulation of the fundamental theorem of natural selection. Moreover the stable equilibria correspond to the local maxima of $\overline{M} \times \overline{F}$ (see, e.g., Kingman, 1961a,b).

CASE 2. $M_{kl} = a - bF_{kl}$ $(a, b > 0)$ *for all* k, l.

In particular, when $M_{kl} = 1 - F_{kl}$ for all k, l, we have a pure sex allocation model (Model I) for which F_{kl} and M_{kl} can be interpreted as probabilities of being female and male, respectively. This multiallele model was introduced in Eshel and Feldman (1982).

In generic cases, the following facts can be proved (Karlin and Lessard, 1983, 1986):

(i) $\overline{M} \times \overline{F}$ increases over two successive attainable equilibria as new alleles are introduced one at a time as long as the maximum value of $\overline{M} \times \overline{F}$ is not achieved. (An attainable equilibrium is an equilibrium that can be reached, i.e., stable, assuming global convergence. Convergence according to this scheme has been proved in many cases including general two-allele models and has always occurred in simulations with any number of alleles.) This optimality property for the maximum value of $\overline{M} \times \overline{F}$ will be called the EGS *property over successive attainable equilibria*.

(ii) The unique point $(\hat{M}, \hat{F})$ where $\overline{M} \times \overline{F}$ is globally maximized corresponds to an equilibrium surface (actually a level surface of a spectral radius functional)[1] with no other equilibrium within the same allele system. In Model I, the equilibrium surface is associated with an even sex ratio since then $(\hat{M}, \hat{F}) = (\frac{1}{2}, \frac{1}{2})$.

(iii) All equilibria with $(\overline{M}, \overline{F}) \neq (\hat{M}, \hat{F})$ are symmetric with $p = q$, i.e., exhibit the same allelic frequencies in males and females, and correspond to equilibria for Case 1 as in Eshel and Feldman (1982). (In such a case the average excesses and average effects of genes coincide and the equilibrium condition is that these quantities vanish for all alleles represented at equilibrium; see Sections 2.2 and 2.3.)

(iv) A tractable criterion for the existence of a non-degenerate equilibrium surface corresponding to $(\hat{M}, \hat{F})$ is the existence of two symmetric equilibria such that $(\hat{M}, \hat{F})$ lies between the mean fitness pairs, (M^*, F^*) and (M^{**}, F^{**}), for the two symmetric equilibria (Lessard, 1986, 1989). The alternative model with fitness determined by the mother's genotype and in particular Model II can be analyzed in the same way and leads to the same evolutionary properties.

CASE 3. (M_{kl}, F_{kl}) *in a general fitness set S for all* k, l.

Consider the unique point $(\hat{M}, \hat{F})$ that maximizes $\overline{M} \times \overline{F}$ on the convex hull $\mathcal{H}(S)$ of S. We have the following properties (see Lessard, 1989, for proofs and further details):

---

[1] The equilibrium surface corresponding to $(\hat{M}, \hat{F})$ with n alleles represented is given by

$$\rho(\text{diag}(Mp) + \text{diag}(p)M) = 2\hat{m}$$

(ρ for spectral radius and diag for diagonal matrix) with corresponding right eigenvector q where

$$p = (p^{(1)}, \ldots, p^{(n)}), \qquad q = (q^{(1)}, \ldots, q^{(n)}), \qquad M = \|M_{kl}\|_{k,l=1}^{n}.$$

(i) The only equilibria that cannot be invaded geometrically fast by a mutant allele are associated with a mean fitness pair $(\hat{M}, \hat{F})$ which is then a candidate for an ES state. (Actually a mutant allele with marginal male and female fitnesses $(\hat{M}, \hat{F})$ invades any other equilibrium.)

(ii) Assuming non-degeneracies in linear approximations, a mutant allele invades at least a class of equilibria[2] if and only if its introduction leads to an increase of $\overline{M} \times \overline{F}$ after the first few generations which is the EGS property for $\hat{M} \times \hat{F}$.

In the event of (ii), if there is a bifurcation to a nearby stable equilibrium as a mutant allele invades, then this equilibrium should exhibit a product $\overline{M} \times \overline{F}$ larger than it was at the original equilibrium. But we have to insist on the fact that a stable equilibrium, even symmetric, may not be a local maximum of $\overline{M} \times \overline{F}$ with respect to small perturbations on the frequencies of the alleles already represented at equilibrium. Nevertheless the product $\overline{M} \times \overline{F}$ might still increase over successive equilibria as new alleles are introduced one at a time. In order to test the validity of this conjecture, we ran simulations with two classes of fitness sets in which the fitness pairs associated with different genotypes were chosen at random. The main conclusions are the following:

(iii) With convex fitness sets, the product $\overline{M} \times \overline{F}$ tends rather rapidly toward its optimum value $\hat{M} \times \hat{F}$ over successive equilibria as new alleles are introduced.

(iv) Non-convex fitness sets favor more alleles represented at equilibrium (i.e., more polymorphism) and more equilibrium changes as new alleles are introduced. Moreover, though rare and small, a decrease of $\overline{M} \times \overline{F}$ over two successive equilibria is possible. (This is an evolutionary attractiveness property for $\overline{M} \times \overline{F}$ with small fluctuations.) But note that such a decrease never occurred from a symmetric equilibrium.

Therefore, even without genetic constraints (e.g., finite number of genes or heterozygote advantages), the evolution toward an EGS state, which is optimal with respect to initial increase properties, may be subject to fluctuations even in one-locus models. Moreover the attainment of an EGS state is quite uncertain for multilocus systems with recombination and frequency-dependent selection (to which initial increase properties can be extended mutatis mutandis) because of the possibility of limit cycles. There may even be a contradiction between initial increase properties with respect to mutant alleles and stability with respect to existing alleles as a consequence of linkage. This will be illustrated in the following section.

3.4. Sex-linked meiotic drive and the evolution of the sex ratio

Maffi and Jayakar (1981) introduced a two-locus model of sex-linked meiotic drive with possible applications to the mosquito *Aedes aegypti* in order to study the influence of recombination on the maintenance of polymorphic equilibria in the

---

[2] Includes every symmetric equilibrium and every equilibrium with (M_{kl}, F_{kl}) represented at equilibrium lying on a straight line.

absence of fitness differences. A primary locus with two possible alleles, M and m, is assumed to be responsible for sex determination: males are Mm while females are mm. A secondary locus allowing multiple alleles $A^{(1)}, \ldots, A^{(n)}$ governs meiotic drive at the primary locus: $MA^{(k)}/mA^{(l)}$ genotypes segregate gametes carrying alleles M and m with frequencies s_{kl} and $1 - s_{kl}$, respectively. Hence s_{kl} is the proportion of males in the progeny. It is assumed that $s_{kl} = s_{lk}$ for all k, l. Moreover a recombination event between the primary locus and the secondary locus occurs with probability r $(0 < r \leqslant \frac{1}{2})$ prior to the meiotic drive effects. Therefore a typical male $MA^{(k)}/mA^{(l)}$ transmits the gametes $MA^{(k)}$, $MA^{(l)}$, $mA^{(k)}$ and $mA^{(l)}$ with frequencies $(1 - r)s_{kl}$, rs_{kl}, $r(1 - s_{kl})$ and $(1 - r)(1 - s_{kl})$, respectively, while the gametes $mA^{(k)}$ and $mA^{(l)}$ are equally represented in the gametic production of a typical female $mA^{(k)}/mA^{(l)}$.

In generic cases, we have the following properties for the dynamics of this system (Lessard, 1987):

(i) For $r = \frac{1}{2}$, the sex ratio, namely, the proportion of females in the population, comes closer to $\frac{1}{2}$ over two successive attainable equilibria following the invasion of a mutant allele at a geometric rate.

(ii) At least for $\frac{1}{4}(3 - \sqrt{3}) < r < \frac{1}{2}$, a Hardy–Weinberg equilibrium (i.e., an equilibrium with frequencies of $A^{(k)}$–carrying gametes for all k independent of the allele present at the M–m locus) associated with a sex ratio $\frac{1}{2}$ is always stable with respect to existing alleles (i.e., always internally stable) but invaded by any mutant allele (i.e., always externally unstable). In general, a Hardy–Weinberg equilibrium associated with a sex ratio closer to $\frac{1}{2}$ is more likely to be internally stable but at the same time more likely to be externally unstable. More precisely, if for example the equilibrium sex ratio is $s^* < \frac{1}{2}$, then the equilibrium is internally stable for $r_1(s^*) < r < r_2(s^*)$ with $r_2(s^*) \to \frac{1}{2}$ as $s^* \to \frac{1}{2}$ and $r_1(s^*) \to r_0 \leqslant (3 - \sqrt{3})/4$ as $s^* \to \frac{1}{2}$, while it is externally unstable if the mutant marginal sex ratio $\tilde{s}$ satisfies $\tilde{s} > s^*$ or $\tilde{s} < s^*$ and $r < (s^* - \tilde{s})/(1 - 2\tilde{s})$ which tends to $\frac{1}{2}$ as $s^* \to \frac{1}{2}$. Moreover, at $r = r_1(s^*)$ there is a Hopf bifurcation to a periodic orbit while at $r = r_2(s^*)$ there is a bifurcation to non-Hardy–Weinberg equilibrium points.

(iii) For r small but positive, limit cycles can exist.

These results can be explained as follows: overdominance at the primary locus promotes the increase in frequency of any new sex ratio modifier at the secondary locus by the hitchhiking effects of linkage while the modifiers themselves still favor the occurrence of an even sex ratio in the population. When $r = \frac{1}{2}$, the hitchhiking effects are inoperant while for r small they are predominant.

Therefore we have to be circumspect with general evolutionary principles since many forces can be acting together or in opposite directions in a population. In particular, evolution cannot be always predicted from initial increase properties sometimes because of genetic mechanisms sometimes because of complex interactions which make vain long-term predictions. At least maximization principles have to be weakened as complexity increases. This is particularly true for two-sex populations.

4. Multifactorial sex determination models

4.1. General description

Many species do not have strong sex determiners (e.g., sex chromosomes) and sex determination by one or several autosomal genes — or a large number of autosomal genes — may be influenced by environmental factors (e.g., incubation temperature, available nutrient, ambiant light, humidity). This is the case for instance for fish, invertebrates (e.g., worms) and reptiles (e.g., turtles, crocodiles).

For such *multifactorial models*, we consider a continuous one-dimensional index variable Z and a *sex determination function* F such that, if $Z = z$ for an offspring, then $F(z)$ is the probability of becoming female and $1 - F(z)$ the probability of becoming male for that offspring. It will be assumed throughout that $F(z)$ is increasing with respect to the index value z.

If $g_t(z)$ is the density of Z in an infinite population at the beginning of generation t, denoted Z_t, such that the fraction of offspring with an index value in the range $(z, z + \Delta z)$ is approximately $g_t(z)\Delta z$, then the densities $g_t^*(z)$ and $g_t^{**}(z)$ of the index variable Z in the female and male populations at the end of generation t, denoted Z_t^* and Z_t^{**}, respectively, are given by

$$g_t^*(z) = F(z)g_t(z)/\overline{F}_t , \tag{4.1a}$$

$$g_t^{**}(z) = [1 - F(z)]g_t(z)/(1 - \overline{F}_t) , \tag{4.1b}$$

where

$$\overline{F}_t = \int F(z)g_t(z)\,\mathrm{d}z . \tag{4.2}$$

The quantity $\overline{F}_t$ is the proportion of females, the sex ratio, in the population at the end of generation t.

The index variable at the beginning of the next generation is assumed to be the result of a 'blending' of the parental index variables plus a residual addend. Formally, we have the model

$$Z_{t+1} = h^*Z_t^* + h^{**}Z_t^{**} + (1 - h^* - h^{**})\varepsilon_t , \tag{4.3}$$

where Z_t^* and Z_t^{**} are the maternal and paternal contributions, respectively, and ε_t is an environmental-mutational term with density function independent of the generation t. The variables Z_t^*, Z_t^{**} and ε_t are assumed to be mutually independent. The coefficients h^* and h^{**} can be viewed as constant heritability coefficients for the female and male parents, respectively. It is assumed that $h^* \geqslant 0$, $h^{**} \geqslant 0$ and $h^* + h^{**} \leqslant 1$.

The formulation of model (4.3) must be understood in distribution. Some of the main difficulties in the analysis of model (4.3) are that the most common distributions are not preserved (e.g., Z_{t+1} is not necessarily Gaussian if Z_t and ε_t are Gaussian) and the sex ratio $\overline{F}_t$ is not generally monotone over successive generations.

4.2. Galtonian model with no residual term

In this section, we consider the Galtonian model

$$Z_{t+1} = \tfrac{1}{2}(Z_t^* + Z_t^{**}),\tag{4.4}$$

with symmetric contributions from the female and male parents and no residual addend. In this case, we have for the expected value of the index variable

$$E(Z_{t+1}) = \tfrac{1}{2}[E(Z_t^*) + E(Z_t^{**})]$$

$$= \frac{1}{2\overline{F}_t}\int zF(z)g_t(z)\,\mathrm{d}z + \frac{1}{2(1-\overline{F}_t)}\int z[1 - F(z)]g_t(z)\,\mathrm{d}z$$

$$= \frac{E(Z_t)}{2(1-\overline{F}_t)} + \left\{\frac{1 - 2\overline{F}_t}{2\overline{F}_t(1-\overline{F}_t)}\right\}\int zF(z)g_t(z)\,\mathrm{d}z$$

$$= E(Z_t) + \left\{\frac{1 - 2\overline{F}_t}{2\overline{F}_t(1-\overline{F}_t)}\right\}\int [z - E(Z_t)]F(z)g_t(z)\,\mathrm{d}z$$

$$\begin{cases} \geqslant E(Z_t) & \text{if } \overline{F}_t < \tfrac{1}{2}, \\ \leqslant E(Z_t) & \text{if } \overline{F}_t > \tfrac{1}{2}, \end{cases}\tag{4.5}$$

since

$$\int [z - E(Z_t)]F(z)g_t(z)\,\mathrm{d}z$$

$$\geqslant \left\{\int [z - E(Z_t)]g_t(z)\,\mathrm{d}z\right\}\left\{\int F(z)g_t(z)\,\mathrm{d}z\right\} = 0,\tag{4.6}$$

by virtue of the Tchebycheff rearrangement inequality provided $F(z)$ is increasing with z (see, e.g., Karlin, 1968). This suggests an evolutionary tendency toward $\overline{F}_t = \tfrac{1}{2}$.

With a threshold criterion for sex determination in the form

$$F(z) = \begin{cases} 0 & \text{for } z < z^*, \\ 1 & \text{for } z \geqslant z^*, \end{cases}\tag{4.7}$$

and an initial distribution which is log-concave, that is, a density for Z_0 in the form

$$g_0(z) = c\,\mathrm{e}^{-\phi(z)},\tag{4.8}$$

where c is a constant and ϕ is a convex function (e.g., a Gaussian density, a gamma density, a Student density, a density of an order statistic associated with a uniform density; see Karlin, 1968), the following result can be proved:

THEOREM 4.1. *Z_t converges in distribution to the threshold value z^*.*

The proof of Theorem 4.1 relies on the facts that log-concave densities are preserved under the transformation (4.4) with the threshold criterion (4.7) and that the variance of a truncated log-concave density is smaller than the variance of the non-truncated variable (see Karlin and Lessard, 1986).

4.3. Purely parental transmission

For the model

$$Z_{t+1} = hZ_t^* + (1-h)Z_t^{**}, \tag{4.9}$$

with index value in $[0, 1]$ and sex determination function

$$F(z) = z \quad \text{for } 0 \leqslant z < 1, \tag{4.10}$$

we have $\overline{F}_t = E(Z_t)$ and

$$\overline{F}_{t+1} = \frac{h}{\overline{F}_t} \int z^2 g_t(z)\, dz + \left\{ \frac{1-h}{1-\overline{F}_t} \right\} \int z[1-z] g_t(z)\, dz, \tag{4.11}$$

from which

$$\overline{F}_{t+1} - \overline{F}_t = \left\{ \frac{h - \overline{F}_t}{\overline{F}_t(1-\overline{F}_t)} \right\} (s_t^2 - \overline{F}_t^2), \tag{4.12}$$

where

$$s_t^2 = \int z^2 g_t(z)\, dz. \tag{4.13}$$

Since

$$s_t^2 - \overline{F}_t^2 = \text{Var}(Z_t) \geqslant 0, \tag{4.14}$$

with equality if and only if Z_t is degenerate to a single value, we have

$$\overline{F}_{t+1} \begin{cases} \geqslant \overline{F}_t, & \text{if } \overline{F}_t < h, \\ \leqslant \overline{F}_t, & \text{if } \overline{F}_t > h, \end{cases} \tag{4.15}$$

with equality if and only if Z_t is degenerate to a single value.

On the other hand, since

$$s_t^2 = \mathrm{E}(Z_t^2) \leqslant \mathrm{E}(Z_t) = \bar{F}_t \tag{4.16}$$

for a variable Z_t taking its values in $[0, 1]$ with equality if and only if Z_t is degenerate to 0 or 1, we have

$$\bar{F}_{t+1} \begin{cases} \leqslant h & \text{if } \bar{F}_t < h, \\ \geqslant h & \text{if } \bar{F}_t > h, \end{cases} \tag{4.17}$$

with equality if and only if Z_t is degenerate to 0 or 1.

Finally we have the following result.

THEOREM 4.2. Z_t *converges in distribution to a single value*

$$z^* \begin{cases} \leqslant h & \text{if } \bar{F}_0 < h, \\ \geqslant h & \text{if } \bar{F}_0 > h, \end{cases} \tag{4.18}$$

as $t \to \infty$.

(See Karlin and Lessard, 1986, for details.)

4.4. Symmetric parental transmission with residual addend

In this section, we consider the model

$$Z_{t+1} = h\{\tfrac{1}{2}(Z_t^* + Z_t^{**})\} + (1 - h)\varepsilon_t, \tag{4.19}$$

with index value in $[0, 1]$ and sex determination function

$$F(z) = z \quad \text{for } 0 \leqslant z \leqslant 1. \tag{4.20}$$

Moreover we assume $0 \leqslant \varepsilon_t \leqslant 1$ with $\mathrm{E}(\varepsilon_t) = \tfrac{1}{2}$ and $\mathrm{Var}(\varepsilon_t) > 0$.

In this model, we find

$$\bar{F}_{t+1} - \bar{F}_t = \left\{ \frac{1 - 2\bar{F}_t}{2\bar{F}_t(1 - \bar{F}_t)} \right\} [h(s_t^2 - \bar{F}_t^2) + (1 - h)(\bar{F}_t - \bar{F}_t^2)], \tag{4.21}$$

where

$$0 < s_t^2 - \bar{F}_t^2 < \bar{F}_t - \bar{F}_t^2, \tag{4.22}$$

since Z_t cannot be degenerate to a single value for $t \geqslant 1$. Therefore

$$\overline{F}_t < \overline{F}_{t+1} < \tfrac{1}{2} \quad \text{if } \overline{F}_t < \tfrac{1}{2}, \tag{4.23a}$$

$$\overline{F}_t > \overline{F}_{t+1} > \tfrac{1}{2} \quad \text{if } \overline{F}_t > \tfrac{1}{2}. \tag{4.23b}$$

Using the method of moments for uniformly bounded variables, the following theorem can be shown (Karlin and Lessard, 1986).

THEOREM 4.3. Z_t *converges in distribution to a variable with mean* $\tfrac{1}{2}$.

5. Multivariate Gaussian models for two-sex populations

5.1. *Formulation of a multivariate Galtonian model*

Multivariate Gaussian models for multifactorial inheritance traits are widespread in quantitative genetics and usually justified when multiple additive and independent factors of small effects are acting on phenotype and/or on fitness. We will focus attention on the two-sex Galtonian model

$$Z_{t+1} = H\{\tfrac{1}{2}(Z_t^* + Z_t^{**})\} + (I - H)\mu_t + \zeta_t, \tag{5.1}$$

where μ_t is the mean of a d-dimensional trait Z_t in offspring of generation t, Z_t^* and Z_t^{**} are the maternal and paternal traits, respectively, obtained from Z_t by applying the female and male fitness functions

$$F(z) = \exp\left[-\tfrac{1}{2}(z - \theta^*)^T \Gamma^{*-1}(z - \theta^*)\right], \tag{5.2a}$$

$$M(z) = \exp\left[-\tfrac{1}{2}(z - \theta^{**})^T \Gamma^{**-1}(z - \theta^{**})\right], \tag{5.2b}$$

with Γ^* and Γ^{**} being symmetric positive definite matrices, H is a mid-parent-offspring trait transmission matrix which is assumed to be a constant positive multiple of I (I for the identity matrix) such that $I - H$ is also a positive multiple of I and ζ_t is an environmental-mutational addend. It will be assumed throughout this section that Z_t^*, Z_t^{**} and ζ_t are independent contributions to Z_{t+1} and ζ_t has a multivariate normal distribution with mean $\mathbf{0}$ and variance–covariance matrix Δ, denoted $\zeta_t \sim N(\mathbf{0}, \Delta)$. Then, if Z_t has a multivariate normal distribution with mean μ_t and variance–covariance matrix Σ_t, we have

$$Z_t^* \sim N(\mu_t^*, \Sigma_t^*), \tag{5.3a}$$

$$Z_t^{**} \sim N(\mu_t^{**}, \Sigma_t^{**}), \tag{5.3b}$$

with

$$\Sigma_t^* = (\Sigma_t^{-1} + \Gamma^{*-1})^{-1}, \tag{5.4a}$$

$$\Sigma_t^{**} = (\Sigma_t^{-1} + \Gamma^{**-1})^{-1}, \tag{5.4b}$$

$$\mu_t^* = (\Sigma_t^{-1} + \Gamma^{*-1})^{-1} (\Sigma_t^{-1}\mu_t + \Gamma^{*-1}\theta^*), \tag{5.4c}$$

$$\mu_t^{**} = (\Sigma_t^{-1} + \Gamma^{**-1})^{-1} (\Sigma_t^{-1}\mu_t + \Gamma^{**-1}\theta^{**}), \tag{5.4d}$$

and

$$Z_{t+1} \sim N(\mu_{t+1}, \Sigma_{t+1}), \tag{5.5}$$

with

$$\Sigma_{t+1} = H\{\tfrac{1}{2}(\Sigma_t^* + \Sigma_t^{**})\}H^T + \Delta, \tag{5.6a}$$

$$\mu_{t+1} = \mu_t + H\{\tfrac{1}{2}(\mu_t^* + \mu_t^{**}) - \mu_t\}. \tag{5.6b}$$

The matrix transformation

$$T(\Sigma_t) = \Sigma_{t+1} \tag{5.7}$$

is *strictly monotone*, that is,

$$T(\Sigma_2) > T(\Sigma_1) \quad \text{if } \Sigma_2 > \Sigma_1 \tag{5.8}$$

$(T(\Sigma_2) - T(\Sigma_1)$ is positive definite if $\Sigma_2 - \Sigma_1$ is positive definite) and *strictly concave*, that is,

$$T\{\tfrac{1}{2}(\Sigma_2 + \Sigma_1)\} > \tfrac{1}{2}[T(\Sigma_2) + T(\Sigma_1)] \quad \text{if } \Sigma_2 > \Sigma_1. \tag{5.9}$$

In such circumstances, we have the following result.

THEOREM 5.1 (Karlin, 1979)

$$\Sigma_t \to \Sigma_\infty \quad as \ t \to \infty, \tag{5.10}$$

where Σ_∞ is a positive definite matrix.

5.2. Increase of the product of the female and male mean fitnesses in Gaussian models

Assuming the Gaussian model (5.1), the mean fitness in the female offspring of generation $t + 1$ is

$$\overline{F}_{t+1} = \frac{1}{(2\pi)^{d/2}\sqrt{\det(\Sigma_{t+1})}} \int \exp\left[-\tfrac{1}{2}(z - \mu_{t+1})^\mathrm{T}\Sigma_{t+1}^{-1}(z - \mu_{t+1})\right]$$

$$\times \exp\left[-\tfrac{1}{2}(z - \theta^*)^\mathrm{T}\Gamma^{*-1}(z - \theta^*)\right]\mathrm{d}z$$
(5.11)

(det for determinant).

Assuming $\Sigma_{t+1} = \Sigma_t = \Sigma_\infty$, a Taylor series expansion of $\ln\overline{F}_{t+1}$ with respect to μ_{t+1} around μ_t yields

$$\ln\overline{F}_{t+1} = \ln\overline{F}_t + (\mu_t^* - \mu_t)^\mathrm{T}\Sigma_\infty^{-1}(\mu_{t+1} - \mu_t)$$

$$- \tfrac{1}{2}(\mu_{t+1} - \mu_t)^\mathrm{T}(\Sigma_\infty + \Gamma^*)^{-1}(\mu_{t+1} - \mu_t)$$
(5.12)

(see, e.g., Lande, 1979).

Similarly, for the mean fitness in male offspring, we have

$$\ln\overline{M}_{t+1} = \ln\overline{M}_t + (\mu_t^{**} - \mu_t)^\mathrm{T}\Sigma_\infty^{-1}(\mu_{t+1} - \mu_t)$$

$$- \tfrac{1}{2}(\mu_{t+1} - \mu_t)^\mathrm{T}(\Sigma_\infty + \Gamma^{**})^{-1}(\mu_{t+1} - \mu_t).$$
(5.13)

Therefore,

$$\Delta\ln(\overline{F}_t \times \overline{M}_t) = \ln(\overline{F}_{t+1} \times \overline{M}_{t+1}) - \ln(\overline{F}_t \times \overline{M}_t)$$

$$= \ln\overline{F}_{t+1} + \ln\overline{M}_{t+1} - \ln\overline{F}_t - \ln\overline{M}_t$$

$$= (\mu_t^* + \mu_t^{**} - 2\mu_t)^\mathrm{T}\Sigma_\infty^{-1}(\mu_{t+1} - \mu_t)$$

$$- \tfrac{1}{2}(\mu_{t+1} - \mu_t)^\mathrm{T}[(\Sigma_\infty + \Gamma^*)^{-1}$$

$$+ (\Sigma_\infty + \Gamma^{**})^{-1}](\mu_{t+1} - \mu_t)$$

$$\geq (\mu_{t+1} - \mu_t)^\mathrm{T}(H\Sigma_\infty)^{-1}(\mu_{t+1} - \mu_t),$$
(5.14)

using

$$\mu_{t+1} - \mu_t = H\{\tfrac{1}{2}(\mu_t^* + \mu_t^{**}) - \mu_t\}$$
(5.15a)

and

$$(\Sigma_\infty + \Gamma^*)^{-1} + (\Sigma_\infty + \Gamma^{**})^{-1}$$

$$\leq 2\Sigma_\infty^{-1} = 2[H\Sigma_\infty + (I - H)\Sigma_\infty]^{-1} \leq 2(H\Sigma_\infty)^{-1}.$$
(5.15b)

Finally, we use the fact that the inverse of a positive multiple of a symmetric positive definite matrix is symmetric positive definite to conclude that

$$\Delta \ln (\overline{F}_t \times \overline{M}_t) \geqslant 0 , \tag{5.16}$$

with equality if and only if

$$\Delta \mu_t = \mu_{t+1} - \mu_t = 0 \tag{5.17}$$

(see Lande, 1980, for a similar result with different assumptions taking into account gene transmission in Mendelian populations). We conclude with the following result.

THEOREM 5.2. *In the multivariate Gaussian model* (5.1) *with an equilibrium variance–covariance structure* (*see Theorem* 5.1), *the product of the female and male mean fitnesses increases over successive generations until equilibrium is reached.*

References

Akin, E. (1979). *The Geometry of Population Genetics*. Lecture Notes in Biomathematics. Vol. 31, Springer-Verlag, Berlin.

Benford, F. A. (1978). Fisher's theory of the sex ratio applied to the social hymenoptera. *J. Theor. Biol.* **72**, 701–727.

Charnov, E. L. (1982). *The Theory of Sex Allocation*. Princeton University Press, Princeton.

Eshel, I. (1975). Selection on sex ratio and the evolution of sex determination. *Heredity* **34**, 351–361.

Eshel, I. and M. W. Feldman (1982). On evolutionary genetic stability of the sex ratio. *Theor. Pop. Biol.* **21**, 430–439.

Ewens, W. J. (1979). *Mathematical Population Genetics*. Springer-Verlag, Berlin.

Fisher, R. A. (1918). The correlation between relatives on the supposition of Mendelian inheritance. *Trans. R. Soc. Edinburgh* **52**, 399–433.

Fisher, R. A. (1930). *The Genetical Theory of Natural Selection*. Clarendon Press, London.

Fisher, R. A. (1941). Average excess and average effect of a gene substitution. *Ann. Eugen. Lond.* **11**, 53–63.

Hamilton, W. D. (1964). The genetical evolution of social behaviours I. *J. Theor. Biol.* **7**, 1–16.

Hamilton, W. D. (1967). Extraordinary sex ratios. *Science* **156**, 477–488.

Hamilton, W. D. (1972). Altruism and related phenomena, mainly in social insects. *Ann. Rev. Ecol. Syst.* **3**, 193–232.

Karlin, S. (1968). *Total Positivity*. Stanford University Press, Stanford.

Karlin, S. (1978). Theoretical aspects of multilocus selection balance I. In: S. A. Levin, ed., *Mathematical Biology, Part II: Populations and Communities*, MAA Studies in Mathematics, Vol. 16, Math. Assoc. of Amer., Washington, DC, 503–587.

Karlin, S. (1979). Models of multifactorial inheritance. I. Multivariate formulations and basic convergence results. *Theor. Pop. Biol.* **15**, 308–355.

Karlin, S. and S. Lessard (1983). On the optimal sex ratio. *Proc. Natl. Acad. Sci. USA* **80**, 5931–5935.

Karlin, S. and S. Lessard (1986). *Theoretical Studies on Sex Ratio Evolution*. Princeton University Press, Princeton.

Karlin, S. and H. M. Taylor (1975). *A First Course in Stochastic Processes*. Academic Press, New York.

Kingman, J. F. C. (1961a). A matrix inequality. *Quart. J. Math.* **12**, 78–80.

Kingman, J. F. C. (1961b). A mathematical problem in population genetics. *Proc. Cambridge Philos. Soc.* **57**, 574–582.

Lande, R. (1979). Quantitative genetic analysis of multivariate evolution, applied to brain: Body size allometry. *Evolution* **33**, 402–416.

Lande, R. (1980). Sexual dimorphism, sexual selection, and adaptation in polygenic characters. *Evolution* **34**, 292–303.

Lessard, S. (1984). Evolutionary dynamics in frequency-dependent two-phenotype models. *Theor. Pop. Biol.* **25**, 210–234.

Lessard, S. (1986). Evolutionary principles for general frequency-dependent two-phenotype models in sexual populations. *J. Theor. Biol.* **119**, 329–344.

Lessard, S. (1987). The role of recombination and selection in the modifier theory of sex-ratio distortion. *Theor. Pop. Biol.* **31**, 339–358.

Lessard, S. (1989). Resource allocation in Mendelian populations: Further in ESS theory. In: M. W. Feldman, ed., *Mathematical Evolutionary Theory,* Princeton University Press, Princeton, 207–246.

Li, C. C. (1967). Fundamental theorem of natural selection. *Nature* **214**, 505–506.

MacArthur, R. H. (1965). Ecological consequences of natural selection. In: T. H. Waterman and H. Morowitz, eds., *Theoretical and Mathematical Biology,* Blaisdell, New York, 388–397.

Maffi, G. and S. D. Jayakar (1981). A two-locus model for polymorphism for sex-linked meiotic drive modifiers with possible applications to *Aedes aegypti. Theor. Pop. Biol.* **19**, 19–36.

Malécot, G. (1948). *Les Mathématiques de l'Hérédité.* Masson, Paris.

Maynard Smith, J. (1982). *Evolution and the Theory of Games.* Cambridge University Press, Cambridge.

Michod, R. E. and W. D. Hamilton (1980). Coefficients of relatedness in sociobiology. *Nature* **288**, 694–697.

Nagylaki, T. (1987). Evolution under fertility and viability selection. *Genetics* **115**, 367–375.

Oster, G., Eshel, I. and D. Cohen (1977). Worker-queen conflict and the evolution of social insects. *Theor. Pop. Biol.* **12**, 49–85.

Price, G. R. (1970). Selection and covariance. *Nature* **227**, 520–521.

Shaw, R. F. and J. D. Mohler (1953). The selective advantage of the sex ratio. *Am. Natur.* **87**, 337–342.

Speith, P. T. (1974). Theoretical considerations of unequal sex ratios. *Am. Natur.* **108**, 834–849.

Taylor, P. D. (1988). Inclusive fitness models with two sexes. *Theor. Pop. Biol.* **34**, 145–168.

Uyenoyama, M. K. and B. O. Bengtsson (1979). Towards a genetic theory for the evolution of the sex ratio. *Genetics* **93**, 721–736.

Uyenoyama, M. K., Feldman, M. W. and L. D. Muller (1981). Population genetic theory of kin selection. I. Multiple alleles at one locus. *Proc. Natl. Acad. Sci. USA* **78**, 5036–5040.

C. R. Rao and R. Chakraborty, eds., *Handbook of Statistics, Vol. 8*
© Elsevier Science Publishers B.V. (1991) 373–393

12

Stochastic Models of Carcinogenesis*

Suresh H. Moolgavkar

1. Introduction

In the four decades since the introduction of the first stochastic models for carcinogenesis, our understanding of the processes underlying malignant transformation has increased considerably. When the first models were proposed by Nordling (1953) and by Armitage and Doll (1954), molecular biology was in its infancy. Some forty years later, molecular biology has triumphantly elucidated the precise mechanism underlying a few cancers. There are tantalizing hints that similar mechanisms underlie many other cancers as well. Yet, the basic assumption — that a malignant tumor arises from a single cell (or from a very few cells) that has sustained a small number of critical insults to its genetic apparatus — on which these early models were predicated remains valid today. It is the fact that cancer is a disease of single cells rather than entire organ systems that makes the mathematical modelling of carcinogenesis feasible.

The first stochastic models (Nordling, 1953; Armitage and Doll, 1954, 1957) were advanced to explain the epidemiologic observation that the age-specific incidence rates of many common adult carcinomas increase roughly with a power of age. Among these models, the multistage model, proposed by Armitage and Doll in 1954, has been widely used for the analysis of epidemiologic data and for cancer risk assessment. From the statistical point of view, this model provides a large and flexible class of hazard functions for the analysis of data. I shall discuss the Armitage–Doll multistage model in some detail.

In 1971, on the basis of epidemiologic observations, Knudson proposed a model for retinoblastoma, a rare tumor of the retina in children. Recent work in molecular biology leaves little doubt that the salient features of Knudson's model for retinoblastoma are correct (Cavanee et al., 1983). This, so-called recessive oncogenesis, model has been generalized and applied to both epidemiologic data in human populations and experimental data in animals by Moolgavkar and collaborators (Moolgavkar, Day and Stevens, 1980; Moolgavkar and Knudson, 1981; Moolgavkar, 1986). In addition to retinoblastoma, the biological under-

* Supported by grant CA-47658 from NIH and contract DOE DE-FG06-88GR60657.

pinnings of this model have been substantiated for some other tumors (e.g., Koufos et al., 1984). These are the triumphs of molecular biology alluded to in the opening paragraph. I shall devote the greater part of this chapter to a discussion of this model.

Why model? There are at least two reasons. First, a model of carcinogenesis that is consistent with all the existing data could have heuristic value by stimulating new and discriminating observations. Second, a biologically realistic mathematical model for carcinogenesis is one component of a scientific approach to cancer risk assessment. Vital regulatory decisions regarding permissible human exposures to various agents need to be made. Because relevant human epidemiologic data are rarely available, permissible levels of exposure are determined on the basis of experiments in which animals are exposed to very high levels of the agent in question. This leads to the problems of inter-species and low-dose extrapolations. Such extrapolations are best done within the framework of a reasonable model of carcinogenesis. The Armitage–Doll model has been extensively used in risk assessment. Indeed the United States Environmental Protection Agency routinely uses the Armitage–Doll model. Recently several investigators have proposed that the recessive oncogenesis model be used on the grounds that it is biologically more realistic (Thorslund, Brown and Charnley, 1987; Wilson, 1986).

In this chapter I shall focus attention on these two models. It is not my intention to give a historical recounting of the literature on stochastic carcinogenesis models. I believe that models are useful only insofar as they are biologically realistic and suggest meaningful experiments and epidemiologic studies. Although many models fit one or another data set they do not satisfy these criteria. In particular, many different models generate age-specific incidence rates that increase roughly with a power of age (Pike, 1966). Thus any model that is incapable of generating such incidence curves must immediately be rejected. However, the ability to generate such curves is not strong evidence in favor of the model. For an excellent review of the multistage model and some estensions of it, I refer the reader to Whittemore and Keller (1977). A less mathematical account is given by Armitage (1985) and fits to various data sets are discussed by Freedman and Navidi (1989). A biological discussion is given in Peto (1977).

Data to which carcinogenesis models can be fit are generally derived from two sources. As stated earlier, the first models were proposed to explain cancer incidence data, and much of the early model fitting was to incidence and mortality data from various population-based cancer registries. More recently models have been fit to other types of epidemiologic data, such as cohort and case-control data. Quantitative data are also available from animal carcinogenesis experiments, particularly from what are called initiation–promotion experiments. I will discuss some of the statistical issues in fitting models to these various data sets. A crucial difference between the human and animal data is that, in human populations, cancer is a rare disease whereas animal experiments are often conducted under conditions such that a large proportion of animals develop tumors and many animals develop multiple tumors of the same site. When models are fit to human data, the analysis is greatly simplified by making a mathematical approximation the adequacy of which depends upon the tumor being rare. This approximation is no longer valid in the experimental situation in which tumors cannot be assumed to be rare. And yet, this approximation continues to be widely used for the analysis of animal data.

The following two fundamental assumptions underlie the models I shall consider here:

(1) Cancers are clonal, i.e., malignant tumors arise from a single malignant progenitor cell.

(2) Each susceptible (stem) cell in a tissue is as likely to become malignant as any other, and the process of malignant transformation in a cell is independent of that in any other cell.

2. The Armitage–Doll multistage model

The Armitage–Doll model, which has been extensively used in the last three decades, was first proposed to explain the observation that, in many human carcinomas, the age-specific incidence rates increase roughly with a power of age. The age-specific incidence rate is a measure of the rate of appearance of tumors in a previously tumor free tissue. The appropriate statistical concept is that of the hazard function.

For the tissue of interest, let T be a random variable representing time to appearance of a malignant tumor, and let

$$P(t) = \Pr\{T \leqslant t\} \, .$$

Then the hazard function $h(t)$ is defined by

$$h(t) = \lim_{\Delta t \to 0} \frac{1}{\Delta t} \Pr\{t < T \leqslant t + \Delta t \,|\, T > t\}$$

$$= P'(t)/[1 - P(t)] \, .$$

Suppose that there are N cells susceptible to malignant transformation in the tissue of interest, and that $T_1, T_2, \ldots, T_N$ are independent identically distributed random variables representing the time to transformation of the individual cells. Then, $T = \min\{T_1, T_2, \ldots, T_N\}$, i.e., T is a minimum order statistic. For a given susceptible cell, let $p(t)$ be the probability of malignant transformation by time t. An easy computation shows that the hazard function $h(t)$ for the tissue is given by the expression

$$h(t) = Np'(t)/[1 - p(t)] \, .$$

The Armitage–Doll model postulates that a malignant tumor arises in a tissue when a single susceptible cell in that tissue undergoes malignant transformation via a finite sequence of intermediate stages, the waiting time between any stage and the subsequent one being exponentially distributed. Schematically, the model may be represented as follows:

$$E_0 \xrightarrow{\lambda_0} E_1 \xrightarrow{\lambda_1} E_2 \xrightarrow{\lambda_2} \cdots \xrightarrow{\lambda_{n-1}} E_n \,.$$

Here E_0 represents the normal cell, E_n represents the malignant cell, and the λ_j represent the parameters of the (exponential) waiting time distributions.

Let $p_j(t)$ represent the probability that a given cell is in stage E_j by time t. Then, $p_n(t) = p(t)$ is the probability that the cell is malignantly transformed by time t, and the expression for the hazard, $h(t)$, can be rewritten as

$$Np'_n(t)/[1 - p_n(t)] \,. \tag{2.1}$$

At the level of the single cell, malignancy is a very rare phenomenon. Thus, for any cell $p_n(t)$ is very close to zero during the life span of an individual, and $h(t)$ is approximately equal to $Np'_n(t)$. Thus, in order to obtain an approximation to $h(t)$, we need only compute $p'_n(t)$.

An explicit expression for $p'_n(t)$ can be obtained in one of two ways. It is not difficult to see that the Kolmogorov forward differential equations for $p_j(t)$ are

$$p'_0(t) = -\lambda_0 p_0(t) \,,$$
$$\vdots$$
$$p'_k(t) = \lambda_{k-1} p_{k-1}(t) - \lambda_k p_k(t) \,, \tag{2.2}$$
$$\vdots$$
$$p'_n(t) = \lambda_{n-1} p_{n-1}(t) \,,$$

with initial condition $p_0(0) = 1$, $p_j(0) = 0$ for $j > 0$. These equations can be explicitly integrated. A second approach is the following. Let S_j represent the waiting time for the cell to go from stage E_j to E_{j+1}, and let $S = S_0 + S_1 + \cdots + S_{n-1}$. Then clearly, $p_n(t) = \Pr\{S \leqslant t\}$, and the problem of computing $p'_n(t)$ reduces to convolving the densities of n independent exponentially distributed random variables. If no two of the transition rates are equal (i.e., if $\lambda_i = \lambda_j \Rightarrow i = j$), then (see, e.g., Feller, 1971, p. 40)

$$p'_n(t) = \lambda_0 \lambda_1 \cdots \lambda_{n-1} \{ \Psi_{0,n-1} \exp(-\lambda_0 t) + \cdots$$
$$+ \Psi_{n-1,n-1} \exp(-\lambda_{n-1} t) \} \,, \tag{2.3}$$

where

$$\Psi_{j,n-1} = [(\lambda_0 - \lambda_j)(\lambda_1 - \lambda_j) \cdots (\lambda_{j-1} - \lambda_j)(\lambda_{j+1} - \lambda_j) \cdots (\lambda_{n-1} - \lambda_j)]^{-1} \,.$$

Now, the expression $\sum_{p=0}^{m} \Psi_{p,m} \exp(-\lambda_p t)$ may be expanded in a Taylor series about $t = 0$. Specifically, we have the following propositions (Moolgavkar, 1978).

PROPOSITION 1. *Let α_k be the coefficient of t^k in the power series about zero for $\sum_{p=0}^{m} \Psi_{p,m} \exp(-\lambda_p t)$, where the $\Psi_{p,m}$ are defined as above. Then,*

$$\alpha_k = \begin{cases} 0 & \text{if } k < m, \\ \dfrac{1}{m!} & \text{if } k = m, \\ \dfrac{(-1)^{k-m}}{k!} \displaystyle\sum_{i_1 < i_2 < \cdots < i_{k-m}} \lambda_{i_1} \lambda_{i_2} \cdots \lambda_{i_{k-m}} & \text{if } k > m. \end{cases}$$

PROOF. The proof proceeds by double induction on k and m. □

PROPOSITION 2. *The coefficients α_k in the power series above are symmetric polynomials in the indeterminants $\lambda_0, \lambda_1, \cdots, \lambda_m$ and can be expressed in terms of he $m + 1$ Newton polynomials, $\sum_{i=0}^{m} \lambda_i^j$; $1 \leqslant j \leqslant m + 1$.*

Proposition 2 says in particular that all the coefficients of the power series can be expressed in terms of the moments of the transition rates, i.e., in terms of the moments of the λ_i, $0 \leqslant i \leqslant n - 1$.

It follows immediately from Proposition 1 that the hazard function $h(t)$ can be written as

$$h(t) \approx N p_n'(t) = \frac{N \lambda_0 \lambda_1 \cdots \lambda_{n-1}}{(n-1)!} \, t^{n-1} \{1 - \bar{\lambda} t + f(\lambda, t)\}, \tag{2.4}$$

where $\bar{\lambda}$ is the mean of the transition rates, $\bar{\lambda} = \frac{1}{n} \sum_{i=0}^{n-1} \lambda_i$, and $f(\lambda, t)$ involves second and higher order products of the transition rates. If only the first non-zero term in this series expansion is used, we obtain the Armitage–Doll approximation, namely

$$h(t) \approx \frac{N \lambda_0 \lambda_1 \cdots \lambda_{n-1} t^{n-1}}{(n-1)!}. \tag{2.5}$$

Thus, with the approximations made, this model predicts an age-specific incidence curve that increases with a power of age that is one less than the number of distinct stages involved in malignant transformation.

It is immediately obvious from the model that any susceptible cell eventually becomes malignant with probability 1. Further, since the waiting time distribution to malignant transformation is the sum of n exponential waiting time distribution, it follows from a well known theorem that $h(t)$ is a monotone increasing function. Moreover, using expressions (2.1) and (2.3), a simple computation shows that $h(t)$ has a finite asymptote: $\lim_{t \to \infty} h(t) = N \lambda_{\min}$, where $\lambda_{\min}$ is the minimum of the

S. H. Moolgavkar

transition rates. Thus, the Armitage–Doll approximation becomes progressively worse with increasing age. If each of the transition rates is small enough, the Armitage–Doll approximation is an adequate description of the hazard during a normal life span.

A simple example serves to illustrate how poor the Armitage–Doll approximation can be if the transition rates are not small enough. Consider a hypothetical malignant tumor that requires seven stages for its genesis, and suppose that the transition rates (per cell per year) are 1×10^{-4}, 2×10^{-4}, 3×10^{-4}, 34×10^{-4}, 7×10^{-3}, 8×10^{-3} and 9×10^{-3} and that the number of susceptible cells in the tissue is 10^9. Table 1 compares the age-specific incidence rates predicted by the Armitage–Doll model with those obtained by retaining the first two terms in the power series expansion above.

Table 1
Comparison of the age-specific incidence rates predicted by the Armitage–Doll approximation and those predicted by retention of the second term in expression (2.4)

Age (yr)	Age-specific incidence rates (per million)		Difference as a percentage of the Armitage–Doll approximation
	Armitage–Doll approximation	Retention of second term	
25	3	3	0
30	10	9	10.0
35	26	23	11.5
40	58	49	15.5
45	119	97	18.5
50	223	178	20.2
55	395	308	22.0
60	666	506	24.0
65	1077	797	26.0
70	1680	1210	28.0
75	2542	1779	30.0
80	3743	2546	32.0

For future reference let me note here that the hazard function can be viewed as follows. Let $X_i(t)$, $i = 1, 2, \ldots, n$, be a sequence of random variables associated with each cell with $X_i(t) = 1$ if the cell is in stage i at time t and 0 otherwise. Then it is clear from the Kolmogorov equations that

$$h(t) = N p_n'(t) / [1 - p_n(t)] = N \, \mathrm{E}[X_{n-1}(t) | X_n(t) = 0] \,, \qquad (2.6)$$

where E denotes the expectation. When $p_n(t)$ is close to zero, or equivalently, the transition rates are small enough,

$$h(t) \approx N p_{n-1}(t) = N \, \mathrm{E}[X_{n-1}(t)] \, .$$

Thus the Armitage–Doll approximation consists of replacing the conditional expectation of X_{n-1} by the unconditional expectation and then retaining only the first non-zero term in the Taylor series expansion of the unconditional expectation.

In order to model the action of environmental carcinogens, one or more of the transition rates can be made functions of the dose of the agent in question, where dose is to be thought of as the effective dose at the target tissue. Usually, the transition rates are modelled as linear functions of the dose, so that $\lambda_i = a_i + b_i d$ for one or more i. There may be justification for assuming first-order kinetics at least for carcinogens that interact directly with DNA to produce mutations. Then, using the Armitage–Doll approximation (2.4), the hazard function at age t and dose d can be written as $h(t, d) = g(d)t^{n-1}$ where $g(d)$ is a polynomial in dose, and the probability of tumor is approximately given by $P(t, d) = 1 - \exp[-\tilde{g}(d)t^n]$. Note that $\tilde{g}(d)$ is a product of linear terms. It is in this form, called the linearized multistage model, that the Armitage–Doll model is applied to the problem of low-dose extrapolation. Generally, the proportion of animals developing tumors at a specified fixed age at each of three different dose levels is known. The linearized multistage model is fitted to the data and the estimated parameters used to extrapolate risk to lower doses. There are formally at least two problems with this procedure. First, as noted above, the Armitage–Doll approximation holds only when the probability of tumor is low and this condition is not satisfied in the usual animals experiments used for risk assessment. Second, in the statistical analysis $\tilde{g}(d)$ is treated as a general polynomial rather than a product of linear terms.

The discussion above applies only when exposure to a carcinogenic agent starts at birth or very early in life, and continues at the same constant level throughout the period of observation. With time-dependent exposures a starting point for the mathematical development is the set of Kolmogorov equations (2.2), which hold even when the transition rates are time dependent. However, the papers in the literature use approximation (2.5) as their starting point (see, e.g., Whittemore, 1977; Day and Brown, 1980; Crump and Howe, 1984; Brown and Chu, 1987; Freedman and Navidi, 1988). As noted above this approximation is inappropriate unless one has reason to believe that each of the transition rates is small enough. The approximation is almost certainly inappropriate when applied to experimental data.

3. A two-stage model

In 1971, Knudson proposed a two-mutation model for retinoblastoma, a rare tumor of the retina in children. A large fraction of retinoblastoma cases occurs in children who have inherited an autosomal dominant gene for the disease. In

many of these cases there is constitutional deletion of a critical piece of chromosome 13. Knudson suggested that retinoblastoma resulted from mutations at a
specific site on both chromosomes 13 leading to homozygous loss of the function
of a critical gene. Further, the first mutation could be either germinal (in the
heriditary cases) or somatic (in the sporadic cases), whereas the second mutation
is always somatic. Thus, according to Knudson's model children with a hereditary
predisposition to retinoblastoma are born with the cells of their retina requiring
only one critical mutation for malignant transformation. The probability that at
least one of the cells will suffer this critical mutation and become malignant is very
high, and hereditary retinoblastoma will appear to segregate in an autosomal
dominant fashion on pedigree analysis, although the disease is recessive at the
cellular level (because homozygous loss of gene function is required). At the time
that Knudson proposed his model, the location of the gene for retinoblastoma was
known to be on the long arm of chromosome 13. Since then spectacular advances
in molecular biology have made it possible to demonstrate conclusively that the
salient features of Knudson's model for retinoblastoma are correct, and indeed,
the gene for retinoblastoma has been cloned (Friend et al., 1986). This, so-called
recessive oncogenesis, mechanisms for malignant transformation has also been
substantiated in some other human malignancies (Koufos et al., 1984; Solomon
et al., 1987). The genes involved in these malignancies, of which the retinoblastoma gene is one example, have been variously called regulator genes, tumor
suppressor genes and antioncogenes. A detailed discussion of the biology of the
recessive oncogenesis model is outside the scope of this chapter. The interested
reader is referred to the literature (e.g., Knudson, 1985; Moolgavkar and
Knudson, 1981).

A weakness of the Armitage–Doll model is that it does not take explicit
account of cell division, cell death and cell differentiation, although these are
known to be of importance in carcinogenesis. The mathematical formulation of
the recessive oncogenesis model proposed by Moolgavkar and colleagues incorporates cell kinetics as an important feature. Thus, the crucial features of the
model are that it views carcinogenesis as the end result of two critical, specific,
irreversible, rate-limiting and hereditary (at the level of the cell) genomic events;
further the kinetics of tissue growth and development are explicitly incorporated.

In addition to the assumptions stated in the introduction, the specific assumptions for the mathematical development are as follows: In a small time interval
$(t, t + \Delta t)$, a normal susceptible cell divides into two normal susceptible cells with
probability $\alpha_1(t)\Delta t + o(\Delta t)$; it dies or differentiates with probability $\beta_1(t) + o(\Delta t)$;
and it divides into one normal cell and one intermediate cell (a cell that has
sustained the first genomic event or mutation) with probability $\mu_1(t)\Delta t + o(\Delta t)$;
the probability of more than one event occurring is $o(\Delta t)$. Likewise at time t, an
intermediate cell divides into two intermediate cells, dies (or differentiates), or
divides into one intermediate and one malignant cell with rate parameters $\alpha_2(t)$,
$\beta_2(t)$, and $\mu_2(t)$, respectively. Biologically, mutation rates per cell division (rather
than per unit of time) are of interest. These rates are $\mu_i(t)/[\alpha_i(t) + \mu_i(t)]$, $i = 1, 2$.
Figure 1 is a schematic representation of the model.

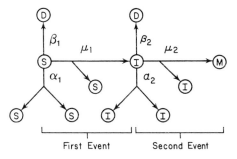

First Event   Second Event

Fig. 1. Two-stage model for carcinogenesis (from Moolgavkar and Knudson, 1981). S = normal stem cell; I = intermediate cell; D = dead or differentiated cell; M = malignant cell. α_1 = rate (per cell per year) of cell division of normal cells; β_2 = rate (per cell per year) of death or differentiation of normal cells; μ_1 = rate (per cell per year) of division into one normal and one intermediate cell. α_2, β_2 and μ_2 are defined similarly. Note that the mutation rate per cell division for normal and intermediate cells is given by $\mu_1/(\alpha_1 + \mu_1)$ and $\mu_2/(\alpha_2 + \mu_2)$, respectively.

Let $X(t)$, $Y(t)$, $Z(t)$ represent, respectively, the number of normal, intermediate, and malignant cells at time t. Let $\Psi(x, y, z; t)$ be the probability generating function at time t starting with a single normal cell at time 0.

$$\Psi(x, y, z; t) = \sum_{i, j, k} \Pr\{X(t) = i, Y(t) = j, Z(t) = k | X(0) = 1,$$
$$Y(0) = 0, Z(0) = 0\} x^i y^j z^k .$$

Then the Kolmogorov forward differential equation for Ψ (Tan and Brown, 1987; Moolgavkar, Dewanji and Venzon, 1988) is

$$\Psi'(t) = \partial \Psi/\partial t = \{\mu_1(t)xy + \alpha_1(t)x^2 + \beta_1(t)$$
$$- [\alpha_1(t) + \beta_1(t) + \mu_1(t)]x\} \partial \Psi/\partial x$$
$$+ \{\mu_2(t)yz + \alpha_2(t)y^2 + \beta_2(t)$$
$$- [\alpha_2(t) + \beta_2(t) + \mu_2(t)]y\} \partial \Psi/\partial y , \qquad (3.1)$$

with the initial condition $\Psi(0) = x$, where the dependence of Ψ and Ψ' on x, y, z and t has been suppressed for notational convenience. The hazard function

$$h(t) = - \Psi'(1, 1, 0; t)/\Psi(1, 1, 0; t) . \qquad (3.2)$$

Because

$$(\partial \Psi/\partial y) (1, 1, 0; t)/\Psi(1, 1, 0; t) = E[Y(t) | Z(t) = 0] ,$$

and, from (3.1),

$$\Psi'(1, 1, 0; t) = - \mu_2(t)(\partial \Psi/\partial y) (1, 1, 0; t) ,$$

it follows that

$$h(t) = \mu_2(t)\, \mathrm{E}[\, Y(t)\,|\, Z(t) = 0]\,. \tag{3.3}$$

Compare this expression with expression (2.6) for the hazard function of he Armitage–Doll model. If the probability of malignancy, $P(t)$, is sufficiently close to zero, then

$$\mathrm{E}[\, Y(t)\,|\, Z(t) = 0] \approx \mathrm{E}[\, Y(t)] \quad \text{and} \quad h(t) \approx \mu_2(t)\, \mathrm{E}[\, Y(t)]\,. \tag{3.4}$$

Recall that this is one of the approximations made in the Armitage–Doll model.

From the Kolmogorov equations, differential equations for the expectations of $X(t)$, $Y(t)$ and $Z(t)$ can be derived immediately (Moolgavkar, Dewanji and Venzon, 1988). In particular, solving the differential equation for $\mathrm{E}[\, Y(t)]$ leads to

$$\mathrm{E}[\, Y(t)] = \int_0^t \left\{ \mu_1(s)\, \mathrm{E}[X(s)] \exp \int_s^t [\alpha_2(u) - \beta_2(u)]\, \mathrm{d}u \right\} \mathrm{d}s\,, \tag{3.5}$$

where

$$\mathrm{E}[X(t)] = \exp \int_0^t [\alpha_1(s) - \beta_1(s)]\, \mathrm{d}s\,.$$

Thus, the approximate hazard function $\mu_2(t)\, \mathrm{E}[\, Y(t)]$ can be easily computed. The use of this hazard function for the analysis of human epidemiologic data is probably appropriate because, in human populations, cancer is a rare disease. However, in many animal experiments, the probability of malignant tumor is very high, and the use of the approximate hazard function may be inappropriate. An approach to computing the exact hazard function is to integrate the characteristic equations associated with Kolmogorov equation (3.1). When the parameters of the model are constant this approach leads to a Riccati equation for Ψ. This equation may also be derived directly (Kendall, 1960; Moolgavkar and Venzon, 1979). I shall not pursue these matters further here because it is more convenient to use a version of the two-mutation model in which the growth of normal tissue is treated as being deterministic.

3.1. Model with growth of normal cells deterministic

Let $X(s)$ represent the number of susceptible stem cells in a tissue at time s, and let $\mu_1(s)$ be the 'rate' of first mutation. To be precise, assume that intermediate cells are generated from normal ones by a nonhomogeneous Poisson process with intensity $\mu_1(s)X(s)$: in a small time interval Δs, the probability that an intermediate cell arises by mutation of a normal cell is $\mu_1(s)X(s)\Delta s + \mathrm{o}(\Delta s)$, and the probability that more than one intermediate cell arises in this fashion is $\mathrm{o}(\Delta s)$. The intermediate cells divide, die (or differentiate) and mutate as described above for the completely stochastic model with parameters $\alpha_2(t)$, $\beta_2(t)$, and $\mu_2(t)$, respectively. The quantity $\int_0^t \mu_1(s)X(s)\, \mathrm{d}s$ may be thought of as the total number of first mutations that have taken place by time t, and this number depends upon the mutation rate and the number of susceptible cells at any instant. Let

$$\Psi(y, z; t) = \sum_{j, k} \Pr[Y(t) = j, Z(t) = k \,|\, Y(0) = 0, Z(0) = 0] \, y^j z^k$$

be the probability generating function for the number of intermediate and malignant cells at time t. The Kolmogorov forward differential equation for Ψ is

$$\Psi' = \partial \Psi/\partial t = (y - 1)\mu_1(t)X(t)\,\Psi(y, z; t)$$
$$+ \{\mu_2(t)yz + \alpha_2(t)y^2 + \beta_2(t)$$
$$- [\alpha_2(t) + \beta_2(t) + \mu_2(t)]y\} \, \partial \Psi/\partial y, \qquad (3.6)$$

with initial condition $\Psi(y, z; 0) = 1$. As in the previous section, one immediately obtains

$$h(t) = -\Psi'(1, 0; t)/\Psi(1, 0; t) = \mu_2(t)\, \mathrm{E}[Y(t)\,|\,Z(t) = 0]\,.$$

The differential equation for $\mathrm{E}[Y(t)]$ can be easily derived and integrated to yield an approximate hazard function

$$h(t) \approx \mu_2(t)\, \mathrm{E}[Y(t)]$$
$$= \mu_2(t) \int_0^t \left\{ \mu_1(s)X(s) \exp \int_s^t [\alpha_2(u) - \beta_2(u)]\, \mathrm{d}u \right\} \mathrm{d}s\,. \qquad (3.7)$$

When the parameters of the model are constant, this expression reduces to

$$h(t) \approx \mu_1\mu_2 \int_0^t X(s) \exp[(\alpha_2 - \beta_2)(t - s)]\, \mathrm{d}s\,. \qquad (3.8)$$

Expression (3.8) for the hazard function has been used for the analysis of human incidence data. It can be shown that expressions (3.7) and (3.8) over-estimate the exact hazard function (Moolgavkar, Dewanji and Venzon, 1988).

When the parameters are constant, an exact expression for the hazard function can be easily computed as follows. Let $\phi(y, z; t)$ be the probability generating function for the number of intermediate and malignant cells at time t starting with 1 intermediate cell at time 0,

$$\phi(y, z; t) = \sum_{j, k} \Pr\{Y(t) = j, Z(t) = k \,|\, Y(0) = 1, Z(0) = 0\} \, y^j z^k\,.$$

Then

$$\Psi(y, z; t) = \exp\left\{ \mu_1 \int_0^t X(s)[\phi(y, z; t - s) - 1]\, \mathrm{d}s \right\} \qquad (3.9)$$

and

$$h(t) = -\Psi'(1, 0; t)/\Psi(1, 0; t) = -\mu_1 \int_0^t X(s)\phi'(1, 0; t - s)\, \mathrm{d}s\,. \qquad (3.10)$$

Further $\phi(y, z; t)$ satisfies the Riccati equation

$$\phi' = \alpha_2 \phi^2 + [\mu_1 z - (\alpha_2 + \beta_2 + \mu_2)] \phi + \beta_2, \tag{3.11}$$

with initial condition $\phi(y, z; 0) = y$. This equation for ϕ can be readily integrated using standard techniques and, in particular, leads to an expression for $\phi(1, 0; t)$:

$$\phi(1, 0; t) = \frac{c_1 - c_2[\{(1 - c_1)/(1 - c_2)\} \exp[\alpha_2(c_1 - c_2)t]]}{1 - \{(1 - c_1)/(1 - c_2)\} \exp[\alpha_2(c_1 - c_2)t]},$$

where $c_1 < 1 < c_2$ are distinct roots of the polynomial $\alpha_2 q^2 - (\alpha_2 + \beta_2 + \mu_2)q + \beta_2$. This expression for $\phi(1, 0; t)$ can be used in (3.9) and (3.10) to yield exact expressions for $\Psi(1, 0; t)$ and $h(t)$, although a numerical integration is still necessary.

As noted above, the approximate hazard function can be shown to overestimate the exact hazard function, and numerical work indicates that for a given fixed value of $\alpha_2 - \beta_2$, the adequacy of the approximation is very sensitive to α_2 and β_2 individually. Thus, for fixed $\alpha_2 - \beta_2$, the approximation gets progressively worse with increasing values of α_2 and β_2. Intuitively, the reason is that with increasing β_2, the probability of extinction of individual clones of intermediate cells increases.

When the parameters of the model are time dependent, exact expressions for $\Psi(1, 0; t)$ and $h(t)$ are more difficult to compute. One approach to computing these is to solve the characteristic equations associated with the Kolmogorov equation for Ψ. I shall not go into the details here. These may be found in a recent paper in which the model is applied to radon-induced lung tumors in rats (Moolgavkar et al., 1990; Moolgavkar and Luebeck, 1990). Suffice it to say here that expressions (3.9) and (3.10) are replaced by

$$\Psi(1, 0; t) = \exp\left\{ \int_0^t \mu_1(s) X(s) [\phi(1, 0; s, t) - 1] \, ds \right\}$$

and

$$h(t) = - \int_0^t \mu_1(s) X(s) \phi_t(1, 0; s, t) \, ds,$$

where $\phi(y, z; s, t)$ represents the generating function at time t for the process starting out with one intermediate cell at time s, and ϕ_t represents its derivative with respect to t. The point is that, because of the time dependence of the parameters, one does not have 'origin invariance', i.e., ϕ depends upon both s and t and not just on $t - s$.

If the two-mutation model of carcinogenesis is a reasonable approximation to reality, then it would be clearly of interest to identify and study the intermediate cells. This appears to be possible for some human tumors and in some initiation–promotion experiments in animals. In particular, there is now considerable data available on the number and size of putative intermediate clones in rat hepato-carcinogenesis experiments. Mathematical expressions for these quantities have recently been derived within the framework of the two-mutation model. Some of the implications of the model for initiation–promotion experiments will be discussed in a later section. The reader is referred to the recent paper by Dewanji, Venzon and Moolgavkar (1989) for mathematical details.

Armitage and Doll (1957) proposed a two-stage model for carcinogenesis in which cells in the intermediate stage grow deterministically and exponentially. Suppose that the intermediate cells are allowed to grow with parameter $\alpha = \alpha_2 - \beta_2$, and that malignant cells are derived from intermediate cells as a Poisson process with intensity $\mu_2 Y(t)$, where $Y(t)$ is the number of intermediate cells at time t. Then the probability that a single intermediate cell generated at time 0 (and its progeny) does not give rise to a malignant cell by time t is easily seen to be $\exp[\mu_2/\alpha - (\mu_2/\alpha)\exp(\alpha t)]$. If this expression is used in (3.10) in place of $\phi(1, 0; t)$ and further, if it is assumed that $X(s) = M$ is constant, then

$$h(t) = \mu_1 M\{1 - \exp[\mu_2/\alpha - (\mu_2/\alpha)\exp(\alpha t)]\},$$

which is the expression obtained by Armitage and Doll.

4. Analysis of human epidemiologic data

4.1. Population-based incidence data

As I noted earlier, the first carcinogenesis models were developed to explain the observation that the age-specific incidence rates of many human carcinomas increase approximately with a power of age. With the Armitage–Doll multistage model, age-specific incidence is simply a multiple of expression (2.4) for the hazard function. With the two-mutation model, age-specific incidence can be modelled by a multiple of (3.8). Note that, in the Armitage–Doll model, the 'shape' of the age-specific incidence curve (i.e., the power of age with which age-specific incidence rates increase) is determined by the number of stages. In contrast, in the two-mutation model, the shape of the age-specific incidence curve is determined by the kinetics of growth of normal and intermediate cells; the mutation rates μ_1 and μ_2 simply determine the overall rates in a population. Because it incorporates tissue kinetics explicitly the two-mutation model can be used to describe the age-specific incidence curves of cancers that do not exhibit the power law behaviour. For example, retinoblastoma does not occur in children after the age of ten years and this is thought to be a consequence of the fact that susceptible normal cells (the retinoblasts) either differentiate or die out by that age. This behaviour of the retinoblasts can easily be explicitly incorporated into the two-mutation model. The growth curves of human tissue show three main patterns:

(1) Most tissues show a steady increase in size during childhood and adolescence. The growth curve of such tissues, e.g., the lung and the colon, can be reasonably well represented by a Gompertz curve. Once adult size is reached, the epithelia of these tissues continue to shed and replenish themselves.

(2) Certain tissues show a sudden burst of growth in early life followed by a greatly decreased rate of cell division in later life, as in lymphoid tissue, or by virtually no cell division in later life, as in neural tissue.

(3) Some tissues, such as the sex organs, show relatively little growth before puberty, followed by a spurt of growth during adolescence. The growth of such tissues may be represented by a logistic curve. In addition, these tissues are sensitive to changes in physiologic state. Thus, the breast in females grows during puberty and varies in size in response to hormonal influences during pregnancy and after menopause.

Reflecting these growth patterns, cancer incidence rates in human populations also show three main patterns.

A steady increase in incidence rates with age
Cancers that exhibit a steady increase in incidence rate with age include the carcinomas in which the age-specific incidence rates increase roughly with a power of age, the log/log cancers. These include many of the commonest cancers such as carcinomas of the lung and colon and, as discussed above, the Armitage–Doll model was developed to explain the age-specific incidence of these cancers. In 1969, Cook, Doll and Fellingham (1969) undertook a comprehensive test of the approximate incidence function generated by the Armitage–Doll model. Specifically, they fitted the relationship $I(t) = bt^k$, where $I(t)$ represents the age-specific incidence, to 31 types of cancer in 11 populations. They concluded that the constant b varied from population to population but that the constant k 'might be a biological constant characteristic of the tissue in which the cancer is produced'. Recall that k is one less than the number of stages required for malignant transformation. However, Cook et al. found that many of the cancers showed departures from simple log/log behaviour.

Even if the simple multistage model presented above is 'correct', there are several reasons for which specific cancers could show departures from log/log behaviour. These have been discussed by Peto (1977), and include changes in transition rates with age as a result of a changing environment; strong cohort effects which distort cross-sectional age-specific incidence curves; transition rates in the model not small enough as is discussed in the first part of this chapter.

With the growth of normal tissue represented by a Gompertz curve (i.e., $X(s)$ in (3.8) represented by a Gompertz curve), the two-mutation model generates incidence curves that describe well the age-specific incidence rates of the cancers in this group. The two-mutation model has been explicitly fitted to the age-specific incidence of lung cancer among non-smokers and gives an excellent description of the data. An analysis, within the framework of the two-mutation model, of lung cancer incidence in a large cohort study is briefly discussed below.

Incidence rates that exhibit a peak sometime in life followed by a decline
Childhood cancers, such as retinoblastoma, Wilms' tumor, and acute lymphocytic leukemia, provide examples of cancers with incidence rates that exhibit a peak sometime in life followed by a decline. As described in Moolgavkar and Knudson (1981), the two-mutation model provides an excellent description of these incidence curves.

Exceptional incidence curves
Here I mean incidence curves such as those of cancers of the sex organs. Carcinoma of the female breast provides an interesting and much-studied example. The age-specific incidence rates of female breast cancer vary widely in different populations: Incidence rates are approximately 4–5 times as high in Western populations (e.g., the United States, Denmark) as in Eastern populations (e.g., Japan). The age-specific incidence rates rise until about menopause, level off, and then continue to rise again, albeit more slowly (it is often stated that in Eastern populations, post-menopausal rates actually decrease; but this is probably an artefact of cohort effects). The two-mutation model provides an excellent description of the age-specific incidence curve of female breast cancer when the hormone induced changes in the kinetics of the mammary epithelium at menarche and menopause are incorporated into the model. Other features of the epidemiology of breast cancer, such as the protective effect of an early full-term pregnancy, are also well described (Moolgavkar, Day and Stevens, 1980).

4.2. Cohort and case-control studies

Relative risk regression models, which are generalizations of the logistic regression model and the Cox proportional hazards model, have become ubiquitous in the analysis of cohort and case-control data. More recently, however, a few epidemiologic studies of cancer have been analyzed within the framework of carcinogenesis models. For the analysis of a cohort study, the hazard function generated by the carcinogenesis model can be used directly with the parameters of the model taken to be functions of the covariates of interest. One approach to analyzing a case-control study is to view it as a synthetic case-control (case-control within a cohort) study, and then the hazard function generated by the carcinogenesis model can be used to define a partial likelihood. Parameter estimation then proceeds by maximization of this partial likelihood. Krailo et al. (1987) and Kampert et al. (1988) have used this approach to analyze case-control studies of breast cancer.

Among the prospective (cohort) studies that have contributed to our understanding of smoking-related diseases, the British doctors' study (Doll and Peto, 1978) stands out for the completeness of its follow-up and the care and precision with which the cigarette smoking information was collected. Doll and Peto showed that lung cancer incidence in this data set was well described by

$$I(t) = 0.273 \times 10^{-12}(d + 6)^2(t - 22.5)^{4.5},$$

where $I(t)$ is the annual incidence (hazard) at age t, for an individual smoking d cigarettes a day. Doll and Peto considered $t - 22.5$ to be a surrogate measure of duration of smoking, and an often repeated conclusion of their analysis is that duration of smoking is more important than daily dose of smoking in determining lung cancer risk.

The lung cancer incidence data are cross-classified into 72 categories (8 age groups, 9 smoking-level categories). Suppose that N_{ij} is the person-years (population) at risk in age group i and smoking category j, and $h(t_i, d_j)$ is the hazard function (appropriately adjusted for lag time, i.e., for time to development of tumor from a single malignant cell) generated by a carcinogenesis model. Then under standard assumptions the likelihood of the data is proportional to that of 72 mutually independent Poisson variates with expectations $N_{ij}h(t_i, d_j)$. Whittemore (1988) analyzed the British doctors' data with h the hazard function from the Armitage–Doll multistage model and concluded that the model described the data well with the first and penultimate stage being dose dependent. Moolgavkar, Dewanji and Luebeck (1989) analyzed the data using the hazard function from the two-mutation model and concluded that this model also described the data well. Further the data are consistent with dose-response relationships quite different from those inferred by Doll and Peto. In particular, the data are consistent with the view that daily level of smoking is at least as important as duration of smoking in determining lung cancer risk (see also Gaffney and Altshuler, 1988).

5. Analysis of experimental data

Much of our understanding of carcinogenesis derives from animal experiments. Particularly in the area of chemical carcinogenesis, animal experiments have led to the current paradigm that there are two broad classes of agents — initiators and promoters — that facilitate carcinogenesis. Moreover, good quantitative data are now becoming available from the so-called initiation–promotion experiments, and these data offer an opportunity and a challenge to the modeller. A discussion of the biological aspects of initiation–promotion experiments, and their interpretation within the framework of carcinogenesis models, is outside the scope of this chapter. The interested reader is referred to some recent papers (Armitage, 1985; Moolgavkar and Knudson, 1981; Moolgavkar, 1986; Peto, 1977). Suffice it to say that the results of recent experiments, involving the so-called IPI (initiation–promotion–initiation) protocol are not easy to reconcile with the multistage model (Freedman and Zeisel, 1988). In contrast, Moolgavkar and Knudson (1981) proposed the IPI protocol on the basis of the two-mutation model and predicted the results that were later obtained.

I would like to concentrate here on a brief description of the mathematical consequences of the two-mutation model for the analysis of initiation–promotion experiments. Within the framework of the two-mutation model, initiators are agents that increase the mutation rates μ_1 and μ_2, whereas the main action of promoters is to increase $\alpha_2 - \beta_2$, i.e., to increase the net growth rate of intermediate cells. In one experimental system, the rat liver, precise quantitative information is now becoming available on the number and size of the so-called enzyme altered foci. These foci are premalignant lesions and can be identified with clones of intermediate cells within the framework of the two-mutation model. A clone of intermediate cells is the collection of all intermediate cells descended from a single intermediate cell that has arisen by direct mutation from the normal cells.

An immediate consequence of the fact that intermediate cells are undergoing a birth and death process is that there is a non-zero probability that the descendants of any intermediate cell will become extinct. In fact, the probability of extinction is just β_2/α_2. Thus, if the tissue of interest is examined one should expect to see different types of intermediate lesions (assuming that intermediate lesions can be identified). Some lesions will have few dead and dying cells, others will appear to be regressing with a large number of dead cells. This is exactly what is observed with the enzyme altered foci in rat liver. In a recent paper Dewanji et al. (1989) have worked out the mathematical expressions for the number and size of intermediate lesions within the framework of the two-mutation model. I will not discuss these further here. They provide a starting point for the analysis of the data on enzyme altered foci. To the best of my knowledge a detailed analysis of such data within the framework of a carcinogenesis model has not yet been undertaken. Figure 2, taken from Dewanji et al. (1989), illustrates some results.

Recall that $E[Y(t)]$, the expectation of the number of intermediate cells, is given by the expression

$$E[Y(t)] = \int_0^t \left\{ \mu_1(s) X(s) \exp \int_s^t [\alpha_2(u) - \beta_2(u)] \, du \right\} ds \,,$$

and thus $E[Y(t)]$ depends only upon the difference $\alpha_2 - \beta_2$ and not on the parameters individually. Figure 2 shows that with increasing values of α_2, the expected number of nonextinct clones decreases, but their expected size increases. An immediate consequence is that for fixed μ_1 and $\alpha_2 - \beta_2$, i.e., for a constant expected number of intermediate cells, a large value of α_2 implies a small number of large clones of intermediate cells; a small value of α_2 implies a large number of small clones of intermediate cells. Further, if the second mutation rate per cell division, μ_2/α_2, is constant the probability of malignant tumor is higher for large α_2 at any time t (see Figure 3 in Moolgavkar et al., 1988). These are simply mathematical predictions of the model. However, it is interesting to note that Schwarz et al. (1984) report on two rat hepatocarcinogens, 4-dimethylaminoazo-benzene (4-DAB) and N-nitrosodiethanolamine (NDEOL), which give rise to the same (mean) number of enzyme altered cells (intermediate cells) in the rat liver. However, 4-DAB treatment yields a small number of large clones, whereas NDEOL treatment yields a larger number of small clones. An interpretation is that both these agents affect α_2 and β_2 in such a way that $\alpha_2 - \beta_2$ is constant but

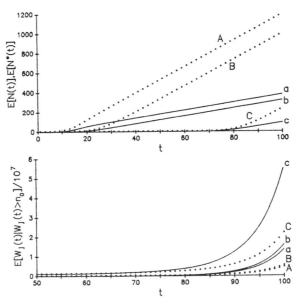

Fig. 2. The expected (mean) number of non-extinct ($E[N(t)]$) and detectable premalignant clones and their expected (mean) size ($E[W_j(t)|W_j(t) > n_0]$) plotted against time (from Dewanji, Venzon and Moolgavkar, 1988). For all plots, $\mu_1 = 10^{-6}$, $X(s)$ is modelled by a Makeham distribution, the value of $\alpha_2 - \beta_2 = 0.2$. The solid line represents $\alpha_2 = 1.0$ ($\beta_2 = 0.8$); the dotted line represents $\alpha_2 = 0.3$ ($\beta_2 = 0.1$). In the upper panel, curves A and a represent the mean number of non-extinct clones; curves B and b represent the mean number of clones with more than 10 cells; curves C and c represent the mean number of clones with more than 10^6 cells. Note that the mean number of clones is smaller with $\alpha_2 = 1.0$ than with $\alpha_2 = 0.3$. In the lower panel curves A and a represent the average size of non-extinct clones; curves B and b represent the average size of clones that are larger than a threshold size of 10 cells; curves C and c represent the average size of clones that are larger than a threshold size of 10^6 cells. Note that the average size of the clones is larger with $\alpha_2 = 1.0$ than with $\alpha_2 = 0.3$.

α_2 is larger with 4-DAB treatment than with NDEOL treatment. Also, as predicted by the model the probability of *malignant* tumor is higher with 4-DAB than with NDEOL treatment.

Finally, recent work (Cohen and Ellwein, 1988) indicates that a two-mutation model is consistent with data from experimental bladder carcinogenesis. Connolly et al. (1988) have proposed a comprehensive model for chemical carcinogenesis by putting the two-mutation model described here back-to-back with a physiologically-based pharmacokinetic model to take into account carcinogen pharmacokinetics.

6. Concluding remarks

In this chapter I have presented a rather personal perspective on carcinogenesis modelling. My bias in favor of the two-mutation model is obvious. I believe that this is the most parsimonious model consistent with the data. In many human and animal tumors one intermediate stage on the pathway to malignancy is clearly observable. Further, the model suggests experiments, such as the IPI protocol, and focuses attention on the biological parameters that need to be measured for a rational approach to risk assessment. Some biologists 'see' several stages on the pathway to malignancy (e.g., Weinstein, 1988). It is not at all clear, however, that these stages are *necessary*, nor that they are rate-limiting. The process of model building inevitably entails simplification and the making of choices. The model builder must decide which features of the biological process must be incorporated into the models and which features can be safely ignored. Thus, most models ignore the growth of a malignant cell into a clinically detectable tumor or, at most, account for it by the introduction of a suitable lag time. Additionally, in the two-mutation model, the cell kinetics of intermediate cells are modelled in primitive fashion. There are entire tomes on the mathematical modelling of the cell cycle and it is clear that cells do not divide or die with exponential waiting times. Nevertheless, in the context of carcinogenesis modelling these simplifications appear to be entirely appropriate in the interests of parsimony.

From the statistical point of view, stochastic carcinogenesis models provide a large and flexible class of hazard functions for the analysis of epidemiologic and experimental data. Their use for this purpose is gradually increasing. One of the virtues of this approach is that time-dependent covariates can be accommodated in a very natural way via their postulated effects on the parameters of the models.

Note added in proof

This chapter was written in late 1988. I would say some things differently today. Parts of the chapter have appeared in expanded form (Moolgavkar and Luebeck, 1990).

References

Armitage, P. (1985). Multistage models of carcinogenesis. *Environmental Health Perspectives* **63**, 195–201.

Armitage, P. and Doll, R. (1954). The age distribution of cancer and a multistage theory of carcinogenesis. *Br. J. Cancer* **8**, 1–12.

Armitage, P. and Doll, R. (1957). A two-stage theory of carcinogenesis in relation to the age distributions of human cancer. *Br. J. Cancer* **11**, 161–169.

Brown, C. and Chu, K. (1987). Use of multistage models to infer stage affected by carcinogenic exposure: Example of lung cancer and cigarette smoking. *J. Chronic Disease* **40** (Suppl. 2), 171S–179S.

Cavanee, W. K., Dryja, T. P., Phillips, R. A., Benedict, W. F., Godbout, R. et al. (1983). Expression of recessive alleles by chromosomal mechanisms in retinoblastoma. *Nature* **305**, 779–784.

Cohen, S. M. and Ellwein, L. B. (1988). Cell growth dynamics in long-term bladder carcinogenesis. *Toxicology Letters* **43**, 151–173.

Connolly, R. B., Reitz, R. H., Clewell, H. J. and Andersen, M. E. (1988). Pharmacokinetics, bio-chemical mechanism and mutation accumulation: A comprehensive model of chemical carcino-genesis. *Toxicology Letters* **43**, 189–200.

Cook, P., Doll, R. and Fellingham, S. A. (1969). A mathematical model for the age distribution of cancer in man. *Int. J. Cancer* **4**, 93–112.

Crump, K. and Howe, R. (1984). The multistage model with a time-dependent dose pattern: Applica-tion to carcinogenic risk assessment. *Risk Analysis* **4**, 163–176.

Day, N. and Brown, C. (1980). Multistage models and primary prevention of cancer. *J. Natl. Cancer Inst.* **64**, 977–989.

Dewanji, A., Venzon, D. and Moolgavkar, S. (1989). A stochastic two-stage model for cancer risk assessment II: The number and size of pre-malignant clones. *Risk Analysis* **9**, 179–186.

Doll, R. and Peto, R. (1978). Cigarette smoking and bronchial carcinoma: Dose and time relation-ships among regular smokers and life-long non-smokers. *J. Epidemiol. Community Health* **32**, 303–313.

Feller, W. (1971). *An Introduction to Probability Theory and its Applications*, Vol. 2, 2nd ed. Wiley, New York.

Freedman, D. A. and Navidi, W. (1989). Multistage models for carcinogenesis. *Environmental Health Perspectives* **81**, 169–188.

Freedman, D. A. and Zeisel, H. (1988). From mouse-to-man: The quantitative assessment of cancer risks. *Statistical Science* **3**, 3–56.

Friend, S. H., Bernards, R., Rogelj, S., Weinberg, R. A., Rapaport, J. M., Albert, D. M. and Dryja, T. P. (1986). A human DNA segment with properties of the gene that predisposes to retino-blastoma and osteosarcoma. *Nature* **323**, 643–646.

Gaffney, M. and Altshuler, B. (1988). Examination of the role of cigarette smoke in lung carcino-genesis using multistage models. *J. Natl. Cancer Inst.* **80**, 925–931.

Kampert, J. B., Whittemore, A. S. and Paffenbarger, R. S. (1988). Combined effect of childbearing, menstrual events, and body size on age-specific breast cancer risk. *Am. J. Epidemiol.* **128**, 962–979.

Kendall, D. G. (1960). Birth-and-death processes, and the theory of carcinogenesis. *Biometrika* **47**, 13–21.

Knudson, A. (1971). Mutation and cancer: Statistical study of retinoblastoma. *Proc. Nat. Acad. Sci. U.S.A.* **68**, 820–823.

Knudson, A. (1985). Hereditary cancer, oncogenes, and antioncogenes. *Cancer Research* **45**, 1437–1443.

Koufos, A., Hansen, M. F., Lampkin, B. C., Workman, M. L., Copeland, N. G. et al. (1984). Loss of alleles at loci on human chromosome 11 during genesis of Wilms' tumor. *Nature* **309**, 170–172.

Krailo, M., Thomas, D. C. and Pike, M. C. (1987). Fitting models of carcinogenesis to a case-control study of breast cancer. *J. Chron. Dis.* **40** (Suppl. 2), 181S–189S.

Moolgavkar, S. (1978). The multistage theory of carcinogenesis and the age distribution of cancer in man. *J. Natl. Cancer Inst.* **61**, 49–52.

Moolgavkar, S. (1986). Carcinogenesis modelling: From molecular biology to epidemiology. *Annual Review of Public Health* **7**, 151–169.

Moolgavkar, S. and Venzon, D. (1979). Two-event models for carcinogenesis: Incidence curves for childhood and adult tumors. *Math. Biosci.* **47**, 55–77.

Moolgavkar, S. and Knudson, A. (1981). Mutation and cancer: A model for human carcinogenesis. *J. Natl. Cancer Inst.* **66**, 1037–1052.

Moolgavkar, S. and Dewanji, A. (1988). Biologically-based models for cancer risk assessment: A cautionary note. *Risk Analysis* **8**, 5–6.

Moolgavkar, S. and Luebeck, G. (1990). Two-event model for carcinogenesis: Biological, mathemati-cal and statistical considerations. *Risk Analysis* **10**, 323–341.

Moolgavkar, S., Day, N. and Stevens, R. (1980). Two-stage model for carcinogenesis: Epidemiology of breast cancer in females. *J. Natl. Cancer Inst.* **65**, 559–569.

Moolgavkar, S., Dewanji, A. and Venzon, D. (1988). A stochastic two-stage model for cancer risk assessment I: The hazard function and the probability of tumor. *Risk Analysis* **8**, 383–392.

Moolgavkar, S., Dewanji, A. and Luebeck, G. (1989). Cigarette smoking and lung cancer: Reanalysis of the British doctors' data. *J. Natl. Cancer Inst.* **81**, 415–420.

Moolgavkar, S. H., Cross, F. T., Luebeck, G. and Dagle, G. E. (1990). A two-mutation model for radon-induced lungtumors in rats. *Radiation Research* **121**, 28–37.

Nordling, C. O. (1953). A new theory of the cancer inducing mechanicm. *Br. J. Cancer* **7**, 68–72.

Peto, R. (1977). Epidemiology, multistage models and short-term mutagenicity tests. In: H. H. Hiatt, J. D. Watson and H. A. Winston (eds.), *Origins of Human Cancer, Book C: Human Risk Assessment*, Cold Spring Harbor Laboratory, Cold Spring Harbor.

Pike, M. C. (1966). A method of analysis of a certain class of experiments in carcinogenesis. *Biometrics* **22**, 142–161.

Schwarz, M., Pearson, D., Port, R. and Kunz, W. (1984). Promoting effect of 4-dimethylaminoazo-benzene on enzyme altered foci induced in rat liver by *N*-nitrosodiethanolamine. *Carcinogenesis* **5**, 725–730.

Solomon, E., Voss, R., Hall, V., Bodmer, W. F., Jass, J. R. et al. (1987). Chromosome 5 allele loss in human colorectal carcinomas. *Nature* **328**, 616–619.

Tan, W. Y. and Brown, C. C. (1987). A nonhomogeneous two-stage model of carcinogenesis. *Mathematical Modelling* **9**, 631–642.

Thorslund, T., Brown, C. and Charnley, G. (1987). Biologically motivated cancer risk models. *Risk Analysis* **7**, 109–119.

Weinstein, I. B. (1988). The origins of human cancer: Molecular mechanisms of carcinogenesis and their implications for cancer prevention and treatment. *Cancer Research* **48**, 4135–4143.

Whittemore, A. S. (1977). The age distribution of human cancer for carcinogenic exposures of varying intensity. *Am. J. Epidemiol.* **106**, 418–432.

Whittemore, A. (1988). Effect of cigarette smoking in epidemiological studies of lung cancer. *Statistics in Medicine* **7**, 223–238.

Whittemore, A. S. and Keller, J. B. (1978). Quantitative theories of carcinogenesis. *SIAM Rev.* **20**, 1–30.

Wilson, J. D. (1986). Time for a change: Guest editorial. *Risk Analysis* **6**, 111–112.

C. R. Rao and R. Chakraborty, eds., *Handbook of Statistics, Vol. 8*
© Elsevier Science Publishers B.V. (1991) 395–406

13

An Application of Score Methodology: Confidence Intervals and Tests of Fit for One-Hit Curves

John J. Gart

1. The single-parameter case

We consider a series of mutually independent components of an experiment, $i = 1, 2, \ldots, I$, yielding the independent variates x_i, based on samples of size n_i. The variates are not necessarily identically distributed, but they all depend on a single unknown parameter λ. The log-likelihood is denoted by

$$\ln L(\lambda) = \sum \ln L_i(\lambda) .$$

Using the classical theory of maximum likelihood (ML) of R. A. Fisher (see Rao, 1973, Ch. 5) one may usually find the ML estimates of λ, $\hat{\lambda}$, from the equation,

$$\mathrm{d} \ln L(\hat{\lambda})/\mathrm{d}\lambda = 0 , \tag{1.1}$$

which may require an iterative solution. The estimator's asymptotic mean is λ and its asymptotic variance is

$$\mathrm{Var}(\hat{\lambda}) = 1/I(\lambda) ,$$

where

$$I(\lambda) = \mathrm{E}\{ - \mathrm{d}^2 \ln L(\lambda)/\mathrm{d}\lambda^2 \} .$$

As $\hat{\lambda}$ is asymptotically normal, subject to modest regularity conditions, approximate $1 - \alpha$ confidence limits are found to be

$$\lambda_{\mathrm{U, L}} = \hat{\lambda} \pm z_{\alpha/2}/\{I(\hat{\lambda})\}^{1/2} , \tag{1.2}$$

where $z_{\alpha/2}$ is the $\alpha/2$ upper percentile of the normal distribution.
    The bias of $\hat{\lambda}$ may be corrected for, at least to terms of order n_i^{-1}. Using the

results of Shenton and Bowman (1977, Ch. 2) and Bartlett (1955), Thomas and Gart (1971), show that this may be simply written as

$$B(\lambda) = -\{2\, dI(\lambda)/d\lambda + E\,(d^3 \ln L/d\lambda^3)\}/[2\{I(\lambda)\}^2]\,. \qquad (1.3)$$

Thus an unbiased estimator, except for terms of $O(n_i^{-2})$, is $\hat{\lambda}_c = \hat{\lambda} - B(\hat{\lambda})$. This estimator may be substituted for $\hat{\lambda}$ in (1.2) to improve these limits.

An alternate method may be based directly on the score (see Bartlett, 1953), that is, $S(\lambda) = d \ln L(\lambda)/d\lambda$. For all λ,

$$E\{S(\lambda)\} = 0 \quad \text{and} \quad \text{Var}\{S(\lambda)\} = I(\lambda)\,,$$

both relations being exact under mild regularity conditions. Furthermore,

$$S(\lambda) = \sum d \ln L_i(\lambda)/d\lambda\,,$$

is a sum of independent variates which, by the central limit theorem, will often be more nearly normally distributed in small sample sizes than the solution to (1.1), which may be a complex function of the independent variates. Thus the confidence limits may be approximated from the formula

$$z(\lambda) = S(\lambda)/\{I(\lambda)\}^{1/2} = \pm z_{\alpha/2}\,, \qquad (1.4)$$

which, like (1.1), may require an iterative solution. As we shall see in the specific application, the secant method will often prove to be an efficient method of solution.

A priori, there is good reason to prefer (1.4) to (1.2) as an approximate confidence interval method. The posited standard normal deviate of (1.4) has exactly a mean of zero and a variance of unity while the approximate normal deviate leading to (1.2), even in the bias-corrected form, has only asymptotically the posited values. More importantly (1.4) adjusts for the possible variation in variances at the two limits, i.e. at λ_u, $\text{Var}\{S(\lambda_u)\} = I(\lambda_u)$. This adjustment might also be done in (1.2), but it would require still further iterative solutions.

As noted above, both (1.2) and (1.4) are in general only asymptotically normal. Skewness is the principal problem in finding valid confidence intervals. Fortunately the third central moment of $S(\lambda)$ is easily found either directly, or from the result of Bartlett (1953a), to be

$$\mu_3\{S(\lambda)\} = 3\, dI(\lambda)/d\lambda + 2E\,(d^3 \ln L/d\lambda^3)\,, \qquad (1.5)$$

always supposing these functions are well defined in the particular application. The skewness is defined to be

$$\gamma_1(\lambda) = \mu_3\{S(\lambda)\}/\{I(\lambda)\}^{3/2}\,,$$

which is a term of order $n_i^{-1/2}$, but is again an exact result. The adjustment of the confidence limits for skewness may be based on the Cornish–Fisher expansion (Bartlett, 1953). The approximate limits are the solution to the equation

$$z_s(\lambda) = z(\lambda) - \gamma_1(\lambda)(z_{\alpha/2}^2 - 1)/6 = \pm z_{\alpha/2},$$

which typically involves an iterative solution. The adjustment depends not only on the magnitude of $\gamma_1(\lambda)$ at the limits but also on the confidence level. For $z_{\alpha/2} = 1.645$, 90% limits, the factor $\frac{1}{6}(z_{\alpha/2}^2 - 1) = 0.284$ while for $z_{\alpha/2} = 2.326$, 99% limits, this factor is 0.735.

Finally one may wish to test the fit of the model, particularly the constancy of λ over the i. A homogeneity test (see Rao, 1973, Ch. 6) may be based on the components of the score,

$$S_i(\lambda) = d \ln L_i(\lambda)/d\lambda,$$

and the components of the information,

$$I_i(\lambda) = E\{-d^2 \ln L_i(\lambda)/d\lambda^2\}.$$

The approximate χ^2 statistic with $I - 1$ degrees of freedom is

$$X_{I-1}^2 = \sum z_i^2(\hat{\lambda}),$$

where

$$z_i^2(\hat{\lambda}) = S_i^2(\hat{\lambda})/I_i(\hat{\lambda}),$$

and $\hat{\lambda}$ is the solution to (1.1).

2. Application to the one-hit curve

The one-hit curve is based on an underlying Poisson distribution of the particles of an agent causing the effect, say a tumor, an infection, or a transformation in the host or target organism. As a single particle is sufficient to cause the effect, its probability is the sum of terms of one or more of a Poisson distribution or, equivalently, unity less the probability of zero, that is,

$$P_i = 1 - \exp(-\lambda d_i),$$

where d_i is the known dose or dilution factor of the agent and λ is the measure of its effectiveness, sometimes called the infectivity in the context of viral or bacterial agents. We shall be concerned with estimating λ. At d_i, n_i independent trials are performed, each having the probability of effect of P_i. Let x_i be the number of such responses, which necessarily has a binomial distribution. The total likelihood is thus

$$\ln L = \sum \ln L_i = \sum \{x_i \ln P_i\} + \sum \{(n_i - x_i) \ln Q_i\} ,$$

where $Q_i = 1 - P_i = \exp(-\lambda d_i)$ and the constant term has been omitted. The score is easily seen to be

$$S(\lambda) = \sum \{(x_i - n_i P_i)d_i/P_i\} \tag{2.1}$$

and

$$I(\lambda) = \sum w_i d_i^2 , \tag{2.2}$$

where $w_i = n_i Q_i/P_i$, $i = 1, 2, \ldots, I$. The ML estimator, $\hat{\lambda}$, is found by iterative solution of $S(\hat{\lambda}) = 0$, say by the Newton–Raphson techniques or the method of scoring (Rao, 1973, Ch. 5), see Thomas (1972) for a computer algorithm.

The test of homogeneity, which is based on the components of the sums in $S(\lambda)$ and $I(\lambda)$, is easily seen to be

$$X_I^2 - 1 = \sum \{S_i^2(\hat{\lambda})/I_i(\hat{\lambda})\} = \sum \{(x_i - n_i \hat{P}_i)^2/(n_i \hat{P}_i \hat{Q}_i)\} , \tag{2.3}$$

where $\hat{P}_i = 1 - \exp(-\hat{\lambda} d_i) = 1 - \hat{Q}_i$. This is simply the usual Pearson χ^2 formula for a $2 \times I$ with the ML estimator of λ used in the expected values.

Further explicit expressions of practical use are

$$dI(\lambda)/d\lambda = -\sum (w_i d_i^3/P_i) ,$$

and

$$E(d^3 \ln L/d\lambda^3) = \sum \{w_i d_i^3 (1 + Q_i)/P_i\} .$$

From these we find, from (1.3),

$$B(\lambda) = (\sum w_i d_i^3)/[2\{\sum w_i d_i^2\}^2] , \tag{2.4}$$

which is always positive, and from (1.5) we have

$$\mu_3\{S(\lambda)\} = \sum \{w_i d_i^3 (Q_i - P_i)/P_i\} . \tag{2.5}$$

Example 1: Adenovirus in HeLa cell cultures. Consider the data of Pereira and Kelly (1957). The infective agent is adenovirus, the host or target is HeLa cell cultures, and the response is a cytopathic effect. The data are displayed in Table 1. Solving (2.1) iteratively, we find $\hat{\lambda} = 0.413$ and $I(\hat{\lambda}) = 386.64$. The goodness of fit test (2.3) yields $\chi_4^2 = 3.91$ ($P = 0.42$). The approximate 95% limits

Table 1
Example 1: Adenovirus in tissue culture

Data[a]

i	d_i	x_i	n_i
1	0.5	3	32
2	1	10	32
3	2	19	30
4	4	27	31
5	8	30	32

Analysis: $\hat{\lambda}$ = 0.413 $\pm$ 0.0509

Homogeneity test: χ_4^2 = 3.91 (P = 0.42)

95% confidence limits:
ML:	(0.313, 0.512)
Corrected ML:	(0.309, 0.509)
Score:	(0.319, 0.519)
Corrected score:	(0.322, 0.522)

[a] Source of data: Pereira and Kelly (1957).

calculated from (1.2) are (0.313, 0.512). The $B(\lambda)$ for $\hat{\lambda}$ = 0.413 is 0.004. Thus $\hat{\lambda}_c$ = 0.409 and the corrected ML limits are (0.309, 0.509).

The solution of the score method, (1.4), for the confidence limits is easily done iteratively by starting with two trial values for a given limit, say the numbers provided by the two ML methods given above. Compute $z(\lambda_0)$ and $z(\lambda_1)$ from (1.4). Call λ_0 the value yielding the $z(\lambda_0)$ closest to the appropriate $\pm z_{\alpha/2}$. Using the secant method, we find the third trial value, λ_2, by the relation,

$$\ln \lambda_2 = \ln \lambda_0 + \{z(\lambda_0) \pm z_{\alpha/2}\} \{\ln(\lambda_1/\lambda_0)\}/\{z(\lambda_0) - z(\lambda_1)\} . \qquad (2.6)$$

The plus sign is used for the upper limit and the minus sign for the lower limit. The iteration proceeds by retaining the two trial values that bracket $\pm z_{\alpha/2}$, if such is the case, and proceeding until convergence is achieved. Doing this, we find the score method yields the 95% interval (0.319, 0.519), an interval shifted to the right of those involving the ML estimator.

Adjustment for skewness proceeds by using $z_s(\lambda)$ in (2.6) and solving iteratively. This yields the 95% interval (0.322, 0.522).

These results are summarized in Table 1. The variation in the various methods is not too great in this application, which has a relatively large sample sizes ($n \sim 30$) at each of the doses. In other examples with smaller sample sizes and where, unlike this example, the P_i do not vary from roughly zero to unity, the variation in the methods can be much greater.

3. The nuisance-parameter case

Consider now the situation where we are concerned with a single parameter of interest, say ρ, when the likelihood also depends on another parameter, say λ. We shall limit our discussion to a single nuisance parameter λ, although the score method extends easily to case where λ is a vector (Bartlett, 1953b). We extend the notation to include a series of mutually independent parallel samples, x_{0i}, n_{0i}; $i = 1, 2, \ldots, I$; and x_{1j}, n_{1j}; $j = 1, 2, \ldots, J$. This gives rise to the log-likelihood

$$\ln L = \sum_i \ln L_{0i}(\rho, \lambda) + \sum_j \ln L_{1j}(\rho, \lambda) \,.$$

For inference on ρ, we first consider the score,

$$S_1(\rho, \lambda) = \partial \ln L(\rho, \lambda)/\partial \rho \,. \tag{3.1}$$

This will in general depend not only on ρ but also on λ. Thus we must also consider

$$S_2(\rho, \lambda) = \partial \ln L(\rho, \lambda)/\partial \lambda \,.$$

The ML estimator of λ for a given ρ is the solution, $\hat{\lambda}$, to the equation,

$$S_2(\rho, \hat{\lambda}) = 0 \,, \tag{3.2}$$

which might be termed the auxiliary equation. This may be a simple linear or quadratic equation, but it may also require an iterative solution. Substituting $\hat{\lambda}$ in (3.1) yields the studentized score, $S_1(\rho, \hat{\lambda})$. We base inferences on ρ on this studentized score.

Unlike the single-parameter case it is not possible to find, in general, the exact moments of $S_1(\rho, \hat{\lambda})$. Bartlett (1953b) shows that asymptotically its moments may be found from the approximate relationship,

$$S_1(\rho, \hat{\lambda}) \sim S_1(\rho, \lambda) - I_{\rho\lambda} S_2(\rho, \lambda)/I_{\lambda\lambda} \,, \tag{3.3}$$

where the I's are the components of the information matrix, that is,

$$I_{\lambda\lambda} = \mathrm{E}(-\partial^2 \ln L/\partial \lambda^2) \,,$$

$$I_{\lambda\rho} = \mathrm{E}(-\partial^2 \ln L/\partial \lambda \, \partial \rho) \,,$$

$$I_{\rho\rho} = \mathrm{E}(-\partial^2 \ln L/\partial \rho^2) \,.$$

Thus we find asymptotically

$$E\{S_1(\rho, \hat{\lambda})\} = 0 ,$$

$$\mathrm{Var}\{S_1(\rho, \hat{\lambda})\} = I_{\rho\rho} - I_{\lambda\rho}^2/I_{\lambda\lambda} = I_{\rho \cdot \lambda}(\rho, \lambda) ,$$

which, except for $I_{\lambda\rho} = 0$, is always less than $I_{\rho\rho}$. The studentized score method is based on the approximate standard normal variate

$$z(\rho) = S_1(\rho, \hat{\lambda})/\{I_{\rho \cdot \lambda}(\rho, \hat{\lambda})\}^{1/2} . \tag{3.4}$$

Letting $z(\rho) = \pm z_{\alpha/2}$ and solving yields the approximate confidence limits. This is done similarly to the single parameter case as in (2.6) with the exception that, in the general case, at each iteration we must solve the auxiliary equation, (3.2), for a new value of $\hat{\lambda}$.

Although the right-hand side of (3.3) has exactly a mean of zero, the $E\{z(\rho)\}$ can sometimes have a bias of order $n_i^{-1/2}$.

Bartlett (1955) shows that its bias, to terms of this order, is in general

$$
\begin{aligned}
B(\rho, \lambda) = & -\{E(\partial^3 \ln L/\partial\rho \, \partial\lambda^2) + 2 \, \partial I_{\rho\lambda}/\partial\lambda\}/\{2I_{\lambda\lambda}(I_{\rho \cdot \lambda})^{1/2}\} \\
& + I_{\rho\lambda}\{E(\partial^3 \ln L/\partial\lambda^3) + 2 \, \partial I_{\lambda\lambda}/\partial\lambda\}/\{2I_{\lambda\lambda}^2(I_{\rho \cdot \lambda})^{1/2}\} .
\end{aligned} \tag{3.5}
$$

Although in many situations this term vanishes, in others it is not negligible. One such situation is the important problem of the estimation of the odds ratio in 2×2 binomial tables (see Gart, 1985, 1987).

The asymptotic third central moment of $S_1(\rho, \hat{\lambda})$ can be found from the third moment of the expression on the right of (3.3). Defining $D = I_{\lambda\rho}/I_{\rho\rho}$ and letting $u = S_1(\rho, \lambda)$ and $v = S_2(\rho, \lambda)$, we have

$$
\begin{aligned}
\mu_3\{S_1(\rho, \hat{\lambda})\} &= \mu_3(u - Dv), \\
&= \mu_3(u) - D^3\mu_3(v) - 3D \, \mathrm{Cov}(u^2, v) + 3D^2 \, \mathrm{Cov}(u, v^2) .
\end{aligned} \tag{3.6}
$$

The third central moments $\mu_3(u)$ and $\mu_3(v)$ may be expressed in terms of the partial derivatives of the components of the information and the $E\{\partial^3 \ln L/\partial\lambda^3\}$ and $E(\partial^3 \ln L/\partial\rho^3)$, analogously to (1.5). The covariances can similarly be found to be

$$\mathrm{Cov}(u^2, v) = 2 \, \partial I_{\rho\lambda}/\partial\rho + \partial I_{\rho\rho}/\partial\lambda + 2E(\partial^3 \ln L/\partial\rho^2 \, \partial\lambda)$$

and

$$\mathrm{Cov}(u, v^2) = 2 \, \partial I_{\rho\lambda}/\partial\lambda + \partial I_{\lambda\lambda}/\partial\rho + 2E(\partial^3 \ln L/\partial\rho \, \partial\lambda^2) . \tag{3.7}$$

The skewness is then, to order $n_i^{-1/2}$,

$$\gamma_1(\rho, \hat{\lambda}) = \mu_3 \{S(\rho, \hat{\lambda})\}/\{I_{\rho \cdot \lambda}(\rho, \hat{\lambda})\}^{3/2} .$$

Thus the bias and skewness-corrected statistic is

$$z_s(\rho) = z(\rho) - B(\rho, \hat{\lambda}) - \gamma_1(\rho, \hat{\lambda})(z_{\alpha/2}^2 - 1)/6 , \tag{3.8}$$

and the bias and skewness-corrected confidence limits, to terms of order $n_i^{-1/2}$, is given by the solutions to $z_s(\rho) = \pm z_{\alpha/2}$. These may be found by an iterative solution using the secant method as in (2.6). Here, again, one must also solve the auxiliary equation, (3.2), for each trial value of ρ.

In the case where the two series of mutually independent samples are matched we may wish to test the model for the homogeneity of ρ over the samples. Letting $j = i = 1, 2, \ldots, I$, we define

$$S_{1i}(\rho, \lambda) = \partial \ln L_i(\rho, \lambda)/\partial \rho ,$$

$$I_{\rho \rho i} = E(-\partial^2 \ln L_i(\rho, \lambda)/\partial \rho^2) ,$$

$$I_{\rho \lambda i} = E(-\partial^2 \ln L_i(\rho, \lambda)/\partial \rho \, \partial \lambda) .$$

The test for homogeneity is based on the approximate χ^2 statistic with $I - 1$ degrees of freedom,

$$\chi_{I-1}^2 = \sum \{z_i(\hat{\rho})\}^2 ,$$

where

$$z_i(\hat{\rho}) = S_{1i}(\hat{\rho}, \hat{\lambda})/\{I_{\lambda \cdot \rho i}(\hat{\rho}, \hat{\lambda})\}^{1/2}$$

and

$$I_{\lambda \cdot \rho i}(\rho, \lambda) = I_{\rho \rho i} - I_{\lambda \rho i}^2/I_{\lambda \lambda} ,$$

and $\hat{\rho}$ and $\hat{\lambda}$ are the joint ML estimators of both ρ and λ.

4. Application to the comparison of one-hit curves

Consider now the comparison of two one-hit curves $P_{0i} = 1 - \exp(-\lambda_0 d_i)$ and $P_{1j} = 1 - \exp(-\lambda_1 d_j)$ where $i = 1, 2, \ldots, I$, and $j = 1, 2, \ldots, J$. We wish to estimate $\rho = \lambda_1/\lambda_0$. The individual maximum likelihood estimators of λ_0 and λ_1 may be found by methods discussed in Section 2. Denote these estimates by $\hat{\lambda}_0$ and $\hat{\lambda}_1$ and their standard errors by $\{I_k(\hat{\lambda}_k)\}^{-1/2}$, $k = 0, 1$, as defined in

(2.2). The ML estimator of ρ is $\hat{\rho} = \hat{\lambda}_1/\hat{\lambda}_0$ and its asymptotic variance may be estimated by

$$\text{Var}(\hat{\rho}) = [1/\{I_1(\hat{\lambda}_1)\} + \hat{\rho}^2/\{I_0(\hat{\lambda}_0)\}]/\hat{\lambda}_0^2,$$

and crude confidence limits are given by

$$\hat{\rho} \pm z_{\alpha/2}\{\text{Var}(\hat{\rho})\}^{1/2}. \tag{4.1}$$

These may be greatly effected by the bias of $\hat{\rho}$ and the estimation and variation of its standard error over the range of values of ρ. Some of these aspects, such as the bias, may be corrected by general methods, see Shenton and Bowman (1977, Ch. 3). Usually, however, it is easier to adjust for these by using the score method.

Turning to the score method, we reparameterize, letting $\rho = \lambda_1/\lambda_0$ and $\lambda = \lambda_0$, so our notation is consistent with that of Section 3. The relevant score statistics are

$$S_1(\rho, \lambda) = \lambda \sum \{(x_{1j} - n_{1j}P_{1j})d_j/P_{1j}\}$$

and

$$S_2(\rho, \lambda) = \sum \{(x_{0i} - n_{0i}P_{0i})d_i/P_{0i}\} + \rho \sum \{(x_{1j} - n_{1j}P_{1j})d_j/P_{1j}\}. \tag{4.2}$$

The elements of the information matrix are

$$I_{\rho\rho} = \lambda^2 \sum w_{1j}d_j^2,$$

$$I_{\lambda\lambda} = \sum w_{0i}d_i^2 + \rho^2 \sum w_{1j}d_j^2,$$

$$I_{\lambda\rho} = \lambda\rho \sum w_{1j}d_j^2,$$

where

$$w_{0i} = n_{0i}Q_{0i}/P_{0i}, \quad i = 1, 2, \ldots, I,$$

and

$$w_{1j} = n_{1j}Q_{1j}/P_{1j}, \quad j = 1, 2, \ldots, J.$$

Thus,

$$I_{\lambda \cdot \rho}(\rho, \lambda) = (\lambda/\rho)^2/V,$$

where

$$V = 1/\{\sum (w_{0i}d_i^2)\} + 1/\{\rho^2 \sum (w_{1j}d_j^2)\}.$$

For a given ρ, we must solve $S_2(\rho, \hat{\lambda}) = 0$. But we see from (4.2) that this itself will require an iterative solution. In many cases this may be simplified. For P_{0i} and P_{1j} relatively small, we have that

$$1/P = 1/(\lambda d) + \tfrac{1}{2} + \tfrac{1}{12}(\lambda d) + O(\lambda d)^2 .$$

The accuracy of this three term approximation is such that even for $P = \tfrac{1}{2}$, where $1/P = 2$ exactly, the approximation yields 2.0005. Substituting this approximation for both P_{1j} and P_{0i} in (4.2) and setting $S_2(\lambda, \rho) = 0$, we obtain the quadratic equation in $\hat{\lambda}$ for a given ρ to be

$$A\hat{\lambda}^2 - B\hat{\lambda} + C = 0 , \tag{4.3}$$

where

$$A = \tfrac{1}{12}(\textstyle\sum x_{0i}d_i^2 + \rho^2 \sum x_{1j}d_j^2) ,$$

$$B = \textstyle\sum n_{0i}d_i + \rho \sum n_{1j}d_j - \tfrac{1}{2}(\sum x_{0i}d_i + \rho \sum x_{1j}d_j) ,$$

$$C = \textstyle\sum x_{0i} + \sum x_{1j} .$$

Thus the approximate normal deviate corresponding to (3.4) is

$$z(\rho) = \rho \hat{V}^{1/2} \textstyle\sum \{(x_{1j} - n_{1j}\hat{P}_{1j})d_j/\hat{P}_{1j}\} , \tag{4.4}$$

where $\hat{V}$ and $\hat{P}_{1j}$ are evaluated from either (4.2) or (4.3) as is appropriate for the given values of ρ and λ.

The bias and third moments require evaluation of partial derivatives of the components of the information matrix and expectations of third partials of the likelihood. Thus we find from (3.5),

$$B(\rho, \lambda) = - \frac{\{(\sum w_{0i}d_i^3)/(\sum w_{0i}d_i^2) - \rho(\sum w_{1j}d_j^3)/(\sum w_{1j}d_j^2)\}}{\{2V^{3/2}\rho^2(\sum w_{0i}d_i^2)(\sum w_{1j}d_j^2)\}} , \tag{4.5}$$

and, from (3.6) and (3.7),

$$\gamma_1(\rho, \lambda) = [\textstyle\sum \{w_{0i}(P_{0i} - Q_{0i})d_i^3/P_{0i}\}/(\sum w_{0i}d_i^2)^3$$
$$+ \textstyle\sum \{(w_{1j}(Q_{1j} - P_{1j})d_j^3/P_{1j}\}/(\rho \sum w_{1j}d_j^2)^3]/V^{3/2} . \tag{4.6}$$

The corrected normal deviate is then given by substituting (4.4), (4.5), and (4.6) and (3.8).

In the case where $d_i = d_j$ for all $i, j = 1, 2, \ldots, J$; a homogeneity test for ρ is given by

$$\chi_{I-1}^2 = \textstyle\sum_i \{z_i(\hat{\rho}, \hat{\lambda})\}^2 , \tag{4.7}$$

where

$$\{z_i(\hat{\rho}, \hat{\lambda})\}^2 = \frac{(x_{1i} - n_{1i}\hat{P}_{1i})^2 \{1 + \hat{\rho}^2(\sum w_{1i}d_i^2)/(\sum w_{0i}d_i^2)\}}{(n_{1i}\hat{P}_{1i}\hat{Q}_{1i})} \ ,$$

and $\hat{\rho}$ is the solution to $S_1(\hat{\rho}, \hat{\lambda}) = 0$, that is, the ML estimator of ρ. It is seen that each component of the χ^2 statistic has a factor similar to that of the Pearson statistic for the first series of binomial variates.

Example 2: Radiation enhancement of chemically induced cell transformations. Consider now a set of two dose response curves where the hosts or targets are Syrian hamster cells and the agent is benzo(a)pyrene. One series of doses is also exposed to X-ray at a dose of 250 R. Each series is exposed to the same doses of the chemicals. The data are given in detail in Table 2.

Table 2
Example 2: Chemically induced transformation with X-ray enhancement

Data[a]

		Non-irradiated		X-ray (250 R)	
i	d_i	x_{0i}	n_{0i}	x_{1i}	n_{1i}
1	0.1	3	2758	12	1235
2	0.5	7	2408	33	1085
3	1.0	9	1234	30	601
4	2.5	16	1003	93	605
5	5.0	24	787	99	422

Analysis: $\hat{\rho} = 9.14 \pm 1.32$

Homogeneity test: $\chi_4^2 = 7.23$ ($P = 0.12$)

95% confidence limits:
  ML: (6.55, 11.7)
  Score: (6.91, 12.1)
  Corrected score: (6.94, 12.2)

[a] As given in Gart, DiPaolo, and Donovan (1979).

We consider the analysis with number of cells at each dose being fixed. An analysis considering their variation over dishes is given in Gart, DiPaolo, and Donovan (1979). The original data are given in DiPaolo, Donovan, and Popescu (1976); the data in Table 2 are the corrected form given in Gart et al. (1979).

If we fit the one-hit curve to each series separately, for the unirradiated series $\hat{\lambda}_0 = 0.00651$ and $\hat{\lambda}_1 = 0.0595$, thus $\hat{\rho} = 9.14$, termed the enhancement in this context. As the doses of the chemical are identical in the two series we may test for the homogeneity of ρ over the doses by employing (4.7). This yields

$$\chi_4^2 = 7.23 , \qquad P = 0.12 .$$

The 95% confidence interval based on the maximum likelihood estimator is, from (4.4), (6.55, 11.7).

Turning to the interval from the score method, we reparameterize $\lambda = \lambda_0$ and $\rho = \lambda_1/\lambda_0$ so that we may write the likelihood in the form of Section 4. As the P_{1i}'s and P_{0i}'s are small we can use the quadratic approximation to the auxiliary equation as in (4.3). Using (4.4), we find the score method yields the 95% limits of (6.91, 12.1). Employing the bias and skewness correction, we find the interval (6.94, 12.2). In this example wherein the sample sizes are quite large, the score intervals for the ratio ρ are substantially different from that based on the ML estimator. In smaller samples the difference in the methods can be quite appreciable.

It might be noted that the difference in score intervals and those based on ML estimators is analogous to the difference between score tests and Wald tests. As noted by several authors score tests are usually preferable.

References

Bartlett, M. S. (1953a). Approximate confidence intervals. *Biometrika* **40**, 12–19.

Bartlett, M. S. (1953b). Approximate confidence intervals II. More than one unknown parameter. *Biometrika* **40**, 306–317.

Bartlett, M. S. (1955). Approximate confidence intervals III. A bias correction. *Biometrika* **42**, 201–204.

DiPaolo, J. A., Donovan, P. J. and Popescu, N. C. (1976). Kinetics of Syrian hamster cells during X-irradiation enhancement of transformation *in vitro* by chemical carcinogen. *Radiat. Res.* **66**, 310–325.

Gart, J. J. (1985). Analysis of the common odds ratio: corrections for bias and skewness. *Bull. Int. Stat. Inst.* **45**, Book 1, 175–176.

Gart, J. J. (1987). The equivalence of two corrections to the approximate mean of an entry in a contingency table. *Biometrika* **74**, 661–663.

Gart, J. J., DiPaolo, J. A. and Donovan, P. J. (1979). Mathematical models and the statistical analyses of cell transformation experiments. *Cancer Res.* **39**, 5069–5075.

Pereira, H. G. and Kelly, B. (1957). Dose-response curves of toxic and infective actions of adenovirus in HeLa cell cultures. *J. Gen. Microbiol.* **17**, 517–524.

Rao, C. R. (1973). *Linear Statistical Inference and Its Applications.* 2nd ed., Wiley, New York.

Shenton, L. R. and Bowman, K. O. (1977). *Maximum Likelihood Estimation in Small Samples.* MacMillan, New York.

Thomas, D. G. (1972). Tests of fit for one-hit vs. two-hit curves (Algorithm AS50). *Appl. Stat.* **21**, 103–112.

Thomas, D. G. and Gart, J. J. (1971). Small sample performance of some estimators of the truncated binomial distribution. *J. Am. Stat. Assoc.* **66**, 169–177.

C. R. Rao and R. Chakraborty, eds., *Handbook of Statistics, Vol. 8*
© Elsevier Science Publishers B.V. (1991) 407–459

14

Kidney-Survival Analysis of IgA-Nephropathy Patients: A Case Study

O. J. W. F. Kardaun *

Λέγειν τὰ προγενόμενα, γινώσκειν τὰ παρεόντα,
προλέγειν τὰ ἐσόμενα μελετᾶν ταῦτα.

HIPPOCRATES

Introduction and summary

In this chapter, a practical case study on survival-time analysis with censored data is presented. An observation is called censored if one knows that the true survival time, which cannot (yet) be observed, is larger than a certain lower bound. If, for instance, a patient is still alive at the closing date of some study, then the true survival time (measured, for example, from the date of diagnosis onwards) is not observable, but one knows that this time is larger than the time between diagnosis and the closing date of the study. Simply omitting the censored observations from the analysis will clearly introduce a bias from the fact that in that case individuals with a relatively short lifetime are selected from the available sample.

The origins of statistical survival analysis date back to the Dark Ages of the history of statistics, i.e. to before the beginning of the last century. Early investigations on this subject are De Witt (1671) on 'Waerdije van lijfrenten' (see Smid, 1947), Bernoulli (1760) on estimating the effects of smallpox inoculation on death rate, and Nightingale (1858), who collected and evaluated mortality data of British troops during the Crimean War. She devised some original graphical representations in her attempts to convince the authorities that hygienic circumstances in the field hospitals were liable to improvement. Life tables have been used since some centuries by demographers and actuaries. In modern terminology, a life table estimates, from censored data and (nearly) without distributional assumptions, the survival function $S(t) = \Pr\{T > t\}$ of an individual taken at random from the population under investigation. (If there are no censored observa-

* Address correspondence to: Max-Planck Institut für Plasmaphysik, Boltzmannstrasse 2, D-8046 Garching bei München, Germany.

tions, a distribution-free estimator of $S(t)$ is obviously the number of survivors at time t divided by the sample size.) In 1926, Greenwood derived a simple formula to estimate the variance of a life table. Renewed interest in the analysis of censored data arose after Kaplan and Meier (1958), in which some properties of various (almost) distribution-free estimators for $S(t)$ were thoroughly discussed. In the period from 1958 to 1972, the 'Renaissance' in the history of survival time analysis, this interest was centered around generalising two- (and k-) sample rank tests to the domain of censored data. These methods were applied to compare, for example, the effects of several treatments on survival time in a specified population of individuals. The influence of other covariables (sometimes called nuisance covariates), such as age, or stage of the cancer, was accounted for 'by stratification': the sample is stratified (i.e. divided into homogeneous subgroups) with respect to the nuisance covariates, a particular k-sample test is used within each stratum, and these within-stratum tests are combined into one global test-statistic. In 1972, an epochal paper (Cox, 1972) appeared, in which the influence of a vector $z = (z_1, \ldots, z_k) \in \mathbb{R}^k$ of covariates on survival time was modelled by

$$S(t|z) = S(t|0)^{h(t, \, \beta^\mathrm{T} z)} ,$$

where $\beta = (\beta_1, \ldots, \beta_k)^\mathrm{T}$ is a vector of regression coefficients, and h a specified function. This model is called the (general) proportional hazards model. In simple situations, h is assumed to be independent of time and one often takes $h(t, \, \beta^\mathrm{T} z) = e^{\beta^\mathrm{T} z}$. If, for example, we have one covariate ($k = 1$) which assumes only two values, we are back in a two-sample situation. An important aspect of this paper was the introduction of a generalised type of likelihood, suitable for (numerically) estimating β without any distributional assumption about $S(t|z)$. It was argued that usual maximum likelihood properties should hold for estimators based on this 'partial' likelihood. The history of the modern era is harder to describe. Many papers and investigations followed Cox's article. Some tendencies are: (1) to complete the set of generalisations of classical procedures to censored data situations, (2) to apply censored data techniques, with the help of suitable computer programs, to more complex data sets, (3) to develop tests for assessing the goodness-of-fit of a proportional hazards model, and (4) to recast the description of censoring as well as Cox's model into the framework of stochastic processes and martingale theory. In this last approach, developed by Aalen, Gill and others, rigorous theorems on Cox's model (and the k-sample problem) are derived under fairly general censoring schemes. Moreover, these results are easily extended to the situation of competing risks (of which ordinary censoring is a special case) and, more generally, to (semi-) Markov processes. A semi-Markov process can be used, for example, to model several (transient) health states, and several causes of death.

    In this chapter, a case study on survival analysis is presented, which arose from a collaboration between nephrologists (Beukhof and Van der Hem) and statisticians (Kardaun and Schaafsma). From a didactical point of view, the reader will be confronted with a variety of statistical aspects, centered around the proportional hazards regression model. More on the medical aspects of the study can be found in Beukhof's thesis (Beukhof, 1986). One is interested in the survival time (measured from the date of diagnosis until terminal renal failure) of patients with a certain kidney disease (IgA nephropathy). Several correlated risk factors, such as the glomerular filtration rate of the kidneys, hypertension, proteinuria and hematuria (macroscopic as well as microscopic) seem to influence this survival time. Besides, there are kidney biopsy variables (with some missing observations), which are possibly related to the seriousness of the disease. The type of hypertension treatment changed rather drastically during the period of investigation. After a descriptive analysis of the data, the problem of selecting a suitable set of predicting variables is explicitly addressed. A practical result of the chapter consists in the construction of two nomograms, from which one can read, for an arbitrary patient (with risk-factor scores z), the estimated survival function $\hat{S}(t|z)$ and (for a fixed time t) the involved statistical uncertainty. Furthermore, a residual analysis is discussed for checking the proportional hazards assumption. This, however, leads to easily interpretable results only in the simple case that there are no censored observations. Finally, attention is paid to the practical handling of the missing values. The nomograms and the residual analysis provide some additional insight into the structure of the proportional hazards model. A rather general interactive computer program has been developed to support, in supplement to the package BMDP, the analysis of this chapter.

Some remarks about notation. Basic random variables are generally denoted by (roman) capitals, their actual outcomes by the corresponding lower-case characters. Parameters in a model are usually denoted by greek characters, their estimates (which are also random variables) by a circonflex ($^\wedge$), incidentally by a haček ($^\vee$). Vectors and matrices are written in bold-face characters. The symbol I with a subscript in brackets denotes the indicator function. For example, $I_{\{X>Y\}}$ means: 1 if $X > Y$ and 0 otherwise (i.e. if $X \leqslant Y$). A vertical bar is used in conditional probabilities. The expression $\Pr\{T > t | Z = z\}$ stands for the probability that T is larger than t, given that the vector of covariates (considered as a random variable) assumes the value z. Frequently, this is abbreviated to $\Pr\{T > t | z\}$ and interpreted as: the probability that a patient with a given vector of covariates z survives t years.

The list of references at the end of the chapter sometimes refers to work of related interest not explicitly mentioned in the main text.

1. Case description

1.1. Data set

Beukhof et al. (1986) performed a follow-up study of a group of patients who, in the period 1967–1982, had undergone renal biopsy at the Nephrology Department of the Groningen University Hospital. They selected patients fulfilling the following inclusion criteria:

Table 1
Kidney-survival data of 75 IgA-nephropathy patients[a]

x_0	x_1	x_2	x_3	x_4	x_5	x_6	x_7	T_{birth}	T_{man}	T_{biop}	δ	T_{term}	y_1	y_2	y_3	y_4
16	070	05	1	1	1	50	1	090742	250180	180380	1	010883				
17	109	01	2	9	2	50	1	120638	010769	240774	1	010883	3	2	2	3
86	165	03	2	9	4	50	1	040460	010176	190178	1	010883	2	5	2	3
41	071	02	2	9	1	11	1	150122	010772	270978	1	010883	2	2	4	5
04	049	05	1	1	1	02	1	080624	150974	300176	2	200180	5	4	4	5
18	118	00	2	9	3	50	1	050563	010773	250675	1	010883	2	2	2	2
73	075	03	1	1	2	50	1	300744	010757	220981	1	010883	3	2	4	4
48	037	00	1	1	2	50	1	300647	151172	200979	1	010883	2	2	2	1
78	080	04	1	1	1	10	1	080151	080880	101280	1	010883	2	2	2	3
21	103	02	2	9	3	16	2	291157	010180	211180	1	010883	3	2	2	2
03	077	01	1	0	1	50	1	010343	190680	260181	1	010883	3	2	4	4
44	025	02	1	1	1	01	1	170653	151280	020381	1	010883	5	2	4	5
72	114	04	2	9	3	50	1	090267	251278	120581	1	010883	2	4	2	4
71	101	00	2	9	3	50	1	040656	300781	060782	1	010883				
34	116	01	2	9	4	50	1	160163	150882	170982	1	010883	2	2	2	3
08	093	01	2	9	1	30	1	150248	151169	230370	2	141174	4	2	4	4
27	040	05	1	0	4	25	1	070845	300156	251067	2	201268				
24	036	03	1	1	1	50	1	211044	191065	171078	2	191280	3	2	4	5
50	125	02	1	1	1	22	1	030728	010769	240374	2	150780	4	2	3	5
89	076	02	2	9	1	50	1	050835	270280	140181	1	010883				
90	089	02	2	9	1	50	1	141160	010581	050382	2	300583				
52	102	02	1	1	1	50	1	191049	010775	091176	2	190783				
25	114	00	2	9	1	11	1	210336	101165	220775	1	010883	4	2	2	2
22	130	01	1	1	2	50	1	291251	011173	240674	1	010883	2	2	2	3
13	134	0	2	9	3	07	1	310563	010169	051177	1	010883	3	2	2	3
14	094	01	2	9	4	50	1	260756	010175	050776	1	010883	2	2	4	4
45	014	4	1	0	1	20	1	210139	011165	280967	2	230169				
77	103	00	2	9	1	11	1	150656	010174	260674	1	010883	2	2	2	3
23	128	03	2	9	1	50	1	260357	010176	260876	1	010883	2	2	4	5
66	094	00	2	9	2	50	1	150437	010752	231281	1	010883	3	2	2	3
39	099	01	2	9	4	50	1	130556	010771	230474	1	010883	2	2	2	2
40	095	01	1	1	4	02	1	061256	151071	171075	1	010883	3	2	4	5
37	138	01	2	9	3	50	1	040163	151280	300781	1	010883	2	2	2	4
76	008	03	1	0	1	05	1	210632	150170	220273	2	040673	1	1	4	5
80	091	02	2	9	1	10	1	270743	150869	040270	1	010883	2	2	2	2
84	058	03	2	9	1	13	1	070935	010773	200478	1	010883	4	5	4	5
47	061	02	1	1	1	02	2	200831	010770	190274	1	010883	5	2	4	5
75	127	01	2	9	4	30	1	300655	010370	120570	1	010883	2	2	2	2
36	106	02	2	9	1	15	2	170753	150768	171268	1	010883				
02	088	02	2	9	1	00	1	300134	010756	240767	3	130273				
38	127	01	2	9	1	02	1	040442	151171	140872	3	250274	2	2	2	2
57	036	04	1	1	1	12	1	020321	010763	250376	1	010883				
01	123	01	2	9	2	02	1	070551	070370	210472	1	010883	2	2	2	2
43	093	00	2	9	2	03	1	260247	010765	130978	1	010883	2	2	2	3
05	080	08	1	1	3	07	1	130951	010776	201180	1	010883	1	1	4	5
81	150	01	2	9	1	40	1	280844	210665	290867	1	010883				
42	113	06	1	1	1	02	1	191241	150977	260380	1	010883	3	2	4	4
51	093	02	2	9	2	15	1	111141	010759	040477	1	010883	4	2	4	5
33	061	01	1	1	1	02	1	200624	020575	060276	1	010883	4	2	4	4
88	016	05	1	1	1	03	1	260251	050879	070482	2	220682	1	1	5	5
15	134	00	2	9	1	20	1	140256	090675	131075	1	010883	2	2	2	3

Table 1 (*Continued*)

x_0	x_1	x_2	x_3	x_4	x_5	x_6	x_7	T_{birth}	T_{man}	T_{biop}	δ	T_{term}	y_1	y_2	y_3	y_4
70	130	05	1	1	1	10	1	161145	150971	190472	2	050279	4	2	4	4
20	089	00	2	9	4	50	2	270559	010769	100478	1	010883	2	2	2	2
74	089	00	1	1	1	02	1	010320	290168	290168	3	281275				
69	103	00	2	9	4	10	2	240754	171075	280676	1	010883	2	2	2	2
32	167	06	2	9	1	00	1	120258	110178	260578	1	010883	2	2	2	2
59	110	00	2	9	2	07	1	300347	240577	150378	1	010883	2	2	2	2
35	111	01	2	9	2	50	2	021252	150281	281081	1	010883	2	2	2	2
82	089	04	2	9	1	12	2	090943	150170	010670	2	090483	3	2	2	2
53	122	00	2	9	4	50	1	121244	190876	031276	1	010883	2	2	2	4
62	114	0	2	9	2	10	1	091240	120577	201277	1	010883	3	2	2	4
63	090	02	1	1	2	50	1	071144	010756	291277	1	010883	5	1	5	5
85	132	02	2	9	2	50	1	060249	241177	121277	1	010883	2	2	2	2
64	064	02	1	0	1	03	1	160829	151268	151069	2	220674	4	2	4	5
26	135	03	2	9	4	50	1	060160	010778	220982	1	010883	4	2	3	4
58	087	01	2	9	1	10	1	300835	290474	210275	1	010883	5	2	2	3
56	074	02	1	1	1	07	2	080158	150178	081079	1	010883				
12	069	04	2	9	1	50	1	080306	131080	261180	1	010883	3	2	3	4
31	135	00	2	9	4	50	2	170830	081080	030481	1	010883	2	2	2	2
09	131	02	2	9	1	07	2	090563	010773	151176	1	010883	2	2	2	2
29	146	01	2	9	1	05	2	170553	210272	190472	1	010883	2	2	2	2
67	111	00	2	9	2	50	2	051262	051075	160180	1	010883	2	2	3	3
49	107	00	2	9	4	11	1	070941	121262	150579	1	010883	3	2	2	3
30	060	00	2	9	1	50	2	150312	120881	140981	1	010883	4	2	2	4
07	073	04	1	1	1	15	1	250356	030180	250180	1	010883	3	3	2	4

[a] The data were collected by Beukhof, van der Hem et al., Department of Nephrology, University of Groningen. All variables denoted by x or y (except x_4) have been measured at the time of renal biopsy. Definitions:

x_0: patient identification number.

x_1: initial creatinine clearance (ml/min).

x_2: amount of proteinuria (rounded to g/24 hr).

x_3: hypertension; 1 = hypertension was present (RR > x/97 mmHg), 2 = hypertension was not present at the moment of biopsy.

x_4: hypertension treatment; 0 = 'not effectively controlled', 1 = 'effectively controlled', 9 = 'no hypertension was present'.

x_5: macroscopic hematuria; 1 = absent, 2, 3, 4 = present.

x_6: microscopic hematuria; number of red blood cell counts per unit square, '50' means 'more than 50'.

x_7: sex; 1 = male, 2 = female.

T_{birth}: birth date (day/month/year).

T_{man}: date of first manifestation (day/month/year).

T_{biop}: date of renal biopsy (day/month/year).

δ: status of the patient at T_{term}; 1 = not yet under dialysis (at 1–8–1983), 2 = under dialysis at T_{term}, 3 = at T_{term} lost during follow-up.

T_{term}: termination date, i.e. closing date of the study if $\delta = 1$, date of dialysis if $\delta = 2$, date of loss during follow-up if $\delta = 3$.

y: bioptic measurements (ranked in increasing order of seriousness in the same week (January, 1983) by the pathologist Hoedemaeker); 1 = uncertain, 2 = −, 3 = + −, 4 = +, 5 = + +.

y_1: global sclerosis.

y_2: extracapillary proliferation.

y_3: interstitial infiltration.

y_4: tubular atrophy.

(1) prominent deposits of IgA antibody complexes in the mesangium of the kidney,

(2) no clinical signs of Henoch Schönlein purpura, systemic lupus erythematosus or cirrhosis of the liver,

(3) no morphological evidence of other types of primary glomerular disease,

(4) older than 12 years,

(5) transplantation or dialysis was not yet necessary.

The first three criteria delineate a disease known as primary IgA nephropathy, which we abbreviate by IgAN. Note that this diagnosis cannot be made without renal biopsy.

From 75 out of the 82 patients sufficient intake and follow-up data were available, the remaining seven being referred to the hospital from outside the region of Groningen. The data of these 75 patients are displayed in Table 1. Although the explanations below the table should generally suffice, we will make some further remarks on the methods of measurement. All intake variables $(x_1, \ldots, x_7)$ are measured just before biopsy. The variables x_1, x_2, x_3, and x_6 are based on the median values of 3 to 10 measurements.

Creatinine clearance (x_1) is a measure of the glomerular filtration rate (GFR) of the kidneys. From its value at biopsy the clearance will globally decrease in time. The rate of decrease characterises the speed of the progression of the disease. If the clearance drops below 5 ml/min, the patient has to be dialysed.

Proteinuria (x_2) was measured from 24-hours urine samples by a biuret reaction and is expressed in (albumine equivalent) grams per 24 hr.

Hypertension (x_3) was assumed to be present if the (uncontrolled) diastolic blood pressure was measured to be above 97 mm Hg. If the diastolic blood pressure of a hypertension patient (under the influence of antihypertensive drugs) decreased below 97 mm Hg within 3 months from biopsy and remained below this limit, the hypertension is considered to be 'effectively controlled'. Otherwise the hypertension is called 'uncontrolled'. Uncontrolled hypertension occurred because effective anti-hypertensives were not available (before 1975), or because patients refused to use these drugs.

Macroscopic hematuria (x_5) was determined by the fact whether or not the patient had a visible amount of blood in the urine. All patients with macroscopic hematuria happened to be biopsied at least six weeks after the last bout of visible hematuria.

Microscopic hematuria (x_6) was measured by the number of red blood cells in a unit square under the microscope and is proportional to the concentration of these cells in urine. (The number 50 was recorded when *at least* 50 cells were counted in the square.)

The variables y_1 to y_4 denote microscopic abnormalities in the biopsied piece of kidney. Semi-quantitative scores were assigned, January 1983, by the pathologist Hoedemaeker; scores 2, ..., 5 indicate the degrees of abnormality, and score 1 indicates that no reliable assignment was possible.

The *termination date* T_{term} for a patient is the date of first hemodialysis (i.e. artificial dialysis of the blood) if it is known that this dialysis occurred before the closing date of the study (1–8–1983); otherwise, T_{term} is either the closing date of the study, or the date of emigration or actual death of the patient. Of almost every patient it was checked (August 1983) whether or not he had been treated with dialysis before 1–8–1983, by contacting the patient himself or his general practitioner. Only three of the 75 patients were lost during follow-up: two because of emigration, one because of actual death unrelated to renal failure.

1.2. Problem overview

If one would ask the clinicians to formulate the purpose of their work, they might approximately say that they want to get a better understanding of IgAN, its pathogenesis, its prognosis, etc., in order to know better what to expect and to do if a particular patient is presented to them. Hence, there are several interesting problem areas connected with the analysis of these data. We shall give here a brief general description. One can roughly distinguish between:

(1) *'Descriptive analysis'*. Summary statistics and/or graphical representations of the data are given, from which the reader quickly can get an impression of the most important features and associations present in the data.

(2) *'Risk analysis'*. One is interested in the influence of several 'risk' factors (e.g. proteinuria, hypertension) on a response variable (for instance time from biopsy until renal failure). One can distinguish between situations with and without special interest in detecting causal relations between the risk factors and the response variable. In *non-causal* risk analysis, for reasons of simplicity, questions of causality (usually involving interpretation of the results of the analysis in a detailed, e.g. physiological, model) are ignored. In *causal* risk analysis main interest is in increasing scientific (e.g. physiological) knowledge, which often implies a close connection between object-science considerations and statistics. For an interesting article about causal inference in the biological sciences, see Rothman in: Schottenfeld and Fraumeni (1982). See also Robins and Greenland (1989). Causal as well as non-causal risk analysis often constitutes a preliminary stage to prediction analysis. Benefit from the labour of revealing a causal structure may consist (among others) in a better generalisability of the corresponding predictions. However, knowledge about the causal structure is usually very incomplete, and tools to reveal it are, at least in the human life sciences, rather limited.

(3) *'Prediction analysis'*. One is interested in predicting the outcome of a response variable for a particular individual from the outcomes of a selection of the predicting variables. In our case these variables are x_1 to x_7, y_1 to y_4, T_{birth} and T_{biop}. The analysis is usually based on a preliminary investigation of (causal or non-causal) risk factors. The selection of an appropriate subset of the predicting variables is by no means easy and constitutes an interesting part of the analysis. The inclusion of more variables generally increases the statistical uncertainty of the prediction and tends to overfit the data, which decreases validity (i.e. generalisability of the results to other, similar, situations). On the other hand, if relevant variables are not included, predictions for individual patients may be seriously biased with respect to the (hypothetical) ideal situation in which all

variables relevant to prediction are known, as well as their influences. An essential topic of concern is the performance of the prediction, a concept that can be defined in various ways.

(4) 'Interdependence analysis'. One is interested in the associative structure of the explanatory variables. Also here one can distinguish between situations with and without particular interest in causal structure. Purpose of the analysis may be increase of physiological knowledge and/or reduction of the complexity associated with a large number of variables or parameters. In our case there are two natural clusters of explanatory variables: (a) the measurements y_1 to y_4 from the biopsied piece of kidney and (b) the measurements (x_1 to x_7, etc.) which are carried out without biopsy. We will call them 'bioptic' measurements and 'non-bioptic' measurements, respectively. (Note that for the diagnosis IgAN a small amount of biopsied kidney is needed.) The area of interest is naturally divided into (i) the interdependence between the variables within each of the two clusters and (ii) the interdependence between the variables of cluster (a) with those of cluster (b). A slight amount of interdependence analysis (which may also be classified as belonging to area (1)) is useful before starting an analysis of risks.

(5) 'Decision analysis'. After a (preliminary) prediction analysis, it is natural to consider the actions that could be taken. In a medical context, these actions may for example consist in (1) doing nothing, (2) performing further investigations, or (3) prescribing a certain treatment. For a population at large one might issue a warning or further investigate groups of people who are at high risk. Most actions will have medical or economical drawbacks. Real medical decision situations are complicated. They may be modelled by probabilistic decision trees consisting of 'diagnostic tests', 'treatments', 'events', etc. Sequential aspects and influences of time usually play an important role. Proper minimisation (according to some criterion) of the expected losses that may occur from the various actions are an important part of the analysis. The area is connected with ethics (determination of the losses associated with the various events) and prediction analysis (estimation of the probability of future events under various treatments). Some aspects of decision analysis are treated in Chapter 15.

There are several particular aspects associated with our data that complicate the situation. We shall discuss them briefly.

(1) 'Censoring'. The response variable of main interest is the time until renal failure (from biopsy or from date of first manifestation onwards). For most patients either the closing date of the study ($\delta_j = 1$), or emigration or actual death ($\delta_j = 3$) preceded renal failure, which prevented the exact measurement of this response variable. In these cases one only knows that the actual time until renal failure is larger than the observed time until one of the just mentioned other end-points, which are (nearly) unrelated to renal failure. The actual times are then called censored. Note that emigration and actual death, which are in this study considered as censoring events, can be looked upon as competing risks. The complication of censoring gives the analysis a particular flavour. Methods to deal with this complication are available if the censoring times are unrelated to the hazard of renal failure.

(2) '*Missing values*'. Some values of the variables y_1 to y_4 are missing. As one can notice from the data, no bioptic measurements are available from before 1969. Discussion with Beukhof revealed that complete absence of bioptic data was due to the fact that either not enough tissue was extracted from the kidney, or the extracted tissue was not available any more at the time (1981) the pathologist Hoedemaeker was classifying the bioptic specimens. (Before 1969, biopsied tissue was not preserved after examination.) Sometimes only one or two of the bioptic variables could not be determined because of the bad condition of the extracted tissue. These measurements (indicated by a score 1) alo have to be considered as missing, and one can see that they are generally associated with unfavourable scores for the other bioptic variables. Of course, it is a great pity that for 13 of the 75 cases the bioptic measurements are (entirely) missing. If these cases can be considered as a random sub-sample from all cases, then there are some ways to repair partially the occurred damage.

(3) '*Treatment*'. Treatment enters in at least two ways. Firstly, one would like to estimate, for a particular patient, the effect of a certain treatment on the response variable. Secondly, the type of treatment applied to the various patients may have been changed during the period for which the data have been collected. In our case, treatments consisted chiefly of (a) the observation of a diet (recommended when serum ureum increased beyond 30 mmol/litre) and (b) the use of anti-hypertensive drugs. The first kind of treatment did not change very much during the period and is not taken into account. The second kind of treatment altered drastically about 1975. This is incorporated as indicated in the description of the data set. (It is remarked that after 1984 a more severe type of low-protein diet was prescribed, while blood pressure was controlled to a lower target value. These measures, however, do not affect our data.)

Attention in this chapter will be mainly restricted to the first three types of analysis. It will be good, however, to reflect occasionally on the relation with the other areas. In Chapter 15 we will discuss some connections with decision analysis.

1.3. Model formulation

In this study we shall analyse the effects of several predicting variables upon the response variable T^0, which measures the time from biopsy until renal failure. The analysis is based on Cox's proportional hazards (PH) model. The values of the covariates will be all regarded as measured at the time of biopsy. Let $T_1^0, \ldots, T_n^0$ be the true survival times (i.e. times between biopsy and renal failure) of the patients. Given the values of the covariates, they are assumed to be independently distributed with unknown (differentiable) survival functions $S(t \mid z_j) = \Pr\{T_j^0 > t \mid z_j\}$, where $z_j = (z_{1j}, \ldots, z_{kj})^T$ denotes the vector of covariates of patient j ($j = 1, \ldots, n$). The survival times are assumed to be censored by independent and identically distributed (i.i.d.) random variables $U_1, \ldots, U_n$ (times between biopsy and the last day of follow-up, regularly the end point of the study) with (unknown) survival function $G(t)$. (Roughly speaking, this assumption follows from the assumption that the patients arrive at the Department according to a, possibly inhomogeneous, Poisson process.) The censoring variables are assumed to be independent from the true survival times. The variables that can be observed and are needed for estimating survival functions, are $T_j = \min(T_j^0, U_j)$ and $\Delta_j = I_{\{T_j^0 \leqslant U_j\}}$. (The last expression means that Δ_j is a binary random variable assuming the value 1 if $T_j^0 \leqslant U_j$, i.e. if renal failure did occur to patient j before the date of censoring, and assuming the value 0 if $T_j^0 > U_j$.) The assumptions so far imply that T_j has survival function $S(t\,|\,z_j)G(t)$, and

$$\Pr(\Delta_j = 1) = -\int_0^\infty G(t)\,\mathrm{d}S(t\,|\,z_j)\,.$$

Instead of using a survival function $S(t)$ in order to express the distribution of survival times, one may consider the corresponding probability density $f(t) = -S'(t)$, or, alternatively, the cumulative hazard rate $\Lambda(t) = -\ln(S(t))$ (in a chemist's notation: $\Lambda = pS$), or its derivative $\lambda(t) = \Lambda'(t)$, which is called the hazard rate. The latter can be interpreted as the probability 'density' of dying at time t, given that one has lived up to time t.

The proportional hazards model specifies that for a patient with covariate vector z,

$$\ln(-\ln[S(t\,|\,z)]) = \ln(-\ln[S(t\,|\,\mathbf{0})]) + \boldsymbol{\beta}^{\mathrm{T}}z\,, \tag{1.1}$$

where $\boldsymbol{\beta} = (\beta_1, \ldots, \beta_k)^{\mathrm{T}}$ is a vector of regression coefficients. In practice, approximate standardisation of the covariates is performed, so that $z = \mathbf{0}$ roughly corresponds to patients under average conditions. Note that the link function $\ln(-\ln S)$ in formula (1.1) is a relative of the logistic link function $\ln[S/(1-S)]$, which is frequently used for proportions. Of course, other link functions are possible (see, e.g., Aranda-Ordaz, 1983; Tibshirani and Ciampi, 1983). However, for simplicity, we shall only use link function (1.1), and perform some (general) model checks after the data have been fitted.

Cox's model will be used for both continuous and discrete explanatory variables. If, however, some discrete covariate with only a few levels has a large estimated effect on survival, we will relax at times the PH assumption for that covariate by considering the (obvious) extension

$$\ln(-\ln[S_i(t\,|\,z)]) = \ln(-\ln[S_i(t\,|\,\mathbf{0})]) + \boldsymbol{\beta}_i^{\mathrm{T}}z\,, \quad i = 1, \ldots, s\,. \tag{1.2}$$

Each level i of the covariate represents a separate stratum. In order to reduce the number of parameters, we usually assume in addition that $\boldsymbol{\beta}_1 = \cdots = \boldsymbol{\beta}_s$. This is equivalent to the assumption of no interaction (on the $\ln(-\ln)$ scale) between the pertaining discrete covariate and the other covariates. Model (1.2) with common $\boldsymbol{\beta}$ is briefly referred to as the 'stratified PH model'.

Complication 1.   In order to estimate survival functions, indicator variables are needed that tell us whether ($\Delta_j = 0$) or not ($\Delta_j = 1$) the survival time of the j-th patient is censored. We will dwell a little on the extraction of that information from the original measurements. As the glomerular filtration rate was measured several times during the period of follow up, one could consider the function GFR(t) as response variable. In that case (see Figure 1) the full measurement region of the response variable for patient j consists of (neglecting for the moment measurement errors in the determination of the GFR):

$$\{(t, \text{GFR}) \mid t \in (0, U_j), \text{GFR} > 5 \text{ ml/min}\} \, .$$

Between 5 and 20 measurements are taken within this region (the frequency is enhanced if the patient approaches the hazardous limit of 5 ml/min). The times U_j between biopsy and the closing date of the study are measured exactly (± 1 day). For survival time analysis one retains the data (T_j, Δ_j), $j = 1, \ldots, n$, where

$$T_j = \begin{cases} \text{GFR}_j^{-1} (5 \text{ ml/min}) & \text{if } T_j \leqslant U_j, \\ U_j & \text{if } \text{GFR}_j(U_j) > 5 \text{ ml/min}, \end{cases} \tag{1.3}$$

and

$$\Delta_j = I_{\{T_j \leqslant U_j\}} \, .$$

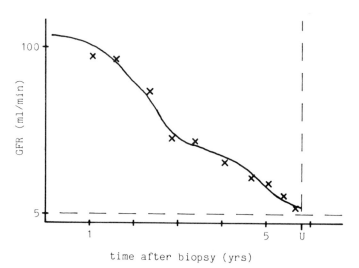

time after biopsy (yrs)

Fig. 1. The creatinine clearance (glomerular filtration rate) as a function of time. (Schematically for an imaginary patient.) The symbol U denotes the censoring time (the time between diagnosis and the closing date of the study). The crosses are measured values. If the filtration rate drops below 5 ml/min, the patient has to be connected to the dialysis equipment, or treated by renal transplantation.

The expression GFR_j^{-1} (5 ml/min) is interpreted as the time (measured from biopsy onwards) at which the true GFR has dropped to 5 ml/min. Because of (1) measurement errors, (2) non-continuous monitoring of the GFR, and (3) occasional hemodialysis before or after the glomerular filtration rate had reached the level of 5 ml/min, the variables T_j and Δ_j are somewhat fuzzy. This means that after follow-up, the outcomes of T_j are still not precisely determined and may again be modelled as random variables, which we shall denote by $T_{j,a}$ and $\Delta_{j,a}$, respectively. In fact, in cases of doubt one could try to estimate the probability $\Pr\{T_{j,a} \leqslant U_{j,a}\}$. The problem of fuzzy variables can be circumvented by considering, instead of a 5 ml/min filtration capacity, the first hemodialysis as final event. This, however, makes a generalisation to other situations more difficult. The corresponding difference in definition of T_j and Δ_j is neglected in the rest of this study. Instead of using the framework of censored survival time analysis, it may be interesting to attempt an analysis which considers

$$(T_j^0 \Delta_j, G_j(1 - \Delta_j), U_j, \Delta_j), \quad j = 1, \ldots, n,$$

as the basic data, where

$$T_j^0 = GFR_j^{-1}(5 \text{ ml/min}), \qquad G_j = GFR_j(U_j),$$

and Δ_j indicates, as before, whether or not $T_j^0 \leqslant U_j$.

The assumption of independent censoring can be relaxed somewhat to so-called non-prognostic censoring (see Lagakos, 1979), which means that $\Pr\{T_j^0 \in (t, t + dt) \mid T_j \in (u, u + du) \text{ and } \Delta_j = 0\} = -dS_j(t)/S_j(u)$ for all $0 < u \leqslant t$, i.e. if a patient is censored at time u, the act of censoring does not influence the probability distribution of the remaining (true) survival time. If a slightly weaker assumption (with no simple physical interpretation) does not hold, the type of censoring is called informative. In that case, identification problems arise and the usual methods for estimating β break down. In our case, however, non-prognostic or even independent censoring seems to be a reasonable assumption. The reader is referred to Wu and Caroll (1988), Wu and Bailey (1989) and Emoto and Matthews (1990) for informative censoring models in a growth curve context.

[It is remarked that, when a patient resides in the hospital (because of a low GFR and/or unfavourable other covariates), the glomerular filtration rate will be measured more frequently. We assume, however, that if the GFR drops below a certain 'warning level' (15 ml/min, say), the patient is regularly seen at the outpatient unit anyhow, as he develops general complaints if untreated, and that, at least if the GFR is above the warning level, the (random) measurement error of the filtration rate is much smaller than its actual value. Accordingly, the probability that a patient, without knowledge of the physicians, will have a GFR < 5 ml/min is quite small. Of course, unless impeded by emigration or actual death, it was checked for all patients whether or not renal failure had occurred before 1–8–1983.]

Complication 2. A lot of the recorded variables are essentially time-dependent.

The main response variable of interest being $X_1(t) = \text{GFR}(t)$, it would be interesting, from a scientific point of view, to select, 'identify', and interpret models of the form

$$\text{GFR}(t) = H(\text{GFR}(s), x_2(s), x_4(s), x_5(s), x_6(s), x_7, \text{age}(s), y_1, y_2, y_3, y_4, \text{error}(t)),$$

$$0 \leqslant s < t \leqslant T_{\max},$$

s and t denoting the time from biopsy onwards and H being an appropriate operator, by using methods of (stochastic) system theory and/or time series analysis. However, in order to reduce complexity, and since no sufficiently accurate data were available for such a time-dependent analysis, we will restrict attention to the special case that all explanatory variables are evaluated at $t = 0$ (the date of biopsy) and the response variable $\text{GFR}(t)$ is simplified to the scalar variable $T = \text{GFR}^{-1}(5 \text{ ml/min})$, which is essentially the time from biopsy until renal failure.

2. Inventory

2.1. Summary statistics

In Table 2 a descriptive summary of the data with respect to the most relevant variables is given. It may be useful to have a look at it before attempting to select an appropriate set of predicting variables. From the table one can see that creatinine clearance can be considered to have approximately a symmetrical, continuous distribution. Proteinuria is skewly distributed to the right and more or less discrete (actually divided into 9 ordered categories). From some points of view this is a difficult distribution and one could consider to simplify it to a factor with only three levels (for example, 'low': 0, 'medium': 1–2, and 'high': 3–8). Microscopic hematuria is truncated at 50 counts, which too makes a division into 2 or 3 classes attractive. Age at biopsy is also skewly distributed to the right, while half of the patients are aged between 20 and 38 years. During the last 5 years of the study, about the same number of patients were referred to the Nephrology Department (and diagnosed to have IgAN) as were during the first 10 years. The time between first manifestation and biopsy is usually short compared to the duration of the study (and accordingly to the mean survival time of the patients), but some values are extremely large. Concerning the discrete variables one can see that about half of the patients had macroscopic hematuria, and that only a few (5) patients had hypertension which was not effectively controlled. This situation occurred mainly before 1975, when the modern 'triple drug regimen' (β-blockers, vaso-dilators and diuretics) was not yet in use. Most of the patients (about $\frac{4}{5}$) are males, while only a smaller fraction (about $\frac{1}{5}$) got into dialysis before the end of the study.

With respect to the bioptic variables the table shows the marginal frequencies of an original $4 \times 4 \times 4 \times 4$ table, together with (for each variable) the numbers

Table 2
Descriptive statistics of 75 IgA-nephropathy patients

(A) Continuous variables[a]

Variable	(min, L, M, H, max)	Stand.skewness	Units
Creatinine clearance	(8, 74, 95, 122, 167)	− 1.6	ml/min
Proteinuria	(0, 0, 2, 3, 8)	3.4	g/24 hr (rounded)
Microscopic hematuria	(0, 7, 16, > 50, > 50)	–	counts/unit square
Age at biopsy	(12, 20, 29, 38, 75)	3.7	years
Date of biopsy	(67.6, 74.4, 77.9, 81.0, 82.8)	− 2.2	years – 1900
$T_{biop} - T_{man}$	(0.0, 0.5, 1.4, 4.4, 29.5)	8.0	years

(B) Discrete variables (non-bioptic)

Variable	Percentage	Category
Macroscopic hematuria	45	Yes
	55	No
Hypertension treatment	64	No hypertension
	29	Hypertension, effectively controlled
	7	Hypertension, not effectively controlled
Sex	83	Male
	17	Female
Censoring status	17	Dead (i.e. renal failure before 1–8–1983)
	79	Censored at 1–8–1983
	4	Lost during follow-up

(C) Discrete variables (bioptic)[b]

Variable	−	+ −	+	+ +	**	Ø	Total
Global sclerosis	30	14	10	5	3	13	75
Tubular atrophy	18	13	15	15	1	13	75
Interstitial infiltration	37	4	19	2	0	13	75
Extracapillary proliferation	53	1	2	2	4	13	75

[a] The characters L, M, and H stand for lower quartile, median, and higher quartile, respectively. The abbreviation Stand.skewness is used for the sample coefficient of skewness divided by its estimated standard error (under the hypothesis of a normal distribution).

[b] The scores − to + + denote the seriousness of the microscopic aberration (determined by one pathologist); ** means that classification was uncertain, and Ø that no bioptic specimen was available or available any more at the date of classification.

of uncertain and missing values. One can see that only a smaller fraction of the patients had advanced microscopic abnormalities of the kidney (score + −, +, or + +). A very few patients had these scores for extracapillary proliferation. Reasons for the occurrence of missing values were already discussed in the previous section.

2.2. Dependencies between covariables

There may exist various associations between the covariables, which were all measured (except for censoring status and hypertension treatment) at the time of biopsy. In Table 3 the sample correlation matrix of 8 predicting variables is displayed. The measurements of microscopic hematuria, being truncated at 50 counts, were divided into three classes, 'low' (< 10 counts), 'medium' (10–30 counts) and 'high' (> 30 counts), with scores -1, 0, and $+1$, respectively. [Note that, except for one measurement of 40 counts, there are no measurements of microscopic hematuria between 30 and 50 counts, while 50 counts never occurs before 1974. The suggestion that before 1974 the measurements were (officially) truncated at 30 counts, while after 1974 this habit was inofficially continued, was not considerd implausible by the medical doctors. In any case, 30 counts seems to be the highest reliable cutting point for grouping the observations.] As we are interested in the dependencies of several variables at the time of biopsy, 'hypertension' (present $= 1$, absent $= 0$), instead of 'hypertension treatment', is considered as a variable. The other variables are the same as in Table 2. Under the hypothesis of independence between two variables, the standard error of the sample correlation coefficient is equal to $1/(n-1)^{1/2} \simeq 0.115$, regardless of the fact whether the joint distribution of these two variables is continuous, discrete, or 'mixed'. Sample correlations which are at least twice this standard error are typed in bold face, while those at least 3 times (in Table 3(C) 4 times) the standard error are in addition underscored. This is to be considered as an informal procedure for sieving out the important correlations from the less important ones. Note that it is difficult to determine which population correlation coefficients are larger than other ones. Besides, it is difficult to give a clear interpretation of such an assertion if the corresponding joint distributions are of different type.

From the table one can see that at least the following associations seem to be manifestly present in the population. Creatinine clearance at biopsy is lower as (1) hypertension, or (2) age is higher ($r = -0.57$ and -0.45, respectively). (3) Proteinuria is positively associated with hypertension ($r = 0.46$), as is (4) macro hematuria with micro hematuria ($r = 0.34$). The table suggests, though less clearly, that there may be some more correlations in the population. (5) Proteinuria, (6) hypertension, and (7) age are negatively correlated with macro hematuria ($r = -0.33$, -0.29 and -0.33). Besides hypertension and age, also (8) proteinuria is negatively correlated with (i.e. is a bad indication for) creatinine clearance ($r = -0.27$), while (9) macroscopic hematuria is positively correlated with it (i.e. a good indication; $r = 0.32$). This last correlation is in accordance with the correlations (5), (6), and (7). The correlations between creatinine clearance and the other variables are interesting, because one may expect that good (bad) indicators for the clearance at biopsy (at least if they are not alarming to the patient) are also good (bad) indicators for the survival time, as this is the time (from biopsy onwards) until the clearance has dropped further to 5 ml/min. We shall see in the next section whether or not this expectation holds true. (10)

Table 3
Sample correlation matrices

(A) Pearson correlation matrix of non-bioptic variables[a]

		1	2	3	4	5	6	7	8
1	Proteinuria	1							
2	Hypertension	**0.46**	1						
3	Macro hematuria	**−0.33**	**−0.29**	1					
4	Micro hematuria	−0.21	**−0.25**	**0.34**	1				
5	Creatinine clearance	**−0.27**	**−0.57**	0.32	0.16	1			
6	Age	0.08	**0.23**	**−0.33**	−0.05	**−0.45**	1		
7	Sex	−0.18	−0.20	0.01	−0.04	0.09	−0.09	1	
8	Date of biopsy	0.04	−0.04	**0.28**	**0.29**	−0.02	0.14	0.04	1

(B) Pearson correlation matrix of bioptic variables[b]

		9	10	11	12
9	Global sclerosis	1			
10	Tubular atrophy	**0.57**	1		
11	Interstitial infiltration	**0.53**	**0.79**	1	
12	Extracapillary proliferation	0.11	0.22	0.08	1

(C) Pearson correlation matrix between bioptic and non-bioptic variables[c]

		1	2	3	4	5	6	7	8
9	Global sclerosis	0.22	**0.45**	**−0.42**	**−0.27**	**−0.55**	**0.42**	−0.09	−0.07
10	Tubular atrophy	**0.31**	**0.54**	**−0.30**	−0.05	**−0.52**	**0.33**	**−0.35**	0.15
11	Interstitial infiltration	**0.37**	**0.57**	**−0.36**	−0.13	**−0.53**	0.24	−0.22	−0.03
12	Extracapillary proliferation	**0.35**	0.01	−0.03	0.03	−0.05	0.03	−0.14	0.11

[a] Sample correlations that differ at least two standard errors from zero (under the hypothesis of zero correlation in the population) are typed in boldface, those which differ at least three standard errors are in addition underscored. The variables 1, 5, 6, and 8 can be considered to be continuous, the other variables are discrete.

[b] The correlation matrix is based on 57 complete cases, the four ordered categories of each variable being scored as 2, 3, 4 and 5, respectively. The underscored, boldface correlations differ at least 4 standard errors from zero.

[c] The matrix is based on 57 complete cases. The boldface correlations differ at least two, the correlations that are in addition underscored at least three standard errors from zero. The 'scientifically most interesting' correlations are those of each of the variables 1, 2, ..., 5 with the variables 9, 10, 11.

Among the patients who are biopsied at a later date there is a relatively larger fraction with hematuria ($r = 0.28$ and 0.29, respectively). Note that old age seems to be associated with higher blood pressure and with a less likely occurrence of macroscopic hematuria.

With respect to the bioptic variables one can see that (11) global sclerosis, tubular atrophy, and interstitial infiltration are clearly correlated with each other and (12) uncorrelated with extracapillary proliferation. Note that, as one can see from Table 2, extracapillary proliferation is somewhat peculiarly distributed.

Have a look at Table 3(C) for the correlations between the bioptic and the non-bioptic variables. Discarding the 'odd' variable extracapillary proliferation and the less interesting non-bioptic variables age, sex and date of biopsy, one can see that (13) hypertension and (14) creatinine clearance are strongly correlated with the three remaining bioptic variables. Though no formal statistical justification is given, it is tempting to say that (15) proteinuria and (16) macroscopic hematuria are somewhat less strongly correlated with these variables. It is interesting to note that the presence of macroscopic hematuria is associated with favourable bioptic scores. Finally, (17) old age is associated with unfavourable bioptic scores, which may be 'explained' by the associations of age with hypertension, creatinine clearance and macroscopic hematuria.

Of course, because of multiple testing, the chance that at least one of these correlations is spurious will be larger than 5%, the actual value of this probability depending on the true correlations in the population. As some variables are clearly correlated and the corresponding tests are accordingly dependent, a simple multiple decision rule based on Bonferoni's inequality seems to be too conservative. It is not easy to construct simultaneous 95% confidence intervals for the various population correlation coefficients, and we shall not attempt such an approach here.

Naturally, it is important to know whether there existed prior knowledge about the various correlations. For most of them it did. Associations (2), (3), (4), (7), (8) and (11) are known among nephrologists and (at least) based on empirical evidence. Associations (1), (13), and (14) fit into a physiological model: an elevated blood pressure destroys the glomeruli of the kidney, thereby reducing the glomerular filtration rate (creatinine clearance); a low GFR implies a diminished capacity of the kidney to excrete excess salts from the blood, which entails an elevated blood pressure, unless rigorous dietary measurements are taken.

3. Towards individual prognosis of IgA nephropathy

3.1. Selecting a set of predicting variables

3.1.1. Theory

The selection of an appropriate set of predicting variables is an important step before entering the stage of making (any reliable) predictions. There is of course much similarity with selecting variables in ordinary regression problems, for which an extensive literature exists. Main complications are the occurrence of censoring and the presence of an infinite dimensional nuisance parameter $S(t \mid \mathbf{0})$. These complications are dealt with by considering for asymptotic theory the partial likelihood function

$$L(\boldsymbol{\beta}) = \prod_{j=1}^{n} [\exp(\boldsymbol{\beta}^{\mathrm{T}} z_j)/Q_j(\boldsymbol{\beta})]^{\Delta_j}, \qquad (3.1)$$

with

$$Q_j(\boldsymbol{\beta}) = \sum_{h \in R(t_j)} \exp(\boldsymbol{\beta}^{\mathrm{T}} z_h) . \tag{3.2}$$

Here, t_j is the observed lifetime (i.e. time to censoring or renal failure), and $z_j = (z_{1j}, \ldots, z_{kj})^{\mathrm{T}}$ the vector of covariates of patient j, while Δ_j indicates whether ($\Delta_j = 1$) or not ($\Delta_j = 0$) renal failure occurred before censoring. As usual, $R(t_j)$ denotes the risk set at time t_j, i.e. the set of patients with life-time larger than or equal to t_j. The denominator $Q_j(\boldsymbol{\beta})$ will be called the partition function (indicating some analogy with models used in statistical mechanics). The dependence of $L(\boldsymbol{\beta})$ and $Q_j(\boldsymbol{\beta})$ on $z_1, \ldots, z_n$ and $\Delta_1, \ldots, \Delta_n$ is suppressed in the notation. Several investigators have argued or proven that this partial likelihood may be treated as an ordinary likelihood (see Cox, 1972, 1975; Johansen, 1983; Andersen and Gill, 1983; Wong, 1986; Doksum, 1987; Arjas and Haara, 1988).

There are at least two separate points of view to assess the relative importance of the various covariates in predicting the survival time.

Viewpoint 1. A covariable z_i ($i = 1, \ldots, k$) is considered to be statistically signifi-cant, if the null hypothesis $H_0 : \beta_i = 0$ cannot be rejected at the 5% level in a model where possibly already some covariables, which are deemed important, are included. No exact theory being available, one resorts to asymptotic methods for testing this hypothesis. Let $\hat{\boldsymbol{\beta}} = (\hat{\beta}_1, \ldots, \hat{\beta}_i)$ and $\check{\boldsymbol{\beta}} = (\check{\beta}_1, \ldots, \check{\beta}_{i-1})$ be the parameter vectors maximising the partial likelihood function for models based on the covariables $z_1, \ldots, z_i$ and $z_1, \ldots, z_{i-1}$, respectively. The 'information matrix' $I(\boldsymbol{\beta}) = -(\partial^2/\partial\boldsymbol{\beta}\,\partial\boldsymbol{\beta}^{\mathrm{T}}) \ln L(\boldsymbol{\beta})$ is assumed to be nonsingular in a sufficiently large neighbourhood of $\hat{\boldsymbol{\beta}}$. The null-hypothesis H_0 is rejected

(a) if $|\hat{\beta}_i/\hat{\mathrm{SE}}(\hat{\beta}_i)| > u_{\alpha/2}$ (or u_α if one-sided testing is allowed), the standard error being estimated by the square root of $(I^{-1})_{ii}(\hat{\boldsymbol{\beta}})$;

(b) if the score statistic $U_i(\check{\boldsymbol{\beta}}, 0) = (\partial/\partial\beta_i) \ln L(\boldsymbol{\beta})|_{\boldsymbol{\beta} = (\check{\boldsymbol{\beta}}, 0)}$ is large with respect to its estimated standard error $\{(I^{-1})_{ii}(\check{\boldsymbol{\beta}}, 0)\}^{-1/2}$;

(c) if $-2\{\ln L(\check{\boldsymbol{\beta}}, 0) - \ln L(\hat{\boldsymbol{\beta}})\} > \chi^2_{1;\,\alpha}$.

(For criterion (b), see e.g. Rao (1973), Peduzzi et al. (1980).) Although, under suitable regularity conditions, these three methods are asymptotically equivalent, for finite sample sizes they may lead to different results, and in a concrete situation it is sometimes useful to apply more than one of them. In order to determine the relative importance of two covariables from Viewpoint 1, the *P*-values of the corresponding test statistics are compared, though it may be difficult to relate a *P*-value to the magnitude of some clearly interpreted effect.

Viewpoint 2. Considering the patients as a sample from a larger population the covariates can be looked upon as random variables and will be denoted by capitals. A covariate Z_i is considered to be 'more relevant for prediction' than another covariate Z_h, if $|\hat{\beta}_i| \hat{\mathrm{SD}}(Z_i) > |\hat{\beta}_h| \hat{\mathrm{SD}}(Z_h)$, where $\hat{\mathrm{SD}}(Z_i)$ denotes the estimated standard deviation of covariate Z_i. The motivation of this point of view is as follows. Let us call patients for whom the i-th covariate differs one sample standard deviation from the (estimated) mean value $\bar{Z}_i$ 'moderately deviant (with respect to the covariate Z_i)'. Then, a moderately deviant patient will have a point-estimated hazard rate which differs a factor $\exp(|\hat{\beta}_i|\hat{SD}(Z_i))$ from that of a 'typical' patient with mean covariate $\overline{Z}_i$ and the same other covariates. (Note that for normally distributed covariates about $\frac{1}{3}$ of the patients are at least moderately deviant.) The viewpoint is concisely expressed by stating that the standard deviation is a natural unit to scale the measurements of the different covariates.

REMARK.   For dichotomous covariates ($Z_i = 0$ or 1), it is convenient to look at the relative risk associated with the two possible values of the covariate, which is $\exp(\hat{\beta}_i)$. If this covariable is approximately symmetrically distributed (i.e. $E(Z_i) \simeq 0.5$), then, to a close approximation, $SD(Z_i) = 0.5$. Therefore, in practice, if some covariates are dichotomous (with $E(Z_i)$ not too close to 0 or 1), while others are not, it is natural to compare the quantities $\exp(\hat{\beta}_i D_i)$, where $D_i = 0$ or 1 for the dichotomous covariates and $D_i = 2\hat{SD}(Z_i)$ for the other covariates.

The two points of view concentrate on different aspects. Viewpoint 1 relates the estimated effects $\hat{\beta}_i$ to their statistical uncertainty, which is strongly dependent on the 'effective' sample size, whereas Viewpoint 2 roughly evaluates, neglecting the uncertainty in $\hat{\beta}_i$, whether one risk factor is likely to be a more important predictor for the survival time of new patients arriving at the hospital than some other factor.

If, for some covariate, adequacy of the proportional hazards model is considered dubious, the stratified model may be used. In that case, with $r = 1, \ldots, s$ denoting the level of the stratifying variable, $L(\boldsymbol{\beta}) = \prod_{r=1}^{s} L_r(\boldsymbol{\beta})$, where $L_r(\boldsymbol{\beta})$ is based on all observations in the r-th stratum. The relative importance of the covariates corresponding to $\boldsymbol{\beta}$ is assessed in just the same way as in the unstratified proportional hazards model.

Based on Viewpoint 1, there are automatic procedures to select a suitable subset from a given set of possibly relevant factors. The reader is referred to the BMDP manual (Dixon et al., 1988), Peduzzi et al. (1980), and the references therein. Instead of using such an automatic procedure, we shall make, in the next sub-section, a selection 'by hand', explicitly writing down the considerations which lead from one step to the other.

3.1.2. Practice

The actual calculations were performed by using the Bio-Medical Data Processing package (Edition 1983) on a DEC-10 computer. The results of a *univariate analysis* are shown in Table 4, where the separate effects of the various covariates (each one transformed into a two-level factor) are indicated by giving approximately the two-sided *P*-values of the Gehan and the logrank test, which are censored data analogues of the two-sample rank tests with Wilcoxon and Savage scores, respectively. It is tested, for example, whether patients with proteinuria (i.e. more than 2.5 g/24 hr) have the same survival curve as patients without proteinuria (i.e. less than 2.5 g/24 hr). The test of Gehan attaches more weight to early differences in

Table 4
Influence of single (dichotomous) factors on the survival of IgA-nephropathy patients[a]

Variable	P-value		Effect	Smallest subsample	
	Gehan	Logrank		Number of patients	Number of renal failures
Non-bioptic				total 75	(13)
1 Proteinuria (>2.5 g/24 hr)	***	***	↓	22	(8)
2 Hypertension	***	***	↓	27	(10)
3 Initial GFR (<80 ml/min)	***	***	↓	22	(7)
4 Macroscopic hematuria	*	*	↑	34	(1)
5 Microscopic hematuria	·	·	0	33	(4)
6 Age (>25 years)	·	–	0	30	(3)
7 Sex (male)	·	·	–	13	(1)
Bioptic				total 57	(7)
8 Global sclerosis	**	**	↓	14	(5)
9 Tubular atrophy	**	**	↓	26	(6)
10 Interstitial infiltration	**	**	↓	17	(5)
11 Extracapillary proliferation	–	·	–	4	(1)

[a] Two-sided P-values: ·: $0.15 < P$; –: $0.05 < P < 0.15$; *: $0.01 < P < 0.05$; **: $0.001 < P < 0.01$; ***: $P < 0.001$.
Effects: ↑: patients with this characteristic have a better prognosis; 0: effect on prognosis not significant; –: test uninformative because of sub-sample sizes; ↓: patients with this characteristic have a worse prognosis.

the samples than the logrank test does. The direction of the (significant) effects can be seen from the arrows in the fourth column, which refer to the sub-samples indicated in the first column. The sample size of the *smaller* of the two sub-samples, together with the corresponding number of renal failures, is displayed in the last column. We take again the first line as example: patients with proteinuria (i.e. > 2.5 g/24 hr) have a significantly worse survival than patients without proteinuria. The smallest sub-sample consists of 22 patients with 8 renal failures, and, consequently, the largest sub-sample of 53 patients with 5 renal failures. From this it should be clear that here the smallest sub-sample corresponds to the group of patients with proteinuria. From the information about the sub-samples one can, of course, construct familiar 2 × 2 tables, which have to be considered, however, with extreme caution because of censoring.

One can see that, when considered separately, initial creatinine clearance, proteinuria, and hypertension have a strong effect on survival. Macroscopic hematuria seems to be favourable, but for microscopic hematuria no effect at all can be determined. Also age is not important as a single predictor. [In the table, age is roughly divided into two classes. However, even when it is treated as a continuous variable, using Cox's model, its estimated effect is small (a relative risk of 1.5 for a patient who is 20 years older) and not significant ($P = 33\%$, two-sided).] All bioptic variables, except for extracapillary proliferation, are significant (when based on the 57 complete cases with 7 renal failures). There is a common belief among nephrologists that extracapillary proliferation is a bad indication, and that females shoud have a better prognosis than males. Because of the very small sizes of the corresponding sub-samples, the data should be regarded as uninformative about these points.

For the *multiple regression analysis* based on Cox's model we decided to consider initially only the non-bioptic variables in order to avoid excessive multidimensionality and because bioptic data are entirely missing for 13 of the 75 cases. Also sex was left out of consideration as the number of females was too small (and there was no significant single effect). So, in the *first round* of the analysis, 7 factors were considered: the first 6 factors of Table 4 and additionally a time variable in order to detect a possible global trend throughout the period 1967–1983.

The results of the first round are presented in Table 5. One can see that, acting simultaneously, initial GFR and proteinuria are significant (as they were in the univariate analysis), and that hypertension and microscopic hematuria are only significant at the 5% level if one-sided testing is allowed. Note that, from Viewpoint 2, hypertension and microscopic hematuria are about 'as relevant for prediction as proteinuria'. The direction of these effects is the same as in Table 4. A little more surprising at first sight is the effect of age, which is significant and rather large (a patient who is 20 years older and has the same scores for the other covariates as some other patient, has, with respect to that patient, an estimated relative risk of $\exp(-2 \times 0.84) \simeq 5^{-1}$). In the univariate analysis the effect of age was, though not significant, in the opposite direction! The difference can be explained by the fact that old age is positively correlated with hypertension, the absence of macroscopic hematuria and in particular with a decreased GFR (see Table 3), which all give a bad prognosis. A similar remark can be made for microscopic hematuria, which has no effect at all (in fact, $P \simeq 0.50$) in the

Table 5
Multiple regression analysis based on the proportional hazards model (*first round*)[a]

Covariable	Scale	$\hat{\beta}$	$\hat{\beta}/\hat{SE}(\hat{\beta})$	$\exp(\hat{\beta}D)$
Hypertension	present = 1; not present = 0	1.6	1.7	5.2
Macroscopic hematuria	present = 1; not present = 0	– 1.8	– 1.5	6.0^{-1}
Microscopic hematuria	present = 1; not present = 0	1.5	1.8	4.3
Initial GFR	$\frac{1}{50}$(GFR(ml/min) – 100)	– 1.9	– 2.8	14.9^{-1}
Proteinuria	$\frac{1}{2}$(prot(g/24 hr) – 2)	1.1	2.3	7.3
Age at biopsy	$\frac{1}{10}$(age(yrs) – 30)	– 0.84	– 1.9	8.7^{-1}
Time of biopsy	$\frac{1}{5}$(date of biopsy – 1975)	– 0.25	– 0.6	1.5^{-1}

[a] The scale is defined in such a way that measurements equal to the mean value in the population are approximately zero, while the units, remaining simple multiples of the original ones, are roughly equal to one standard deviation. $D = 2\ \hat{SD}(Z)$ if Z is a continuous covariable, and $D = 1$ if Z is a dichotomous covariable.

univariate analysis, and an adverse effect on survival in the multiple regression approach. This is consistent with the fact that microscopic hematuria is positively correlated with macroscopic hematuria and with the absence of hypertension (see Table 3).

In the *second round* some adjustments were made. First, hypertension treatment was taken into account by dividing the patients into three groups: those without hypertension, those with 'effectively controlled hypertension', and those with 'insufficiently controlled hypertension' (see Sections 1 and 2). In order to apply Cox's model, the 3 groups were indicated by 2 binary covariables (H + C + and H + C −), and scored as (0, 0), (1, 0) and (0, 1), respectively. Second, it was realised that for normal people there is a natural decline in creatinine clearance of about 1 ml/min per year (Dewitt Baines, 1980). Being interested in the excessive decrease of GFR due to IgAN, we decided to use the variable 'age-adjusted GFR' [by definition, aa.GFR = GFR(ml/min) + age(years)] instead of just the glomerular filtration rate. Finally, the variable 'time of biopsy', which had turned out to be unimportant in the first round, was omitted.

The results of the second round are summarised in Table 6. One can see that insufficiently treated hypertension has a strong negative effect on kidney survival, whereas adequately treated hypertension seems to be markedly less harmful. Furthermore, age has lost significance due to the introduction of aa.GFR (the factor $\exp(-0.4) \simeq 1.5^{-1}$ is to be interpreted as the estimated relative risk of a patient who is 10 years older and has the same other covariates, which now implies, as is normal for healthy people, a GFR that is 10 ml/min less).

In the *third round* of the analysis, the variables 'age at biopsy' and 'adequately controlled hypertension' (H + C +) were dropped from the analysis. One can see from Table 6 that they are the least important ones (from both viewpoint 1 and 2) among the non-significant factors. Another reason for dropping the factor H + C + is that measuring the effect of hypertension treatment was not a primary

Table 6
Multiple regression analysis (*second round*)[a]

Covariable	Scale	$\hat{\beta}$	$\hat{\beta}/\hat{SE}(\hat{\beta})$	$\exp(\hat{\beta}D)$
Adequately controlled hypertension (H + C +)	present = 1; not present = 0	1.3	1.3	3.7
Insufficiently controlled hypertension (H + C −)	present = 1; not present = 0	3.0	2.4	19.4
Macroscopic hematuria	present = 1; not present = 0	− 1.9	− 1.7	6.4^{-1}
Microscopic hematuria	present = 1; not present = 0	1.7	2.0	5.3
Age adjusted initial GFR	$\frac{1}{50}$(GFR′ − 130) with GFR′ = GFR(ml/min) + age(yrs)	− 1.6	− 2.3	7.5^{-1}
Proteinuria	$\frac{1}{2}$(prot(g/24 hr) − 2)	1.2	2.4	8.8
Age at biopsy	$\frac{1}{10}$(age(yrs) − 30)	− 0.4	− 1.0	2.5^{-1}

[a] $D = 2 \hat{SD}(Z)$ if Z is a continuous covariable, and $D = 1$ if Z is a dichotomous covariable.

objective of the study. Instead, it is rather to be considered as a 'nuisance effect', which has to be accounted for in order to avoid a source of bias in estimating the other effects. In view of the very large influence of 'uncontrolled hypertension', due to only a few cases, the patients were, with respect to this variable, divided into two strata (corresponding to patients with 'uncontrolled hypertension' and with 'effectively controlled or no hypertension', respectively). This stratified proportional hazards model allows for more freedom in the relative hazard rate of the two (remaining) hypertension groups than the ordinary PH model does. Interest from the medical doctors in the possible effects of macroscopic hematuria led to the idea that we should stop deleting more variables from the analysis.

The results of the third round are summarised in Table 7, from which one can see that the effects of aa.GFR, proteinuria, and microscopic hematuria, common to both strata, are 'statistically significant' from Viewpoint 1, and 'rather large' from Viewpoint 2. Macroscopic hematuria is nearly significant at the 5% level if one-sided testing is alowed, which is disputable here. Nevertheless, from Viewpoint 2, its estimated effect is considerable and more or less comparable to that of aa.GFR or proteinuria (more precisely, the presence versus absence of macroscopic hematuria corresponds to roughly $(1.7/1.6) \times 50 \simeq 50$ ml/min aa.GFR and to $(1.7/1.3) \times 2.0 \simeq 2.5$ g/24 hr proteinuria). Note that microscopic hematuria has an unfavourable effect on survival, whereas macroscopic hematuria seems to be favourable (although the magnitude of the effect of macroscopic hematuria cannot be estimated precisely from the data). A favourable effect of macroscopic hematuria is also suggested from the fact that only 1 out of the 34 patients with macroscopic hematuria had to be dialysed (this patient happened to have also uncontrolled hypertension).

From a *physiological* point of view, the effect of macroscopic hematuria is more surprising than that of microscopic hematuria. The presence of slight amounts of

Table 7
Multiple regression analysis (*third round*)[a]

Strata		Renal failure	Lost	Censored	Total
'No hypertension' or 'Adequately controlled hypertension'		9	3	58	70
Insufficiently controlled hypertension		4	0	1	0

Covariable	Scale	$\hat{\beta}$	$\hat{\beta}/\hat{SE}(\hat{\beta})$	$\exp(\hat{\beta}D)$	D
Macroscopic hematuria	present = 1; not present = 0	-1.7	-1.6	5.5^{-1}	1
Microscopic hematuria	present = 1; not present = 0	1.9	2.2	6.7	1
Age-adjusted initial GFR	$\frac{1}{50}$(GFR′ − 130) with GFR′ = GFR(ml/min) + age(yrs)	-1.6	-2.4	7.7^{-1}	63 ml/min
Proteinuria	$\frac{1}{2}$(prot(g/24 hr) − 2)	1.3	2.6	10.4	3.6 g/24 hr

[a] $D = 2\,\hat{SD}\,(Z)$ if Z is a continuous covariable, and $D = 1$ if Z is a dichotomous covariable.

blood in the urine indicates destruction of the blood vessels and an abnormal permeability of the glomeruli, and one may expect that it is associated with a lower survival probability of the kidney. But also the presence of macroscopic hematuria indicates such an abnormal situation. As the analysis has been performed for the survival time from biopsy onwards, while (age-adjusted) GFR at the time of biopsy has been included as a covariate, the fact that patients with the alarming symptom of macroscopic hematuria are detected at an earlier stage of the disease, is insufficient to explain its favourable effect. A serious possibility is that macroscopic hematuria indicates a different, more benign, subentity within the disease IgAN. This viewpoint seems also to be supported by the finding that patients with macroscopic hematuria differ in several respects from those without macroscopic hematuria (see the correlation coefficients presented in Section 2.2 and, for an overview, Beukhof et al., 1984b). In any case, as the data are not conclusive on this point, it would be interesting to see whether a beneficial effect of macroscopic hematuria (in the presence of the other covariates) would be confirmed by analysing another group of IgAN patients, for example those who were treated at the University Hospital in Leiden.

One can relate the results of the analysis to the correlations between the initial creatinine clearance and the other risk factors (see Table 3), in order to see whether there is some (physiological) consistency. (One expects that unfavourable risk factors, at least if they don't alarm the patient, would in general also be associated with a low initial GFR.) From Table 3 one can see that hypertension is strongly associated with a low initial creatinine clearance, in accordance with the results of the survival analysis. Proteinuria an the absence of macroscopic hematuria are, though less clearly, also associated with a low initial GFR, whereas microscopic hematuria is unrelated to it. This is in accordance with the univariate analysis in Table 4. However, too far-fetched conclusions should not be attached to this. In order to make a comparison with the multiple regression results of Table 7, a multiple (for example linear) regression analysis of the initial GFR on the other covariates would be useful, but we will not present this here.

3.1.3. Methodological remarks

(1) For both discrete and continuous variables we have rather heavily relied on the assumption of proportional hazards. The following distinction between types of tests for checking this assumption is useful: (a) tests for a particular covariate, (b) 'overall' goodness-of-fit tests, which check the assumption for all covariates included in the model. For a discrete covariate, the PH assumption can be informally checked by stratifying for that covariate, and plotting the estimated survival functions corresponding to the strata on an appropriate scale (i.e. $-\ln(-\ln S_i(t\,|\,\mathbf{0}))$ against t, or $\ln t$, for $i = 1, \ldots, s$). If the hazards are proportional, estimates should result that approximately differ by a vertical shift.

Incidentally, looking ahead at the graphical representation of Figure 2 (in the next section), one can see that for these data the hazards of the two hypertension groups (considered in the third round) are approximately proportional. Applying the PH model also for the factor $H + C -$, the estimates for the regression coefficients of the other covariates were very nearly the same as those presented in Table 7. (The stratified model was maintained in the final round, because uncontrolled hypertension will hardly occur for new patients and we did not want to overburden Figure 2 with too many arrows in the left lower corner.) It is remarked that for the factor macroscopic hematuria the method breaks down, because only one patient with macroscopic hematuria entered end-stage renal disease. In fact, for this covariate, the PH assumption seems hardly testable.

Checking the PH assumption for continuous covariates is more difficult. Several methods have been proposed (see Kay, 1984; Moreau, O'Quigley and Meshbah, 1985; Nagelkerke, Oosting and Hart, 1986; Gill and Schumacher, 1987; Barlow and Prentice, 1988; O'Quigley and Pessione, 1989; Gray, 1990). Although interesting, we will, due to time constraints, not apply such goodness-of-fit tests here. Extensive testing of the model will be hampered by the small number of actual renal failures. Nevertheless, some indication of the adequacy of the model fit may also be gained from the residual analysis which is presented in the next section.

(2) Interaction between the influences of the various covariates has been systematically neglected. There were no prior reasons for expecting interactions to be important. (A large amount of interaction would of course spoil the results of Table 7.)

Interactions may be studied from the data by introducing product terms $z_1 z_2$ in the PH model, and by assuming different regression coefficients for the various strata in the stratified model, though the limited number of actual renal failures will hinder an extensive analysis. In particular, the fact that only one patient with macroscopic hematuria entered end-stage renal disease, blocked an investigation of the interesting possible interaction between macroscopic hematuria and the other covariates.

(3) No multiple comparison methods have been used for selecting, according to criterion A, the relevant predicting variables. As the null-hypothesis $\beta_i = 0$ is tested for many covariates (for example, 4 in the third round), one might argue that the significance level for each single test should be lower than 5% in order to avoid, by maintaining an overall significance level of 5%, the inclusion of spurious covariates. On the other hand, prior evidence existed for the effects of initial GFR, proteinuria and hypertension on survival time, which makes the null-hypothesis that all effects are zero very implausible. Furthermore, it should be noticed that hypothesis testing theory was not especially designed for selecting covariates in multiple regression problems. Accordingly, the 5% significance level is quite arbitrary. If hypothesis testing theory is used nevertheless in this context, the significance level should, in the author's opinion, depend on the sample size. [If one strictly adheres to, e.g., the 5% level, then, for large samples many variables may well be included which are practically insignificant, but statistically significant, whereas for small samples (taken from the same population) only a very few variables would be included, since even rather large estimated effects would be statistically insignificant.] Because in this case the effective sample size is small, one may consider to use an overall significance level of, say, 10–15%

so as to have at least some power of detecting effects whose magnitude is clinically relevant. This counterbalances the multiple comparison considerations. Looking at Figure 2, one can see that the estimated relative risks have to be rather large (about 3 or 4) in order to be included by a 5% level for each single statement. Hence, it is felt that in this case the 5% level is a reasonable compromise between the conflicting goals of not including spurious covariates and not missing relevant ones.

3.2. Point estimates and confidence intervals for $S(t \mid z)$

3.2.1. Theory

Identification of the various risk factors and estimation of their joint influence was studied in the previous sub-section. We now proceed by estimating the survival function of a patient with a particular set of k risk-factor scores. We recall that $\hat{\boldsymbol{\beta}} = (\hat{\beta}_1, \ldots, \hat{\beta}_k)^{\mathrm{T}}$ denotes the ML estimator maximising

$$\ln L(\boldsymbol{\beta}) = \sum_{j=1}^{n} \Delta_j(\boldsymbol{\beta}^{\mathrm{T}} z_j - \ln Q_j(\boldsymbol{\beta})), \tag{3.3}$$

where $Q_j(\boldsymbol{\beta}) = \sum_{h \in R(t_j)} \exp(\boldsymbol{\beta}^{\mathrm{T}} z_h)$ is the partition function at time t_j. Remember that $t_1, \ldots, t_n$ denote the observed lifetimes, while Δ_j indicates whether ($\Delta_j = 1$) or not ($\Delta_j = 0$) the observation at t_j is an actual renal failure.

There exist various, asymptotically equivalent, point estimates for the cumulative hazard rate $\Lambda(t \mid z) = -\ln S(t \mid z)$, see Breslow (1979), Kalbfleisch Prentice (1980) and Gill and Johansen (1990), among others.

Let

$$\tilde{\Lambda}(t, \boldsymbol{\beta} \mid z) = \exp(\boldsymbol{\beta}^{\mathrm{T}} z) \tilde{\Lambda}(t, \boldsymbol{\beta} \mid 0) = \exp(\boldsymbol{\beta}^{\mathrm{T}} z) \sum_{t_j < t} \Delta_j \Big/ Q_j(\boldsymbol{\beta}), \tag{3.4}$$

where z denotes the vector of covariates of an arbitrary patient. The summation is to be understood as 'for all j for which $t_j < t$'. Then, $\hat{\Lambda}(t \mid z) = \tilde{\Lambda}(t, \hat{\boldsymbol{\beta}} \mid z)$ is a convenient estimator for $\Lambda(t \mid z)$. It reduces to Nelson's estimator for the cumulative hazard rate if there are no covariates (which is formally equivalent to the distribution of $\hat{\boldsymbol{\beta}}$ being concentrated in $\mathbf{0}$). Of course, $\hat{\Lambda}(t \mid z)$, which is a step function, may be replaced by a broken continuous line, based on some interpolation method.

The asymptotic joint distribution of $\hat{\boldsymbol{\beta}}$ and $\hat{\Lambda}(t \mid 0)$ has been derived, under somewhat different conditions and model assumptions, by Bailey (1979, 1983), Tsiatis (1981), and Andersen and Gill (1982). In this section we need only the following results, see Chapter 15, Section 1.1, for further details.

Under mild assumptions concerning the boundedness of the covariates and regular behaviour of the information matrix $I(\boldsymbol{\beta})$ as $n \to \infty$, the joint distribution of $n^{1/2}(\hat{\boldsymbol{\beta}} - \boldsymbol{\beta})$ and $n^{1/2}(\ln \hat{\Lambda}(t \mid z) - \ln \Lambda(t \mid z))$ is, for all t belonging to a suitable interval $[T_{\min}, T_{\max}]$, asymptotically normal with mean zero and covariance matrix $\Sigma(t)$, which can consistently be estimated by

$$\hat{\Sigma}(t) = n \begin{pmatrix} \hat{V} & \hat{V}\hat{\Gamma}(t) \\ (\hat{V}\hat{\Gamma}(t))^{\mathrm{T}} & \hat{\Psi}(t) + \hat{\Gamma}^{\mathrm{T}}(t)\hat{V}\hat{\Gamma}(t) \end{pmatrix}, \tag{3.5}$$

where

$$\hat{V} = V(\hat{\beta}) = I(\hat{\beta})^{-1}, \tag{3.6}$$

$$\hat{\Psi}(t) = \hat{\Psi}(t, \hat{\beta}) = \hat{\Lambda}(t\,|\,0)^{-2} \sum_{t_j < t} \Delta_j Q_j(\hat{\beta})^{-2}, \tag{3.7}$$

$$\hat{\Gamma}(t) = \tilde{\Gamma}(t, \hat{\beta}, z) = (\partial/\partial\beta)\ln \tilde{\Lambda}(t, \beta\,|\,z)|_{\beta = \hat{\beta}} = z + \tilde{\Gamma}(t, \hat{\beta}, 0). \tag{3.8}$$

A computationally convenient expression for $\tilde{\Gamma}(t, \hat{\beta}, 0)$ is given by

$$\tilde{\Gamma}(t, \hat{\beta}, 0) = -\left[\sum_{t_j < t} \Delta_j \Big/ Q_j(\hat{\beta}) \right]^{-1} \sum_{t_j < t} \Delta_j \bar{z}_j(\hat{\beta})/Q_j(\hat{\beta}). \tag{3.9}$$

Here, $\bar{z}_j(\beta) = (\partial/\partial\beta)\ln Q_j(\beta)$ denotes the 'mean' vector of covariates at t_j (weighted according to the distribution $\hat{p}_h = \exp(\hat{\beta}^{\mathrm{T}}z_h)/Q_j(\hat{\beta})$, $h \in R(t_j)$).

The variance of $n^{1/2}\ln \hat{\Gamma}(t\,|\,z)$ is estimated by $\hat{\Sigma}_{22}(t)$. Thus, a two-sided level-α confidence interval for $\ln \Lambda(t\,|\,z)$ is given by

$$(\ln \underline{\Lambda}(t\,|\,z), \ln \overline{\Lambda}(t\,|\,z)) = (\ln \hat{\Lambda}(t\,|\,z) - u_{\alpha/2}\hat{C}(t), \ln \hat{\Lambda}(t\,|\,z) + u_{\alpha/2}\hat{C}(t)), \tag{3.10}$$

with

$$\hat{C}(t) = \{\hat{\Psi}(t) + [z + \tilde{\Gamma}(t, \hat{\beta}, 0)]^{\mathrm{T}}\hat{V}[z + \tilde{\Gamma}(t, \hat{\beta}, 0)]\}^{1/2}. \tag{3.11}$$

Transforming back to $\Lambda(t\,|\,z)$ and $S(t\,|\,z)$ we have

$$(\underline{\Lambda}(t\,|\,z), \overline{\Lambda}(t\,|\,z)) = (\hat{\Lambda}(t\,|\,z)/\hat{D}(t), \hat{\Lambda}(t\,|\,z)\hat{D}(t)) \tag{3.12}$$

and

$$(\underline{S}(t\,|\,z), \overline{S}(t\,|\,z)) = (\hat{S}(t\,|\,z)^{\hat{D}(t)}, \hat{S}(t\,|\,z)^{1/\hat{D}(t)}), \tag{3.13}$$

respectively, where $\hat{D}(t) = \exp(u_{\alpha/2}\hat{C}(t))$.

In the stratified PH model, β and $I(\hat{\beta})$ are based on the combined log likelihood $\ln L(\beta) = \sum_{i=1}^{s} \ln L_i(\beta)$, whereas the expressions for $\hat{\Psi}(t)$ and $\hat{\Gamma}(t)$ in formulae (3.7), (3.8) and (3.9) pertain to the stratum considered.

The set of confidence intervals given by formula (3.13) (for various t and at a fixed value of z) may be looked upon as an 'instantaneous confidence band' for the unknown survival function $S(t\,|\,z)$. For each fixed value of t $(t \in [0, T_{\max}])$ the band gives a confidence interval for the survival probability, and, conversely, for each fixed survival percentage p $(p \in [p_{\min}, 1]$ for some suitable $p_{\min})$ it gives a confidence interval $(\underline{t}_p, \bar{t}_p)$ for the corresponding percentile. The band is called 'instantaneous', because for each separate interval the asymptotic confidence coefficient is $1 - \alpha$.

Of course, one could consider the construction of (a set of) simultaneous

confidence bands. However, the corresponding overall confidence coefficients are
not very easily calculated, at least not without simulation experiments. As the
hypothesis tests corresponding to intervals for time-points that are not too much
separated are strongly dependent, the simultaneous confidence bands are perhaps
not very much wider than the instantaneous bands of the same level. In this
respect, one is referred to Kardaun (1983b) for a comparison between various types
of confidence bands in the simple case in which there are no covariates, and to
Kardaun (1991, Ch. 15), where a general numerical method is discussed for
calculating the confidence coefficient of a simultaneous band in the presence of
covariates.

3.2.2. Practice

In order to construct an instantaneous confidence band for $\Lambda(t\,|\,\mathbf{0})$ the author has
written, with help from Ronald Kiel, an interactive FORTRAN program, which uses
numerical utility routines from IMSL (1982) and graphical routines from the
package KOMPLOT (Kraak, 1983). Some information about this program, called
'CONCERN', can be found in Appendix B.

Predictions for an arbitrary patient may be performed by the reader himself by
using the nomogram presented in Figure 2, which was made by CONCERN. In this
figure, the estimated survival curves are plotted on the scale $(-\ln(-\ln[S(t\,|\,z)])$
against $\ln t$) on which Weibull distributions correspond to straight lines and, more
importantly, Cox's model is additive. The survival curves (interpolated by joining
the midpoints of the horizontal line segments) correspond to the standard patients
in each of the two remaining hypertension groups ('no or adequately controlled'
(a) and 'uncontrolled' hypertension (b), respectively). A standard patient, who has
a vector of covariates $z = \mathbf{0}$, is defined as a patient with 130-z ml/min GFR at age
z, with 2 g/24 hr proteinuria, and with no microscopic or macroscopic hematuria.
The end-points of (asymptotically) 67% confidence intervals at the times-to-death
are connected by broken lines, which may be considered to delimit approximately
an instantaneous 67% confidence band for the survival curve. The arrows in the
corner represent the estimated effects $\hat{\beta}_i$ (for covariates which are measured in the
units of Table 7: 50 ml/min GFR, 2 g/24 hr proteinuria, etc.). For each $\hat{\beta}_i$, the
accompanying vertical bar indicates ± 1 estimated standard deviation. The distance between two heavy horizontal lines of the grid corresponds to one unit of
$\hat{\beta}_i$ (and accordingly to a factor $e \simeq 2.7$ in relative risk).

The estimated survival function for a patient with scores $z_1, \ldots, z_k$ (in the units
of Table 7) can be found by consecutively multiplying each arrow with the corresponding score and shifting accordingly the base-line curve upwards and downwards. For example, in order to obtain the survival curve for a patient who has
macroscopic hematuria, and standard other covariates, the base-line survival
curve $\hat{S}(t\,|\,\mathbf{0})$ has to be shifted 1.7 units upwards. If the patient has, in addition,
3 instead of 2 g proteinuria per 24 hr, the curve has, in addition, to be shifted
0.65 units downwards.

In order to find a confidence interval for e.g. the 5 year survival probability for
a patient with covariate z, one could turn to the computer and order a new

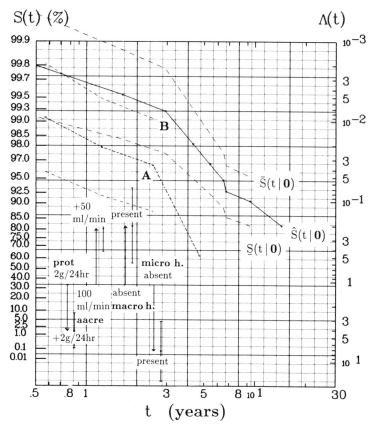

Fig. 2. A nomogram representing base-line survival curves of IgA-nephropathy patients (with approximately 67% instantaneous confidence bands) and the influence on these curves of various covariates. A base-line survival curve corresponds to patients with 100 ml/min GFR at age 30, 2 g/24 hr proteinuria, no macroscopic and no microscopic hematuria. Curve B corresponds to patients with no, or adequately controlled, hypertension, and curve A to patients with not effectively controlled hypertension.

nomogram from the program CONCERN. Alternatively, taking $\hat{V}$, $\Gamma(5, \hat{\beta}, 0)$ and $\hat{\Psi}(5)$ from the non-graphical output for $z = 0$ one could, using formulae (3.5) to (3.8), perform the calculations on a pocket calculator. Of course, it would be nice to construct such intervals directly from a nomogram, without having to resort to a computer program, or to tedious calculations. Actual construction of such a nomogram for a general case (i.e. for some covariates, and for all $t \in [T_{min}, T_{max}]$), however, is not too easy. (The reader is invited to try and invent one.)

Some insight about the widths of the intervals at $t = 5$ years for various values of the covariates can be gained from Table 8, which was calculated by running the program several times. (One unit corresponds, as usual, to the distance between two thick horizontal lines in Figure 2 and to a relative risk of $e \simeq 2.7$.)

From formula (3.5) and (3.8) one can see that, considered as a function of

Table 8
Lengths of approximately 67% confidence intervals for $-\ln(-\ln S(t\,|\,z))$
at $t = 5$ years and for some values of z

No macroscopic or microscopic hematuria ($z_3 = z_4 = 0$)

Proteinuria (g/24 hr)	aa.GFR		
	100 + 30	50 + 30	25 + 30
0	2.4	2.4	2.7
3	1.5	1.5	1.9
6	1.8	1.8	2.2

Proteinuria ($= z_1$) = 2 g/24 hr; age-adjusted GFR ($= z_2$) = 130

Macro hematuria	Micro hematuria	
	0	1
0	1.7	1.4
1	2.6	2.3

$z \in \mathbb{R}^4$, the contours of constant (estimated) variance of $\hat{S}(5\,|\,z)$ are elliptical surfaces in $\mathbb{R}^4$, centered around $-\tilde{\varGamma}(5,\,\hat{\beta},\,0)$. As macroscopic and microscopic hematuria can actually only assume the values 0 and 1, one could draw (e.g. for the main stratum) four sets of two-dimensional elliptical contours, representing, for the four possible combinations of hematuria, the loci of constant (estimated) variance of $\hat{S}(5\,|\,z)$.

Actually, a compact version of such a graphical representation is Figure 3, for which the underlying formulae are derived in the Appendix. The figure corresponds to the stratum consisting of patients with no, or adequately controlled, hypertension. The straight lines are lines of constant relative risk $\exp(\hat{\beta}^{\mathrm{T}}z) = rR$ ($r = 1/4,\ 1,\ 4,\ 16,\ 64$) with respect to $\hat{\varLambda}(5\,|\,0) = 0.03$, or, alternatively, lines of constant $\hat{S}(5\,|\,z)$. [For convenience, the relative risk is split into two factors: $r = \exp(\beta_1 z_1 + \beta_2 z_2)$ and $R = \exp(\beta_3 z_3 + \beta_4 z_4)$. The occurring values of R can be seen in the table at the right lower corner of the figure.]

The ellipses are contours of constant length of the 67% confidence intervals for $-\ln \varLambda(t\,|\,z) = -\ln(-\ln S(t\,|\,z))$. The length of such an interval equals $2\sqrt{D + d}$, $d = 0.2,\ 0.5,\ 1,\ 2$, where one unit corresponds to the distance between 2 heavy horizontal lines in Figure 2, i.e. to a factor $e \simeq 2.7$ in relative risk. [For convenience, the variance of $-\ln\hat{\varLambda}(t\,|\,z)$ is split up into two parts; the part D depends on z only through the covariates z_3 and z_4, while for fixed z_3 and z_4, the remaining part d (being constant on ellipses) depends in a simple way on z_1 and z_2.] The ellipses are drawn for $(z_3,\,z_4) = (0,\,0)$, i.e. for patients without macro-

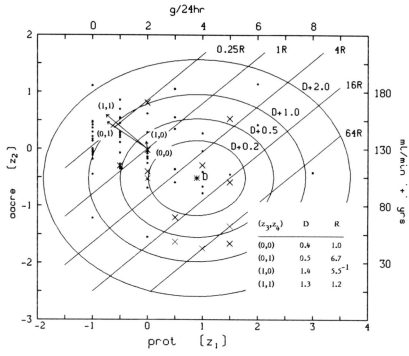

Fig. 3. A nomogram for determining the length of an approximately 67% confidence interval as well as a point estimate of the 5 year survival probability for a patient with no, or effectively controlled, hypertension and with an arbitrary vector $z = (z_1, z_2, z_3, z_4)$ of scores with respect to proteinuria, age-adjusted creatinine clearance, macroscopic hematuria, and microscopic hematuria. For (z_1, z_2) on one of the drawn ellipses, the length of a 67% confidence interval, equals $2\sqrt{D + d}$, $d = 0.2, 0.5,$ 1.0, 2.0, one unit corresponding to the distance between two heavy horizontal lines in Figure 2. For patients with (z_3, z_4) unequal to $(0, 0)$, the center of the ellipses has to be shifted according to one of the arrows. The relative risk with respect to $-\ln \hat{S}(5 \mid \mathbf{0}) = 0.03$, for (z_1, z_2) on one of the straight lines, equals rR ($r = 1/4, 1, 4, 16, 64$). The dots and the crosses represent observations in the sample (corresponding to censored and uncensored lifetimes, respectively). The observations from patients with uncontrolled hypertension are plotted with light-face symbols. The age-adjusted creatinine clearance is defined as aacre = cre (ml/min) + age (yrs). The units at the upper and the right axes are commonly used units. The units at the lower and the left axes are standardised units that are also used in Figure 2 and Table 7.

scopic or microscopic hematuria. For other values of (z_3, z_4), the center of the ellipses has to be shifted according to the arrows (or, alternatively, the value of (z_1, z_2) has to be shifted in opposite direction), and the value of D changes according to the associated table. Note that, in practice, the variables proteinuria and creatinine clearance can only assume positive values. The dots and crosses correspond to observations in the sample. The first ones to patients with censored lifetimes and the second ones to patients who entered end-stage renal disease (ESRD). The observations from the patients with uncontrolled hypertension (for which the nomogram is not designed) are plotted in lightface.

Examples

(1) Consider a patient who has no hypertension, 4 g/24 hr proteinuria, a glomerular filtration rate of 75 ml/min at age 30 (i.e. aa.GFR = 105), and neither macroscopic nor microscopic hematuria. The standard scores for this patient are $z = (z_1, \ldots, z_4) = (1, -0.5, 0, 0)$. From nomogram 2 one can infer by shifting the base-line survival curve that his estimated 5 year survival probability $\hat{S}(5 \mid z) = 78\%$, while the corresponding cumulative hazard rate is $\hat{\Lambda}(5 \mid z) = 0.25$. Alternatively, from Figure 3 one can see that the relative risk with respect to $\hat{\Lambda}(5 \mid \mathbf{0})$ is $rR \simeq 8$, so $\hat{\Lambda}(5 \mid z) \simeq 0.24$. An approximately 67% confidence interval for $S(5 \mid z)$ can be obtained by observing from Figure 3 that $d \simeq 0$, $D = 0.4$, hence the variance of $\ln \hat{\Lambda}(5 \mid z)$ is 0.4, and the width of the confidence interval, on the scale of the first nomogram, is $2\sqrt{0.4} \simeq 1.25$. Using again Figure 2 one can see that $(\underline{S}(5 \mid z),\ \overline{S}(5 \mid z)) = (42\%, 93\%)$, and $(\underline{\Lambda}(5 \mid z),\ \overline{\Lambda}(5 \mid z)) = (0.07, 0.88)$.

(2) Consider a patient with the same covariates as in the first example, except that, in addition, he has also microscopic hematuria. In this case $z = (1, -0.5, 0, 1)$. Now, as inferred from Figure 2, the point prediction is: $\hat{S}(5 \mid z) = 20\%$, $\hat{\Lambda}(5 \mid z) = 1.6$.

From Figure 3 one can verify that $rR \simeq 6.7 \times 8 \simeq 54$, and see that $2\sqrt{D + m} = 2\sqrt{0.5 + 0.25} \simeq 1.7$. Turning again to Figure 2, we get the interval prediction $(\underline{S}(5 \mid z),\ \overline{S}(5 \mid z)) = (0.03\%, 75\%)$. Remember that shifting the center of the ellipses according to the arrow denoted by (0, 1) is equivalent to keeping the ellipses fixed and shifting the point (z_1, z_2) in the opposite direction.

3.3. Analysis of residuals

After having selected a model, one will be interested in some assessment of its adequacy. Several methods have been proposed (see, e.g., Kay, 1977; Schoenfeld, 1982; Andersen, 1982; Kardaun, 1983a; Kay, 1984; Reid and Crèpeau, 1985; Barlow and Prentice, 1988) for inspecting (generalised) residuals in order to obtain an impression of the goodness-of-fit of a proportional hazards model, and to detect outlying and/or influential data points. In this section, we will consider a graphical residual analysis, which can be viewed as an extension of some frequently used methods. A precise assessment of the distribution of the involved residuals, however, turns out to be difficult (except in the special case of no censoring). Hence, improvements of the following analysis, possibly based on simulation or on resampling methods, would be interesting.

A simple and familiar way for physicians to look at residuals is to compare, e.g., the estimated 5-year survival probabilities (of those patients whose diagnosis was established at least five years before the moment of censoring) with the actual outcomes. We will neglect the issue that these probabilities should ideally be calculated from a new sample, i.e. a sample which has not been used for estimating the parameters of the model.

A refinement of this method is as follows. For all patients in the training sample, the estimated probabilities for surviving a fixed period of time (say $c = 5$ years) are plotted against the observed lifetimes, i.e. $\hat{S}(c \,|\, z_j)$ is plotted against t_j for $j = 1, \ldots, n$. One distinguishes between (a) patients who have 'died', (b) patients who were censored and were diagnosed at least c years before the censoring date, and (c) patients who were censored and were diagnosed less then c years before the date of censoring. Look at Figures 4 and 5, which were drawn by the program CONCERN. The residuals are plotted on the scale of the nomograms of Figures 2 and 3. Note that, abbreviating $-\ln(-\ln(x))$ by $\mathrm{lln}(x)$,

$$\mathrm{lln}(S(c \,|\, z)) = \mathrm{lln}(S(c \,|\, \mathbf{0})) - \boldsymbol{\beta}^{\mathrm{T}} z, \qquad (3.14)$$

so the vertical axis is essentially equivalent to (minus the logarithm of) the relative risk $rR = \exp(\boldsymbol{\beta}^{\mathrm{T}} z)$ with respect to the standard patient. Some values of rR, separately for both strata, can be seen on the rightmost scales. For some relevant cases the accuracy of an estimate $\hat{S}(t \,|\, z_j)$ is indicated by ± 1 estimated standard deviation. The same interpolation method as in the previous section was used. (Actually, c was taken to be 4.7 instead of 5 years, as this was the largest time for which reasonable confidence intervals could be drawn for patients from stratum 2.)

For clarity of the picture, observations from group (a) are plotted separately from those of the two other groups. Thus, in Figure 4 the patients are displayed who entered end-stage renal disease (ESRD). For each patient, the time t_j from biopsy until renal failure is indicated by a $\times$, whereas the time u_j from biopsy until the date of censoring is indicated by a $\circ$. Lightface symbols correspond to the patients from stratum 2 ('uncontrolled hypertension'). One expects that for patients who actually went into ESRD after (before) five years, $\hat{S}(5 \,|\, z_j)$ is high (low), so the general trend of the crosses should be upwards.

Under the assumption that the PH model holds, this trend may be compared with so-called 'reference lines'. These are defined in Figures 4 and 5 as lines on which, for a specified stratum, $S(t \,|\, z)$ is constant. In our case we consider the reference lines from stratum 1, which corresponds to 'no' or 'adequately controlled' hypertension. This definition has the following motivation. If the model holds, then, for the specified stratum, $\Lambda(c \,|\, z)/\Lambda(t \,|\, z) = \Lambda(c \,|\, \mathbf{0})/\Lambda(t \,|\, \mathbf{0})$ is independent of the covariate z. Moreover, $\Lambda(t \,|\, \mathbf{0})$ is a monotonic function of t. Accordingly, for each $p \in (0, 1)$, the set $\{(t, \Lambda(c \,|\, z)); \Lambda(t \,|\, z) = -\ln p\}$, which corresponds to a $100p$ reference line, constitutes in Figures 4 and 5 the graph of a monotonic function. Alternatively stated: identifying all triples $(t, \Lambda(c \,|\, z), \Lambda(t \,|\, z))$ with the positive orthant of $\mathbb{R}^3$, Cox's model is easily seen to define a two-dimensional surface (see REMARK 2 below); the reference lines are obtained by intersecting this surface with planes of constant $\Lambda(t \,|\, z)$ and projecting these intersections on the coordinate plane determined by t and $\Lambda(c \,|\, z)$. In Figures 4 and 5 we have drawn some 'empirical' reference lines, for which $\hat{S}(t \,|\, z)$ assumes the values $p = 90, 70, 50, 30, 10\%$ if the proportional hazards assumption holds

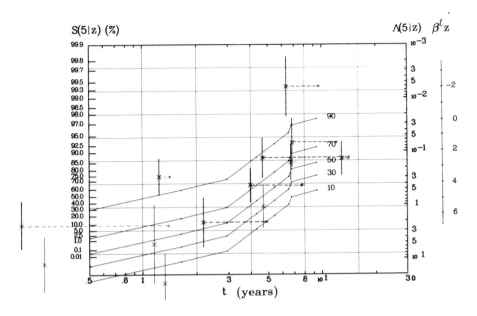

Fig. 4. Residual plot (on the scale of the nomogram of Figure 2) of the patients who entered end-stage renal disease. The vertical axis on the left represents the 5-year survival probability $S(5\,|\,z)$. If the proportional hazards assumption holds, then this axis is essentially equivalent the axis representing $\beta^T z$, which is the logarithm of the relative risk with respect to a standard patient (see the right-most vertical scale). For each patient in the sample who entered end-stage renal disease, the estimated 5 year survival probability (with an approximately 67% confidence interval) is plotted against the observed lifetime (indicated by a cross), and against the time (indicated by a circle) between dialysis and the closing date of the study. A $p\%$ reference line is a line on which $\text{lln}\,S(5\,|\,z) - \text{lln}\,S(5\,|\,0) = p\% - \text{lln}\,S(t\,|\,0)$. Empirical reference lines (for which unknown parameters are replaced by their estimates) are drawn for $p = 10(20)90$. If the proportional hazards assumption holds, then a $p\%$ reference line is precisely a line on which $S(t\,|\,z) = p\%$. If there would have been no censoring at all, and the proportional hazards model fits well, then the data would fluctuate around the 50% reference line.

for the fitted parameters. On the transformed scale of the plot we then have

$$\text{lln}(\hat{S}(c\,|\,z)) = \text{lln}(p) + \text{lln}(\hat{S}(c\,|\,\mathbf{0})) - \text{lln}(\hat{S}(t\,|\,\mathbf{0})).\qquad(3.15)$$

Hence, each reference line in Figures 4 and 5 is just minus the estimated base-line survival function, shifted by an amount determined by p. (In fact, if the PH assumption does not hold, then the empirical reference lines are still defined by the right-hand side of (3.15). In that case they do not converge, as $n \to \infty$, to the 'true' reference lines.) A further interpretation of the (empirical) reference lines is as follows. Draw a horizontal line, e.g. through the data point corresponding to $t = 1.2$ yrs and $\hat{S}(c\,|\,z_j) = 73\%$. This horizontal line intersects the reference lines at the times t_p ($p = 90, 70, \ldots, 10\%$). In the example, $t_{0.90}$ is about 3 years.

According to the fitted PH model, a new patient having a risk-factor score $\exp(\boldsymbol{\beta}^{\mathrm{T}} z_j)$ that corresponds to the horizontal line, has an estimated probability p to survive a period of at least t_p years. (This probability is exactly equal to p if the model holds and the parameters $\boldsymbol{\beta}$ and $S(t\,|\,0)$ are exactly known.)

In a similar way, reference lines might be constructed on which, instead of $\hat{S}(t\,|\,z)$, e.g. $\overline{S}(t\,|\,z)$ is constant. (As usual, $\overline{S}(t\,|\,z)$ denotes the upper boundary of an instantaneous confidence band.) As in general, however, such reference lines depend on z, this would certainly overburden the figure. If in Figures 2 and 3, for some z_j, the confidence bands for $S(t\,|\,z_j)$ *differ only by a vertical shift from* $\hat{S}(t\,|\,z_j)$, then the reference lines corresponding to the confidence bands for $S(t\,|\,z_j)$ coincide with those corresponding to $\hat{S}(t\,|\,z)$, and an approximately 67% confidence interval for e.g. the median survival time (i.e. the 50% quantile) is easily constructed from Figures 4 and 5 by intersecting the two horizontal lines which pass through $\overline{S}(c\,|\,z_j)$ and $\underline{S}(c\,|\,z_j)$, respectively, with the reference line corresponding to $p = 50\%$. This interval, then, holds for all patients having the same vector of covariates z_j.

There are several aspects about which one can possibly obtain some information from the residual plot, such as:

(1) the 'predictive power' of the model,
(2) the 'discriminating power' of the model,
(3) the detection of outliers,
(4) the adequacy of the fitted proportional hazards model.

Of course, these aspects are somewhat interrelated. If there are too many outliers, then the fitted model may not be adequate; several adequate models may exist which differ in predictive and/or discriminating power.

The first two aspects are rather easily observed from Figure 4.

(1) The prediction of survival for a new patient is the more accurate as the horizontal distance between the reference lines is smaller. (The vertical distance between the reference lines being determined by the differences in p only, this is equivalent to the slope of the lines being larger.) As a measure of the predictive power one can take, for example, $R_{\mathrm{pred},\,2/3} = t_{0.83\,:\,q}/t_{0.17\,:\,q}$, where $t_{p\,:\,q}$ stands for the time at which the reference line determined by $S(t\,|\,z) = p$ crosses the horizontal line corresponding to the q-th quantile of the distribution of the risk-factor $\exp(\boldsymbol{\beta}^{\mathrm{T}} Z)$ (where Z is considered as a random variable), or, equivalently, of the distribution of $S(t\,|\,Z)$. In other words, on a logarithmic time scale, $R_{\mathrm{pred},\,2/3}$ is the length of an estimated (equal tails) 67% prediction interval for a patient with the just mentioned risk-factor score. If the reference lines in Figure 4 are approximated by straight lines, then this measure is independent of q. In our case we have $R_{\mathrm{pred},\,2/3} \approx 5^{-1}$, so, for each patient having a risk-factor score in the range where the linear approximation is adequate, the survival time, with a chance of $2/3$, can be determined within a factor of 5. (This holds true if the statistical errors in estimating $\boldsymbol{\beta}$ and $S(t\,|\,0)$ are neglected.)

(2) The discriminating power of a model for a group of patients (assumed to be randomly selected from the population) is the larger as the slope of the reference lines, but also as the range of the distribution of $\boldsymbol{\beta}^{\mathrm{T}} Z$ (the logarithm of

the risk-factor score) is larger. As a measure one can consider $R_{dis, 2/3} = t_{p;\,0.83}/t_{p;\,0.17}$. If the reference lines are again approximated by straight lines, $R_{dis,\,2/3}$ is independent of p. In our case, discarding $33\% = 25$ of the upper and lower extreme risk-factor scores in the combination of the plots 4 and 5, we get $R_{dis,\,2/3} \simeq 8$. Hence, for two patients whose values of $\boldsymbol{\beta}^T z$ differ about 2 SD (estimated from all patients in the sample), the estimated median survival times differ by a factor 8. (This holds also, in the straight line approximation, if the median is replaced by any reasonable quantile.)

The other two aspects are more difficult to assess, at least if part of the observations is censored. Note that if there are no censored observations, the data points in Figure 4 (in arbitrary order, or ordered with respect to $\boldsymbol{\beta}^T z$) should fluctuate around the 50% reference line. This means that between any pair of horizontal lines approximately $x\%$ of the data points should lie to the right of the $x\%$ reference line ($0 < x < 100$). Note also that, if no covariates are used, a degenerate situation arises: the data points run horizontally at $\hat{S}(5)$, as do the reference points. If there are only discrete covariates, the whole scene is concentrated on a number of horizontal lines, each line corresponding to a certain combination of covariate scores.

If, as in our case, part of the data is censored, then patients with a high risk-factor score $\exp(\boldsymbol{\beta}^T z)$ are more likely to be included in the uncensored group (which is plotted in Figure 4) than patients with a low risk-factor score. Hence, the uncensored group is a selected sub-sample and the residuals do not fluctuate regularly around the 50% (or some other) reference line. One property in the presence of censoring that should hold if a model fits well is that $\{1 - \hat{S}(t_j|z_j); j = 1, \ldots, n\}$ constitutes a (censored) sample from approximately, as $S(t_j|z_j)$ is estimated, a uniform $U(0, 1)$ distribution. This is essentially the residual analysis, introduced by Kay (1977), and Crowley and Hu (1977). The distributional property of the residuals is often inspected by plotting the cumulative hazard rate of the transformed residuals $\{-\ln \hat{S}(t_j|z_j); j = 1, \ldots, n\}$ against the residuals themselves. At present it is unclear, however, which deviations from the model can easily be detected from the plot (see also Lagakos (1980), Kardaun (1983a, Sect. 4).

A second property that should hold is that, conditional on the arrival times of the patients and on the fact that the patients represented in Figure 4 are the ones who suffered renal failure, $S(T_i|z_i) \sim U(p_{min,\,i}, 1)$, where $i \in \{1, \ldots, m\}$ indicates the i-th patient with renal failure, $p_{min,\,i} = \hat{S}(u_i|z_i)$, and u_i denotes the censoring time of the i-th patient. As a test for detecting a misfit of the model one could consider to investigate whether

$$V_i = (1 - \hat{S}(t_i|z_i))/(1 - \hat{S}(u_i|z_i)), \quad i = 1, \ldots, m \qquad (3.16)$$

(arbitrarily ordered, or ordered with respect to $\boldsymbol{\beta}^T z_i$) constitutes approximately an independent sample from a uniform $U(0, 1)$ distribution. The points $(u_i, \hat{S}(5|z_i), \hat{S}(u_i|z_i))$ for stratum 1 are indicated in Figures 4 and 5 by a dot. In our case we have, for stratum 1,

$$8^{-1} \sum_{i=1}^{8} V_i \simeq 0.6 \in (0.5 - 2\sqrt{(12 \times 8)^{-1}}, 0.5 + 2\sqrt{(12 \times 8)^{-1}}), \quad (3.17)$$

so, from this point of view, there is no evidence of a substantial misfit. As $\Pr\{\min_{i=1}^{m} V_i < x\} = 1 - (1-x)^m$, $x \in [0, 1]$, any patient for which $V_i = 1 - (1-\alpha)^{1/m}$, $\alpha = 0.05$, can be suspected to have entered end-stage renal disease untimely. This happened to none of the patients in Figure 4. Although the two patients who died at $t = 1.24$ and $t = 6.3$ yrs at first sight seem to have entered ESRD too rapidly ($(\hat{S}(t \mid z) = 95.5\%$ and 97.8%, respectively), this may be due to chance, as $V_i = 90\%$ and 45%, respectively. This observation holds also if one compares for these two patients $\hat{S}(t \mid z)$ with the extreme order statistics of a sample of 70 from the $U(0, 1)$ distribution, which have expectation values $70/71 = 98.6\%$ and $69/71 = 97.2\%$, respectively.

Although it is difficult to obtain a definite answer, from the residual plot of Figure 4, on whether a particular proportional hazards model is adequate, one can compare two different PH models by inspecting each residual plot and looking for desirable features. These are: a large slope of the reference lines, a large range of $\{\hat{S}(c \mid z_j); j = 1, \ldots, n\}$, small confidence intervals for $S(c \mid z_j)$, and a firm and regular association of the observed (uncensored) lifetimes t_i with the risk-factor scores $\exp(\boldsymbol{\beta}^T z_i)$ $(i = 1, \ldots, m)$. Preferably, in Figure 4, the slope of the empirical regression line (with $\ln t$ as dependent variable) should not be much larger than that of the reference lines, as this may indicate (unless there is heavy censoring) an overfitting of the data.

Figure 5 corresponds to the patients who were censored. Again, light-face symbols are used for the patients of stratum 2. For censored lifetimes larger than 5 years, one expects that $\hat{S}(5 \mid z_j)$ will be close to 1. This is seen to be approximately true in our case. If a censored lifetime is smaller than five years, then $\hat{S}(5 \mid z_j)$ may assume any value. Hence, this type of data is entirely uninformative for assessing the adequacy of the model. Nevertheless, if one has already some confidence in the model, the display may be useful to identify those patients (two in our case) who are at high risk of entering ESRD within a short elapse of time.

Just for comparison, in Figure 6 the residual plot, for censored as well as uncensored data, is presented on the ordinary (linear) scale. Although the plot seems somewhat complicated, the reader should now be able to interpret this figure by himself.

In summary, from the residual plots, the model does not seem to fit too badly. Two patients entered ESRD relatively rapidly (however, no statistical significance was found). Two other patients are suspected to be at high risk of entering ESRD soon.

REMARKS. (1) We have used the term residuals in a somewhat loose sense ('that what is left after a model has been fitted'). More precise definitions are, of course, available. For these the reader is referred to Cox and Snell (1968).

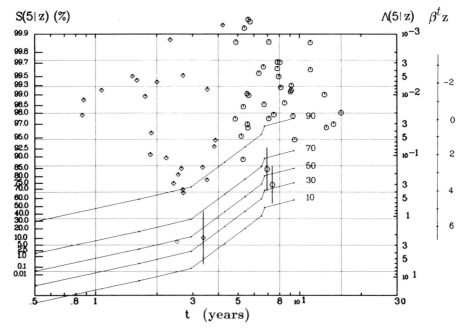

Fig. 5. Residual plot (on the scale of the nomogram of Figure 2) of the patients who did not enter
end-stage renal disease. Estimated 5-year survival probabilities and approximately 67% confidence
intervals are plotted against the observed lifetimes. Diamonds correspond to the patients who arrived
less then $c = 5$ years before the end of the study, and circles to the patients who arrived at least
$c = 5$ years before the end of the study. If the proportional hazards model does not fit too badly,
one expects that the 5 year survival probability for the last group is close to 100%, while for the
first group it may vary. If one is reasonably confident of the model, one can discern 2 diamonds (at
$t \approx 3$ years) corresponding to patients who are at high risk of entering end-stage renal disease
soon.

(2) The residual plot of Figures 4 and 5 can be considered as derived from a
three-dimensional plot, of which the x, y, and z axes are defined by $\ln t$,
$\text{lln } S(c \mid z)$, and $\text{lln } S(t \mid z)$, respectively. (The reader is suggested to draw a
picture.) A proportional hazards model

$$\text{lln } S(t \mid z) = \text{lln } S(t \mid 0) + \text{lln } S(c \mid z) - \text{lln } S(c \mid 0), \qquad (3.18)$$

defines a two-dimensional surface in this plot.

The model being postulated and being applied to the estimates $\hat{\beta}$ and
$\hat{S}(t \mid 0)$ as well, the data points $(\ln t_i, \text{lln } \hat{S}(c \mid z_i), \text{lln } \hat{S}(t \mid z_i))$, $i = 1, \ldots, n$, are lying
on a similar surface. (Statistical errors in $\hat{S}(c \mid z)$ and $\hat{S}(t \mid z)$ lead to a family of
surfaces. For simplicity, we wil neglect this complication here.) The data points
in Figures 4 and 5 are the projection of the data points in $\mathbb{R}^3$ on the $x-y$ plane,
the reference lines being the contours of constant $\text{lln } S(t \mid z)$. Of course, one can
make also projections on the other coordinate planes, and perform a similar

S(5| z)

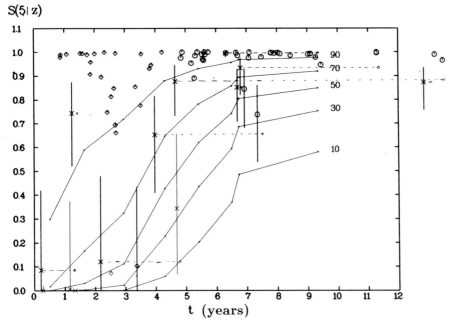

t (years)

Fig. 6. Residual plot for all patients on ordinary (linear) scale. For the legends, see Figures 4 and 5.

residual analysis. Projecting on the x–z plane, one obtains the nomogram of Figure 2 (in which are also plotted the data points, and the survival functions corresponding to a few values of $\boldsymbol{\beta}^T z$). Likewise, one can analyse, without additional difficulty, the third projection (the one on the z–y plane), which essentially plots the residuals $\hat{S}(t_j | z_j)$ considered by Kay (1977) against $\hat{\boldsymbol{\beta}}^T z_j$ ($j = 1, \ldots, n$). It is illustrative to consider the (very simple) projections on the coordinate axes: the x-axis projections are the basic sample of censored survival times, as represented, e.g., by Nelson (1972), projection on the y-axis (almost) yields the familiar residual plot used by Beukhof (1986), and the z-axis projections constitute the censored sample of residuals analysed by Kay (1977).

Evidently, all information about the data points and the estimated model is contained in each of the three two-dimensional projection plots, although some features may be more clearly viewed from one plot than from the other. There is a clear analogy with, for example, an equation of state $f(P, V, T) = 0$ in statistical mechanics. (Here, as for an ideal gas ($\ln T = \ln P + \ln V - \ln C$), there is obviously one phase.) With a capable three-dimensional graphical package it will be possible to inspect interactively the three-dimensional plot as well as the projections on the coordinate planes. Note that the y-axis corresponds to a line perpendicular to the lines of constant relative risk rR in Figure 3, and that, if the relative risk of a vector of covariates can be summarised by a function of $\boldsymbol{\beta}^T z$

only, there exists no useful higher dimensional extension of our three-dimensional plot.

4. Treatment of missing values

4.1. Introduction

One can see from Table 1 that for the bioptic variables quite a few data are missing, which is uncomfortable if one wants to use these variables for predicting survival. The missing values in our situation can be divided into two classes: class a pertains to those cases for which all bioptic variables are missing (because no bioptic specimen was available, or available any more, at the time Hoedemaeker performed the classifications); class b pertains to those cases for which only some of the bioptic variables are missing, which was due to the fact that not enough tissue of sufficient quality was available for a clear assessment of all bioptic scores. Presumably, the missing values do not occur 'at random', as there are six patients with renal failure among the eighteen ones with missing values. So, omitting in a prediction analysis the cases with missing values may bias the results. Evidently, much survival-time information present in the sample is lost by excluding so large a fraction of the total number of renal failures. Hence, it would be interesting to investigate (1) whether (some) bioptic variables are useful as survival predictors, and (2) whether they are so when added to a model which already includes an adequate selection of the most important non-bioptic variables.

4.2. Method

We will attack this problem in four steps.

(1) The missing values of class b will be interpolated by linear regression on 2 or 3 covariates which have a high correlation with the missing variables.

(2) Then, the bioptic variables will be recoded as binary covariates, where a '0' corresponds to the former scores 2 and 3 (i.e. to 'low severity' of the morphological aberration of the bioptic specimen), and '1' to the former scores 4 and 5 (i.e. to 'high severity').

(3) Because we find it somewhat risky to interpolate the missing values of class A, we will introduce, for prediction by Cox's model, an extra binary variable, assigning it the value 1, if the bioptic variables are missing, and the value 0, if they are not missing. The binary variable is added as an extra variable y_5 in Cox's model. This method allows the hazard rate of patients with missing values to be proportional to that of the other patients (within the same stratum). The base-line hazard rate is given by patients without missing values and with low scores on the bioptic variables. (One could also consider leaving out the five missing values with two renal failures from before 1969, as these may be missing at random. Although preferable from a theoretical point of view, we did not do this because of the small total number of patients with renal failure.)

(4) Effects of the bioptic variables on survival time will be investigated by using the (stratified) proportional hazards model.

4.3. Results

The results of interpolating the missing values of class b are displayed in Table 9. Calculations were performed by using the program BMDPAM (Dixon et al., 1988, Version 1983). The program selected predictor variables (from all or from only the bioptic variables) which, stepwisely, passed an F-test indicating (approximately) significance of the regression coefficients. The critical value of the F-statistic was set to 4.0. As the results do not depend appreciably on whether the linear regression was based on a selection of all, or of only the bioptic covariates, we have some confidence in the adequacy of the interpolation and the subsequent dichotomisation. Note that an alternative method would have been to estimate, e.g. by logistic regression or discriminant analysis techniques, the posterior probabilities that the missing values belong to score-class '1' (indicating a 'high severity' of the morphological aberration).

From Table 10 one can get an impression of the influence of the bioptic variables on survival time. Table 10(A) shows the results of a PH analysis where the bioptic variables are the only covariates, and (as usual) treatment of hypertension is used as a stratifying variable. One can see that, probably, only global sclerosis and tubular atrophy may have a substantial 'influence' on the survival

Table 9
Estimation of missing values[a]

Case number	x_0	Variable	Estimate	Predictor variables	Estimate	Predictor variables
8	48	atr	2.6	$y_3 y_2 y_1$	2.7	$y_3 y_2$
34	76	scl	3.8	$x_3 y_4 x_6$	3.6	y_4
		prol	2.3	x_2	2.2	
45	05	scl	3.8	$x_3 y_4 x_6$	3.6	y_4
		prol	2.9	x_2	2.2	
50	88	scl	3.9	$x_3 y_4 x_6$	3.6	y_4
		prol	2.6	x_2	2.2	
62	63	prol	2.2	x_2	2.2	

[a] A missing score for the variable in the third column is estimated ('predicted') by linear regression, using the variables in the 5-th (7-th) column. The corresponding estimates are written in the 4-th (6-th) column. The 5-th column is a selection from all variables, and the 7-th column a selection of the bioptic variables only. The method of selection is briefly explained in the main text. As one can see, both selections of predictor variables yield approximately the same numerical predictions of a missing value. The coding of the predictor variables is the same as in Table 1. For example, case number 50 corresponds to the 50-th patient (from the top) in Table 1, who has identification number $x_0 = 88$. For this patient, the missing score for extracapillary proliferation is estimated to be 2.6 by regression on x_2 (proteinuria), and to be 2.2 if the sample mean of all non-missing proliferation scores is used as predictor.

Table 10

Multiple regression analysis based on the proportional hazards model (bioptic variables included)

(A) Bioptic variables only (with control of hypertension as stratifying variable)[a]

Variable	$\hat{\beta}$	$\hat{\beta}/\hat{SE}(\hat{\beta})$	$\exp(\hat{\beta})$
Global sclerosis	1.5	1.6	4.4
Extracapillary proliferation	0.2	0.2	1.2
Interstitial infiltration	0.3	0.3	1.3
Tubular atrophy	1.5	1.0	4.6
Missing value indicator	2.6	2.3	14.0

(B) Bioptic variables in combination with age-adjusted GFR and proteinuria (and with control of hypertension as stratifying variable)[b]

Variable	$\hat{\beta}$	$\hat{\beta}/\hat{SE}(\hat{\beta})$	$\exp(\hat{\beta})$
Global sclerosis	1.7	1.8	5.2
Extracapillary proliferation	0.4	0.3	1.5
Interstitial infiltration	1.2	1.3	3.3
Tubular atrophy	2.2	1.7	9.2

[a] The bioptic variables are dichotomous, 0 indicating low severity, and 1 indicating high severity of the microscopic aberration. The fifth variable indicates whether (score 1) or not (score 0) all bioptic variables are missing. In order to facilitate comparison with Table 7, the data were stratified with respect to the variable H + C − (control of hypertension).

[b] Each bioptic variable was added *separately* to a model which included aa.GFR, proteinuria, and the missing value indicator as covariates, and (the control of) hypertension as stratifying variable.

time, and that the overall fit is worse than that of the third round (see Section 3.1), which comprised four non-bioptic variables. In fact, the value of $\ln L(\beta)$ is markedly lower than in the third round (-40.6 instead of -24.2). Note that the effect of atrophy can only be measured with low precision. In practice, of course, no one would use just the bioptic variables for predicting the survival time, so in Table 10(B) the effects of the bioptic variables are given when added to a PH model which already includes proteinuria and (age-adjusted) GFR. These results indicate that global sclerosis and tubular atrophy are of predictive value, even in addition to the main non-bioptic variables.

However, when we tried to construct a final model consisting of the two main bioptic variables and all five non-bioptic variables of the third round, some of the estimated effects, as well as their estimated standard deviations, became very large. (Note that the latter phenomenon corresponds to the fact that the observed Fisher information matrix was getting nearly singular.) This unfortunate situation persisted, even after reduction of the number of parameters by means of the following index scores:

$$hematuria\ index = (macroscopic\ hematuria - microscopic\ hematuria),$$

$$bioptic\ index = (global\ sclerosis + tubular\ atrophy).$$

REMARK. The first index, which can assume the values -1, 0, and 1, is motivated by the fact that the estimated effects of macroscopic hematuria and of microscopic hematuria are approximately equal in magnitude, but of opposite sign (see Table 7). A corresponding motivation holds for the second index, which assumes the values 0, 1, or 2, according to whether 0, 1, or 2 of the 'main' bioptic scores indicates a serious aberration.

In summary, combining these results with those of Section 2.2, the following can be concluded.

(1) As one can see from Table 3, the bioptic variables have interesting correlations with the non-bioptic variables: Hypertension is positively correlated with the three 'main' bioptic variables (global sclerosis, tubular atrophy and interstitial infiltration), whereas macroscopic hematuria (!) and initial creatinine clearance are negatively correlated with them. Extracapillary proliferation behaves differently from the other bioptic variables.

(2) Two of the bioptic variables (global sclerosis and tubular atrophy) seem to be an indication of low survival, even if the effects of the three main non-bioptic variables are already taken into account.

(3) The data are insufficient to permit reliable lifetime predictions from a full model which includes all relevant bioptic and non-bioptic variables, at least if proportional hazards regression techniques are used that are based on Cox's partial likelihood.

(4) The remarkable correlations of macroscopic hematuria with the bioptic and non-bioptic variables, together with the fact that it has a favourable effect on survival time, supports the view, expressed by Beukhof et al. (1984b), that macroscopic hematuria indicates a benign form of the disease IgAN.

Appendix A: The nomogram of Figure 3

In the first part of this appendix we derive the underlying formulae for the nomogram in Figure 3, for which we will rely on the asymptotic theory concerning the distribution of $\ln \hat{\Lambda}(t\,|\,z)$ (see Section 3.2 and Chapter 15, Section 1.1).

As we have seen in Section 3.2, for a fixed time t (e.g., five years), the contours of constant estimated variance of $\ln \hat{\Lambda}(t\,|\,z)$, $z = (z_1, z_2, z_3, z_4) \in \mathbb{R}^4$, are ellipses in $\mathbb{R}^4$. For fixed z_3 and z_4, they are two-dimensional ellipses in $\mathbb{R}^2$. (In our case z_1 (proteinuria) and z_2 (age-adjusted creatinine clearance) are continuous covariates, while z_3 (macroscopic hematuria) and z_4 (microscopic hematuria) are binary.)

In order to derive some convenient expressions for these ellipses, we first state a (mathematically simple) lemma, which has some interesting statistical interpretations on its own.

LEMMA. *The family of ellipses in* $\mathbb{R}^2$

$$\{x \in \mathbb{R}^2 \mid x^{\mathrm{T}} A x - 2 a^{\mathrm{T}} x = c\}, \tag{A.1}$$

with

$$a \in \mathbb{R}^2, \quad c \in \mathbb{R}, \quad \text{and} \quad A^{-1} = \Sigma = \begin{pmatrix} \sigma_1^2 & \rho\sigma_1\sigma_2 \\ \rho\sigma_1\sigma_2 & \sigma_2^2 \end{pmatrix},$$

can, alternatively, be represented by the parametric form

$$\begin{pmatrix} x_1(t) \\ x_2(t) \end{pmatrix} = \begin{pmatrix} m_1 \\ m_2 \end{pmatrix} + \begin{pmatrix} k\sigma_1 \sin(\omega t) \\ k\sigma_2 \sin(\omega t + \phi) \end{pmatrix}, \quad 0 \leqslant \omega t < 2\pi, \tag{A.2}$$

where $m = (m_1, m_2)^{\mathrm{T}}$ *denotes the center of the ellipse,* ϕ *is a phase angle, and* k *is a scale factor. The connection between the two representations is*

$$m = \Sigma a, \quad k^2 = c + m^{\mathrm{T}} A m, \quad \rho = \cos\phi. \tag{A.3}$$

(The parametric representation corresponds to a special case of so-called 'Lissajous figures', familiar to physicists and electrical engineers.)

REMARKS. The ellipses are interpreted as the contours of constant probability density of a bivariate normal $N(m, \Sigma)$ distribution. Note the relationship between the phase angle and the correlation coefficient. The term $x_2(t)$ can be expressed as:

$$x_2(t) = m_2 + k\sigma_2(\rho \sin(\omega t) + (1 - \rho^2)^{1/2} \cos(\omega t)), \tag{A.4}$$

which again gives an interpretation of the correlation coefficient. Note also that $a = \Sigma^{-1} m$ are the weights of Fisher's linear discriminant function used for discriminating between $N(0, \Sigma)$ and $N(m, \Sigma)$ as $\{x \mid 2a^{\mathrm{T}} x = m^{\mathrm{T}} A m = a^{\mathrm{T}} \Sigma a\}$ is the intersection line of the corresponding contours of equiprobability (see Figure A.1).

PROOF. Differentiating the quadratic form with respect to x we obtain the equation for the center $m: Am - a = 0$. Hence, instead of (A.1), we can write

$$(x - m)^{\mathrm{T}} A (x - m) = c + m^{\mathrm{T}} A m. \tag{A.5}$$

By using standardised variables $y_i = (x_i - m_i)/\sigma_i$ $(i = 1, 2)$ we get

$$y^{\mathrm{T}} A_\rho y = c + m^{\mathrm{T}} A m, \tag{A.6}$$

where

$$A_\rho = \begin{pmatrix} 1 & \rho \\ \rho & 1 \end{pmatrix}^{-1} = (1 - \rho^2)^{-1} \begin{pmatrix} 1 & -\rho \\ -\rho & 1 \end{pmatrix}$$

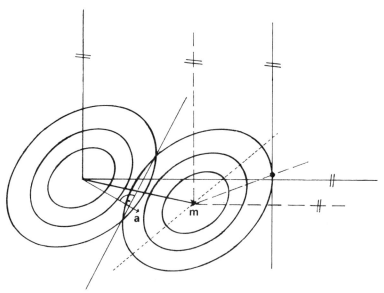

Fig. A.1. A diagram for illustrating various (statistical and geometrical) properties of two shifted bivariate normal distributions. (Notice, among others, the linear discrimination line and the geometrical interpretations of the correlation coefficient.)

is the inverse of the correlation matrix. Substituting $(y_1, y_2) = (k \sin(\omega t), k \sin(\omega t + \phi))$ in (A.6) for $\omega t = 0, \frac{1}{2}\pi$, gives the relations expressed in (A.3). Then, by substitution one can see that, given (A.3), (A.2) satisfies (A.1) for all $t \in [0, 2\pi)$. Both (A.1) and (A.2) represent two-dimensional ellipses, which concludes the proof. $\square$

We now start with our derivation. Let the estimated variance of $\ln \hat{\Lambda}(t \mid z)$ be denoted by α. The contours of constant α are given by, see (3.11),

$$(z + \hat{\Gamma})^{\mathrm{T}} \hat{V}(z + \hat{\Gamma}) + \hat{\Psi}(5) = \alpha, \tag{A.7}$$

where $\hat{\Gamma} = \tilde{\Gamma}(5, \hat{\beta} \mid 0) \in \mathbb{R}^4$. For fixed z_3 and z_4 we want to describe the two-dimensional ellipses that are a function of z_1 and z_2. So, supressing for notational convenience the circonflex in all $n \times n$ matrices ($n \geqslant 2$), let

$$V = \begin{pmatrix} V_{11} & V_{12} \\ V_{21} & V_{22} \end{pmatrix}, \qquad I = V^{-1} = \begin{pmatrix} I_{11} & I_{12} \\ I_{21} & I_{22} \end{pmatrix},$$

$$m = (m_1, m_2)^{\mathrm{T}} = -\hat{\Gamma} = -(\hat{\Gamma}_1, \hat{\Gamma}_2, \hat{\Gamma}_3, \hat{\Gamma}_4)^{\mathrm{T}},$$

$$z = (z_1, z_2)^{\mathrm{T}} = (z_1, z_2, z_3, z_4)^{\mathrm{T}}. \tag{A.8}$$

Here, m denotes the center of a four-dimensional ellipse defined by (A.7). Instead of (A.7) we can now write,

$$(z_1 - m_1)^\mathrm{T} V_{11}(z_1 - m_1) + 2(z_1 - m_1)^\mathrm{T} V_{12}(z_2 - m_2)$$
$$+ (z_2 - m_2)^\mathrm{T} V_{22}(z_2 - m_2) + \hat{\Psi}(5) = \alpha, \tag{A.9}$$

which corresponds to (A.1) if

$$A = V_{11}, \qquad x = z_1 - m_1, \qquad a = -V_{12}(z_2 - m_2),$$

and

$$c = \alpha - (z_2 - m_2)^\mathrm{T} V_{22}(z_2 - m_2) - \hat{\Psi}(5). \tag{A.10}$$

With respect to $x = z_1 - m_1$, the center M of a pertaining two-dimensional ellipse is given by

$$M = (M_1, M_2)^\mathrm{T} = -V_{11}^{-1} V_{12}(z_2 - m_2), \tag{A.11}$$

and an alternative expression for (A.9) is

$$(x - M)^\mathrm{T} V_{11}(x - M) = c + M^\mathrm{T} V_{11} M = \alpha - D,$$

with

$$D = \hat{\Psi}(5) + (z_2 - m_2)^\mathrm{T} I_{22}^{-1}(z_2 - m_2). \tag{A.12}$$

(Recall that $I_{22}^{-1} = V_{22} - V_{12}^\mathrm{T} V_{11}^{-1} V_{12}$.) To obtain a convenient parametric representation, let us denote the *inverse* of the estimated covariance matrix of $\hat{\beta}_1$ and $\hat{\beta}_2$ by

$$V_{11}^{-1} = \begin{pmatrix} \hat{\sigma}_1^2 & \hat{\rho}\,\hat{\sigma}_1\,\hat{\sigma}_2 \\ \hat{\rho}\,\hat{\sigma}_1\,\hat{\sigma}_2 & \hat{\sigma}_2^2 \end{pmatrix}. \tag{A.13}$$

Applying the Lemma, we get

$$z_1(t) = \hat{\Gamma}_1 + M_1 + \hat{\sigma}_1(\alpha - D)^{1/2} \sin(\omega t),$$
$$z_2(t) = \hat{\Gamma}_2 + M_2 + \hat{\sigma}_2(\alpha - D)^{1/2} \sin(\omega t + \phi), \tag{A.14}$$

where $\cos \phi = \hat{\rho}$, D is given by (A.12), and $(M_1, M_2)^\mathrm{T}$ by (A.11).

Notice that at the center of the two-dimensional ellipses the variance α equals D. The ellipses in Figure 3 are drawn for $\alpha = D + d$, $d = 0.2, 0.5, 1, 2$. Note also that the center depends on $(z_3, z_4)^\mathrm{T}$ by

$$(M_1, M_2)^\mathrm{T} = -V_{11}^{-1} V_{12}((z_3, z_4)^\mathrm{T} + (\hat{\Gamma}_3, \hat{\Gamma}_4)^\mathrm{T}).$$

Since in our case there are only four combinations of values of (z_3, z_4): $(0, 0)$, $(0, 1)$, $(1, 0)$, $(1, 1)$, the displacement of the centers with respect to $(z_3, z_4) = (0, 0)$ can be indicated by three arrows.

The straight lines in Figure 3 are simply lines for which $\beta_1 z_1 + \beta_2 z_2$ is constant (1/4, 1, 2, 4, 16, 64, respectively). On these lines, the relative risk $\exp(\boldsymbol{\beta}^{\mathrm{T}} z)$ with respect to the standard patient equals rR, where $R = \beta_3 z_3 + \beta_4 z_4$ can be found (for the above mentioned four combinations of (z_3, z_4)) in the table at the left lower corner of the figure. The straight lines correspond to the lines of constant posterior probability used in discriminant analysis.

Appendix B: The computer program CONCERN

The program CONCERN (CONfidence from CEnsored Regression Nomograms) is an approximately 4000 lines interactive computer program, written in FORTRAN 77, for multiple regression analysis with censored data, based on Cox's proportional hazards model.

Special features are

(1) supply of an (instantaneous) confidence band for the survival function of an individual with a given vector of scores;

(2) representation of the base-line survival function and the effects of the various covariates in a compact nomogram; from this nomogram one can directly estimate, within a certain proportional hazards model, the survival function $S(t \mid z)$ of an individual with an arbitrary vector z of scores;

(3) a nomogram displaying the variation of the standard deviation of $\hat{S}(t \mid z)$ with respect to z, if z has two continuous and some ($\leqslant 3$) discrete components;

(4) a graphical analysis of residuals in order to get an impression of the (overall) adequacy of the model assumptions, and to identify, from the individuals who are still alive, those who are at high risk; this type of analysis is primarily informative in the simple case in which there are no censored observations.

The interactive input/output of the program is largely self-explaining, and will not be discussed here. (Interested readers may contact the author.) The first version of the program was written by the author for analysing the IgAN data. In the second version an interface for arbitrary input data was made by Ronald Kiel, who also contributed a lot to general improvements.

The program makes use of IMSL routines (for sorting, solving a set of simultaneous linear equations, and inverting the observed Fisher information matrix), and of graphical procedures of the package KOMPLOT, written by Kraak. Accordingly, the program will primarily run on computers where these products are available.

Acknowledgement

This chapter corresponds (apart from some minor corrections and an update of the references) to Chapter II of the author's Thesis at the University of Groningen. He is especially indebted to Dr J. R. Beukhof and Prof. Dr G. K. van der Hem from the Nephrology Department, who initiated the subject and from which collaboration the present study emerged, to his supervisor Prof. Dr W. Schaafsma for many stimulating discussions and drastic proposals to improve the manuscript, to Prof. Dr A. J. Stam for carefully reading and commenting the manuscript, and to Drs. R. Kiel for extensive help in making a general version of the program CONCERN.

References

Aalen, O. O. (1978). Nonparametric inference for a family of counting processes. *The Annals of Statistics* **6**, 701–726.

Altman, D. G. and Andersen, P. K. (1986). A note on the uncertainty of a survival probability estimated from Cox's regression model. *Biometrika* **73**, 722–724.

Andersen, P. K. and Gill, R. D. (1981). Cox's regression model for counting processes: a large sample study. *The Annals of Statistics* **10**, 1100–1121.

Andersen, P. K. (1982). Testing goodness-of-fit of Cox's regression and life model. *Biometrics* **38**, 67–77.

Andersen, J. R., Bernstein, L. and Pike, M. C. (1982). Approximate confidence intervals for probabilities of survival and quantiles in life table analysis. *Biometrics* **38**, 407–416.

Andersen, P. K., Borgan, O., Gill, R. D. and Keiding, N. (1982). Linear nonparametric tests for comparison of counting processes, with applications to censored survival data (with discussion). *International Statistical Review* **50**, 219–259.

Andersen, P. K., Christensen, E., Fauerholdt, L. and Schlichting, P. (1983). Measuring prognosis using the proportional hazards model. *Scandinavian Journal of Statistics* **10**, 49–52.

Ansell, J. I. and Phillips, M. J. (1989). Practical problems in the analysis of reliability data (with discussion). *Applied Statistics* **38**, 205–247.

Arjas, E. and Haara, P. (1988). A note on the asymptotic normality in the Cox regression model. *The Annals of Statistics* **8**, 1133–1140.

Bailey, K. R. (1979). The general maximum likelihood approach to the Cox regression model. Ph.D. dissertation, University of Chicago, Chicago, Illinois.

Bailey, K. R. (1983). The asymptotic joint distribution of regression and survival estimates in the Cox regression model. *The Annals of Statistics* **11**, 39–48.

Bailey, K. R. (1984). Asymptotic equivalence between the Cox estimator and the general ML estimators of regression and survival parameters in the Cox model. *The Annals of Statistics* **12**, 730–737.

Barlow, W. E. and Prentice, R. L. (1988). Residuals for relative risk regression. *Biometrika* **75**, 65–74.

Bennett, S. (1983). Analysis of survival data by the proportional odds model. *Statistics in Medicine* **2**, 273–277.

Bernoulli, D. (1760). Essai d'une nouvelle analyse de la mortalité causée par la petite Vérole, et les avantages de l'Inoculation pour la prévenir. *Mémoires de l'Académie Royale de Science, 1760*, 1–45.

Beukhof, J. R. (1986). *Primary IgA Nephropathy*. Thesis, University of Groningen.

Beukhof, J. R., Fleuren, G. J., Hoedemaeker, Ph. J. et al. (1984a). IgA nephropathy: Triumph of immunology or misconception? *Netherlands Journal of Medicine* **27**, 393–404.

Beukhof, J. R., Ockhuizen, Th., Halie, L. M. et al. (1984b). Subentities within adult primary IgA-nephropathy. *Clinical Nephrology* **22**, 195–199.

Beukhof, J. R., Kardaun, O., Schaafsma, W., Poortema, K., Donker, A. J. M., Hoedemaeker, Ph. J. and Van Der Hem, G. K. (1986). Toward individual prognosis of IgA nephropathy. *Kidney International* **29**, 549–556.

Breslow, N. E. (1979). Statistical methods for censored survival data. *Environmental Health Perspectives* **32**, 181–192.

Breslow, N. E. and Crowley, J. (1974). A large sample study of the life table and product limit estimates under random censorship. *The Annals of Statistics* **2**, 437–453.

Brillinger, D. R. (1986). The natural variability of vital rates and associated statistics (with discussion). *Biometrics* **42**, 693–734.

Brookmeyer, R. and Crowley, J. (1982). A confidence interval for the median survival time. *Biometrics* **38**, 29–41.

Brookmeyer, R. (1983). Prediction intervals for survival data. *Statistics in Medicine* **2**, 486–495.

Brown, M. (1984). On the choice of variance for the logrank test. *Biometrika* **71**, 65–74.

Budde, M. (1983). Das Jackknife- und Bootstrap-Verfahren im Cox-Regressionsmodell als Mittel der Verzerrungsreduktion, Modellüberprüfung und Influenzbestimmung. Dissertation der Abteilung Statistik der Universität Dortmund.

Bundschuh, W. (1982). Simultane Konfidenzbänder für die Survivalfunktion. Diplomarbeit, Fakultät für Mathematik, Heidelberg.

Cox, D. R. (1972). Regression models and life tables (with discussion), *Journal of the Royal Statistical Society B* **34**, 187–220.

Cox, D. R. (1975). Partial likelihood. *Biometrika* **62**, 269–276.

Cox, D. R. (1979). A note of the graphical analysis of survival data. *Biometrika* **66**, 188–190.

Cox, D. R. and Hinkley, D. V. (1974). *Theoretical Statistics*. Chapman & Hall, London.

Cox, D. R. and Oakes, D. (1984). *Analysis of Survival Data*. Chapman & Hall, London.

Cox, D. R. and Snell, E. J. (1968). A general definition of residuals. *Journal of the Royal Statistical Society B* **30**, 248–275.

Crowly, J. and Hu, M. (1977). Covariance analysis of heart transplant survival data. *Journal of the American Statistical Association* **72**, 557–565.

Cuzick, J. (1982). The efficiency of the proportions test and the logrank test for censored survival data. *Biometrics* **38**, 1033–1041.

Dabrowska, D. M. and Doksum, K. A. (1987). Estimates and confidence intervals for median and mean life in the proportional hazards model. *Biometrika* **74**, 799–807.

Dewitt Baines, A. (1980). Disorders of the kidney and the urinary tract. In: A. G. Gornall, ed., *Applied Biochemistry of Clinical Disorders*, Harper & Row, London.

De Witt, J. (1671). *Waerdije van lijfrenten naer proportie van losrenten*, 's-Gravenhage, facsimilé Joh. Enschedé & Zonen (1879), Haarlem.

Dixon, W. J. et al., eds. (1988). *BMDP Statistical Software Manual, Edition 1988*, University of California Press, Berkeley, Los Angeles, London.

Doksum, K. A. (1987). An extension of partial likelihood methods for proportional hazards models to general transformation models. *The Annals of Statistics* **15**, 325–345.

Efron, B. (1977). Efficiency of Cox's likelihood function for censored data. *Journal of the American Statistical Association* **72**, 557–565.

Efron, B. (1981). Censored data and the bootstrap. *Journal of the American Statistical Association* **76**, 312–319.

Efron, B. and Hinkley, D. V. (1978). Assessing the accuracy of the maximum likelihood estimator: Observed versus Expected Fisher Information (with discussion). *Biometrika* **65**, 457–489.

Emerson, J. D. (1982). Nonparametric confidence intervals for the median in the presence of right censoring. *Biometrics* **38**, 17–27.

Emoto, S. E. and Matthews, P. C. (1990). A Weibull model for dependent censoring. *The Annals of Statistics* **18**, 1556–1557.

Fleming, T. R., O'Fallon, J. R., O'Brien, P. C. and Harrington, D. P. (1980). Modified Kolmogorov-Smirnov test procedures with application to arbitrarily right censored data. *Biometrics* **36**, 607–625.

Fleming, T. R. and Harrington, D. P. (1981). A class of hypothesis tests for one and two sample censored survival data, *Communications in Statistics A* **10**, 763–794.

Gehan, E. A. (1965). A generalized Wilcoxon test for comparing arbitrarily singly-censored samples. *Biometrika* **52**, 203–223.

Gillespie, M. J. and Fisher, L. (1979). Confidence bands for the Kaplan–Meier survival curve estimate. *The Annals of Statistics* **7**, 920–924.

Gill, R. D. (1980). *Censoring and Stochastic Integrals*. M. C. Tracts 124, Mathematisch Centrum, Amsterdam.

Gill, R. D. (1984). Understanding Cox's regression model: A martingale approach. *Journal of the American Statistical Association* **79**, 441–448.

Gill, R. D. and Schumacher, M. (1987). On a simple test of the proportional hazards assumption. *Biometrika* **74**, 187–197.

Gill, R. D. and Johansen, S. (1990). A survey of product-integration with a view toward application in survival analysis. *The Annals of Statistics* **18**, 1501–1555.

Gore, S. M., Pocock, S. J. and Kerr, G. R. (1984). Regression models and non-proportional hazards in the analysis of breast-cancer survival. *Applied Statistics* **33**, 176–195.

Gray, R. J. (1990). Some diagnostic methods for Cox regression models through hazard smoothing. *Biometrics* **46**, 93–102.

Greenwood, M. (1926). The natural duration of cancer. In: *Reports on Public Health and Medical Subjects 33*. London, Her Majesty's Stationary Office, 1–26.

Hall, W. J. and Wellner, J. A. (1980). Confidence bands for a survival curve from censored data. *Biometrika* **67**, 133–143.

Heckmann, J. J. and Honoré, B. E. (1989). The identifiability of the competing risk model. *Biometrika* **76**, 325–330.

Hempenius, A. L. and Mulder, P. G. H. (1986). Multiple failure causes and observations of time dependent covariables. *Statistica Neerlandica* **40**, 47–64.

Hill, C. (1981). Asymptotic relative efficiency of survival tests with covariates. *Biometrika* **68**, 699–702.

IMSL (1990). *International Mathematical and Statistical Libraries (MATH, SFUN, STAT)*. IMSL Inc., 2500 Park West Tower One, 2500 City West Boulevard, Houston, Texas 77036–5085, USA.

Jennen, C. (1982). Asymptotische Bestimmung von Kenngrössen sequentieller Verfahren. Dissertation, Universität Heidelberg.

Jennrich, R. I. (1984). Some exact tests for comparing survival curves in the presence of unequal right censoring. *Biometrika* **71**, 57–64.

Johansen, S. (1978). The product-limit estimator as maximum likelihood estimator. *Scandinavian Journal of Statistics* **5**, 195–199.

Jones, M. P. and Crowley, J. (1989). A general class of nonparametric tests for survival analysis. *Biometrics* **45**, 157–170.

Kalbfleisch, J. D. and Prentice, R. L. (1973). Marginal likelihoods based on Cox's regression and life model. *Biometrika* **60**, 267–278.

Kalbfleisch, J. D. and Prentice, R. L. (1980). *The Statistical Analysis of Failure time Data*. Wiley, New York, Toronto.

Kaplan, E. L. and Meier, P. (1958). Nonparametric estimation from incomplete observations. *Journal of the American Statistical Association* **53**, 457–481.

Kardaun, O. J. W. F. (1983a). Statistical survival analysis of male larynx-cancer patients – A case study. *Statistica Neerlandia* **37**, 103–125.

Kardaun, O. J. W. F. (1983b). Over betrouwbaarheidsbanden en aanpassingstoetsen bij gecensureerde waarnemingen. *Kwantitatieve Methoden* **4**, 3–20.

Kardaun, O. J. W. F. (1986). *On Statistical Survival Analysis and its Applications in Medicine*. Thesis, University of Groningen.

Kardaun, O. J. W. F. (1991). Confidence bands and the relation with decision analyses: Theory. Chapter 15 in: C. R. Rao and R. Chakraborty, eds., *Statistical Methods in Biological and Medical Sciences*, Handbook of Statistics **8**, North-Holland, Amsterdam (this volume).

Kay, R. (1977). Proportional hazard regression models and the analysis of censored survival data. *Applied Statistics* **26**, 227–237.

Kay, R. (1984). Goodness of fit methods for the proportional hazards regression model: a review. *Revue d'épidémiologie et de santé publique* **32**, 185–199.

Kay, R. (1986). Treatment effects in competing risks analysis of prostate cancer. *Biometrics* **42**, 203–212.

Kay, R. and Schumacher, M. (1983). Unbiased assessment of treatment effects on disease recurrence and survival in clinical trials. *Statistics in Medicine* **2**, 41–58.

Klein, J. P. and Moeschberger, M. L. (1988). Bounds on net survival probabilities for dependent competing risks. *Biometrics* **44**, 529–538.

Koziol, J. A. and Green, S. B. (1976). A Cramér–von Mises statistic for randomly censored data. *Biometrika* **63**, 465–474.

Kraak, J. (1989). *KOMPLOT en GRAFIEK, Programmatuur voor het maken van Grafieken (Mainframes en IBM-compatibele PC's)*. RC publicatie 7, Rekencentrum der Rijksuniversiteit, Groningen.

Kronborg, D. and Aaby, P. (1988). Piecewise comparison of survival functions in stratified proportional hazards models. *Biometrics* **46**, 375–380.

Kuk, A. Y. C. (1984). All subsets regression in a proportional hazards model. *Biometrika* **71**, 587–593.

Lagakos, S. W. (1979). General right censoring and its impact on the analysis of survival data. *Biometrics* **35**, 139–156.

Lagakos, S. W. (1980). The graphical evaluation of explanatory variables in proportional hazard regression models. *Biometrika* **68**, 93–98.

Laird, N. and Olivier, D. (1981). Covariance analysis of censored survival data using log-linear analysis techniques. *Journal of the American Statistical Association* **76**, 231–240.

Landwehr, J. M., Pregibon, D. and Shoemaker, A. C. (1984). Graphical methods for assessing logistic regression models (with discussion). *Journal of the American Statistical Association* **79**, 61–84.

Lee, E. T. and Desu, M. M. (1974). A computer program for comparing K samples with right-censored data. *Computer Programs in Biomedicine* **2**, North-Holland, Amsterdam, 315–321.

Lee, E. T. (1980). *Statistical Methods for Survival Data Analysis*. Lifetime Learning Publications, Belmont, CA.

Leurgans, S. (1984). Asymptotic behaviour of two-sample rank tests in the presence of random censoring. *The Annals of Statistics* **12**, 572–590.

Liang, K. Y., Self, S. G. and Liu, X. (1990). The Cox proportional hazards model with change point: An epidemiologic application. *Biometrics* **46**, 738–793.

Link, C. L. (1984). Confidence intervals for the survival function using Cox's proportional hazards model with covariates. *Biometrics* **40**, 601–610.

McKnight, B. and Crowley, J. (1984). Tests for differences in tumor incidence based on animal carcinogenesis experiments. *Journal of the American Statistical Association* **79**, 639–649.

Miller, R. G. Jr., Efron, B., Brown, B. W. Jr. and Moses, L. E., eds. (1980). *Biostatistics Casebook*. Wiley, New York/Toronto.

Miller, G. Jr. (1981). *Survival Analysis*. Wiley, New York.

Moreau, T., O'Quigley, J. and Meshbah, M. (1985). A global goodness-of-fit statistic for the proportional hazards model. *Applied Statistics* **34**, 212–218.

Moolgavkar, S. H. and Venzon, D. J. (1987). Confidence regions in curved exponential families: Application to matched case-control and survival studies with general relative risk function. *The Annals of Statistics* **15**, 346–359.

Morton, R. (1978). Regression analysis of life tables and related non-parametric tests. *Biometrika* **65**, 329–333.

Nagelkerke, N. J. D., Oosting, J. and Hart, A. A. M. (1984). A simple test for goodness of fit of Cox's proportional hazards model. *Biometrics* **40**, 483–486.

Nagelkerke, N. J. D., Oosting, J. and Hart, A. A. M. (1986). Comparison of several goodness of fit tests for Cox's model. *Biometrical Journal* **28**, 491–494.

Nair, V. N. (1981). Plots and tests for goodness of fit with randomly censored data. *Biometrika* **68**, 99–103.

Nelson, W. (1969). Hazard plotting for incomplete failure data. *Journal of Quality Technology* **1**, 27–52.

Nelson, W. (1972). Theory and application of hazard plotting for censored failure data. *Technometrics* **14**, 945–966.

Nightingale, F. (1858). *Notes on matters affecting the health, efficiency, and hospital administration of the British Army, founded chiefly on the experience of the late war*. Harrison & Sons, London.

Oakes, D. (1981). Survival times: Aspects of partial likelihood (with discussion). *International Statistical Review* **49**, 235–264.

O'Quigley, J. and Pessione, F. (1989). Score tests for homogeneity of regression effect in the proportional hazards model. *Biometrics* **45**, 135–144.

Peduzzi, P. N., Hardy, R. J. and Holford, T. R. (1980). A stepwise variable selection procedure for nonlinear regression model. *Biometrics* **36**, 511–516.

Peto, R., Pike, M. C., Armitage, P., Breslow, N. E., Cox, D. R. et al. (1977). Design and analysis of randomized clinical trials requiring prolonged observation of each patient, Part II analysis and examples. *British Journal of Cancer* **35**, 1–39.

Pettit, A. N. (1984). Proportional odds models for survival data and estimates using ranks. *Applied Statistics* **33**, 169–175.

Pettit, A. N. (1986). Censored observations, repeated measures and mixed effects models: An approach using the EM algorithm and normal errors. *Biometrika* **73**, 635–643.

Pettit, A. N. and Stephens, M. A. (1976). Modified Cramér–von Mises statistics for censored data. *Biometrika* **63**, 291–298.

Rao, C. R. (1973). *Linear Statistical Inference and its Applications*, 2nd ed. Wiley, New York.

Reid, N. (1981a). Influence functions for censored data. *The Annals of Statistics* **9**, 78–92.

Reid, N. (1981b). Estimating the median survival time. *Biometrika* **68**, 188–190.

Reid, N. and Crépeau, H. (1985). Influence functions for proportional hazards regression. *Biometrika* **72**, 1–9.

Robins, J. and Greenland, S. (1989). The probability of causation under a stochastic model for individual risk. *Biometrics* **45**, 1125–1138.

Schoenfeld, D. A. (1982). Partial residuals for the proportional hazards regression model. *Biometrika* **69**, 239–241.

Schoenfeld, D. A. (1983). Sample-size formula for the proportional-hazards regression model. *Biometrics* **39**, 499–505.

Scholz, F. W. (1980). Towards a unified definition of maximum likelihood. *Canadian Journal of Statistics* **8**, 193–203.

Schottenfeld, D. and Fraumeni, J. F. (1982). *Cancer Epidemiology and Prevention*. W. B. Saunders Company, Philadelphia.

Schumacher, M. (1982). *Analyse von Überlebenszeiten bei nichtproportionalen Hazardfunktionen*. Habilitationsschrift, Fakultät für Theoretische Medizin, Heidelberg.

Schumacher, M. (1984). Two-sample tests of Cramér–von Mises and Kolmogorov–Smirnov type for randomly censored data. *International Statistical Review* **52**, 263–283.

Schumacher, M. and Vaeth, M. (1984). On a goodness-of-fit test for the proportional hazards model. *EDP in Biology and Medicine* **15**, 19–23.

Slud, E. V. (1984). Sequential linear rank tests for two-sample censored survival data. *The Annals of Statistics* **12**, 551–572.

Slud, E. V., Byar, D. P. and Green, S. B. (1984). A comparison of reflected versus test-based confidence intervals for the median survival time based on censored data. *Biometrics* **40**, 587–601.

Smid, L. J. (1949). *Levensverzekeringswiskunde*. Servire, Den Haag.

Stephenson, J. L. (1981). Case studies in renal and epithelial physiology. In: *Mathematical Aspects of Physiology, Lecture Notes in Applied Mathematics* **19**, Amer. Mathematical Soc., Providence, RI, 171–212.

Susarla, V. and Van Ryzin, J. (1980). Large sample theory for an estimator of the mean survival time from censored samples. *The Annals of Statistics* **8**, 1002–1016.

Tarone, R. E. (1975). Tests for trend in life table analysis. *Biometrika* **62**, 679–682.

Tarone, R. E. (1981). On the distribution of the maximum of the log-rank statistic and the modified Wilcoxon statistic. *Biometrics* **37**, 79–86.

Taulbee, J. D. (1979). A general model of hazard rate with covariables. *Biometrics* **35**, 439–450.

Taulbee, J. D. and Symons, M. J. (1983). Sample size and duration for cohort studies of survival time with covariables. *Biometrics* **39**, 351–361.

Thomas, D. R. and Grunkemeier, G. L. (1975). Confidence interval estimation of survival probabilities for censored data. *Journal of the American Statistical Association* **70**, 865–871.

Tibshirani, R. J. and Ciampi, A. (1983). A family for proportional- and additive-hazards models for survival data. *Biometrics* **39**, 141–149.

Tsiatis, A. A. (1981). A large sample study of Cox's regression model. *The Annals of Statistics* **9**, 93–108.

Wei, L. J. (1984). Testing goodness of fit for proportional hazards model with censored observations. *Journal of the American Statistical Association* **79**, 649–653.

Wei, L. J. and Lachin, J. M. (1984). Two-sample asymptotically distribution-free tests for incomplete multivariate observations. *Journal of the American Statistical Association* **79**, 653–662.

Wong, W. H. (1986). Theory of partial likelihood. *The Annals of Statistics* **14**, 88–123.

Woolson, R. F. and Lachenbruch, P. A. (1983). Rank analysis of covariance with right censored data. *Biometrics* **39**, 727–735.

Wu, M. C. and Bailey, K. R. (1989). Estimation and comparison of changes the presence of informative right censoring: conditional linear model. *Biometrics* **45**, 939–955.

Wu, M. C. and Caroll, R. J. (1988). Estimation and comparison of changes the presence of informative right censoring by modeling the censoring process. *Biometrics* **44**, 175–188.

C. R. Rao and R. Chakraborty, eds., *Handbook of Statistics, Vol. 8*
© Elsevier Science Publishers B.V. (1991) 461–513

15

Confidence Bands for the Survival Function $S(t\,|\,z)$ and the Relation with Decision Analysis: Theory

O. J. W. F. Kardaun

$Κακοὶ\ μάρτυρες\ ἀνθρώποισιν\ ὀφθαλμοί.$

HERACLITUS

Summary

In this chapter, two theoretical developments, motivated by practice, are discussed. These are (1) the construction of simultaneous confidence bands for the survival function $S(t\,|\,z) = S(t\,|\,z_1, \ldots, z_k)$, and (2) the implementation of the results of multivariable survival analysis, as described in Chapter 14, in decision analysis. This implementation is performed from the medical as well as from the mathematical side.

A (two-sided) simultaneous confidence band is a generalisation of a confidence interval for a fixed time t. It consists of a lower boundary $\underline{S}(t\,|\,z)$ and an upper boundary $\overline{S}(t\,|\,z)$ such that, for a given interval $[t_{\min}, t_{\max}]$ and a given $\alpha \in [0, 1]$, the probability that the unknown survival function lies between $\underline{S}(t\,|\,z)$ and $\overline{S}(t\,|\,z)$ for every $t \in [t_{\min}, t_{\max}]$ is equal to $1 - \alpha$. The calculation of the width of a band with a realistic shape presents a non-trivial mathematical problem, which is solved by using an integral equation approach. Such a confidence band can be used (among others) to test the null hypothesis $H_0: S(t\,|\,z) = S_0(t)$, where $S_0(t)$ is a reference survival function (for example, of the normal population at a certain age).

In Section 2 of this chapter, the following problem is considered. Suppose one is interested in choosing between two treatments, which we shall call A and B. (A may denote 'to treat' and B 'not to treat', according to some method.) We further assume that data are available from patients who have been treated according to A and from (other) patients who have been treated according to B. The question is whether for a new patient with covariables $z_1, \ldots, z_k$, on the basis of the available data, (1) treatment A has to be recommended, (2) treatment B has to be recommended, or (3) no recommendation should be made.

Our analysis of this problem is based on the estimated survival functions of the 'effective lifetime' under each treatment. The effective lifetime is determined by subjective ratios, established in a discussion with the patient, of the qualities of life under various conditions. Sometimes, the difference between two 'effective' survival functions is well expressed by a one-dimensional parameter (such as the ratio of the mean, or median, effective survival times), the estimator of which is asymptotically normally distributed. For such a situation, the (two-sided) Neyman–Pearson testing theory is replaced by a loss-function formulation. This formulation leads to the same partition of the outcome space $\mathbb{R}$ as does some two-sided test. It adequately reflects, however, the three-decision character of the above mentioned problem, and allows a motivated choice of the level α of the corresponding (two-sided) test. In a traditional approach, this level is often set to 5% or 1%.

The same remarks about notation and the list of references apply as in Chapter 14.

1. Expressing statistical uncertainties in estimating $S(t|z)$

> *Misschien is niets geheel waar, en zelfs dat niet*
>
> DOUWES DEKKER

We consider Cox's proportional hazards model, expressed by

$$S(t|z) = S(t|0)^{\exp(\boldsymbol{\beta}^T z)}, \tag{1.1}$$

where $S(t|z) = \Pr\{T > t \,|\, Z = z\}$ is the survival function of an individual with vector of covariates $z = (z_1, \ldots, z_k)$.

In many publications where Cox's model is applied, one encounters both point estimates and confidence intervals for the regression coefficients $\beta_1, \ldots, \beta_k$, but only point estimates for the survival function $S(t|z)$. Consequently, if the selection of covariates is adequate, and if the proportional hazards assumption holds to a reasonable extent, the 'influences' of the main covariates are represented in a statistically satisfactory manner. The inaccuracy in estimating $S(t|z)$, however, is ignored.

Often, in clinical studies, the sample sizes are such that the random error in estimating $S(t|z)$ is considerable. Concentrating only on a point estimate $\hat{S}(t|z)$, medical scientists may be misled, or even induced to pursue a false line of further research, if they read statements such as: for patient A, having a vector of covariates z, the 5 year survival probability is 75%, while for patient B, having another vector of covariates, it is only 60%. Such a statement is deceptive if the standard error in the survival probability is of the order of magnitude of 10%, which is not unusual. So, in many cases, after having carefully selected a 'suitable' set of covariates, one would like to have an impression of the statistical inaccuracy in estimating $S(t|z)$. In order to express this inaccuracy one can try to construct some confidence band for $S(t|z)$. The construction of such confidence bands is the subject of this section.

One can distinguish between 'instantaneous' and 'simultaneous' confidence bands. The first type of band consists of a set of confidence intervals

$$\{(\underline{S}(t|z), \overline{S}(t|z)); \, t \in (t_{\min}, t_{\max})\}, \tag{1.2}$$

for a suitable time interval $(t_{\min}, t_{\max})$. A band is said to have an instantaneous confidence coefficient $1 - \alpha$ if, for each fixed $t \in (t_{\min}, t_{\max})$,

$$\Pr\{S(t|z) \in (\underline{S}(t|z), \overline{S}(t|z))\} = 1 - \alpha. \tag{1.3}$$

(In practice, a band will also be said to have this confidence coefficient if the above equality only holds approximately.)

An important interpretation of such an instantaneous confidence band is that, for each fixed $p \in (p_{\min}, p_{\max})$,

$$A_p = \{t; \, p \in (\underline{S}(t|z), \overline{S}(t|z))\}, \tag{1.4}$$

constitutes a level $1 - \alpha$ confidence set for the p-th survival quantile. In regular cases, this set constitutes an interval, which we shall denote by $(\underline{t}_p, \overline{t}_p)$. The position of the interval $(\underline{S}(t|z), \overline{S}(t|z))$ with respect to $\hat{S}(t|z)$ is, of course, not determined by α. Boundaries of an instantaneous band can be calculated, for example, by considering the asymptotic distribution of

$$Y_{\varphi, n}(t) = \sqrt{n}[\varphi(S(t|z)) - \varphi(\hat{S}_n(t|z))], \tag{1.5}$$

where n denotes the sample size and φ is some appropriate continuously differentiable and monotonic function $[0, 1] \to \mathbb{R}$. A two-sided symmetric (or a one-sided) band for $\varphi(S(t|z))$ is easily transformed back to a band for $S(t|z)$. As the asymptotic distribution is used as an approximation for finite sample sizes, the shape of the band for $S(t|z)$ depends on the choice of the transformation φ. It follows from the theory presented in Section 1.1 that each such $Y_{\varphi, n}(t)$ is asymptotically normal. A Taylor series expansion gives that, as $n \to \infty$,

$$\operatorname{Var} Y_{\varphi, n}(t) = n\{\varphi'(x)|_{x = \hat{S}_n(t|z)}\}^2 \operatorname{Var} \hat{S}_n(t|z), \tag{1.6}$$

so, the choice of $\varphi(x)$, being irrelevant asymptotically, can in practice be based (1) on the ease of interpretation of the form of the confidence intervals, and (2) on the supposed speed of convergence of $Y_{\varphi, n}(t)$. An explicit expression for the variance of $Y_{\varphi, n}(t)$ for the transformation $\varphi(x) = -\ln(-\ln x)$ is given in Chapter 14, Section 3.2, the corresponding instantaneous band being actually drawn there. More details about the pertaining asymptotic theory are presented in Section 1.1.

The second type of band consists of boundaries $\underline{S}(t|z)$ and $\overline{S}(t|z)$ such that, approximately,

$$\Pr\{S(t|z) \in (\underline{S}(t|z), \overline{S}(t|z)) \quad \text{for all } t \in (t_{\min}, t_{\max})\} = 1 - \alpha. \tag{1.7}$$

In general, it is much more difficult to calculate such a boundary, for some fixed $x \in (0, 1)$, than it is in the case of an instantaneous band. The shape of the band, which is of course not determined by α, has to be fixed by imposing additional constraints. An important class of choices consists in making a transformation $\varphi(S(t|z))$, and then constructing a two-sided symmetric [i.e. $\varphi(S(t|z)) \in (\varphi(\hat{S}(t|z)) - d, \varphi(\hat{S}(t|z)) + d)$], or a one-sided [e.g. $\varphi(S(t|z)) < \varphi(\hat{S}(t|z)) + d$] band around $\varphi(\hat{S}(t|z))$. The choice of transformation may be determined by (1) the ease of interpretation, (2) the distributional adequacy of the asymptotic approximation for finite samples, and (3) computational convenience.

A shape which seems reasonable, at least from the first two points of view, is the 'proportional hazards shape':

$$(\underline{S}(t|z), \overline{S}(t|z)) = (\hat{S}(t|z)^c, \hat{S}(t|z)^{1/c}), \tag{1.8}$$

where the constant c is determined by the confidence coefficient $1 - \alpha$. (Evidently, this corresponds to an 'equidistant' band for (1.5) if the transformation $\varphi(x) = -\ln(-\ln x)$ is made. A similar instantaneous band was constructed in Chapter 14, Section 3.2.) A method for actually calculating the one-sided analogue of this type of simultaneous confidence band is presented in Sections 1.2 and 1.3.

1.1. Instantaneous confidence bands

Several authors, among whom Tsiatis (1981), Andersen and Gill (1982) and Bailey (1979, 1983) have derived, using somewhat different models and regularity assumptions, the asymptotic normality of $\hat{A}(t|0)$, and have indicated how to estimate its variance.

Here we shall present, loosely following Bailey (1979, 1983), a derivation of the (estimated) variance of $\ln \hat{A}(t|z)$. From this we can construct an instantaneous confidence band for $A(t|z)$, or, equivalently, for $S(t|z)$.

Bailey's model assumptions are as follows: Let the arbitrary sequence (z_1, u_1), (z_2, u_2), ... denote the k-dimensional (non-random) covariate vectors and the non-random censoring times associated with a sequence of patients. The true survival times $T_1^0, T_2^0, ...$ of the patients are independent random variables with survival function $S(t|z) = S(t|0)^{\exp(\beta^T z)}$, where z denotes the patient's covariate, and the base-line survival function $S(t|0)$ is a differentiable function of time. The hazard rate $-d/dt(\ln S(t|0))$ is denoted by $\lambda_0(t)$. The observable lifetimes are $T_j = \min(T_j^0, u_j)$, $j = 1, 2, ...$ All functions of β are supposed to be sufficiently differentiable in a neighbourhood B of the true value β_0 (which is also denoted by β). The following, quite natural, regularity assumptions are made:

A1: there exists an $M < \infty$, such that $\|z_i\| < M$ for all $i = 1, 2, ...$;

A2: there exists a $c_1 \in (0, \infty)$, such that for all sufficiently large n, $n^{-1}x^T EI(\beta)x > c_1 > 0$ for all $x \in \mathbb{R}^k$ with $\|x\| = 1$, and β in B;

A3: there exists a $c_2 \in (0, \infty)$, such that for all sufficiently large n, $-\int G_n(t)\,dS(t|0) > c_2 > 0$, where $G_n(t) = n^{-1} \# \{j | u_j > t\}$ denotes the 'empirical survival function' of the non-random censoring times.

The second condition roughly means that the eigenvalues of the expected Fisher information matrix $EI(\beta)$, see expression (1.12), increase (at least) proportionally with n, and the third condition that the censoring times do not pile up, as $n \to \infty$, on the set where $S(t|\mathbf{0}) = 1$. (Note that $-\int G_n \, dS(t|\mathbf{0})$ is interpreted as the probability that a random variable with distribution $S(t|\mathbf{0})$ is smaller than a random variable with distribution $G_n(t)$.)

Starting with the logarithm of the partial likelihood

$$\ln L(\boldsymbol{\beta}) = \sum_{j=1}^{n} \Delta_j (\boldsymbol{\beta}^{\mathrm{T}} z_j - \ln Q_j(\boldsymbol{\beta})), \tag{1.9}$$

where

$$Q_j(\boldsymbol{\beta}) = \sum_{h \in R(t_j)} \exp(\boldsymbol{\beta}^{\mathrm{T}} z_h), \tag{1.10}$$

one can introduce, successively,

$$\bar{z} = \partial/\partial \boldsymbol{\beta} \ln Q_j(\boldsymbol{\beta}) = \sum_{h \in R(t_j)} z_h \exp(\boldsymbol{\beta}^{\mathrm{T}} z_h)/Q_j(\boldsymbol{\beta}),$$

$$U(\boldsymbol{\beta}) = \partial/\partial \boldsymbol{\beta} \ln L(\boldsymbol{\beta}) = \sum_{j=1}^{n} \Delta_j(z_j - \bar{z}_j(\boldsymbol{\beta})) = \sum_{j=1}^{n} \Delta_j U_j, \tag{1.11}$$

$$I(\boldsymbol{\beta}) = -\partial/\partial \boldsymbol{\beta} \, U(\boldsymbol{\beta}) = \sum_{j=1}^{n} \Delta_j V_j, \tag{1.12}$$

where

$$V_j = \partial^2/\partial \boldsymbol{\beta} \partial \boldsymbol{\beta}^{\mathrm{T}} \ln Q_j(\boldsymbol{\beta}) = \partial/\partial \boldsymbol{\beta} \, \bar{z}_j^{\mathrm{T}}, \tag{1.13a}$$

equals the 'covariance matrix'

$$\sum_{h \in R(t_j)} z_h z_h^{\mathrm{T}} \exp(\boldsymbol{\beta}^{\mathrm{T}} z_h)/ \sum_{h \in R(t_j)} \exp(\boldsymbol{\beta}^{\mathrm{T}} z_h) - \bar{z}_j \bar{z}_j^{\mathrm{T}}, \tag{1.13b}$$

of the (random) covariate Z, calculated from the distribution which assigns probability $\exp(\boldsymbol{\beta}^{\mathrm{T}} z_h)/Q_j(\boldsymbol{\beta})$ to individual h in the risk set $R(t_j)$ at time t_j. As usual, $\partial/\partial \boldsymbol{\beta}$ indicates that a k-dimensional column vector of partial derivatives is constructed. Note that one may take the expectation of (1.12), as the variables Δ_j and V_j are random ($j = 1, \ldots, n$).

Under suitable regularity conditions with respect to the behaviour of the derivatives of $U(\boldsymbol{\beta})$, the Taylor expansion

$$U(\boldsymbol{\beta}) = U(\hat{\boldsymbol{\beta}}) - I(\boldsymbol{\beta})(\boldsymbol{\beta} - \hat{\boldsymbol{\beta}}) + O_{\mathrm{P}}(\|\boldsymbol{\beta} - \hat{\boldsymbol{\beta}}\|^2), \tag{1.14}$$

where (since $\hat{\boldsymbol{\beta}}$ maximises the partial likelihood) $U(\hat{\boldsymbol{\beta}}) = \mathbf{0}$, shows that

$$\hat{\boldsymbol{\beta}} \sim \mathrm{AN}(\boldsymbol{\beta}, I(\boldsymbol{\beta})^{-1}) \quad \text{is equivalent to} \quad U(\boldsymbol{\beta}) \sim \mathrm{AN}(\mathbf{0}, I(\boldsymbol{\beta})), \tag{1.15}$$

if $\hat{\boldsymbol{\beta}}$ is a consistent estimator for $\boldsymbol{\beta}$. (The notation AN stands for asymptotically normal, and $X_n = O_{\mathrm{P}}(Y_n)$ means that the sequence X_n/Y_n is bounded in probability.) In Bailey (1979, 1983), it is proven that

$$n^{-1} \| I(\hat{\beta}) - EI(\beta) \| \overset{P}{\to} 0 , \tag{1.16}$$

and, by concavity arguments, that this implies the consistency of $\hat{\beta}$ and the asymptotic interchangeability of $I(\beta)$, $I(\hat{\beta})$ and $EI(\beta)$. With some effort, Bailey derives that $U(\beta) \sim \text{AN}(0, I(\beta))$ by approximating, using Hájek's projection lemma, the vectors $\Delta_j U_j$ by the stochastically independent variables $\Delta_j \hat{U}_j = E(U(\beta)|z_j, T_j, \Delta_j)$, $j = 1, \ldots, n$. (Note that the indepencence of $\Delta_1 \hat{U}_1$, $\Delta_2 \hat{U}_2$, ... follows from the definition of the conditional expectation and from the independence of (Δ_1, T_1), (Δ_2, T_2), ...)

As we have seen in Chapter 14, Section 3.2, $\hat{\Lambda}(t|0) = \sum_{t_j < t} \Delta_j / Q_j(\hat{\beta})$ constitutes a reasonable estimator for the cumulative hazard rate $\Lambda(t|0)$. We now want to have a further look at its asymptotic distribution. A basic decomposition is

$$\sqrt{n}(\hat{\Lambda}(t|0) - \Lambda(t|0)) = \sqrt{n}(\hat{\Lambda}(t|0) - \tilde{\Lambda}(t|0))$$
$$+ \sqrt{n}(\tilde{\Lambda}(t|0) - \Lambda(t|0)) , \tag{1.17}$$

where

$$\tilde{\Lambda}(t|0) = \tilde{\Lambda}(t, \beta|0) = \sum_{t_j < t} \Delta_j / Q_j(\beta) , \tag{1.18a}$$

and, as indicated before,

$$\hat{\Lambda}(t|0) = \tilde{\Lambda}(t, \hat{\beta}|0) = \sum_{t_j < t} \Delta_j / Q_j(\hat{\beta}) . \tag{1.18b}$$

Consider the (truncated) Taylor expansion of $\hat{\Lambda}(t|0)$ around $\tilde{\Lambda}(t|0)$:

$$\hat{\Lambda}(t|0) = \tilde{\Lambda}(t, \beta|0) + (\hat{\beta} - \beta)^{\mathsf{T}} \tilde{\Gamma}_1(t, \beta|0) + R_n(\hat{\beta} - \beta, t) \tag{1.19}$$

with

$$\tilde{\Gamma}_1(t, \beta|0) = \partial/\partial\beta \, \tilde{\Lambda}(t, \beta|0) = - \sum_{t_j < t} \Delta_j \bar{z}_j / Q_j(\beta) . \tag{1.20}$$

Using the assumptions A1, A2, and A3, Bailey derives that, at least if attention is restricted to a time-interval $[0, t_{\max}]$, where $S(t_{\max}) = \delta > 0$,

(1) $\hat{\beta}$ and $\tilde{\Lambda}(t, \beta|0)$ are asymptotically independent,

(2) $\tilde{\Gamma}_1(t, \hat{\beta}|0)$ converges in probability to a non-random function $\Gamma_1(t, \beta)$,

(3) the distribution of $\sqrt{n}[\tilde{\Lambda}(t, \beta|0) - \Lambda(t, \beta|0)]$ converges to that of a gaussian process with independent increments and finite variance $n\psi_1(t, \beta)$, which is difficult to express, but can consistently be estimated by $n\hat{\psi}_1(t, \hat{\beta}) = n \sum_{t_j < t} \Delta_j / Q_j^2(\hat{\beta})$,

(4) the remaining terms of the Taylor expansion are negligible, i.e. $o_P(1/\sqrt{n})$, uniformly in t. ($X_n = o_P(Y_n)$ means that $X_n / Y_n \overset{P}{\to} 0$.)

REMARKS. Strictly speaking, in Bailey (1983) the proof of this representation is given only when there is no censoring, whereas in Bailey (1979) it is asserted that the derivation (given in the simple case of one covariate) is not essentially altered by the occurrence of censoring. The details of this are rather complicated to check. Nevertheless, we feel confident about this representation, also in the general situation, for which a weak additional condition about regular occurrence of the (non-random) censoring times may perhaps be necessary. Under slightly more involved, but rather weak, conditions and quite general model assumptions (multiple failure types, time-dependent stochastic covariates, and various types of censoring), Andersen and Gill (1982) have obtained asymptotically the same results, using martingale theory in a stochastic counting-process formulation. Their basic assumption with respect to censoring is that the empirical distribution of the (fixed or random) censoring times converges in norm to some fixed distribution function. The counting-process formulation contains the model with independent failure times and independent censoring as a special case, in which the intensity function $\lambda(t|z)$ has the usual hazard-rate interpretation, see Self and Prentice (1982). Jacobsen (1990) presents, in the case of no covariates, a general discussion of the interplay between the various structures and derives, in some sense, a minimal class of censoring patterns. Based on the approach of Andersen and Gill, Arjas and Haara (1988) have expressed the conditions for the consistency and the asymptotic normality of $\hat{\beta}$ into a simple form, similar to **A1** and **A2**. For results on time-dependent regression coefficients, see Zucker and Karr (1990).

Finally, we consider the distribution of

$$X(t) = \sqrt{n}\,[\ln \hat{\Lambda}(t|z) - \ln \Lambda(t|z)]$$
$$= \sqrt{n}\,[\ln \hat{\Lambda}(t|0) - \ln \Lambda(t|0) + (\hat{\beta} - \beta)^{\mathrm{T}} z]\,. \qquad (1.21)$$

By expanding $\ln \hat{\Lambda}(t|0)$ around $\ln \tilde{\Lambda}(t, \beta|0)$, and using the above results, one can infer that the variance of $X(t)$ can be estimated by

$$n(\hat{\psi}(t) + [z + \tilde{\Gamma}(t, \hat{\beta}|0)]^{\mathrm{T}} I(\hat{\beta})^{-1}[z + \tilde{\Gamma}(t, \hat{\beta}|0)])\,, \qquad (1.22)$$

where now

$$\tilde{\Gamma}(t, \hat{\beta}|0) = \partial/\partial\beta \ln \tilde{\Lambda}(t, \beta|0)|_{\beta = \tilde{\beta}} = \tilde{\Lambda}(t|0)^{-1}\,\tilde{\Gamma}_1(t, \hat{\beta}|0)\,, \qquad (1.23a)$$

and

$$\hat{\psi}(t) = \tilde{\psi}(t, \hat{\beta}) = \hat{\Lambda}(t|0)^{-2}\,\tilde{\psi}_1(t, \hat{\beta})\,. \qquad (1.23b)$$

As usual, the estimated covariance matrix of $\hat{\beta}$ is $I(\hat{\beta})^{-1}$. As $\hat{\beta}$ and $\ln \tilde{\Lambda}(t|0)$ are asymptotically independent, one can derive, using (1.21) and the expansion around $\ln \tilde{\Lambda}(t|0)$, that the covariance between $(\hat{\beta} - \beta)$ and $(\ln \hat{\Lambda}(t|z) - \ln \Lambda(t|z))$ asymptotically equals the covariance matrix of $\hat{\beta}$ times $z + \partial/\partial\beta \times \ln \tilde{\Lambda}(t, \beta|0)$, and can be estimated by

$$I(\hat{\beta})^{-1}[z + \tilde{\Gamma}(t, \hat{\beta}|0)]\,. \qquad (1.24)$$

This gives a motivation of formulae (3.5) to (3.9) in Chapter 14, Section 3.2.

Formula (1.22) shows how the variance of $X(t)$ depends in a simple way on z. Given the expression for $z = \mathbf{0}$, this can alternatively be derived by an argument based on translational invariance: Denote formula (1.22) by $\text{Var}(z, z_1, \ldots, z_n)$, and assume, from Bailey (1983), that it is correct for $z = \mathbf{0}$. Consider the effect of shifting all covariates of the sample to the right by an amount z_0: $z_j' = z_j + z_0$, $j = 1, \ldots, n$. It is nice to observe how the various quantities behave under this shift transformation (we abbreviate $Q_j'(\boldsymbol{\beta}, z_2, \ldots, z_n) = Q_j(\boldsymbol{\beta}, z_1', \ldots, z_n')$ by Q_j', $Q_j(\boldsymbol{\beta}, z_1, \ldots, z_n)$ by Q_j, and so on).

We have, successively

$$Q_j' = Q_j \exp(\boldsymbol{\beta}^\mathrm{T} z_0)\,, \qquad\qquad \tilde{\Lambda}' = \tilde{\Lambda} \exp(-\boldsymbol{\beta}^\mathrm{T} z_0)\,, \qquad (1.25)$$

$$\bar{z}_j' = \bar{z}_j + z_0\,, \qquad\qquad z' = z + z_0\,, \qquad I' = I\,, \qquad (1.26)$$

$$\tilde{\Gamma}_1' = (\tilde{\Gamma}_1 - z_0 \tilde{\Lambda}) \exp(-\boldsymbol{\beta}^\mathrm{T} z_0)\,, \qquad \tilde{\psi}_1'(t, \boldsymbol{\beta}) = \exp(-2\boldsymbol{\beta}^\mathrm{T} z_0) \tilde{\psi}_1(t, \boldsymbol{\beta})\,, \qquad (1.27)$$

$$\tilde{\Gamma}' = \tilde{\Gamma} - z_0\,, \qquad\qquad \tilde{\psi}'(t) = \tilde{\psi}(t)\,. \qquad (1.28)$$

Now, $\text{Var}(z', z_1', \ldots, z_n')$ at $z' = z_0$ should, by invariance, be equal to $\text{Var}(z, z_1, \ldots, z_n)$ at $z = \mathbf{0}$. Using (1.25) to (1.28), formula (1.21) with $z = \mathbf{0}$ can be expressed in primed quantities, or equivalently, in primed coordinates $z_1', \ldots, z_n'$. If we do so, we obtain a shift of $+(\hat{\boldsymbol{\beta}} - \boldsymbol{\beta})^\mathrm{T} z_0$. Applying the same procedure to formula (1.22), i.e. to $\text{Var}(\mathbf{0}, z_1, \ldots, z_n)$, we see (in primed coordinates) that the variance of (1.21) with $z = z_0$ is given by (1.22) with $z = z_0$. Note that the simple shift in the expression of the variance only occurs after the 'proportional hazards transform', i.e. on the scale of $\ln \Lambda(t|z)$.

1.2. Simultaneous confidence bands

From theoretical as well as practical points of view, it would be interesting to construct, in addition to the instantaneous confidence band described in the previous section, a simultaneous confidence band for the unknown survival function $S(t|z)$. As we have seen, such a band should ideally have the property that, for some suitable $t_- = t_{\min}$ and $t_+ = t_{\max}$, and for some prescribed $\alpha \in (0, 1)$,

$$\Pr\{S(t|z) \in (\underline{S}(t|z), \bar{S}(t|z)) \quad \text{for all } t \in [t_-, t_+]\} = 1 - \alpha\,. \qquad (1.29)$$

In practice, we will have to be content with approximate results.

A simultaneous band can be used, among others, for testing the goodness-of-fit hypothesis

$$\mathrm{H}_0\colon S(t|z) = S_{\text{ref}}(t) \quad \text{for all } t \in [t_-, t_+]\,, \qquad (1.30)$$

where $S_{\text{ref}}(t)$ denotes an arbitrary reference survival function. (Conversely, by indicating the set of reference survival functions for which H_0 is not rejected, each (type of) goodness-of-fit test generates a particular confidence band. In general, the relation between the alternative for which the test has some 'optimality' property, and the shape of the corresponding confidence band is complicated.)

From the variety of possible choices for the shape of the band, we want to select the 'proportional hazards band'

$$(\underline{S}(t|z), \overline{S}(t|z)) = (\hat{S}(t|z)^c, \hat{S}(t|z)^{1/c}),$$
(1.31)

for further investigation (c is some constant $\geqslant 1$). Transforming to $\ln \Lambda(t|z) = \ln(-\ln S(t|z))$, we get

$$(\ln \underline{\Lambda}(t|z), \ln \overline{\Lambda}(t|z)) = (\ln \hat{\Lambda}(t|z) - d, \ln \hat{\Lambda}(t|z) + d),$$
(1.32)

with $d = \ln c$, i.e. on the scale of the nomogram of Figure 2 in Chapter 14, Section 3.2.2, the bounds correspond to parallel shifts of the estimated survival functions. This type of band is preferred because of its simple interpretation from a practical point of view. Moreover, the approximation based on asymptotic normality is expected to hold reasonably well on this scale. The 'only' problem, which we will consider now, is the determination of c or d as a function of α. For simplicity, we will restrict the attention to the construction of a one-sided band. (Once the relation between c and α is determined for one-sided bands, the problem for two-sided bands can in principle be attacked by using inclusion–exclusion arguments. However, it is felt that for $1 - \alpha = 95\%$ only a few terms of the associated series expansion are needed, while for practical purposes, instead of a two-sided band, one can also draw, and suitably interpret, one-sided upper and lower bands, each with confidence coefficient $1 - \frac{1}{2}\alpha$.)

To construct the required one-sided band, we need an approximation for

$$F_n(d) = \Pr\{\sqrt{n}(\ln \Lambda(t|z) - \ln \hat{\Lambda}(t|z)) < d \quad \text{for all } t \in [t_-, t_+]\}.$$
(1.33)

(For convenience, we have added a factor $\sqrt{n}$ to the definition of d, which we shall use from now on.) From the theory presented in Section 1.1 one can infer, at least if attention is restricted to a time-interval $[0, t_{max}]$ such that $\Lambda(t_{max}|z) \leqslant M \exp(\boldsymbol{\beta}^T z) < \infty$, that

$$\sqrt{n}(\ln \Lambda(t|z) - \ln \hat{\Lambda}(t|z))$$
(1.34)

can asymptotically be represented as

$$(\Lambda(t|z))^{-1} W(\psi_n(t)) + \sqrt{n}(\boldsymbol{\beta} - \hat{\boldsymbol{\beta}})^T \boldsymbol{\Gamma}(t) + o_P(1),$$
(1.35)

where (1) $\hat{\boldsymbol{\beta}}$ and the time-transformed Wiener process $W(\psi_n(t))$ are independent, and (2) $\psi_n = n\psi_1(t, \boldsymbol{\beta}|z)$ and $\boldsymbol{\Gamma}(t) = \boldsymbol{\Gamma}(t, \boldsymbol{\beta}|z) = (\Lambda(t|z))^{-1} \boldsymbol{\Gamma}_1(t, \boldsymbol{\beta}|z)$ are unknown deterministic functions which can be estimated from the data. (The explicit expressions for $\psi_1(t, \boldsymbol{\beta}|z)$ and $\boldsymbol{\Gamma}_1(t, \boldsymbol{\beta}|z)$ are rather complicated, see Bailey, 1983). It is noted that the first term in (1.35), not being simply a time-transformed Wiener process, is a (nonstationary) Gauss–Markov process (see, e.g., Mehr and McFadden, 1965).

From this representation, one can derive that for $n \to \infty$ the deterministic sequence of functions $F_n(d)$ converges, for fixed $\Lambda(t|z)$, t_- and t_+, to

$$
F(d) = (2\pi)^{-k/2} |\boldsymbol{\Sigma}|^{-1/2} \int_{-\infty}^{+\infty} \cdots \int_{-\infty}^{+\infty} dy_1 \cdots dy_k \, e^{-(1/2)y^{\mathrm{T}} \boldsymbol{\Sigma}^{-1} y}
$$
$$
\times \Pr\{W(\psi_n(t)) < \Lambda(t|z)(d - y^{\mathrm{T}}\boldsymbol{\Gamma}(t)) \quad \text{for all } t \in [t_-, t_+]\},
$$
(1.36)

where $\boldsymbol{\Sigma}$ denotes the covariance matrix of the asymptotic (normal) distribution of $\sqrt{n}(\boldsymbol{\beta} - \hat{\boldsymbol{\beta}})$. The expression for $F(d)$ contains the unknown parameters $\boldsymbol{\Sigma}$, $\psi_n(t)$, and $\boldsymbol{\Gamma}(t)$. Therefore, in practice, $F(d)$ is estimated by substituting the following estimates:

$$
\hat{\boldsymbol{\Sigma}}^{-1} = n^{-1} I(\hat{\boldsymbol{\beta}}) = -n^{-1} \partial^2/\partial \boldsymbol{\beta} \partial \boldsymbol{\beta}^{\mathrm{T}} \ln L(\boldsymbol{\beta})|_{\boldsymbol{\beta} = \hat{\boldsymbol{\beta}}},
\tag{1.37}
$$

$$
\hat{\boldsymbol{\Gamma}}(t) = \tilde{\Lambda}(t, \hat{\boldsymbol{\beta}}|z)^{-1} (\partial/\partial \boldsymbol{\beta}) \tilde{\Lambda}(t, \boldsymbol{\beta}|z)|_{\boldsymbol{\beta} = \hat{\boldsymbol{\beta}}},
\tag{1.38}
$$

$$
\hat{\psi}_n(t) = n \tilde{\psi}_1(t, \hat{\boldsymbol{\beta}}|z) = n \, e^{2\hat{\boldsymbol{\beta}}^{\mathrm{T}} z} \sum_{t_j < t} \Delta_j / Q_j(\hat{\boldsymbol{\beta}})^2,
\tag{1.39}
$$

$$
\hat{\Lambda}(t|z) = \tilde{\Lambda}(t, \hat{\boldsymbol{\beta}}|z),
\tag{1.40}
$$

where

$$
\tilde{\Lambda}(t, \boldsymbol{\beta}|z) = e^{\boldsymbol{\beta}^{\mathrm{T}} z} \sum_{t_j < t} \Delta_j / Q_j(\boldsymbol{\beta}),
\tag{1.41}
$$

$$
\ln L(\boldsymbol{\beta}) = \sum_{j=1}^{n} \Delta_j (\boldsymbol{\beta}^{\mathrm{T}} z_j - \ln Q_j(\boldsymbol{\beta})),
\tag{1.42}
$$

$$
Q_j(\boldsymbol{\beta}) = \sum_{h \in R(t_j)} e^{\boldsymbol{\beta}^{\mathrm{T}} z_h} \quad \text{and} \quad |\boldsymbol{\Sigma}| = \det \boldsymbol{\Sigma}.
\tag{1.43}
$$

As both $-\ln \hat{\Lambda}(t|z)$ and its estimated variance grow to infinity as t approaches 0, we want to have our band from t_- onwards. Therefore, we first let the stochastic process run from 0 to t_-, and notice that the marginal distribution of $W(\psi_n(t_-))$ is gaussian with variance $\psi_n(t_-)$. Accordingly, making use of the Markov property of the Wiener process we can write (see Figure 1)

$$
\hat{F}(d) = (2\pi)^{-(k+1)/2} n^{-k/2} |I(\hat{\boldsymbol{\beta}})|^{1/2} \int_{-\infty}^{+\infty} \cdots \int_{-\infty}^{+\infty} dy_1 \cdots dy_k \int_{-\infty}^{\hat{a}(t_-)} dx
$$
$$
\times e^{-(1/2)x^2} e^{-(1/2)ny^{\mathrm{T}} I(\hat{\boldsymbol{\beta}})y} \Pr\{W(\tau) < \hat{\Lambda}(t|z)[d - y^{\mathrm{T}} \hat{\boldsymbol{\Gamma}}(t)] - x\sqrt{\hat{\psi}_n(t_-)}
$$
$$
\text{for all } 0 \leqslant \tau \leqslant \hat{\psi}_n(t_+) - \hat{\psi}_n(t_-)\},
$$
(1.44)

where

$$\hat{u}(t_-) = \hat{\Lambda}(t_-|z)[d - y^{\mathrm{T}} \hat{\Gamma}(t_-)]/\sqrt{\hat{\psi}_n(t_-)}, \tag{1.45}$$

$$\tau = \hat{\psi}_n(t) - \hat{\psi}_n(t_-) \quad \text{and} \quad t = \hat{\psi}_n^{-1}(\tau + \hat{\psi}_n(t_-)). \tag{1.46}$$

The core of this expression consists of a 'boundary crossing problem' for the standard Wiener process $W(\tau)$ with boundary

$$f_{d,x,y,z}(\tau) = \hat{\Lambda}(t|z)[d - y^{\mathrm{T}} \hat{\Gamma}(t)] - x\sqrt{\hat{\psi}_n(t_-)}, \tag{1.47}$$

where the relation between t and τ is given by (1.46).

(Because of the upper limit in integrating over x we have $f_{d,x,y,z}(0) \geqslant 0$.) Note that when the usual step-function estimators are inserted, the boundary is also a step-function, jumping at the transformed times-to-death $\tau_1, \ldots, \tau_m$ only. If $\hat{\Gamma}(t)$, $\hat{\psi}_n(t)$ and $\hat{\Lambda}(t|z)$ are interpolated in the same way as was $\hat{\Lambda}(t|z)$ in Figure 2 of Chapter 14, the derivative of the boundary is jumping at the transformed time points that interpolate the times-to-death.

1.2.1. Auxiliary problem: Boundary crossing probabilities for a Wiener process
To the author's knowledge, although the subject has been extensively investigated, no closed form expression has been derived for the probability that a Wiener process remains below a general boundary $f(\tau) \geqslant 0$ for $0 < \tau \leqslant \tau_0$. The special

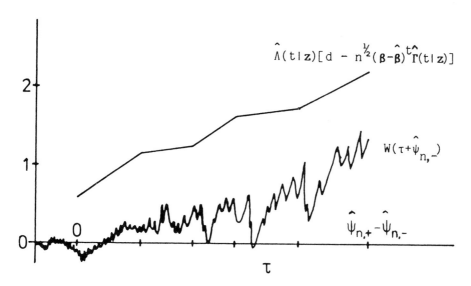

Fig. 1. Illustration of a one-sided boundary crossing problem for the stochastic process $\sqrt{n}(\ln \Lambda(t|z) - \ln \hat{\Lambda}(t|z)) = \Lambda(t|z)^{-1}[W(\psi_n(t))] + \sqrt{n}(\beta - \hat{\beta})^{\mathrm{T}} \Gamma(t|z)$. The (estimated) time scale for the Wiener process is $\tau = \hat{\psi}_n(t) - \hat{\psi}_{n,-}$. The functions $\hat{\psi}_n(t)$ and $\hat{\Gamma}(t|z)$ are data-based estimates of $\psi_n(t)$ and $\Gamma(t|z)$, respectively; $\psi_n(t)$ and its estimate are increasing functions of t. The crossing of the boundary is considered on the time interval $[t_-, t_+]$, which is transformed to the proper-time interval $[0, \hat{\psi}_{n,+} - \hat{\psi}_{n,-}]$.

case $f(\tau) = a + b\tau$ is well known. The case $f(\tau) = A(\tau)\sqrt{\tau}$, where $A(\tau)$ is asymptotically equivalent to $\sqrt{2 \ln \ln \tau}$ for $\tau \to \infty$, was studied, in an early stage, by Robbins and Siegmund (1968). For arbitrary 'smooth' (i.e. continuously differentiable and 'remote' boundaries, asymptotic approximations have recently been derived (see Ferebee, 1982; Jennen and Lerche, 1981, and the references therein; Siegmund, 1986, is a general survey). However, as is known for some years (see Durbin, 1971), an integral equation for this probability can be derived, even for boundaries which do not have to be smooth or remote. It is such an integral equation that we want to consider here.

Denoting the (random) first exit time by T, and its distribution function by $P_T(s) = \Pr\{T < s\}$, we have, for the event $A_\tau = \{W(\tau) > f(\tau)\}$, that at time τ the process exceeds the boundary:

$$\Pr(A_\tau) = \int_0^\tau \Pr(A_\tau | T = s)\, dP_T(s),\qquad (1.48)$$

or, equivalently,

$$1 - \Phi(f(\tau)/\sqrt{\tau}) = \int_0^\tau (1 - \Phi[(f(\tau) - f(s))/\sqrt{\tau - s}])\, dP_T(s).\qquad (1.49)$$

Integrating by parts, we obtain

$$1 - \Phi(f(\tau)/\sqrt{\tau})$$
$$= 0.5 P_T(\tau) + 0.5(2\pi)^{-1/2} \int_0^\tau e^{-(f(\tau) - f(s))^2/2(\tau - s)} (\tau - s)^{-3/2}$$
$$\times [f(\tau) - f(s) - 2f'(s)(\tau - s)] P_T(s)\, ds.\qquad (1.50)$$

This equation is at least valid if f is piecewise smooth and has finite right- and left-hand derivatives at the (finitely many) points of nondifferentiability. (We have used the fact that $P_T(0) = 0$.) The equation can be written in the form

$$g(\tau) = y(\tau) + \int_0^\tau (\tau - s)^{-1/2} K(\tau, s)\, y(s)\, ds,\qquad (1.51)$$

where

$$y(\tau) = P_T(\tau),\qquad (1.52)$$

$$g(\tau) = 2(1 - \Phi[f(\tau)/\sqrt{\tau}]),\qquad (1.53)$$

and

$$K(\tau, s) = (2\pi)^{-1/2} e^{-(f(\tau) - f(s))^2/2(\tau - s)} [(f(\tau) - f(s))/(\tau - s) - 2f'(s)].\qquad (1.54)$$

As one can see, this is a second kind Volterra integral equation (of Abel's type, as the kernel of the equation has a square root singularity).

Equation (1.49), which is an integral equation of the first kind for the density $p_T(s) = \mathrm{d}/\mathrm{d}s\,P_T(s)$, was first derived by Durbin (1971). It is much harder to solve numerically than (1.50). Independently, Ferebee (1982) and Durbin (1985) have derived a second kind integral equation similar to (1.50), but for the first-exit density $p_T(s)$, and have stressed its importance for asymptotic expansions and for numerical purposes, respectively. Since the derivative of our boundary is discontinuous, and we primarily interested in $P_T(t_{\max})$, it is more convenient in our case to solve directly the (second kind) integral equation for the first-exit distribution function $P_T(s)$.

In order to evaluate $\hat{F}(d)$ in practice, equation (1.50) has to be solved numerically. This can, for example, be done by adapting a method of Branca (see Te Riele and Schoevers, 1983). The solution will have to be incorporated in the $(k + 1)$ fold integral, for which a straightforward Monte Carlo method will be employed. (Using this method one can restrict the number of costly integrand evaluations to a moderately large, prefixed number (say, 1000), irrespective of the dimension k (running from 1 to 7 for many applications), and a statistical estimate of the integration error is obtained. Note that, as $\hat{F}(d)$ is a crude approximation to $F(d)$ anyhow, the required accuracy for calculating $\hat{F}(d)$ is low.) An elaboration of this approach for solving and incorporating the boundary crossing problem can be found in the next section.

The actual calculations were performed by SIMBAND, a FORTRAN program, which uses output from CONCERN. In the program $\hat{\Gamma}(t)$, $\hat{\psi}_n(t)$ and $\hat{A}(t|z)$ are interpolated in the same way as $\hat{A}(t, z)$ in Figure 2 of Chapter 14. The results are shown in Figure 2, where the linear scale of the vertical axis represents the difference $d/\sqrt{n}$ $(= \ln c)$ between $-\ln \hat{A} = -\ln(-\ln \hat{S})$ and $-\ln \overline{A} = -\ln(-\ln \underline{S})$. One unit corresponds to a relative risk of a factor e. As a function of c, the horizontal axis gives the estimated probability that $\underline{S}(t) = \hat{S}(t)^c < S(t)$ for all t between $t_- = 0.5$ yr and $t_+ = 8$ yrs (lower curve) and, just for comparison, the estimated probability that $\underline{S}(t) < S(t)$ just at $t = 6$ yrs (upper curves). The scale of the horizontal axis is such that normal distributions are represented by straight lines. As one can see, at least for $\alpha \approx 90\%$, the simultaneous band is not very much wider than the instantaneous band at $t = 6$ yrs. The normal approximations on the two different scales (see the legends of Figure 2) differ appreciably with respect to their tails.

1.2.2. A crude approximation

A relatively easy expression for $F(d)$ is derived by approximating the upper boundary $f_{d, x, y, z}(\tau)$ for the Wiener process by a straight line, see (1.47). The expression becomes even more simple if t_+ can be assumed to be effectively infinite. The first approximation may be adequate if $\psi_n(t|z)$ and $\Lambda(t|z)$ can reasonably be represented by linear functions of time and $\Gamma_i(t|z)$ by time-independent constants $\Gamma_{i, z}$ $(i = 1, \ldots, k)$. This last property is equivalent to the fact that, for each component $i = 1, \ldots, k$, the quantities $\overline{z}_j(\boldsymbol{\beta})$, $j = 1, \ldots, m$ (j denotes the j-th uncensored observation), are approximately constant, i.e. independent of j. (Note that the mean value $m^{-1} \sum_{j=1}^{m} \overline{z}_j(\hat{\boldsymbol{\beta}})$ is fixed by the

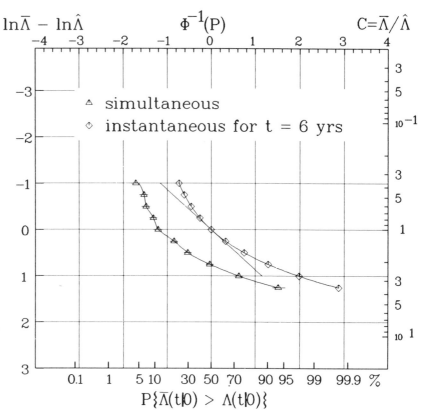

Fig. 2. The estimated asymptotic distribution of a one-sided simultaneous confidence band (with shape $\underline{S}(t|0) = \hat{S}(t|0)^c$) for the base-line survival function $S(t|0)$. The diagram is to be used in conjunction with Figure 2 of Chapter 14: the quantity c on the right-hand side denotes the 'relative risk of $\underline{S}(t|0)$ with respect to $\hat{S}(t|0)'$, and its logarithm on the left-hand axis measures the parallel vertical shift from $\hat{S}(t|0)$ to $\underline{S}(t|0)$ in linear units of that figure. Using the lower curve in this diagram, one can read off at the horizontal axis the estimated probability that $\underline{S}(t|0)$ (being equal to $\hat{S}(t|0)^c$) is smaller than $S(t|0)$ for all $t \in [t_-, t_+] = [0.5 \text{ yr}, 8 \text{ yrs}]$. From the upper curves one reads off the estimated probability that $\underline{S}(t|0) < S(t|0)$ just at $t = 6$ yrs (the straight line corresponds to a normal approximation for $\ln[-\ln \hat{S}(t|0)]$, and the curved line to a normal approximation for $-\ln \hat{S}(t|0)$).

maximum likelihood equation for β, which reads $\sum_{j=1}^{m} (z_j - \bar{z}_j(\beta)) = 0$.) The simplifying assumptions about $\psi_n(t|z)$ and $\Lambda(t|z)$ and $\Gamma_i(t|z)$ should be checked from the data. The assumption that t_+ is effectively infinite can be dropped, see (1.60).

Let, for fixed z, the linear approximation of $\Lambda(t|z)$ be denoted by $c_0 e^{\beta^T z} t = c_z t$. Then, the analogue of equation (1.36) can be written as

$$F(d) = E_U \Pr\{W(\psi_n(t|z)) < Ut \quad \text{for all } t \in [t_-, \infty)\}, \qquad (1.55)$$

where the expectation is taken with respect to $U = c_z(d - Y^T \Gamma_z) \sim$ N$(c_z d, c_z^2 \sigma_z^2)$, $\sigma_z^2 = \Gamma_z^T \Sigma \Gamma_z$, and Σ is the asymptotic covariance matrix of

$Y = \sqrt{n}(\beta - \hat{\beta})$. The dependence of $\psi_n(t|z)$, c_z, Γ_z, and σ_z on z (evidently, $\psi_n(t|z) = \exp(2\beta^T z)\psi_n(t|0)$, $c_z = c_0 \exp(\beta^T z)$, and $\Gamma_z = z + \Gamma_0$) will from now on be suppressed in the notation, and we shall abbreviate $\psi_n(t_-)$ by ψ_-, and $\psi_n(t_+)$ by ψ_+. Approximating $\psi_n(t)$ by $a + bt$, and transforming to $\tau = \psi_n(t) - \psi_- = b(t - t_-)$, we get

$$F(d) = E_U \Pr\{W(\psi_- + \tau) < U(t_- + \tau/b) \quad \text{for all } \tau \in [0, \infty)\}. \quad (1.56)$$

Realising that the marginal distribution of $W(\psi_-)$ is normal with mean zero and variance ψ_-, we can write

$$F(d) = E_X E_V \Pr\{W(\tau) < V(bt_- + \tau) - X\sqrt{\psi_-} \quad \text{for all } \tau \in [0, \infty)\}, \quad (1.57)$$

where X is standard normal and independent of $V = U/b$ as well as of $W(\tau)$. As is well known (see e.g. Pollard, 1984, Ch. V.6.), the probability that a Wiener process $W(\tau)$ remains, for $0 < \tau < \infty$, below a linear boundary $a + b\tau$ is equal to $I_{\{a > 0, b > 0\}}(1 - \exp(-2ab))$. (Note that our situation can be considered as some 'doubly stochastic' linear boundary problem for the Wiener process.) Inserting this into expression (1.57), and writing Y instead of $bt_- V - X\sqrt{\psi_-}$, we get

$$F(d) = E_{VY}(1 - \exp(-2VY))I_{\{V > 0, Y > 0\}}, \quad (1.58)$$

where the pair (V, Y) has a bivariate normal distribution with its parameters $(\mu_v, \mu_y, \sigma_v^2, \sigma_y^2, \rho_{vy})$ equal to

$$(cd/b, ct_- d, (c\sigma/b)^2, \psi_- + (c\sigma t_-)^2, c\sigma t_-/\sqrt{\psi_- + (c\sigma t_-)^2}). \quad (1.59)$$

Note that $\rho_{vy} = (\mu_y/\sigma_y)(\mu_v/\sigma_v)^{-1}$, so only four parameters are functionally independent.

In practice, $F(d)$ is evaluated by inserting estimates for the parameters in equation (1.58): $\hat{c}$ is the estimated linear slope of $\hat{\Lambda}(t|z)$, $\hat{b}$ the estimated slope of $\hat{\psi}_n(t|z)$, $\hat{\psi}_- = \hat{\psi}_n(t_-|z)$ (or $\hat{a} + \hat{b}t_-$), $\hat{\sigma} = \sqrt{(\hat{\Gamma}^T \hat{\Sigma} \hat{\Gamma})}$, and $\hat{\Gamma} = (\hat{\Gamma}_1, \ldots, \hat{\Gamma}_k)$, see equations (1.37) to (1.39), where Γ_i ($i = 1, \ldots, k$) is estimated by fitting a straight horizontal line through the calculated values $\hat{\Gamma}_i(t_j)$, $j = 1, \ldots, m$. Expression (1.58) may be cast in some other form for numerical convenience. In any case, its evaluation, by numerical quadrature or by Monte Carlo simulation, is quite easier than in the general situation.

Since for a Wiener process the probability density of the first exit time to a linear boundary is known to be the inverse gaussian distribution (see, e.g., Mehr and McFadden, 1965; Jørgensen, 1981), the assumption that t_+ is effectively infinite can be dropped. In that case, $\exp(-2YV)$ in (1.58) has to be replaced by

$$\exp(-2YV)\Phi(V\sqrt{\tau_r} - Y/\sqrt{\tau_r}) + \Phi(-V\sqrt{\tau_r} - Y/\sqrt{\tau_r}), \quad (1.60)$$

where $\tau_r = \psi_+ - \psi_-$. In many practical situations, it will be more realistic to approximate each $\Gamma_i(t)$ by a straight line $a_i + b_i t$, but then the relatively simple approximation becomes considerably more difficult.

REMARK. Evidently, under the model assumption $\Lambda(t|z) = c_0 t \exp(\boldsymbol{\beta}^{\mathrm{T}} z)$ of exponentially distributed lifetimes, the parameter $(c_0, \boldsymbol{\beta})$ can be estimated by full maximum likelihood, and a confidence band for the exponential survival function $S(t|z)$ can be directly constructed from the asymptotic distribution of $(\hat{c}_0, \hat{\boldsymbol{\beta}})$. In this case, making the same choice for the form of the (one-sided) band as in the rest of this section, we have

$$F_e(d) = \Pr\{\sqrt{n}(\ln c_0 - \ln \hat{c}_0 + (\boldsymbol{\beta} - \hat{\boldsymbol{\beta}})^{\mathrm{T}} z) < d\} \tag{1.61}$$

and

$$\hat{F}_e(d) = \Phi(d/\hat{\sigma}_e), \tag{1.62}$$

where $\hat{\sigma}_e^2 = (1, z)^{\mathrm{T}} \hat{\boldsymbol{\Sigma}}_e (1, z)$, $\hat{\boldsymbol{\Sigma}}_e$ being the estimated covariance matrix of $\sqrt{n}(\ln \hat{c}_0, \hat{\boldsymbol{\beta}})$. Note that here a simultaneous band coincides with an instantaneous band having the same confidence coefficient, and that this property is lost if the exponential lifetime distribution is replaced by a Weibull distribution (for which $\ln \Lambda(t|z) = \ln c_0 + \boldsymbol{\beta}^{\mathrm{T}} z + \gamma \ln t$) with unknown γ. The purpose of equations (1.58) and (1.60) was not to construct a confidence band for an exponential-lifetime regression model (see Kalbfleisch and Prentice, 1980; Moller, 1988), but to make a technical approximation for $F(d)$ by replacing in equation (1.44), for each x and y, the upper boundary of the Wiener process by a straight line.

1.3. Numerical aspects

Integral equations of various types have extensively been studied by many mathematicians in the fields of functional analysis as well as numerical analysis (see, e.g., Bourbaki, 1973; Riesz and Sz.-Nagy; 1956, Ljusternik and Sobolew, 1960; Miller, 1971; Halmos and Sunder, 1978; Baker, 1981; Te Riele, 1982, 1983; Brunner, 1982, 1990 and the references therein). From the previous section, we have an integral equation of the form

$$g(t) = -\lambda y(t) + \int_0^t (t - s)^{-1/2} K(t, s) y(s)\, \mathrm{d}s, \tag{1.63}$$

where λ is a numerical parameter, and the kernel $K(s, t)$ as well as the known function g depend on the boundary f for the Wiener process, see (1.53) and (1.54). We assume that the boundary is continuous, and that its derivative is piecewise continuous with only finitely many jumps of finite size. (Note that, although asymptotic theory is used, the boundary crossing problem is applied to a finite sample-size situation.) The kernel $K(s, t)$, then, is a bounded, measurable (and even piecewise continuous) function on $[0, T]^2$.

1.3.1. Functional analytic digression

In abstract form, equation (1.63) can be written as

$$g = (K_0 - \lambda I)y, \tag{1.64}$$

where, for some suitable Banach space E of (at least) measurable functions, $K_0 \colon E \to E$ is a bounded integral operator, and $I \colon E \to E$ is the identity operator. (We use the word operator for linear operator. It is relatively easy to verify that, although $(t - s)^{-1/2} K(t, s)$ is an unbounded kernel, K_0 is bounded as an operator $C([0, T]) \to C([0, T])$, and also as an operator $L_p([0, T]) \to L_p([0, T])$, $1 \leqslant p \leqslant \infty$, see, e.g., Folland (1983, Ch. 1).) If $K(t, s)$ is continuous on the compact set $S = \{(t, s) | 0 \leqslant s \leqslant t \leqslant T\}$, then K_0 is compact in the space $C[0, T]$ of continuous functions with uniform norm (see, e.g., Jörgens, 1970, par. 8).

The property that K_0 is compact means, by definition, that it maps bounded subsets of E onto subsets which are relatively compact with respect to the strong topology. (Note that a bounded operator maps bounded subsets into bounded subsets.) For reflexive Banach spaces $(E = E^{**})$, such as $L_p([0, T])$ for $1 < p < \infty$, the compactness property is equivalent to the fact that weakly convergent sequences are mapped into strongly convergent ones. (Note that in an arbitrary Banach space, a bounded operator maps weakly convergent sequences into weakly convergent sequences and strongly convergent sequences into strongly convergent ones. It is recalled that for $C[(0, T)]$, which is not reflexive, weak convergence of a sequence is equivalent to pointwise convergence of a sequence of functions which are uniformly bounded by a constant M, while strong convergence of a sequence is equivalent to uniform convergence of a sequence of functions.)

A compact operator has desirable properties. Outside an arbitrarily small neighbourhood of 0, its spectrum is simple and consists of a finite number of points. A compact operator can be approximated in norm by finite-dimensional operators. We will explain these properties somewhat further. The spectrum of an operator is called simple (see e.g. Grossmann, 1988, Ch. 14) if it consists only of a point spectrum (i.e. the set of those λ for which $(K_0 - \lambda I)^{-1}$ does nowhere exist) with no accumulation point, while the dimension of the eigenspace corresponding to each isolated eigenvalue is finite. If the Banach space E is infinite dimensional, the inverse of a compact operator, if it exists at all, cannot be bounded, so in that case the point 0 has to belong to the spectrum of the compact operator, but it is an exceptional point.

The point 0 may be an element of the point spectrum. In that case, the dimension of the associated eigenspace may be finite or infinite, and 0 may be or not be accumulation point of the point spectrum. Alternatively, the point 0 may be element (and if so, it is the only element) of the continuous spectrum (i.e. the set of points for which $(K_0 - \lambda I)^{-1}$ exists, and is defined on a dense subset of E, but is not bounded) or of the residual spectrum (i.e. the set of points for which $(K_0 - \lambda I)^{-1}$ exists, but is only defined on a non-dense subset of E). The above spectral properties of the point 0 explain why integral equations of the first type (where $\lambda = 0$) are in general numerically much harder to solve than integral equations of the second type. Note that if λ is not an element of the spectrum, then for each $g \in E$ there exists a unique solution $y \in E$, which is a special case of one part of the Fredholm alternative. The compact operators form a closed (with respect to the uniform operator topology) two-sided ideal within the Banach algebra of all bounded linear operators on E. This means that if an operator is not compact, it cannot be uniformly approximated by finite-dimensional bounded linear operators (which are compact). If the operator is a compact operator on $L_p([0, T])$, $1 \leqslant p \leqslant \infty$, or on $C([0, T])$, uniform approximation by finite-dimensional operators is possible. One may show that, e.g. for $E = C([0, T])$, the finite-dimensional operators that arise by discretising equation (1.63) will provide reasonable approximations to the solution.

A slight complication in our case is that, for fixed t, the kernel is only piece-wisely continuous in s, but this does not lead to real problems. From Zabreyko et al. (1975, Section V.1.2), one can deduce that, for a piecewisely smooth boundary f, our kernel corresponds to a bounded, and even compact, integral operator K_0 which maps $L_p([0, T])$ into $C([0, T])$ for $p > 2$. [The crucial conditions

$$\int_0^t ((t - s)^{-1/2} |K(t, s)|)^{p'} \, ds < \infty \quad \text{for all } t \in [0, T] \tag{1.65}$$

and

$$\lim_{t_0 \to t} \int_0^t ((t - s)^{-1/2} |K(t, s) - K(t_0, s)|)^{p'} \, ds = 0 \quad \text{for all } t \in [0, T] , \tag{1.66}$$

where $1/p' = 1 - 1/p$, are not affected by finitely many jumps of finite size in $K(t, \cdot)$.] Note that a bounded set in $C([0, T])$ is also a bounded set in $L_\infty([0, T])$, so K_0 is also compact $C([0, T]) \to C([0, T])$. As the derivative of the solution $y(t)$ is discontinuous at a finite number of points, it will be suitable to tailor the discretisation of (1.63) to these discontinuities. This is worked out further in the next sub-section.

The spectrum of a compact Volterra operator, i.e. a compact operator of the above 'triangular' form, consists of 0 only (Zabreyko, 1975). So, for $\lambda \neq 0$, the existence and uniqueness of a solution of (1.63) is guaranteed for all $g \in E = C([0, T])$. Alternatively, without referring to compactness, this can also be derived by observing that the iterated kernel

$$K_2(s, t) = \int_s^t ((t - u)(u - s))^{-1/2} K(t, u) K(u, s) \, du$$

(corresponding to K_0^2) is bounded, which entails, for $E = C([0, T])$, that $\|K_0^{2n}\| \leqslant T^n M^n / n!$, where M is an upper bound for $K_2(s, t)$ on $[0, T] \times [0, T]$, and that $\|K_0^{2n+1}\| \leqslant \|K_0\| \|K_0^{2n}\|$. Hence,

$$\sum_{j=1}^{\infty} \|(\lambda^{-1} K_0)^j\| < \infty ,$$

which implies that the series expansion for $(\lambda^{-1} K_0 - I)^{-1}$ converges for all $\lambda \neq 0$.

The compactness of K_0 implies that the operator can well be approximated by finite dimensional operators arising from discretisation methods. The stability of a particular numerical method for slight perturbations of the known function g (and/or the kernel K) is another matter. In Eggermont (1984) this was investigated for collocation methods (in which the solution is piecewise approximated by polynomials) applied to compact Abel-type integral equations.

1.3.2. Method of solution

Despite intense attention from numerical analysts to numerical approaches for solving integral equations, implementation of such methods in general software packages, such as IMSL and NAG, is still limited to some special smooth cases. Stoutmyer (1975) discusses an interesting approach based on symbolic computation with the package REDUCE. In the rest of this section we will, instead, construct a numerical solution tailored to the discontinuities in our case.

As one can see from (1.63) and (1.47), the derivative of the solution may be discontinuous at the interpolated and transformed times-to-death $(\tau_1, \ldots, \tau_{m-2}) = (\hat{\psi}_n(\tilde{t}_2) - \hat{\psi}_n(\tilde{t}_1), \ldots, \hat{\psi}_n(\tilde{t}_{m-1}) - \hat{\psi}_n(\tilde{t}_1))$, where we take $\tilde{t}_j = (t_{j+1} t_j)^{1/2}$ for $j = 1, \ldots, m - 1$ (corresponding to midpoint interpolation on the logarithmic scale), $t_- = \tilde{t}_1$, and $t_+ = \tilde{t}_{m-1}$. So a collocation method, where the solution is approximated by a low order polynomial between these points of discontinuous derivative, seems to be appropriate. Accordingly, (1.51) is replaced by

$$g(\tau) = y(\tau) + \sum_{j=0}^{k-1} \int_{\tau_j}^{\tau_{j+1}} (\tau - s)^{1/2} K(s, \tau) y(s) \, ds$$

$$+ \int_{\tau_k}^{\tau} (\tau - s)^{-1/2} K(s, \tau) y(s) \, ds, \quad \tau \in [\tau_k, \tau_{k+1}), \quad 0 \leqslant k \leqslant m - 3. \tag{1.67}$$

The lengths of the intervals $(\tau_0, \tau_1) = (0, \tau_1)$, (τ_1, τ_2), $\ldots$, (τ_{m-3}, τ_{m-2}) are denoted by $h_1, h_2, \ldots, h_{m-2}$, respectively. We approximate the solution by a first degree polynomial between the collocation points:

$$y(s) \approx h_{k+1}^{-1}[(\tau_{k+1} - s) y(\tau_k) + (s - \tau_k) y(\tau_{k+1})] \tag{1.68}$$

for $s \in (\tau_k, \tau_{k+1})$, $0 \leqslant k \leqslant m - 3$. With (1.68), equation (1.67), restricted to the points $(\tau_1, \ldots, \tau_{m-2})$, can be rewritten as

$$g_k = y_k + \sum_{j=0}^{k-1} \int_0^1 (\tau_k - \tau_j - u h_{j+1})^{-1/2} (y_j(1 - u) + u y_{j+1}) h_{j+1}$$

$$\times K(\tau_j + u h_{j+1}, \tau_k) \, du, \quad 1 \leqslant k \leqslant m - 2, \tag{1.69}$$

where the transformation $u = h_{j+1}^{-1}(s - \tau_j)$ has been made, y_k stands for $y(\tau_k)$, and g_k for $g(\tau_k)$. The integrals have to be approximated numerically. The square root singularity makes trapezoidal or Simpson rules inefficient. Instead, as a variation of a method mentioned in Te Riele and Schoevers (1983), we like to approximate the integrals by applying a (generalised) two-point Radau integration rule,

$$I = \int_0^1 (c_2 + c_1 u)^{-1/2} G(u)\, \mathrm{d}u = w_1 G(a) + w_2 G(0^+),   \tag{1.70}$$

where the weights w_1, w_2 and the abcissa a are chosen such that the integration is exact for $G(u) = D_1 + D_2 u + D_3 u^2$, i.e. for a kernel in (1.69) which is piecewisely linear in u. Inserting the quadratic function we get an equation of the following form:

$$I = \sum_{j=1}^3 D_j t_j = D_1(w_1 + w_2) + D_2 w_1 a + D_3 w_1 a^2 \quad \text{for all } D_1, D_2, D_3 \in \mathbb{R},$$

which is indeed satisfied if

$$a = t_3/t_2,   \qquad w_1 = t_2^2/t_3,   \qquad w_2 = t_1 - w_1,   \tag{1.71}$$

where $t_j = \int_0^1 (c_2 - c_1 u)^{-1/2} u^{j-1}\, \mathrm{d}u$ ($j = 1, \ldots, 3$). Determination of the terms t_1, t_2, t_3 with $0 < c_1 < c_2$, performed by substituting $v = (c_2 - c_1 u)^{1/2}$, is straightforward, but tedious. The results, actually obtained by using the formula-manipulation package ALTRAN (with assistance from J. Hollenberg), are

$$t_1 = 2/B,   \qquad t_2 = 2C/3B^2,   \qquad t_3 = 2D/15B^3,   \tag{1.72}$$

where

$$B = S_1 + S_2,   \qquad C = 2S_1 + S_2,   \qquad D = 8S_1^2 + 9S_1 S_2 + 3S_2,$$

$$S_1 = c_2^{1/2},   \qquad S_2 = c_2^{1/2} - c_1^{1/2}.   \tag{1.73}$$

So, the integrals in (1.69) are replaced by the simple right-hand side expressions in (1.70), where the weights and the abcissae are readily calculated by a subroutine of the program. Note that $0 < a < 1$ as long as $0 < c_1 < c_2$, and that for $c_2 = 1$ and $c_1 = 0$ we have the familiar Radau formula (cf. Hildebrand, 1956, p. 343)

$$\int_0^1 G(u)\, \mathrm{d}u = \tfrac{1}{4} G(0) + \tfrac{3}{4} G(\tfrac{2}{3}).   \tag{1.74}$$

Replacement of the integrals in (1.69) by the two-point (generalised) Radau sums (1.70) leads to a triangular matrix equation for the unknown vector $y = (y_1, \ldots, y_{m-2})$, which is easily derived as follows. Note that a, w_1, and w_2 depend on c_1 and c_2. We abbreviate $a(h_{j+1}, \tau_k - \tau_j)$ by $a(j, k)$, and likewise for

w_1 and w_2. Substituting (1.70) in (1.69), we get, for $1 \leqslant k \leqslant m - 2$,

$$g_k = y_k + \sum_{j=0}^{k-1} \{w_2(j, k) G(0^+) + w_1(j, k) G(a(j, k))\}, \tag{1.75}$$

where

$$G(0^+) = y_j K(\tau_j, \tau_k) h_{j+1}$$

and

$$G(a(j, k)) = \{(1 - a(j, k)) y_j + a(j, k) y_{j+1}\} h_{j+1}$$
$$\times K(\tau_j + a(j, k) h_{j+1}, \tau_k). \tag{1.76}$$

Accordingly,

$$g_k = y_k + \sum_{j=0}^{k-1} (y_j B_{k, j} + y_{j+1} A_{k, j}) \tag{1.77}$$

with

$$B_{k, j} = w_2 h_{j+1} K(\tau_j, \tau_k) + (1 - a) w_1 h_{j+1} K(\tau_j + ah_{j+1}, \tau_k)$$

and

$$A_{k, j} = aw_1 h_{j+1} K(\tau_j + ah_{j+1}, \tau_k). \tag{1.78}$$

For simplicity, we have written a for $a(j, k)$, and w_p for $w_p(j, k)$ ($p = 1, 2$). Thus, the integral equation is approximated by the matrix equation

$$
\begin{bmatrix}
1 + A_{1,0} & & & \\
B_{2,1} + A_{2,0} & 1 + A_{2,1} & & \\
\cdot & \cdot & & \\
\cdot & \cdot & \cdot & \\
\cdot & \cdot & \cdot & \\
B_{m-2,1} + A_{m-2,0} & B_{m-2,2} + A_{m-2,1} & \cdots & 1 + A_{m-2,m-3}
\end{bmatrix}
\begin{bmatrix}
y_1 \\
y_2 \\
\cdot \\
\cdot \\
\cdot \\
y_{m-2}
\end{bmatrix}
=
\begin{bmatrix}
g_1 \\
g_2 \\
\cdot \\
\cdot \\
\cdot \\
g_{m-2}
\end{bmatrix},
\tag{1.79}
$$

which can be solved by direct substitution. As $y_0 = P_T(0)$ is the probability that the stochastic process $W(\tau)$ already exceeds the boundary at $\tau = 0$, we have, obviously, $y_0 = 0$ if $f_{d,x,y,z}(0) > 0$. Note from (1.53) that $g_0 = 0$ if $f_{d,x,y,z}(0) > 0$. (In fact, because of this initial condition, the first column in the matrix of eq. (1.79), the one corresponding to g_0, was deleted.) For $f_{d,x,y,z}(0) < 0$ the integral equation does not hold, and we get immediately $y_0 = y_1 = \cdots = y_{m-2} = 1$.

Te Riele (1982) proposes to improve the method if the solution has the following type of singularity near the origin

$$y(t) = f_1(t) + t^{1/2} f_2(t), \quad f_1 \text{ and } f_2 \text{ smooth}, \tag{1.80}$$

by adjusting the weights and the abcissa for the first interval so that (1.70) becomes exact for $G(u) = D_1 + D_2 u^{1/2} + D_3 u$. However, numerical examples in Te Riele and Schoevers (1983) suggest that a relative error of less than 10^{-3} can be achieved by using (for 10 equidistant collocation points) a straight (1 point) interpolation method, even if the solution has the form (1.80). As this accuracy is high enough for statistical purposes, we did not attempt to adapt the weights and the abcissa for the first interval more precisely to the behaviour of the solution in our case, which, from (1.60), is inferred to be of the type $t^{1/2} e^{(-c_3/t)} f_3(t)$, for some positive constant c_3 and for some smooth function $f_3(t)$.

2. Connection with decision analysis

> *Τὴν δὲ θεραπείην ἄριστα ἄν ποιέοιτο προειδὼς*
> *τὰ ἐσόμενα ἐκ τῶν παρεόντων παθημάτων.*
>
> HIPPOCRATES

In the previous sections we constructed procedures to express statistical un-certainties associated with the estimation of the survival function of a patient who has a vector of features (covariates) $z \in \mathbb{R}^k$, and receives a certain treatment. It is, then, natural to ask whether and how these uncertainties can be incorporated in the actual process of making a medical decision. In this section we will consider some of the pertaining decision theory, which, by making simplified abstractions from real situations, aims at clarifying and quantifying some aspects which the physician is confronted with when he has to make decisions in the face of uncertainty.

It is sometimes thought that the patient should not be confronted with scientific uncertainties, because uncertainty has a 'negative utility'. This attitude is, however, not always appropriate. Exaggerated firmness of opinion may lead to dis-agreement and difficult discussions, between, for instance, patients, physicians, and life insurance companies. Moreover, it may lead to a loss of credibility of the individual physician, or even to less confidence in the medical profession as a whole.

One recognises a difference of interest between the scientist and the practising physician. Evaluating past experiences, the scientist's responsibility with respect to science requires that uncertainties are expressed and no false pretensions are made. The physician bears responsibility for the patient: adequate decisions have to be made and knots to be cut in quite uncertain situations. This distinction between the two types of responsibility parallels the distinction, clearly made by Schwartz, Flammand and Lellouch (1980), between an 'explanatory' and a 'prag-matic' objective in a clinical trial. In the first case, the goal is to increase scientific (e.g. physiological) knowledge, and in the second case to reach a practical de-cision concerning the treatment to be given to a certain (group of) patient(s). The above difference of (main) objective may have a strong influence on the set-up as well as on the analysis of the trial.

We will exemplify our general discussion by considering the pragmatic question whether or not treatment with antihypertensive drugs should be advised to a particular, hypertensive, IgA-nephropathy patient. Occasionally, we will also touch upon the 'scientific version' of this question, i.e. whether or not effective control of hypertension is 'proven to be' beneficial, maleficial, or indifferent for certain groups of IgA-nephropathy patients. Although the answer to these questions may seem to be relatively clear from a practical medical point of view, our theory illustrates various issues that may be important in this, or in similar situations. For definiteness, the methodological discussion will be based upon the availability of a retrospective data set similar to the one described in Kardaun (1991). (Note that this data set accomodates an involuntary, historical trial, as before 1975 sufficiently effective anti-hypertensive drugs were not yet available.)

2.1. From survival time to effective lifetime

Summum nec metuas diem nec optes

MARTIALIS

Consider two treatments, A and B, for a disease (like IgAN) which has two competing risks: (1) the risk of a serious, non-lethal failure caused by the disease, and (2) the risk of actual death, which is assumed to be almost independent from the previous risk, i.e., it is supposed that the probabilities of actual death (due to all causes) are not much influenced by the just mentioned (non-lethal) failure. It is presumed that at the date of diagnosis and prognosis, also a decision between treatment A and treatment B is made. Of course, the prognosis may depend on the treatment that will be given. Let $y_i(t)$ denote some gain function for a patient. This function describes the cumulative gain associated with living t time-units after prognosis if treatment i is used ($i = $ A, B). The gain per time-unit of living depends on side-effects associated with treatment, and, in the case of IgA-nephropathy, also upon whether or not the patient has entered end-stage renal disease, an event which necessitates renal dialysis and/or renal transplantation. Sometimes, the side-effects of one treatment may be neglected with respect to those of the other treatment. In our case, we will look at the situation in which treatment A consists of keeping diet, and treatment B of keeping diet combined with taking antihypertensive drugs. We will not consider an evaluation of the decisions involved in choosing between dialysis and renal transplantation.

A basic question is, of course, how to specify the functions $y_i(t)$. There is an ample range of possible choises. It seems obvious that, for each treatment, $y_i(t)$ is an increasing function of time, and that $y_i(0) = 0$. If the patient has no clear preference with respect to the choice of the gain function, we propose to adopt the following principle: for each time-unit (e.g. day) from the date of biopsy until the date of dialysis, the gain is b_A units if treatment A is applied, and b_B units if treatment B is used. For each time-unit between dialysis and actual death, the gain is taken equal to c units. The values of b_i and c will have to be specified in a discussion with the patient. We assume that the value of c does not depend on the type (A or B) of treatment, which is given before dialysis. For concreteness, we also assume that $0 \leqslant c \leqslant b_i \leqslant 1$ (i = A, B). (If necessary, the mathematical analysis is easily extended to the case $0 \leqslant b_i \leqslant c \leqslant 1$.) For a patient with some vector z of covariates and treated with a certain treatment i, the (random) time from diagnosis until dialysis will be denoted by T_i, and the (random) time between diagnosis and actual death by T_D. The associated gain

$$Y_i = \begin{cases} b_i T_D & \text{if } T_D < T_i, \\ b_i T_i + c(T_D - T_i) & \text{if } T_D > T_i, \end{cases} \tag{2.1}$$

is a random variable depending upon treatment (i = A, B), and, which is suppressed in the notation, the vector of covariates. Note that the gain can be interpreted as the 'effective duration of life' after diagnosis.

The just described model resembles somewhat the accelerated life model, discussed, for example, in Cox and Oakes (1984) and Kalbfleisch and Prentice (1980). Here, the acceleration factors arise from a subjective evaluation by the patient of his future health states, which are possibly affected by treatment. Note that differences in economic costs between the treatments can be incorporated in a similar fashion. It may be difficult, however, to assess the relative importance of economic costs (largely) paid by the community, and the quality-of-life of the patient. In this study, we shall pay no further attention to the economic costs of treatments.

For a patient with vector of covariates z, let $S_i(t|z)$ denote the survival function of the time-to-dialysis associated with treatment i (i = A or B). The survival function for the time-to-actual-death, depending upon a (possibly different) vector of covariates z_D, is denoted by $S_D(t|z_D)$, or, to indicate a possible dependence on treatment and on the vector z, by $S_{D,i}(t|z_D, z)$. All survival functions are presupposed to be absolutely continuous, those pertaining to the time-to-dialysis with possibly a positive absorption probability $\lim_{t \to \infty} S_i(t|z)$ at infinity. For simplicity, we suppose that, given the vectors z and z_D of covariates, the time-to-actual-death does not depend on treatment and is independent of the time-to-dialysis. For actual practice, we consider z_D = (sex, age), and make the crude assumption that $S_D(t|z_D)$ taken from national lifetables constitutes an adequate approximation to the corresponding survival function of the patients. In summary, our assumptions are that for z_D = (sex, age), $S_{D,i}(t|z_D, z) \approx S_D(t|z_D)$ and $S_i(t|z_D, z) \simeq S_i(t|z)$, while the corresponding (conditional) random variables are independent.

From (2.1), we see that the survival function $S_{Y,i}(y) = \Pr\{Y_i > y\}$ equals

$$\Pr\{(b_i T_D > y) \,\&\, (T_D < T_i)\} + \mathrm{P}\{(b_i T_i + c(T_D - T_i) > y) \,\&\, (T_D > T_i)\}. \tag{2.2}$$

This can be rewritten (see Figure 3) as

$$S_{Y,i}(y) = \{S_D(c^{-1}y|z_D) + \int_{t=0}^{y/b_i} S_i(t|z)\,dS_D(c^{-1}[y - (b_i - c)t]|z_D)\},$$

(2.3)

for $0 \leqslant c \leqslant b_i \leqslant 1$, and $i = A, B$. (For simplicity, the dependence of $S_{Y,i}(y)$ on z and z_D is suppressed in the notation.) Note that, obviously,

$$S_{Y,i}(y) = S_D(c^{-1}y|z_D) \quad \text{if } b_i = c,$$

(2.4)

and, as is directly inferred from (2.1),

$$S_{Y,i}(y) = S_D(b_i^{-1}y|z_D)S_i(b_i^{-1}y|z) \quad \text{if } c = 0.$$

(2.5)

A point estimate $\hat{S}_{Y,i}(y)$ is obtained by substituting a usual step function estimate for $S_i(t|z)$, taking S_D from a national lifetable. (See, e.g., Kardaun, 1983,

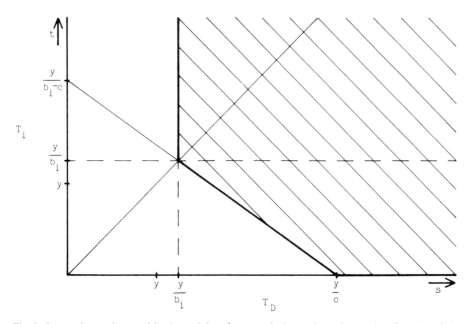

Fig. 3. Integration region used in determining, for a particular patient, the survival function of the effective lifetime under treatment i ($i = A, B$). Integration over the shaded area of the joint distribution of (T_D, T_i) yields the survival function $S_{Y,i}(y) = \text{Pr}\{Y_i > y\}$, where Y_i (being equal to b_iT_D if $T_D < T_i$, and to $(b_iT_i + c(T_D - T_i)$ if $T_i < T_D$) is the effective lifetime after the date of prognosis for a patient using treatment i. The numbers c and b_i, ranging from 0% to 100%, are subjective assessments by the patient of the quality-of-life after renal failure, and before renal failure under treatment i, respectively ($i = A, B$).

Figure 7, for graphs of $S_D(t | \text{age})$ for the Dutch male population, and Brillinger, 1986, for a thorough discussion on lifetable statistics.)

2.1.1. *Three models of increasing complexity*

Unless the sample of patients is quite homogeneous (or can be divided into homogeneous sub-samples of a reasonable size), a model has to be specified to account for the influence of the vector z of nuisance covariates on the lifetimes, or on the effective lifetimes. Although there are several possibilities, we consider here the case that the (stratified) proportional hazards model,

$$S_i(t|z) = S_i(t|0)^{\exp(\beta_i^T z)}, \quad i = A, B, \tag{2.6}$$

is applied to the original lifetimes. Unless stated differently, we assume that there is no interaction (on the proportional hazards scale) between treatment and the nuisance covariates, i.e., $\beta_A = \beta_B$.

A very special case of (2.3) combined with (2.6), arises if $c = 0$, and, for the time interval of interest, $S_D = 1$. These simplifications are justified if the quality-of-life after renal failure is considered to be very low, and if the age of the patient is such that actual death is very unlikely to occur in the time interval considered (e.g., the first ten years after diagnosis). In that case, the gain Y_i, interpreted as the 'effective lifetime' under treatment i, is simply $b_i T_i$ (i = A, B), i.e. only the time until renal failure counts, and each day is worth b_i units. The corresponding survival function $S_{Y,i}(y) = S_i(b_i^{-1} y | z)$. This mimics the accelerated life model. Note, however, that in the present context the factor b_i is not an estimated quantity, but a subjective measure of the quality of life, determined in a discussion with the patient. We shall refer to this model as the *Sphald* (simple proportional-hazards accelerated-life decision) model. It is remarked that, on the scale of Figure 2 of Chapter 14, acceleration with a factor b corresponds to a *horizontal* shift of the survival function by an amount $\ln b$ to the left, whereas the effect of a covariate in the proportional hazards model is expressed by a *vertical* shift of the survival function. (As is well known, the accelerated life model and the proportional hazards model coincide if the survival functions are represented by straight lines in that figure, i.e. if Weibull distributions are assumed for the survival times.)

A slightly more complicated model is the *Phald* (proportional-hazards accelerated-life decision) model, which arises if the quality-of-life after renal failure cannot be considered as effectively 0 (i.e. if there is still 'life after death') whereas S_D is effectively 1 on the time-interval of interest. Here, the gain is given by

$$Y_i = (b_i - c)T_i, \tag{2.7}$$

which may be called the 'additional effective lifetime' (being a shorthand expression for 'the effective lifetime in addition to the effective lifetime when the patient enters immediately end-stage renal disease'). The corresponding survival function is, of course,

$$S_{Y,i}(y) = S_i(y/(b_i - c)) . \qquad (2.8)$$

To emphasize the difference with 'effective lifetime', one should perhaps write $Y_{a,i}$ instead of Y_i (subscript a standing for additional), but we will not do so in order to keep the subsequent notation as simple as possible.

REMARK. We will investigate how the gain in the 'Phald' model relates to the 'limit' of the gain in (2.1). Conceptually, it is convenient to restrict first attention to a fixed period of time (say the first t_m years after the date of diagnosis, $t_m = 1, 2, 3, \ldots$). We assume that $S_D(t)$ is effectively 1 on the intervals $[0, t_{m,i}]$ of interest, which are, commonly, the intervals on which reliable semiparametric estimates of $S_i(t\,|\,z)$ are available $(i = A, B)$. The gain corresponding to equation (2.1) is now given by

$$Y_i = \begin{cases} b_i t_m & \text{if } T_i > t_m , \\ b_i T_i + c(t_m - T_i) & \text{if } T_i < t_m , \end{cases} \qquad (2.9)$$

which can be rewritten as

$$Y_i = ct_m + (b_i - c) \min(T_i, t_m) , \qquad (2.10)$$

and is interpreted as the 'effective lifetime' during the first t_m years after diagnosis. The corresponding survival function $S_{Y,i}(y)$ equals

$$S_{Y,i}(y) = \begin{cases} 1 & \text{for } 0 \leqslant y < ct_m , \\ S_i((y - ct_m)/(b_i - c)) & \text{for } ct_m \leqslant y \leqslant b_i t_m , \\ 0 & \text{for } b_i t_m < y < \infty . \end{cases} \qquad (2.11)$$

One could be inclined to consider (2.10) or (2.11) as an alternative definition of the 'Phald' model. However, if we compare the two effective-lifetime distributions by stochastic inequality (a subject which will be discussed further in the next sub-section), the jump in $S_{Y,i}$ at $b_i t_m$ leads to certain difficulties provoked by the finite horizon t_m, which are inessential to the problem. (This was pointed out to me by Professor Dr W. Schaafsma.) These difficulties are removed by discarding, for both treatments, the identical terms ct_m and letting $t_m \to \infty$. One thus obtains a simpler model with only a slightly more difficult interpretation of the gain.

The most general model to be considered here, which we will call the *Gphald* model, just consists of (2.3) combined with (2.6), with $S_D(t\,|\,z)$ taken from national lifetables. It is noted that 'Gphald' may be appropriate if the simplifying condition on S_D in the model 'Phald' does not hold in practice.

REMARKS. (1) In the models 'Sphald' and 'Phald' one may consider to transform first all measured lifetimes (censored and uncensored) into (additional) effective lifetimes and then postulate the proportional hazards (PH) model for the (additional) effective lifetimes. This is identical to postulating the PH model for the original lifetimes. (Note that the ordering of the observations is not disturbed by the scale transformation, so all estimates based on the partial likelihood remain the same.) In the 'Gphald' model it is not possible to transform the original lifetimes directly into effective lifetimes. Transformation of the associated survival functions is given by (2.3). The PH model, here, is applied to $S(t|z)$, which is quite different from a PH model for $S_{Y,i}(y|z)$.

(2) Equal random censorship in the stratified PH model for the original lifetimes is destroyed by the transformation of the scale. If $G(u)$ is the censoring distribution function for the original lifetimes, we get (in 'Sphald') for the effective lifetimes: $G_A(u) = G(b_A^{-1} u)$ for sample A, and $G_B(u) = G(b_B^{-1} u)$ for sample B. Unequal censorship, however, is not unusual. It may be postulated for the original stratified PH model as well. In the k-sample situation without covariates it has extensively been studied (see, e.g., Breslow, 1979, and its references). Here we expect that not too wildly different censoring in the samples will not spoil the asymptotic approximations referred to in Section 1. (Note that in Bailey (1983) fixed censoring is considered, and that in Andersen and Gill (1982) the asymptotic results are derived, under suitable regularity conditions, for more general random censoring schemes.)

(3) For the IgA-nephropathy case, the 'Phald' model is perhaps an appropriate choice, being a reasonable compromise between simplicity and adequacy. Accordingly, the reader primarily interested in a simple evaluation of this case may, in the following sub-sections, quickly pass through the considerations about the more complicated model 'Gphald'. The model 'Sphald', being (in two respects) a simplified version of 'Phald', will henceforth not be considered.

2.1.2. Construction of confidence bands

In the model 'Phald', an asymptotic simultaneous confidence band for the 'additional effective lifetime' can directly be obtained by shifting the simultaneous band for $S_i(t|z)$, as constructed according to Section 1, an amount $-\ln(b_i - c)$ to the left. The same applies to an instantaneous confidence band.

In the model 'Gphald' the construction of confidence bands is more complicated. In general the variance of $\hat{S}_{Y,i}(y)$ is estimated by

$$
\hat{\text{Var}}(\hat{S}_{Y,i}(y)) = \int_{s=0}^{y/b_i} \int_{t=0}^{y/b_i} dS_D(h_i(s)|z_D)\, dS_D(h_i(t)|z_D)\, \hat{S}_i(s|z)\hat{S}_i(t|z)
$$
$$
\times \{ \tilde{\psi}_{1,i}(s \wedge t, \hat{\beta}_i|z) + \tilde{\Gamma}_{1,i}(s, \hat{\beta}_i|z)^{\mathrm{T}} \hat{\Sigma} \, \tilde{\Gamma}_{1,i}(\cdot t, \hat{\beta}_i|z) \},
$$

$$
\tag{2.12}
$$

$0 \leqslant c \leqslant b_i \leqslant 1$, where $h_i(s) = c^{-1}(y - (b_i - c)s)$, $i = $ A, B denotes the treatment, and $s \wedge t$ the minimum of s and t. The function $\tilde{\psi}_{1,i}(s, \hat{\beta}_i|z)$ is given by (1.39). Obviously, $\tilde{\Gamma}_{1,i}(s, \hat{\beta}_i|z) = (\partial/\partial\beta) \tilde{\Lambda}_i(s,\beta|z)|_{\beta=\hat{\beta}_i}$, where the right-hand side can be determined from (1.41) and (1.43). A similar expression can be used if there is no interaction between treatment and the nuisance covariates. In that case, $\hat{\beta}_A = \hat{\beta}_B$ and $\hat{\Sigma}_A = \hat{\Sigma}_B$ in (2.12) are determined from a joint partial likelihood function. If $S_i(t|z)$ is estimated, semiparametrically, for $t \in [0, t_m]$, then $\hat{S}_{Y,i}(y)$ and $\text{Vâr}(\hat{S}_{Y,i}(y))$ can be determined for $y \in [0, b_i t_m]$. Step functions for $\hat{S}_i(s|z)$ and $\hat{S}_i(t|z)$ entail that the double integral is converted into a double summation. Because of symmetry, the first term in (2.12), the one which remains (in a degenerate form) if there are no covariates, can be converted into two times an integral on a triangular region.

Using formula (2.12), the construction of an instantaneous confidence band for $S_{Y,i}(y)$ is relatively easy, once values for b_i and c have been established in a discussion with the patient. The covariance structure of the process $X_i(y) = \sqrt{n_i}(\hat{S}_{Y,i}(y) - S_{Y,i}(y))$, obtained by integrating the integrand in (2.10) over a rectangular region $[0, y_1/b_i] \times [0, y_2/b_i]$, however, is (for general b_i and c) not particularly nice for making simultaneous confidence bands based on asymptotic theory. This can be seen from the fact that the term containing $\hat{\psi}_{1,i}(s \wedge t, \hat{\beta}|z)$ is not the covariance structure of a Gauss–Markov process (i.e. cannot be written in the form $g_1(y)W(g_2(y))$ for monotonic g_1 and g_2), which means that the derivation, as presented in Section 1.2, of the integral equation for the distribution function of the first-exit time with respect to a general boundary breaks down.

If the limits of a simultaneous confidence band for $S_i(t|z)$ have been calculated, e.g. according to the theory of Section 1.2, a *conservative* band for $S_{Y,i}(y)$ is simply obtained by substituting the limits of the band for $S_i(t|z)$ into (2.3). It is easily verified that the confidence coefficient of the band for $S_{Y,i}(y)$ is at least as large as the coefficient of the corresponding band for the survival curve. Although this band for $S_{Y,i}(y)$ is conservative, the fact that its construction is easy (once a band for $S_i(t|z)$ is established), and that the lower bound for the confidence coefficient holds for all values of b_i and c, and for all monotonic functions S_D, means that the procedure is a useful complement to the construction of an instantaneous band.

Alternatively, one might attempt making simultaneous confidence bands by applying a Monte Carlo approach, or using bootstrap (or other resampling) methods. See, e.g., Hall (1988). These approaches, however, will not be discussed in this study.

2.2. Comparison of effective-lifetime distributions

> *Tute hoc intristi; tibi omne est exedendum.*
>
> TERENTIUS

2.2.1. Preliminary observations

We want to compare the two treatments A and B on the basis of their effective-lifetime distributions $S_{Y,A}$ and $S_{Y,B}$. It is interesting to recall that these distributions are partially ordered by 'stochastic inequality'. By definition,

$$S_{Y,A} \succcurlyeq S_{Y,B} \quad \text{if } S_{Y,A}(y) \geqslant S_{Y,B}(y) \quad \text{for all } y \in [0, \infty). \tag{2.13}$$

Of course, the survival functions are defined with respect to a certain vector z of covariates. Sometimes, the interval $[0, \infty)$ in this definition will be replaced by a suitable sub-interval $[y_1, y_2]$. In the notation we suppress the dependence of $\geqslant$ on y_1 and y_2, and also the dependence of the survival function on z. Note that the stochastic ordering leads to precisely four possible relationships between $S_{A,Y}$ and $S_{B,Y}$. We abbreviate (suppressing for a moment even subscript Y in the notation for the survival function):

$$(S_A \succcurlyeq S_B, \text{ and not } S_B \succcurlyeq S_A) \qquad \text{by } S_A \succ S_B, \tag{2.14a}$$

$$(S_B \succcurlyeq S_A, \text{ and not } S_A \succcurlyeq S_B) \qquad \text{by } S_A \prec S_B, \tag{2.14b}$$

$$(S_A \succcurlyeq S_B \text{ and } S_B \succcurlyeq S_A) \qquad \text{by } S_A = S_B, \tag{2.14c}$$

$$(\text{neither } S_A \succcurlyeq S_B \text{ nor } S_B \succcurlyeq S_A) \quad \text{by } S_A \prec \succ S_B. \tag{2.14d}$$

For a patient with a certain vector of covariables, we say that, on the interval $[y_1, y_2]$,
   (a) treatment A is 'strictly better' than treatment B,
   (b) treatment A is 'strictly worse' than treatment B,
   (c) treatment A is 'equivalent' to treatment B,
   (d) treatment A and treatment B are 'incomparable'.
Notice that in the models 'Phald' and 'Gphald', stochastic inequality is generally not preserved by a transformation of the original lifetime to the effective lifetime, and that we want to compare the treatments on the basis of the effective lifetime.

In a concrete situation, $S_{Y,A}$ and $S_{Y,B}$ are estimated from the data. A useful preliminary procedure for comparing the two treatments consists in drawing a picture like Figure 4, which displays the estimated survival functions for the (additional) effective lifetime under treatment A and treatment B, respectively, together with instantaneous confidence bands. (For illustrative purposes only, the two survival curves of Figure 2 of Chapter 14 have been transformed with $c = 0$, $b_A = 1$, and $b_B = 0.75$.) A careful look at such a picture gives already much information.

First we concentrate on the point estimates. If $\hat{S}_{Y,A}(y) > \hat{S}_{Y,B}(y)$ for all $y \in [y_1, y_2]$ such that the ranges $[\hat{S}_{Y,i}(y_1), \hat{S}_{Y,i}(y_2)]$, $i = A, B$, cover the interval $[0, 1]$ to a large extent, one may say that, for a patient with vectors of covariates z and z_D, treatment A is 'estimated to be strictly better than treatment B'. If, for example, $[y_1, y_2]$ is the transformation (see Section 2) of the time interval $[0.5 \text{ yrs}, 10 \text{ yrs}]$, and $\hat{S}_{Y,i}(y_1)$ is close to 1, but $\hat{S}_{Y,i}(y_2)$ is not close to 0 $(i = A, B)$, then we may say that 'A is estimated to be better than B for the first 10 years'.

Next, we want to incorporate confidence statements in comparing the two treatments. This will be done in the next sub-section for the following specific

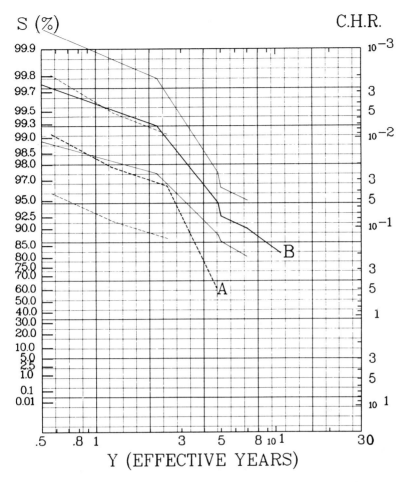

Fig. 4. Estimated survival functions and approximate 67% confidence intervals for the effective lifetime under treatment A and treatment B, respectively. For illustrative purposes only, the figure consists of the two survival functions presented in Figure 2 of Chapter 14, but with curve B shifted a factor $\frac{4}{3}$ to the left. This corresponds to $b_B = 0.75$, $b_A = 1$ and $c = 0$ (i.e. to a relative quality-of-life of 75% during effective hypertension treatment, and of 0% after renal failure).

models. We use the proportional hazards model for the effect of the nuisance covariates on the original (or effective) survival times:

$$S_A(t|z) = S_A(t|0)^{\exp(\beta_A^T z)} \quad \text{and} \quad S_B(t|z) = S_B(t|0)^{\exp(\beta_B^T z)}, \qquad (2.15)$$

where, as usual, $S_i(t|0)$ denotes the base-line survival function for the time-to-renal-failure under treatment i, $i = A, B$. We will consider the following model specifications for the effect of treatment:

(a) the 'stratified model with interaction', for which there is neither a connection between β_A and β_B, nor between S_A and S_B,

(b) the 'stratified model without interaction', for which $\boldsymbol{\beta}_A = \boldsymbol{\beta}_B$, whereas S_A and S_B are arbitrary,

(c) the 'unstratified model (without interaction)', for which $\boldsymbol{\beta}_A = \boldsymbol{\beta}_B$ and $S_B(t|\mathbf{0}) = S_A(t|\mathbf{0})^{\exp(\gamma)}$,

(d) the 'proportional-hazards accelerated-life' model, for which $\boldsymbol{\beta}_A = \boldsymbol{\beta}_B$ and $S_B(t|\mathbf{0}) = S_A(b_e^{-1}t|\mathbf{0})$, where b_e is an accelerating factor measuring the effect of the difference in treatment on the actual survival times. (Note the difference between the subjective parameter b and the model parameter b_e, the latter of which is estimated from the data.)

REMARKS. (1) In the first three models, the term 'stratified' is shorthand for 'stratified with respect to treatment', while the term 'with(out) interaction' stands for 'interaction is (not) modelled between treatment and the nuisance covariates'.

(2) In model (d) a partial likelihood profile can be drawn:

$$L_{\text{profile}}(b_e) = \max_{\boldsymbol{\beta}} L(b_e, \boldsymbol{\beta}),\qquad\qquad (2.16)$$

from which one can derive a point estimate and a confidence interval for b_e. Inference for b_e may be less complicated than in a full accelerated-life model (see, e.g., Kalbfleisch and Prentice, 1980, Ch. 6). Etezadi-Amoli and Ciampi (1987), among others, consider model (d) in a more general formulation, using splines for the base-line survival function.

In the two stratified models (a) and (b), it is natural to try to base the comparison of the treatments on stochastic inequality. Note that when the 'Phald' model is used in these cases, the PH model applied to the original lifetimes leads to the same estimation procedure of the regression coefficient β as the PH model applied to the additional effective lifetimes. In the two unstratified models (c) and (d), comparison of the treatments is, instead, described by a one-dimensional parameter (which is consistent, of course, with stochastic inequality).

In model (c), estimation of the regression coefficient β is influenced by the fact whether the PH model is postulated for the original, or for the additional effective lifetime. For a proper treatment comparison, we like to apply the model to the additional effective lifetime. This can be done, as soon as it seems reasonable from the plot of the stratified model (b) for the additional effective lifetime.

Model (d) with 'Phald' is a natural combination, having a clear and simple interpretation. Postulating model (d) for the original lifetime is equivalent to postulating it for the additional effective lifetime. Whether or not model (d) applies to the data can be seen from the plot of the stratified model. (The two base-line survival curves should now approximately coincide after a *horizontal* shift.)

If 'Gphald' is used, the proportional hazards assumption for the covariates and the model assumptions (a) to (d) will always be applied to the original lifetimes (otherwise an evaluation seems to be too difficult).

2.2.2. *Confidence band considerations*

We now want to discuss incorporating confidence statements in comparing the gains of the two treatments in some specific situations. Firstly, we will make some remarks on the general case 'Gphald', which is rather complicated, and subsequently, we shall give a more detailed elaboration of the model 'Phald'.

Note that in 'Gphald' as well as in 'Phald', comparison by stochastic inequality is hampered by the fact that the two intervals on which reliable estimates for $S_{Y,A}$ and $S_{Y,B}$ are available, may not be the same. This difficulty can be removed by restricting the interval $[y_1, y_2]$ to the intersection of these two intervals, although ideally the choice of this restricted interval should not depend on the data.

In 'Gphald' we mainly consider the 'stratified model with interaction'. In this situation estimates of $S_A(t|z)$ and $S_B(t|z)$ are independent. Suppose there is some prior knowledge that treatment A will not be strictly better than treatment B (i.e., either $S_{Y,A} \prec \succ S_{Y,B}$, or $S_{Y,A} \preccurlyeq S_{Y,B}$). Then, some insight is gained by constructing, according to the conservative method explained in Section 2.1, a one-sided, simultaneous upper confidence band for $S_{Y,A}$ and a one-sided, simultaneous lower confidence band for $S_{Y,B}$. If both bands have confidence coefficient $1 - \alpha$, and if the first band nowhere exceeds the second band, then, with confidence of at least $1 - 2\alpha + \alpha^2$, $S_{Y,A}$ is stochastically smaller than $S_{Y,B}$. In that case one may say that, for the type of patient considered, treatment A is 'likely to be strictly worse' than treatment B, a lower bound for the confidence coefficient being approximated by $1 - 2\alpha$. The same procedure can formally also be applied in situations (b) and (c), but then $1 - 2\alpha + \alpha^2$ may be a quite conservative approximation to the actual confidence level. In the absence of prior knowledge about the relation between $S_{Y,A}$ and $S_{Y,B}$, the same procedure should be carried out with two-sided instead of one-sided confidence bands. (For small α, a two-sided band with confidence coefficient $1 - \alpha$, which is complicated to construct, can be approximated by two one-sided bands, each with confidence coefficient $1 - \frac{1}{2}\alpha$.) If a simultaneous band for $S_{Y,A}$ is, for some $y \in [y_1, y_2]$, above the simultaneous band for $S_{Y,B}$, then it may be said that treatment B is 'likely to be somewhere better' than treatment A (with the appropriate confidence coefficient).

In the 'Phald' model, we will consider all four situations. (Note that in situation (c) it will be convenient to transform the original lifetimes into the additional effective lifetimes, before the proportional hazards model is applied. In the other situations one can, equivalently, apply the models to the original lifetimes, and then make appropriate shifts.)

In situation (a) ('stratified model with interaction') one may apply the just described method based on a simultaneous confidence band for each survival function. Each band has now asymptotically the correct confidence coefficient, although the combining procedure is, of course, conservative.

In the second situation ('stratified model without interaction') the above method may be too conservative. One would like, perhaps, to make a simultaneous confidence band of constant width for the 'log-effect' function

$$\beta_{AB}(y|z) = \ln \Lambda_A(y|z) - \ln \Lambda_B(y|z), \tag{2.17}$$

where, as usual, $\Lambda_i(y|z) = -\ln S_i(y|z)$ (i = A, B). This, however, runs into difficulties, as $\sqrt{n}(\beta_{AB}(y|z) - \hat{\beta}_{AB}(y|z))$ does not converge to a Gauss–Markov process. It is easier to consider the difference $D_{AB}(y|z) = \Lambda_A(y|z) - \Lambda_B(y|z)$ between the cumulative hazard rates, rather than to investigate their quotient. One observes that $\sqrt{n}(D_{AB}(y|z) - \hat{D}_{AB}(y|z))$ can asymptotically be represented by

$$\sqrt{n}\,\mathrm{W}([\psi_{1,A} + \psi_{1,B}](y)) + \sqrt{n}(\beta - \hat{\beta})^{\mathrm{T}}(\Gamma_{1,A}(y) - \Gamma_{1,B}(y)) + o_{\mathrm{P}}(1),$$

$$(2.18)$$

where $[\psi_{1,A} + \psi_{1,B}](y)$ stands for $\psi_{1,A}(y) + \psi_{1,B}(y)$. Estimates of $\psi_{1,i}(y)$ = $\psi_{1,i}(y, \beta|z)$ and $\Gamma_{1,i}(y) = \Gamma_{1,i}(y, \beta|z) = \Lambda(t, \beta|z)\Gamma_i(y, \beta|z)$, i = A or B, are given in equations (1.38) to (1.43). Each estimate is based on the ordered effective lifetimes of sub-sample i, and on the common estimate of β. As $\psi_{1,i}(y)$ and its estimate are monotonic functions of y, the structure of the process described by (2.18) is similar to that for just one survival function. So, the integral equation approach for the first-exit time of the Wiener process with respect to a general boundary may be combined with a Monte Carlo integration for the regression term in (2.18) in order to find the confidence coefficient of a general type of band for $D_{AB}(y|z)$.

If $\Gamma_{1,i}(y)$ can reasonably be approximated by a straight line, and if simple boundaries for $D_{AB}(y|z)$, such as straight lines, are used, then the integral equation need not to be solved.

In Schumacher (1982), which is a pleasure to read, several tractable simultaneous band constructions for the log-effect function are discussed in the case that there are no covariates. They are based on the (time-transformed) Wiener process, i.e. on the first term in (2.18).

In situation (c) it makes a difference whether the model assumptions apply to the original lifetimes or to the additional effective lifetimes. For an easy evaluation of the effect of treatment, the PH model with respect to the additional effective lifetimes is preferable. From a plot, on the scale of Figure 4, of the stratified model applied to the additional effective lifetimes one can observe whether this model is reasonable. If so, the estimate of γ and of its variance give a clearly interpretable (one-dimensional) evaluation of the treatment effect. For effects of misspecification see Solomon (1984).

The model 'Phald' in combination with (d) is quite natural. Whether this model holds can be inferred from a plot of the stratified model (for the original lifetimes). When model (d) is used, only one base-line survival curve (for the original lifetime) is drawn. The estimated effect of not using adequate hypertension treatment is represented by a horizontal arrow with length $|\ln \hat{b}_e|$ and, in our case, pointing to the left. However, if treatment i is used, the value of each day, for a particular patient and excess to its value when the patient has entered end-stage renal disease, is $b_i - c$ units (i = A, B). So, for treatment i, the transition from original lifetime to additional effective lifetime is indicated by an arrow which shifts the base-line hazard survival curve to the left by an amount $|\ln(b_i - c)|$. It may be illustrative to discuss with the patient the ratio between b_e (with its estimated uncertainty) and his subjective factor $(b_A - c)/(b_B - c)$. A scientific version of such a discussion will be presented in the next sub-section.

2.3. Decision problems when treatment difference is estimated by $X \sim N(\theta, 1)$

Simplex munditiis

In a substantial number of cases, no definite conclusions will be reached by applying the procedures of the previous section. Important reasons are, firstly, that the ordering of the survival functions is only partial, and, secondly, that simultaneous confidence bands tend to be large, at least if popular values of α, such as 5%, are applied to a rather limited number, say 50 to 100, of patients.

In a scientifically oriented situation, one is inclined to state that two treatments have a different effect on survival only if the data are sufficiently incompatible with the null-hypothesis of 'no difference'. This is reflected by the usual choice in hypothesis testing theory of a significance level of $\alpha = 5\%$. On the other hand, in pragmatic decision situations, e.g. when one out of two treatments has to be recommended to a particular patient, one will be inclined to cut the Gordian knot, even when there is considerable scientific doubt.

Here, we shall discuss aspects of pragmatic decision problems in the simplified situation that the two treatments are compared by a one-dimensional summary characteristic, such as

(a) the ratio of the median effective lifetimes,

(b) the ratio $\Lambda_A(y|z)/\Lambda_B(y|z)$ of the cumulative hazard rates at a certain effective lifetime,

(c) the 'acceleration factor' between the treatments, which is a postulated scale factor for the effective-lifetime distributions $(S_A(y|z) = S_B(b_a y|z))$, or

(d) the 'relative risk' between the treatments, which is a postulated constant of proportionality between the two cumulative hazard rates $(\Lambda_A(y|z) = c_a \Lambda_B(y|z))$.

It is noted that such one-dimensional measures cannot reflect every aspect of the difference between two survival functions, and are sometimes even rather misleading. In particular, the following procedures should only be applied to situation (c) or (d) if inspection of the separate survival functions reveals approximate adequacy of the utilised one-dimensional measure of discrepancy.

In many cases, the estimate $\hat{\theta}$ of the one-dimensional parameter will be asymptotically normal, and the asymptotic variance σ_θ^2 of $\sqrt{n}(\hat{\theta}_n - \theta)$ can be estimated using large sample theory (for example by inspection of a partial likelihood profile). For moderately large sample sizes we will use the asymptotic distribution as an approximation to the true probability distribution. We consider situations in which σ_θ^2 varies only slightly with θ, at least in the region $(\theta_{\min}, \theta_{\max})$ of interest. Within this region we approximate the distribution of $\sqrt{n}(\hat{\theta}_n - \theta)/\sigma_0$ by the standard normal distribution, independently of the true value θ, and we assume that σ_θ^2 is known (i.e. $\hat{\sigma}_\theta^2$ is regarded as the true value σ_θ^2). By making these, sometimes strong, assumptions, we have reduced the problem to a deceptively simple one: statistical inference and decision making about a one-dimensional parameter θ on the basis of one observation $X = \hat{\theta}_n$, of which the distribution is approximated by $N(\theta, 1)$. (It is understood that all quantities, parameters as well as estimates, are measured in units of $\sigma_\theta / \sqrt{n}$.)

Pure significance testing theory, which has its roots in the work of Fisher and others before him, considers the exceedance probability $P = P_0\{X > |x|\} = 2\Phi(-|x|)$ as a measure of agreement between the data, summarised by a statistic X, and the null-hypothesis H_0: $\theta = 0$. If P is small, then the agreement is low and either the null-hypothesis is false, or an improbable event has taken place. The null-hypothesis is rejected if the exceedance probability is sufficiently small. For 'ordinary' sample sizes a rather arbitrary, but commonly accepted, cut-off value is 5%.

Classical Neyman–Pearson theory provides clear-cut procedures for testing H_0: $\theta = 0$ against one-sided (A_1: $\theta > 0$, e.g.), or two-sided (A_2: $\theta \neq 0$) alternatives. For the one-sided problem, a uniformly most powerful (UMP) test exists among all tests of which the size, i.e. the (maximum) probability of falsely rejecting H_0, does not exceed some predetermined level α. This test uses as rejection region (c_α, ∞), where $c_\alpha > 0$ is determined by requiring that the size of the test is equal to α. If $\alpha = 5\%$, $c_\alpha = 1.65$. For the two-sided problem, no UMP level-α test exists. Restricting attention, however, to unbiased procedures (for which the probability of rejecting H_0 is larger if $\theta \neq 0$ than if $\theta = 0$), it is shown that the test which rejects H_0 if $X \notin (-c_{\alpha/2}, c_{\alpha/2})$, is UMP among unbiased level-α tests. For $\alpha = 5\%$, we have $c_{\alpha/2} = 1.96$. Restriction to the class of unbiased procedures makes the optimality criterion somewhat less compelling.

Under the stimulus of Wald, hypothesis testing theory was put into a decision-theoretic framework. In the usual formulation of the just mentioned testing problems, the (non-randomised) decision space consists of only two actions:

a_0: both H_0 and A are neither rejected nor accepted ,

a_1: H_0 is rejected, and A is accepted .   (2.19)

A loss function L can be defined (see, e.g., Lehmann, 1986, Ch. 1) by

$$L(\theta, a_0) = 0; \quad L(\theta, a_1) = a \quad (\theta \in H_0) ,$$
$$L(\theta, a_0) = b; \quad L(\theta, a_1) = 0 \quad (\theta \in \text{A}) . \tag{2.20}$$

Note that this loss function is assumed to be constant on the parameter regions defined by H_0 and A. Protection against falsely rejecting the null-hypothesis is obtained by choosing a/b large. For the one-sided testing problem, the UMP level-α test corresponds to the decision procedure which has uniformly minimum risk among the minimax risk procedures, provided that $\alpha = b/(a + b)$. For the two-sided problem, the same decision-theoretic property holds for the UMP unbiased level-α test. The loss function formulation attaches some interpretation to the level α, and thus provides a loose guide to its choice.

In a number of two-sided situations, however, the asymmetric two-decision formulation of hypothesis-testing theory is not particularly attractive. One would prefer an action space consisting of three decisions. Such a situation may arise, for example, in drug testing experiments when attention is focussed on 3 interesting states of nature: $\theta < -\theta_1$: drug A is better, $\theta > \theta_1$: drug B is better, $\theta \in [-\theta_0, +\theta_0]$: the two drugs are essentially equivalent (θ_0 is supposed to be smaller than, or equal to, θ_1). It arises also in certain decision problems about treating a particular patient. If we suppose that either treatment A or treatment B has to be given (treatment A may be a conservative treatment, which is applied anyhow, or simply consist in 'doing nothing'), then the decisions of the physician with respect to the patient are that, on the basis of the considered information,

$$a_1 \quad : \text{treatment B is recommended},$$

$$a_{-1} \quad : \text{treatment A is recommended},$$

$$a_0 \quad : \text{no recommendation is made}. \tag{2.19'}$$

In the absence of prior knowledge about the direction of the difference between the treatment effects, we are faced with a two-sided situation.

With respect to the second problem, about recommending one of two treatments to a particular patient, we shall, in the next sub-sections, investigate some fixed-sample, three-decision procedures in a setting which can be considered to complement the three-decision formulation studied in Schaafsma (1966, 1969). The whole is based on general theory of statistical decision functions, developed by Wald and others, which will be described first.

The need for a three-decision formulation of two-sided testing problems was already felt by Neyman and Pearson (1933). It is noted that for the first type of problem, a sequential analysis was carried out by Sobel and Wald (1949), see also Govindarajulu (1981). Such an approach may be useful, for example, in quality control or in short-term toxicity experiments.

2.3.1. Theory of statistical decision functions

To refresh the reader's memory we will review some elements of the theory of statistical decision functions. Of special interest to us are its applications to multiple decision problems, which are decision problems for which the space $\mathscr{A}$ of actions for the statistician is finite, $\mathscr{A} = \{a_1, \ldots, a_m\}$. The decisions are made with respect to some parameter $\theta \in \Omega \subset \mathbb{R}^p$. Hypothesis testing problems can be looked upon as multiple decision problems with $m = 2$. Note that in estimation theory, the action space will be infinite (for example equal to the parameter space).

Motivated by the theory of Von Neumann and Morgenstern (1944), Wald considered multiple decision problems as two-person zero-sum games, in which player I (Nature) 'chooses' the value of a parameter θ, and player II (the statistician) chooses an action a_j. For each θ and a_j, a loss $L(\theta, a_j)$ occurs to player II. (As we have a zero-sum game, the gain for player I is supposed to be precisely the loss for player II. See Rauhut, Schmitz and Zachow (1979), and Szép and Forgó (1983), among others, for more background on game theory.) The game is denoted by $(\Omega, \mathcal{A}, L)$. Of course, player II tries to minimise his losses and will usually base his action on the outcome of a statistical experiment, which is modelled by a random variable X taking values in some outcome space $\mathcal{X}$, whereas the probability distribution of X depends on $\theta \in \Omega$. (In our case we will be interested in $\mathcal{X} = \mathbb{R}$, the probability distribution being given by the density $f(x|\theta) = (2\pi)^{-1/2} e^{-1/2(x-\theta)^2}$.) A non-randomised decision rule is a function d: $\mathcal{X} \to \mathcal{A}$, and can be identified with a partition $(\mathcal{X}_1, \ldots, \mathcal{X}_m)$ of the outcome space: action a_j is taken if X falls in the subset $\mathcal{X}_j$. A so-called behavioural decision rule is a function φ: $\mathcal{X} \to \mathcal{A}^*$, where $\mathcal{A}^*$ is the set of probability distributions on $\mathcal{A}$. We use the notation $\varphi(x) = (\varphi_1(x), \ldots, \varphi_m(x))$: if x is the outcome of the experiment, then action a_j is chosen with probability $\varphi_j(x)$. Of course, $\sum_{j=1}^{m} \varphi_j(x) = 1$. For a non-randomised decision rule d, the risk function is defined by

$$R(\theta, d) = E_\theta L(\theta, d(X)) = \int_{\mathcal{X}} L(\theta, d(x)) f(x|\theta) \, dx \, . \tag{2.21}$$

Denoting the set of decision rules by D, we now have another zero-sum game, with 'imperfect spying', (Ω, D, R), which is of course related to the previous game. For any prior distribution g of the parameter θ, the Bayes risk is

$$R(g, d) = EE_\theta L(\theta, d(X)) = \int_\Omega \int_{\mathcal{X}} L(\theta, d(x)) f(x|\theta) g(\theta) \, dx \, d\theta \, . \tag{2.22}$$

Obviously, for a degenerate decision rule d_a, which chooses action a, regardless of the outcome of X, we have

$$R(g, d_a) = \int_\Omega L(\theta, a) g(\theta) \, d\theta \, . \tag{2.23}$$

REMARK. We do not want to exclude discrete prior probabilities. Therefore, in general, using Dirac's notation, g is a distribution of order zero (see, e.g., Schwartz, 1966; Roddier, 1978). As g is the derivative of a probability measure, one might, alternatively, use the notation $g(\theta) \, d\theta = d\tau(\theta)$. (If one wants to include discrete probabilities for the random variable X as well, a similar observation can be made for $f(x|\theta) \, dx$, which is sometimes notated as $dP_\theta(x)$.)

A rule d_g which minimises the Bayes risk, i.e. which satisfies

$$R(g, d_g) = \inf_d R(g, d) \, , \tag{2.24}$$

is called a Bayes rule with respect to the prior distribution g. A decision rule d^* is said to have minimax risk if

$$\sup_\theta R(\theta, d^*) = \inf_d \sup_\theta R(\theta, d) \qquad (= \bar{R}) \, . \tag{2.25}$$

A prior distribution $g*$ is said to be least favourable, if

$$\inf_d R(g*, d) = \sup_g \inf_d R(g, d) \qquad (=\underline{R}). \qquad (2.26)$$

Note that, for each $d \in D$, $\sup_g R(g, d) = \sup_\theta R(\theta, d)$, and that the maximin value $\underline{R}$ is always smaller than or equal to the minimax value $\overline{R}$. Under certain regularity conditions (see, e.g., Ferguson, 1967, Ch. 2; Schaafsma, 1971) $\underline{R} = \overline{R}$ and the game is said to have an equilibrium point. A decision rule d_0 is called an equalizer rule if $R(\theta, d_0)$ is constant for all values of θ. There are several relations between minimax rules, equalizer rules, Bayes rules and least favourable distributions. Briefly stated, under some weak conditions, minimax rules are Bayes with respect to a least favourable distribution, and under certain additional conditions, minimax rules are precisely those equalizer rules which are also Bayes.

Some more precise formulations are as follows (see, e.g., Ferguson, 1967):

(1) If the game (Ω, D, R) has an equilibrium point, then any minimax rule is extended Bayes (i.e. for every $\varepsilon > 0$ there exists a prior distribution g_ε such that $R(g_\varepsilon, d_0) \leqslant \inf_d R(g_\varepsilon, d) + \varepsilon$).

(2) An equalizer rule which is extended Bayes is also minimax.

(3) If the game has an equilibrium point, and there exists a least favourable distribution g_0, then every minimax rule is Bayes with respect to g_0.

(4) If the game has an equilibrium point, the parameter space consists of a finite number of points, and there exists a least favourable distribution which gives a positive weight to each point of the parameter space, then every minimax rule is an equalizer rule. (In fact, in that case the minimax rules are precisely those equalizer rules which are also Bayes.)

(5) If there exists a Bayes rule d_0 with respect to some distribution g_0, which satisfies $R(\theta, d_0) \leqslant R(g_0, d_0)$ for all θ, then the game has an equilibrium point, g_0 is least favourable, and d_0 is minimax.

A class of decision rules C is called essentially complete if for each d not in C there exists a d_0 in C, which is at least as good as d, i.e. $R(\theta, d_0) \leqslant R(\theta, d)$ for all $\theta \in \Omega$. If $\Omega = \{\theta_1, \ldots, \theta_q\}$, the 'risk set' is defined as the set of all vectors $(R(\theta_1, d), \ldots, R(\theta_q, d))$ in $\mathbb{R}^q$, where d runs through the decision space D. If, for finite Ω the risk set is closed from below and bounded from below, then the class of extended Bayes rules is essentially complete. Clearly, if an essentially complete class is identified, then all decision rules outside C can be disregarded, and the problem is reduced to choosing a particular element of C.

2.3.2. Application to the three-decision problem
For our decision problem 'treatment A is recommended, treatment B is recommended, no recommendation is made', the two-sided hypothesis testing formulation (2.19) and (2.20) is not attractive, because here we clearly have three decisions, and, in a symmetric situation, three (instead of two) types of error may occur. An error occurs if

(I) a recommendation is made, whereas no recommendation would be appropriate,

(II) no recommendation is made, whereas a recommendation would be appropriate,

(III) a wrong recommendation is made.

Following terminology of Schwartz et al. (1970), these are called errors of the first, second, and third kind; respectively. The losses corresponding to these errors will be denoted by a, b, and c. The last type of error is the most serious one. It will usually be reasonable to assume that, for our type of problem, $c > b > a > 0$.

In a concrete situation, we have to specify for which values of θ the various recommendations would be appropriate. The parameter space is divided into $\Omega_0 = \mathbb{R} \backslash (\Omega_{-1} \cup \Omega_1 \cup \Omega_2)$, $\Omega_{-1} = (-\infty, -\theta_1]$, $\Omega_1 = [\theta_1, +\infty)$, and $\Omega_2 = [-\theta_0, +\theta_0]$ for some small value θ_0 and larger value θ_1. (Please draw a figure.) If $\theta \in \Omega_{-1}$ (Ω_1), treatment A (B) should be recommended, and if $\theta \in \Omega_2$, it would be appropriate to make no recommendation. We denote these actions by a_{-1}, a_1, and a_0, respectively. The region Ω_0 is an indefiniteness zone, in which we do not have a clear opinion. For $\theta \in \Omega_0$ the loss function is undefined. Instead of (2.20), we now have the loss matrix

$L(\theta, d)$	a_{-1}	a_0	a_1
$\theta \in \Omega_{-1}$	0	b	c
$\theta \in \Omega_2$	a	0	a
$\theta \in \Omega_1$	c	b	0

At the beginning of a testing procedure, a joint specification has to be made of θ_0, θ_1, and the ratio $a : b : c$. (Note that the values of the parameter θ have an intuitively clear interpretation, such as the logarithm of the relative risk or of the ratio of the median effective lifetimes.)

REMARKS. (1) In Schaafsma (1966, 1969) and Ferguson (1967) a similar loss matrix formulation for the three-decision problem is used. (The reader is cautioned against a slightly different notation for the losses and a different terminology for the various types of error.) Considering the situation that $\Omega_2 = \{0\}$ and discarding the possibility that $\theta \in \Omega_2$, Schaafsma develops and evaluates various decision rules in the special case that $\Omega_0 = \emptyset$. We, however, consider the use of a non-empty indefiniteness zone and allowance for the situation that $\theta \in \Omega_2$ as a physically meaningful extension. Parts of Schaafma's results, and of the two-sided hypothesis-testing theory can formally be recovered by putting (θ_0 and θ_1 being 0) $a = 0$ and $c = 0$, respectively.

(2) It is, of course, somewhat unnatural to consider the loss to be constant for all θ in a whole region. This simplification is, however, also made in Schaafsma (1966, 1969) and in some other texts on the decision-theoretic foundation of the Neyman–Pearson theory. In our formulation, there is quite some flexibility in choosing appropriate values of θ_0 and θ_1. It may be shown that some classes of non-constant loss functions lead to the same least favourable distribution and the same minimax procedure, as in the simplified situation with constant losses.

(3) One may argue that, although we did not define them, there actually will be some losses associated with each decision if $\theta \in \Omega_0$. One possible physically meaningful interpretation of Ω_0 is that the proposed procedures will be the same for a fairly large class of loss functions on Ω_0 that satisfy the boundary conditions on θ_0 and θ_1 (and which do not behave too wildly). We will not attempt to find out such classes more precisely here.

In general, a (behavioural) decision rule for our problem is defined by a set $\{\varphi_j: \mathbb{R} \to [0, 1], j = -1, 0, 1, \sum_{j=-1}^{1} \varphi_j(x) = 1\}$ of three decision functions. They have the interpretation that action a_j is chosen with probability $\varphi_j(x)$ if x is the outcome of the experiment. If θ is the true state of nature, then the unconditional probability that action a_j will be chosen is $E_\theta(\varphi_j(X)) = \int_\mathbb{R} \varphi_j(x) f(x \mid \theta) \, dx$.

The set of all behavioural rules is rather large. It is convenient to restrict attention to a smaller, essentially complete, class of decision rules. This is done by applying a procedure from Karlin and Rubin (1956). As, in our case, $\Omega = \mathbb{R}$ and $c > b$ (while a, b and c are positive), we have a so-called monotone multiple decision problem. Furthermore, our class of normal distributions for the random variable X has monotone likelihood ratio. From these two properties it follows that attention may be restricted to the monotone decision rules, as these form an essentially complete class. By definition, for a monotone decison problem, a rule $\varphi = (\varphi_1, \ldots, \varphi_m)$ is called monotone if each $\varphi_j(x)$ is the indicator function of an interval $C_j = (x_{j-1}, x_j) \subset \mathbb{R}$, where $-\infty = x_0 \leqslant x_1 \leqslant \ldots \leqslant x_m = \infty$ is a non-decreasing sequence of points. (If the outcome of X happens to be x_{j-1}, which has probability zero in our case, then a randomisation between action a_{j-1} and a_j may be required.) So, in our case we may restrict attention to

$$\varphi_{-1}(x) = I_{\{(-\infty, c_0)\}}, \qquad \varphi_0(x) = I_{\{[c_0, c_1]\}}, \qquad \varphi_1(x) = I_{\{(c_1, \infty)\}},$$

$$(2.27)$$

with $c_1 > c_0$. The problem is symmetric under reflection with respect to the origin. Therefore, we further restrict attention to procedures of the form (2.27) with $-c_0 = c_1 = k > 0$. (See, e.g., Ferguson, 1967, Ch. 6 for a detailed invariance argument.) With a slight abuse of notation the risk function of such a procedure will be denoted by $R(\theta, k)$.

For $X \sim N(\theta, 1)$, we have

$$E_\theta \varphi_{-1} = P_\theta\{X < -k\} = \Phi(-\theta - k),$$

$$E_\theta \varphi_0 \ = P_\theta\{-k < X < k\} = 1 - \Phi(-\theta - k) - \Phi(\theta - k), \qquad (2.28)$$

$$E_\theta \varphi_1 \ = P_\theta\{X > k\} = \Phi(\theta - k),$$

where $E_\theta \varphi_j$ is a shorthand notation for $E_\theta \varphi_j(X)$. Note that, as $\sum_{j=-1}^{1} E_\theta \varphi_j = 1$, a vector $(E_\theta \varphi_{-1}, E_\theta \varphi_0, E_\theta \varphi_1)$ of decision probabilities can be represented as a point in a ternary diagram (see Figure 5). Each vector is determined by the value

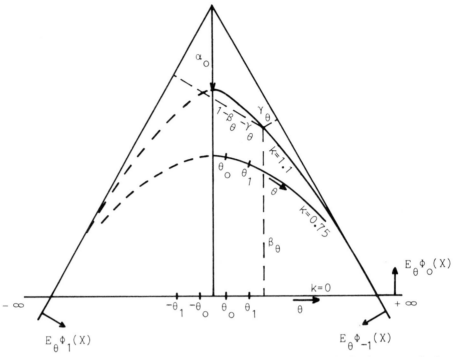

Fig. 5. A ternary diagram displaying properties of monotone decision rules for the symmetric three-decision problem associated with $X \sim N(\theta, 1)$. The distances of a point inside the triangle to the edges of the triangle correspond to $E_\theta \varphi_1(X)$, $E_\theta \varphi_{-1}(X)$, and $E_\theta \varphi_0(X)$, which are the probabilities of recommending treatment A, of recommending treatment B, and of making no recommendation, respectively. For a non-randomised decision rule based on a symmetric interval $[-k, k]$ we have, correspondingly, $P_\theta\{X > k\}$, $P_\theta\{X < -k\}$, and $P_\theta\{X \in [-k, k]\}$. The position of such a point is uniquely determined by the parameters θ and k. Each line of constant k corresponds to a particular decision procedure, and is symmetric with respect to $\theta = 0$. Note that for $|\theta| < \theta_0$, $1 - P_\theta\{X \in [-k, k]\} = \alpha_0$, whereas for $|\theta| > \theta_1$, $P_\theta\{X \in [-k, k]\} = \beta_0$, and $P_\theta\{X < -k\} = \gamma_0$. These are called the probabilities of error of the first, second, and third kind, respectively. As $X \sim N(\theta, 1)$,
$$\text{we have } k = \Phi^{-1}(1 - \tfrac{1}{2}\alpha_0).$$

of k and θ. For each fixed value of k, the set $\{(E_\theta \varphi_{-1}, E_\theta \varphi_0, E_\theta \varphi_1), \theta \in \Omega\}$, characterising a particular decision rule, constitutes a curve in the diagram. Notice the exceptional values $k = 0$, and $k = \infty$. Because of the assumed symmetry, we have, for all θ, $E_\theta \varphi_1 = E_{-\theta} \varphi_{-1}$, and $E_\theta \varphi_0 = E_{-\theta} \varphi_0$, so the whole plot is symmetric around the vertical axis. Another simple representation is given in Figure 6. Here, for some fixed k, $E_\theta \varphi_{-1}$, $E_\theta \varphi_0$, and $E_\theta \varphi_1$ are plotted as a function of θ. Of course, in our symmetric situation, the properties of a monotone decision rule are determined by only one function, e.g. $E_\theta \varphi_1 = \Phi(\theta - k)$, $\theta \in \mathbb{R}$, which is (for $\theta > \theta_1$) the power function of a one-sided test with critical value k. The problem is to find a suitable value for the cut-off point k, which minimises, according to some criterion, the expected losses. One can immediately see that the risk

prob.

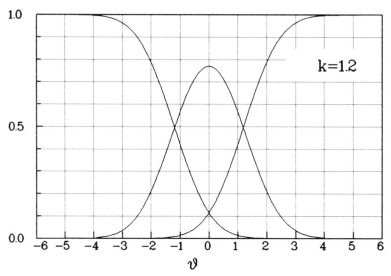

Fig. 6. The three-decision probabilities $P_\theta\{X > k\}$, $P_\theta\{X \in [-k, k]\}$ and $P_\theta\{X < -k\}$ as a function of θ, for X standard normal and for a fixed value of $k = \Phi^{-1}(1 - \alpha_0/2)$. This representation is alternative to that of Figure 5. Note the probabilities of the different types of error for various values of θ. Because of the symmetry, all information is contained in one of the two monotonic functions.

$$R(\theta, k) = \begin{cases} bE_\theta\varphi_0 + cE_\theta\varphi_1 & \text{if } \theta \in \Omega_{-1}, \\ a(E_\theta\varphi_1 + E_\theta\varphi_{-1}) & \text{if } \theta \in \Omega_2, \\ bE_\theta\varphi_0 + cE_\theta\varphi_{-1} & \text{if } \theta \in \Omega_1. \end{cases} \tag{2.29}$$

As $R(\theta, k) = R(-\theta, k)$ for all θ, one can restrict attention to positive values of θ. A minimax rule minimises $\sup_\theta R(\theta, k)$. It is easy to verify by differentiating the three decision function with respect to θ, that the suprema of the probabilities of incorrect decision occur at the boundaries of the parameter regions. For correspondence with the familiar testing theory, we call

$$\alpha_{\theta_1}(k) = \sup_{\theta \in \Omega_2} (E_\theta\varphi_1 + E_\theta\varphi_{-1}) = \Phi(-\theta_0 - k) + \Phi(\theta_0 - k),$$

$$\beta_{\theta_1}(k) = \sup_{\theta \in \Omega_1} E_\theta\varphi_0 = 1 - \Phi(-\theta_1 - k) - \Phi(\theta_1 - k), \tag{2.30}$$

$$\gamma_{\theta_1}(k) = \sup_{\theta \in \Omega_1} E_\theta\varphi_{-1} = \Phi(-\theta_1 - k)$$

the (maximum) probability of the error of the first, second, and third kind, respectively. (Note that these are easily identified in the ternary diagram.) If, instead of considering the whole risk function, we restrict attention to the maximum risks $\sup_{\theta \in \Omega_j} R(\theta, k)$ ($j = -1, 0, 1$), then (because of the assumed symmetry) we only have to look at the two parameter points θ_0 and θ_1. In this way, the parameter space is effectively reduced, which is a considerable simplification. (We might call our problem a 'quasi two-state, three-decision problem'.)

We now are in a position to express the critical values according to various optimality criteria in an easily interpretable form, liable to simple numerical evaluation. The *minimax rule* chooses the critical value k_{mm}, such that

$$\max\left(a\alpha_{\theta_0}(k), b\beta_{\theta_1}(k) + c\gamma_{\theta_1}(k)\right) \tag{2.31}$$

is minimised. The *equalizer rule* (for the reduced problem) determines a value k_{eq}, such that

$$a\alpha_{\theta_0}(k) = b\beta_{\theta_1}(k) + c\gamma_{\theta_1}(k). \tag{2.32}$$

A *Bayes rule* (for the reduced problem), which gives a prior weight q to state θ_0 (and a weight $1 - q$ to state θ_1), minimises

$$qa\alpha_{\theta_0}(k) + (1 - q)\{b\beta_{\theta_1}(k) + c\gamma_{\theta_1}(k)\}. \tag{2.33}$$

In particular, the *Bayes-1/2 rule* minimises $a\alpha_{\theta_0}(k) + b\beta_{\theta_1}(k) + c\gamma_{\theta_1}(k)$, which is a weighted sum of the maximum probabilities of the three types of error. The envelope risk $R^*(\theta)$ is defined as $\inf_d R(\theta, d)$. In our case, we can restrict attention to the rules $R(\theta, k)$, $k > 0$, which are given by (2.29). Hence, it is easily shown that, while $R^*(\theta) = R^*(-\theta)$ for all θ, $R^*(\theta) = 0$ for $\theta \in \Omega_0$ and $R^*(\theta) = b\Phi(d/\theta - \theta) + (c - b)\Phi(-d/\theta - \theta)$ for $\theta \in \Omega_1$ with $d = 0.5 \ln((c/b) - 1)$. The *minimax regret rule* chooses a critical value k_{mmr} such that

$$\max_\theta(R(\theta, k) - R^*(\theta)) \tag{2.3.4}$$

is minimised. The minimax rule seems attractive in a number of situations (at least if θ_1 is not too small: in the special situation $\theta_0 = \theta_1 = 0$, the rule gives, for $c > a + b$, the degenerate result 'never make a recommendation'; in that case the minimax regret rule will be preferred). A comparison between the minimax and the minimax regret rule in a specified situation is presented in example (2) of Section 2.3.4. For the minimax rule to be reasonably adequate, it seems advisable to choose the value of θ_1 such that θ_1 is either of the same order of magnitude (say, within a factor 2 or 3) as $\sigma_\theta/\sqrt{n}$, or near the true value of θ, and to specify $a : b : c$ ($a < b < c$) after the choice of θ_1.

For $a : b : c = 1 : 3 : 9$ the critical value k_{mm} of the minimax rule can be read off from Figure 7, which displays k_{mm} as a function of θ_1 for various values of $r_0 = \theta_0/\theta_1$. Note the logarithmic scale of the axes. The numbers indexing the various graphs represent the values of $^2\log r_0$. The plot is drawn for $X \sim N(\theta, 1)$. (When the plot is applied to $X' \sim N(\theta, \sigma_\theta^2/n)$, it is to be understood that θ_0, θ_1, and the outcome of X' are measured in units of $\sigma_\theta/\sqrt{n}$.) Only results for $\theta_0 \leqslant \theta_1$ are meaningful.

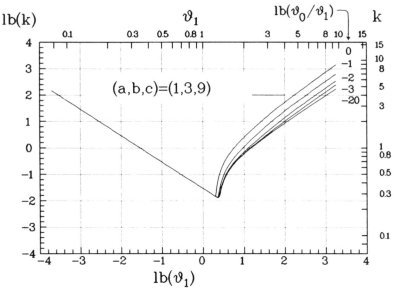

Fig. 7. Critical values $k = \Phi^{-1}(1 - \frac{1}{2}\alpha_0)$ as a function of θ_1 for various values of $r_0 = \theta_0/\theta_1$, according to the minimax rule for the three-decision problem associated with $X \sim N(\theta, 1)$. Notice the logarithmic scale for θ_0 and θ_1. The values of $\mathrm{lb}\,r_0 = {}^2\log r_0$ are plotted to the right of the curves. (Only $r_0 \in [0, 1]$ leads to a physically meaningful interpretation of k.) If $\theta_1 \leqslant \tilde{\theta}_1 \simeq 1.2$, then $k = d/\theta_1 \simeq 0.35/\theta_1$ for all $\theta_0 \leqslant \theta_1$. If $\theta_1 \to 0$ or $\theta_1 \to \infty$, then $k \to \infty$.

As one can see, for θ_1 smaller than a certain value $\tilde{\theta}_1$, the critical value k_{mm} does not depend on θ_0 (for all $\theta_0 \leqslant \theta_1$), while k_{mm} decreases as θ_1 increases (linearly, on the logarithmic scale). In this region, the minimax rule minimises $b\beta_{\theta_1}(k) + c\gamma_{\theta_1}(k)$, which is the second term in (2.31).

[A short analytic elaboration is as follows: By differentiation one can show that $b\beta_{\theta_1}(k) + c\gamma_{\theta_1}(k)$ is minimised (for $c > b$) by $k_{\mathrm{mm}2} = d/\theta_1$, with $d = 0.5 \ln((c/b) - 1)$. This minimising value $k_{\mathrm{mm}2}$ (being positive for $c > 2b$) is for all $\theta_0 \leqslant \theta_1$ equal to the minimax value k_{mm} if, for all $\theta_0 \leqslant \theta_1$, $a\alpha_{\theta_0}(k_{\mathrm{mm}2}) \leqslant b\beta_{\theta_1}(k_{\mathrm{mm}2}) + c\gamma_{\theta_1}(k_{\mathrm{mm}2})$. As $a\alpha_{\theta_0}(k_{\mathrm{mm}2})$ is an increasing function of θ_0, this is equivalent to the condition $(b - a)\Phi(d/\theta_1 - \theta_1) + (c - b - a)\Phi(-d/\theta_1 - \theta_1) \geqslant a$, which means, at least for the common situation $b > a$, $c > 2b$, and $c > (a + b)$, that θ_1 is not greater than the above defined boundary value $\tilde{\theta}_1$, where $\tilde{\theta}_1 = \tilde{\theta}_1(a, b, c)$ can be found by numerical means.

More generally, if θ_1 is smaller than some function $\bar{\theta}_1(r_0)$, $r_0 \in [0, 1]$, then k_{mm} does not depend on θ_0 for all $\theta_0 \leqslant r_0\theta_1$, while k_{mm} decreases as θ_1 increases. In this situation, the minimax rule minimises $b\beta_{\theta_1}(k) + c\gamma_{\theta_1}(k)$. The function $\bar{\theta}_1(r_0)$ can be determined numerically. Note that $\tilde{\theta}_1 = \bar{\theta}_1(1)$.

If $\theta_1 \geqslant \bar{\theta}_1(r_0)$ and $\theta_0 \geqslant r_0\theta_1$, then the minimax critical value k_{mm} equals k_{eq}. This can be inferred from the facts that (1) if $k_{\mathrm{mm}2}$ minimises $b\beta_{\theta_1}(k) + c\gamma_{\theta_1}(k)$, then $a\alpha_{\theta_0}(k_{\mathrm{mm}2})$ is now at least $b\beta_{\theta_1}(k_{\mathrm{mm}2}) + c\gamma_{\theta_1}(k_{\mathrm{mm}2})$, and (2) $b\beta_{\theta_1}(k) + c\gamma_{\theta_1}$ tends to $b > 0$, whereas $a\alpha_{\theta_0}(k)$ is continuously decreasing to 0 as $k \to \infty$.]

The q-values (as a function of θ_0 and θ_1) such that a Bayes-q rule coincides with the minimax rule are shown in Figure 8 ($a:b:c = 1:3:9$). Note that if $q \in (0, 1)$, the equalizer rule coincides with the minimax rule, whereas for $q = 0$ the minimax rule is determined by the dominating term $b\beta_{\theta_1}(k) + c\gamma_{\theta_1}(k)$.

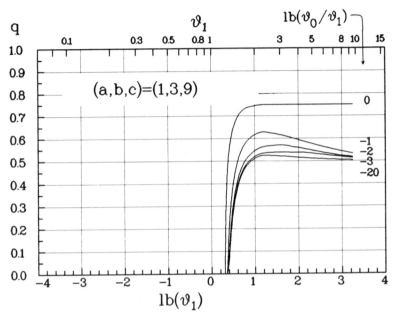

Fig. 8. Prior weights of the Bayes rule that corresponds to the minimax rule. The quantity q is plotted as a function of θ_1 for various values of $r_0 = \theta_0/\theta_1$. This quantity is defined such that the Bayes rule minimising (as a function of k) $q a\alpha_{\theta_0}(k) + (1 - q)(b\beta_{\theta_1}(k) + c\gamma_{\theta_1}(k))$ corresponds to the minimax rule, which minimises $\max\{a\alpha_{\theta_0}(k), b\beta_{\theta_1}(k) + c\gamma_{\theta_1}(k)\}$. If $q = 0$, then the minimax rule minimises $b\beta_{\theta_1}(k) + c\gamma_{\theta_1}(k)$. If $q \in (0, 1)$, then the minimax rule is an equalizer rule (which determines k such that $a\alpha_{\theta_0}(k) = b\beta_{\theta_1}(k) + c\gamma_{\theta_1}(k)$).

2.3.3. Conclusion

In summary, we state that for the three-decision problem considered in this section (on the basis of a one-dimensional estimate of treatment difference, it is to be decided whether a_{-1}: treatment A is recommended; a_1: treatment B is recommended, or a_0: no recommendation is made) the two-sided testing formulation is inadequate. A three-decision formulation, such as the one described above, is to be preferred. The associated procedure is briefly described as follows: Apply twice a one-sided size-$\alpha/2$ test (H_0: $\theta = 0$ against A_1: $\theta > 0$, and against A_{-1}: $\theta < 0$, respectively), where α is determined by the loss specification and the chosen criterion of optimality. Applying the minimax rule, which is sensible if θ_1 is not too small, α or, equivalently, the cut-off point $k = -\Phi^{-1}(\alpha/2)$, can graphically be found from a plot as in Figure 7. If one wants to consider the situation

$\theta_0 = \theta_1 = 0$ (or θ_1 very small), in some instances a minimax regret procedure, as described in Schaafsma (1966, 1969), may be preferable. Three types of error may occur. The associated probabilities, as a function of θ, can easily be read off from a graph of the 'power function' of such a one-sided test. (If, for example, H_0 is tested against A_1, the 'power function', being defined as the probability of rejecting H_0 as a function of θ for all $\theta \in R \backslash \{0\}$, is $\Phi(\theta - k)$.)

It is remarked that the above described choice of α, induced by the three-decision nature of the problem, may perhaps also be applied to the comparison of the (effective-) lifetime distributions on the basis of simultaneous confidence bands, as described in Section 2.2. (Note that the particular bands used are of a 'proportional hazards' type, which fixes the scale for θ_0 and θ_1.) A further investigation would be needed to cover this more complicated case. A band comparison of the lifetime distributions may be appropriate if e.g. $\hat{S}_{Y,A} < \hat{S}_{Y,B}$, while a proportional hazards (or an accelerated lifetime) assumption between $S_{Y,A}$ and $S_{Y,B}$ does not seem appropriate.

2.3.4. Discussion of two examples

(1) Consider $a:b:c = 1:3:9$, $\theta_0 = 0.05$, and $\theta_1 = 0.7$. This means that a relative risk of $e^{0.05} = 1.05$ between the two treatments is considered as negligible, whereas a relative risk of $e^{0.70} \simeq 2$ is looked upon as a serious one, which should not be overlooked. (In fact, if the hazard rate associated with treatment A is consistently a factor 2 higher than that of treatment B, then advising the wrong treatment is considered to give a loss which is 9 times higher than the loss incurred if the two treatments have essentially the same effect on the (effective) lifetime and one of the two is explicitly recommended.) Suppose that, by applying Cox's model to a particular data set, an estimate $\hat{\theta} = 1.5$ is obtained for the regression coefficient measuring (in the presence of other covariates) the effect of treatment difference, with a standard error of 1.25. (The estimate $\hat{\theta}$ corresponds roughly to the effect of hypertension treatment on the distribution of the original lifetimes in the IgAN case study, see Chapter 14, Figure 2.) Measuring all effects in units of the standard error of $\hat{\theta}$, we get $\hat{\theta}_{st} = 1.5/1.25 = 1.17$, $\theta_{0,st} = 0.04$, and $\theta_{1,st} = 0.56$. From Figure 7 we see that if $\theta_{1,st}$ is slightly larger than $\frac{1}{2}$, $k \simeq 0.65$, irrespective of the value of $\theta_{0,st}$. (In this region, of course, the expression $k = d/\theta_{1,st}$ holds, with $d = 0.5 \ln((c/b) - 1) = 0.35$, which gives the more precise value $k = 0.63$.) As $1.17 > k \simeq 0.65$, according to the minimax rule, recommendation of the treatment with the lowest hazard rate would be appropriate. (Note that, according to the classical (one- or two-sided) level-α test, the estimated treatment difference is far from being statistically significant at 5%.)

(2) The second example is meant to expose some weaknesses of the minimax risk rule if θ_1 has to be chosen such that $\theta_{1,st} = \theta_1(/SE(\hat{\theta})$ is substantially smaller than 1. Consider the same situation as in the previous example, but now with $\theta_0 = \theta_1 = 5\%$. In units of the standard error, we have $\theta_{1,st} = 0.04$. Consequently, $k = 0.35/0.04 = 8.75$. In this situation, according to the minimax rule, no recommendation should be made.

This result is not in contradiction with the previous example. If effects that are an order of magnitude smaller than the noise (standard error of $\hat{\theta}$) are deemed important, and must not be overlooked, then an error of the third kind is readily made, and one has to be very cautious in making a recommendation. In fact, in this situation one is reluctant to make a recommendation, because the available sample is uninformative about effects of such a small magnitude (in comparison with the accuracy of measurement).

[If on the other hand θ_1 larger than, say 2 or 3 times the standard error of $\hat{\theta}$, then, if no recommendation is made, the difference in effect between the treatments is likely to be small. The distinction between these two types of 'no recommendation' is made explicitly in a scientifically oriented approach, alluded to in the beginning of this sub-section. See, e.g., Sobel and Wald (1949) for a sequential analysis, in which, at each stage, a choice between four decisions is made. The experiment is stopped when there is sufficient evidence that 'A is better than B', 'B is better than A', or 'A and B are equivalent', and sampling is continued as long as 'evidence is not yet sufficient'. In a pragmatic approach, the two types of 'no recommendation' are mixed up in the final decision. Nevertheless, large values of k in this region are provoked by small values of $\theta_1/\text{SE}(\hat{\theta})$, and consequently are an indication that the available sample is uninformative in discriminating between two treatments with a relative risk $\text{RR} \approx \exp(\theta_1)$, which, at the onset, was considered an important one.]

One may be bothered, however, about the extremely large value $k = 8.75$ following from the minimax risk principle. For this value of k, hardly ever a definite recommendation is made. As an (objectivistic) alternative to the minimax rule, we consider the minimax regret rule, developed in Schaafsma (1966, 1969) for a general class of problems with restricted alternatives. Applied to our one-dimensional situation, this procedure gives for $b:c = 3:9$ (and $\theta_0 = \theta_1 = 0$) a cut-off value $k_2 \approx 1.0$. For illustration, in Figure 9, the risk functions of the minimax risk and of the minimax regret procedure (corresponding to $k_1 = 8.75$ and $k_2 = 1.0$, respectively) are plotted as a function of θ, together with the envelope risk with respect to all symmetric interval procedures. [It is not easy to find a scale which is satisfying in every respect. The relative risk $\text{RR} = \exp(\theta)$ is a familiar concept with a clear interpretation. The problem is how to compare two relative risks throughout the entire 'spectrum' of small and large relative risks. The present scale $Q = {}^2\log(\text{RR} - 1)$ is chosen in such a way that 'RR_1 is twice as serious as RR_2' is, for large as well as for small values of the relative risk, approximately represented by the same horizontal distance. The transformation from Q to RR and to θ is simple, and can be seen from the diagram.] In our essentially complete class of decision rules based on a symmetric interval, there exists no uniformly best procedure. The minimax regret procedure has a higher risk if $\theta \in (0.04, 0.25)$, but a lower risk in a large region of the parameter space ($\theta > 0.25$) consisting of relatively large values of θ. Which of the two procedures should be preferred depends, among others, on vague prior information about the order of magnitude of the effects. (Such vague information is often available, in which case there is no reason not to use it. Not being subjectivists, we do not

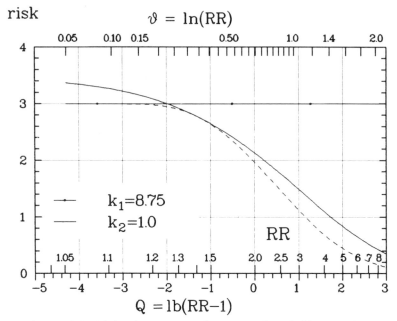

Fig. 9. Comparison of the risk functions corresponding to some three-decision procedures associated with $X \sim N(\theta, 1)$ and based on a symmetric interval $[-k, k]$. The broken line represents the envelope risk $R^*(\theta) = \min_k R(\theta, k)$. The risk function $R(\theta, k_1)$ corresponds to the minimax risk rule (with $\theta_1 = 0.05$, and $\theta_0 \leqslant \theta_1$), and $R(\theta, k_2)$ to the minimax regret rule. The last procedure minimises $\max_{\theta \geqslant 0.5}\{R(\theta, k) - R^*(\theta)\}$. The relative risk between the two treatments is denoted by $RR = \exp(\theta)$. The scale $Q = {}^2\log(RR - 1)$ is chosen such that if a relative risk is 'twice as serious' as another relative risk, then the corresponding distance on the plot is approximately one unit, for small as well as for large values of RR.

like to specify a precise prior distribution for θ. Usually, we do not need to do so.) Other considerations in choosing between minimax risk and minimax regret may be the following. In contrast to the minimax risk procedure, the minimax regret procedure is robust against misspecification of the losses occurring at $\theta = \theta_1$. On the other hand, reasonable loss functions for the errors of type 2 and type 3 will, in our problem, be strictly increasing for $\theta > \theta_1$. Constancy of the loss functions was only assumed as a mathematically convenient simplification. Provided the loss function is correctly specified at $\theta = \theta_1$, the minimax risk procedure will, in contrast to the minimax regret procedure, be robust against (slightly) increasing loss functions.

Acknowledgement

Apart from some minor corrections and an update of the references, this chapter corresponds to Chapter III of the author's Thesis at the University of Groningen. He is especially indebted to Prof. Dr W. Schaafsma for his inspiration and guidance throughout the study, to Prof. Dr A. J. Stam and Professor Dr B. L. J. Braaksma for valuable comments on the section on confidence bands, to Prof. Dr E. G. F. Thomas for his comments on the properties of the Volterra operator, and to Dr H. J. Buurema for discussions on numerical analysis.

References

Andersen, R. S., De Hoog, F. R. and Weiss, R. (1973). On the numerical solution of Brownian motion processes. *Journal of Applied Probability* **10**, 409–418.

Andersen, P. K. and Gill, R. D. (1982). Cox's regression model for counting processes: a large sample study. *The Annals of Statistics* **10**, 1100–1121.

Andersen, P. K. and Vaeth, M. (1989). Simple parametric and nonparametric models for excess and relative mortality. *Biometrics* **45**, 523–535.

Arjas, E. and Haara, P. (1988). A note on the asymptotic normality in the Cox regression model. *The Annals of Statistics* **8**, 1133–1140.

Arnold, L. (1974). *Stochastic Differential Equations: Theory and Applications.* Wiley, New York, London.

Bailey, K. R. (1979). *The general maximum likelihood approach to the Cox regression model.* Ph.D. dissertation, University of Chicago, Chicago, IL.

Bailey, K. R. (1983). The asymptotic joint distribution of regression and survival estimates in the Cox regression model. *The Annals of Statistics* **11**, 39–48.

Baker, C. T. H. (1981). An introduction to the numerical treatment of Volterra and Abel-type integral equations. In: *Lecture Notes in Mathematics, Vol 965*, 1–38.

Bismut, J. M. (1985). Transformations différentiables du mouvement Brownien. In: *Colloque en l'Honneur de Laurent Schwartz, Vol 1, Astérique* **131**, Société Mathématique de France, Paris.

Blake, I. F. and Lindsey, W. C. (1973). Level-crossing problems for random processes, *IEEE Transactions on Information Theory* **IT-19**, 295–316.

Bourbaki, N. (1967). *Théories Spectrales.* Éléments de mathématique, fascicule XXXII, actualités scientifiques et industrielles 1332, Hermann, Paris.

Bourbaki, N. (1973). *Espaces Vectorielles Topologiques.* Éléments de mathématique, fascicule XVIII, actualités scientifiques et industrielles 1229 (nouveau tirage), Hermann, Paris.

Breslow, N. E. (1979). Statistical methods for censored survival data. *Environmental Health Perspectives* **32**, 181–192.

Brillinger, D. R. (1986). The natural variability of vital rates and associated statistics (with discussion). *Biometrics* **42**, 693–734.

Brunner, H. (1982). A survey of recent advances in the numerical treatment of Volterra integral and integro-differential equations. *Journal of Computational and Applied Mathematics* **8**, 213–229.

Brunner, H. (1990). On the numerical solution of non-linear Volterra-Fredholm integral equations by collocation methods. *Siam Journal on Numerical Analysis* **27**, 987–1000.

Cox, D. R. and Oakes, D. (1984). *Analysis of Survival Data.* Monographs on statistics and applied probability, Chapman and Hall, London.

Curtain, R. F. and Pritchard, A. J. (1977). *Functional Analysis in Modern Applied Mathematics.* Mathematics in Science and Engineering, Vol. 132, Academic Press, London.

Dabrowska, D. M., Doksum, K. A. and Song J. K. (1989). Graphical comparison of cumulative hazards for two populations. *Biometrika* **76**, 763–773.

Delves, L. M. and Walsh, J., ed. (1974). *Numerical Solution of Integral Equations.* Clarendon Press, Oxford.

Durbin, J. (1971). Boundary-crossing probabilities for the Brownian motion and Poisson processes and techniques for computing the power of the Kolmogorov–Smirnov test. *Journal of Applied Probability* **8**, 431–453.

Durbin, J. (1985). The first passage density of a continuous gaussian process to a general boundary. *Journal of Applied Probability* **22**, 99–122.

Efron, B. and Johnstone, I. M. (1990). Fisher's information in terms of the hazard rate. *The Annals of Statistics* **18**, 38–62.

Eggermont, P. P. B. (1984). Stability and robustness of collocation methods for Abel-type integral equations, *Numerische Mathematik* **45**, 432–447.

Etezadi-Amoli, J. and Ciampi, A. (1987). Extended hazard regression for censored survival data with covariates: A spline approximation for the baseline hazard function. *Biometrics* **43**, 181–192.

Ferebee, B. (1982). The tangent approximation to one-sided Brownian exit densities. *Zeitschrift für Wahrscheinlichkeitstheorie und verwandte Gebiete* **61**, 309–327.

Ferguson, T. S. (1967). *Mathematical Statistics: A Decision Theoretic Approach*. Academic Press, New York.

Folland, G. B. (1983). *Lectures on Partial Differential Equations*. Springer-Verlag, New York/Heidelberg.

Friedrichs, K. O. (1973). *Spectral Theory of Operators in Hilbert Space*. Applied Mathematical Sciences **9**, Springer-Verlag, New York/Heidelberg.

Gill, P. E., Murray, W. and Wright, M. H. (1981). *Practical Optimization*. Academic Press, London/New York.

Gill, R. D. (1984). Understanding Cox's regression model. *Journal of the American Statistical Association* **79**, 441–447.

Govindarajulu, Z. (1981). *The sequential Statistical Analysis of Hypothesis Testing, Point and Interval Estimation, and Decision Theory*. American Sciences Press, Columbus, OH.

Grossmann, S. (1988). *Funktionalanalysis (im Hinblick auf Anwendungen in der Physik)* (4th ed.). AULA-Verlag, Wiesbaden.

Gupta, S. S. and Huang, D. Y. (1980). *Multiple Statistical Decision Theory: Recent Developments*. Lecture Notes in Statistics **6**, Springer-Verlag, New York/Heidelberg.

Hall, P. (1988). Theoretical comparison of bootstrap confidence intervals (with discussion). *The Annals of Statistics* **16**, 925–981.

Halmos, P. R. and Sunder, V. S. (1978). *Bounded Integral Operators on L^2 Spaces*. Ergebnisse der Mathematik und ihre Grenzgebiete **96**, Springer-Verlag, New York/Heidelberg.

Hildebrand, F. B. (1974). *Introduction to Numerical Analysis* (2nd ed.), Constable and Company, London. (Dover edition: 1987, New York.)

Intriligator, M. D. (1973). *Optimización Matemática y Teoria Economica*. Prentice-Hall, Bogotá, Madrid.

Jacobsen, M. (1982). *Statistical Analysis of Counting Processes*. Lecture Notes in Statistics **12**, Springer-Verlag, New York/Heidelberg.

Jacobsen, M. (1990). Right censoring and martingale methods for failure time data. *The Annals of Statistics* **17**, 1133–1156.

Jennen, C. and Lerche, H. R. (1981). First exit densities of Brownian motion through one-sided moving boundaries, *Zeitschrift für Wahrscheinlichkeitstheorie und verwandte Gebiete* **55**, 133–148.

Jörgens, K. (1970). *Lineaire Integraloperatoren*. Teubner, Stuttgart.

Jørgensen, B. (1981). *Statistical Properties of the Generalized Inverse Gaussian Distribution*. Lecture Notes in Statistics **9**, Springer-Verlag, New York/Heidelberg.

Kalbfleisch, J. D. and Prentice, R. L. (1980). *The Statistical Analysis of Failure Time Data*. Wiley, New York.

Kamakura, T. and Yanagimoto, T. (1983). Evaluation of the regression parameter estimators in the proportional hazards model. *Biometrika* **70**, 530–532.

Kardaun, O. J. W. F. (1983). Statistical survival analysis of male larynx-cancer patients—A case study. *Statistica Neerlandica* **37**, 103–125.

Kardaun, O. J. W. F. (1986). *On Statistical Survival Analysis and its Applications in Medicine*. Thesis, University of Groningen.

Kardaun, O. J. W. F. (1991). Kidney-survival analysis of IgA-nephropathy patients: A case study. In: *Statistical Methods in Biological and Medical Sciences*, Ch. 14, Handbook of Statistics **8**, North-Holland, Amsterdam, (*this volume*), pp. 407–458.

Karlin, S. and Rubin, D. B. (1956). The theory of decision procedures for distributions with monotone likelihood ratio. *The Annals of Mathematical Statistics* **27**, 272–229.

Kolmogorov, A. and Fomine, S. (1974). *Éléments de la Théorie des Fonctions et de l'Analyse Fonctionelle*. Éditions Mir, Moscou (traduction française).

Lehmann, E. L. (1957). A theory of some multiple decision problems I, II. *The Annals of Mathematical Statistics* **28**, 1–25, 547–572.

Lehmann, E. L. (1986). *Testing Statistical Hypotheses* (2nd ed.). Wiley, New York.

Ljusternik, L. A. and Sobolew, W. I. (1960). *Elemente der Funktionalanalysis*. Akademie-Verlag, Berlin.

Mehr, C. B. and McFadden, J. A. (1965). Certain properties of Gaussian processes and their first passage times. *Journal of the Royal Statistical Society* **8**, 505–522.

Miller, R. G., Jr. (1980). *Simultaneous Statistical Inferende* (2nd ed.). Springer Series in Statistics, Springer-Verlag, New York/Heidelberg.

Miller, R. K. (1971). *Nonlinear Volterra Integral Equations*. Mathematics Lecture Note Series. W. A. Benjamin Inc., Philippines/Menlo Park, California.

Moller, R. A. (1988). On the exponential model for survival. *Biometrika* **75**, 582–586.

Nelson, E. (1967). *Dynamical Theories of Brownian Motion*. Princeton University Press, Princeton.

Neyman, J. and Pearson, E. S. (1933). The theory of statistical hypotheses in relation to probabilities a priori. *Proceedings of the Cambridge Philosophical Society* **29**, 492–510.

Pollard, D. (1984). *Convergence of Stochastic Processes*. Springer-Verlag, New York/Heidelberg.

Prentice, R. L. and Mason, M. W. (1986). On the application of linear relative risk regression models. *Biometrics* **45**, 523–535.

Rauhut, B., Schmitz, N. and Zachow, E. W. (1979). *Spieltheorie*. Teubner, Stuttgart.

Te Riele, H. J. J. (1982). Collocation methods for weakly singular second kind Volterra integral equations with non-smooth solution. *IMA Journal of Numerical Analysis* **2**, 437–449.

Te Riele, H. J. J. and Schoevers, Ph. (1983). A comparative survey of numerical methods for the linear generalized Abel integral equation. Report 155/83, Mathematical Centre, Amsterdam.

Riesz, F. and Sz.-Nagy, B. (1952). *Leçons d'Analyse Fonctionelle*. Académie des Sciences de Hongrie, Budapest.

Robbins, H. and Siegmund, D. (1969). Probability distributions related to the law of the iterated logarithm. *Proceedings of the National Academy of Science* **62**, 11–14.

Roddier, F. (1978). *Distributions et Transformations de Fourier*. McGraw-Hill, Paris.

Sancho-Garnier, H., Benhamou, E., Avril, M.-F. et al. (1984). An example of a pragmatic trial in the treatment of basocellular carcinoma of the skin. *Revue d'Epidémiologie et de Santé Publique* **32**, 249–253.

Schaafsma, W. (1966). *Hypothesis Testing Problems with the Alternative Restricted by a Number of Inequalities*. Noordhoff, Groningen.

Schaafsma, W. (1969). Minimax risk and unbiasedness for multiple decision problems of type I. *The Annals of Mathematical Statistics* **40**, 1648–1720.

Schaafsma, W. (1971). The Neyman–Pearson theory of testing statistical hypotheses. *Statistica Neerlandica* **25**, *1*, 1–28.

Schumacher, M. (1982). *Analyse von Überlebenszeiten bei nichtproportionalen Hazardfunktionen*. Habilitationsschrift. Fakultät für Theoretische Medizin, Heidelberg.

Schwartz, L. (1966). *Théorie des Distributions*. Hermann, Paris.

Schwartz, D., Flamant, R. and Lellouch, J. (1980). *Clinical Trials*. Academic Press, London, New York.

Self, S. G. and Prentice, R. L. (1982). Commentary on Andersen and Gill's "Cox's regression model for counting processes: A large sample study". *The Annals of Statistics* **10**, 1121–1124.

Siegmund, D. (1986). Boundary crossing probabilities and statistical applications. *The Annals of Statistics* **14**. 361–404.

Sobel, M. and Wald, A. (1949). A sequential decision procedure for choosing one of three decision hypotheses concerning the unknown mean of a normal distribution. *The Annals of Mathematical Statistics* **20**, 502–523.

Solomon, P. J. (1984). Effect of misspecification of regression models in the analysis of survival data. *Biometrika* **71**, 291–298.

Stoutmyer, D. R. (1977). Analytically solving integral equations by using computer algebra. *ACM Transactions on Mathematical Software* **3**, 128–146.

Stratonovich, R. L. (1967). *Topics in the Theory of Random Noise, Vols. I & II.* Gordon and Breach, New York/London.

Szép, J. and Forgó, F. (1983). *Einführung in die Spieltheorie.* Verlag Harri Deutsch, Thun/Frankfurt am Main.

Tricomi, F. G. (1985). *Integral Equations.* Dover Publications, New York.

Von Neumann, J. and Morgenstern, O. (1944). *Theory of Games and Economic Behaviour.* Princeton University Press, Princeton.

Wald, A. (1947). *Sequential Analysis.* Wiley, New York (Dover edition: 1973, New York).

Wald, A. (1950). *Statistical Decision Functions.* Wiley, New York.

Wald, A. (1955). *Selected Papers in Statistics and Probability.* Edited by T. W. Anderson et al., McGraw-Hill, New York/London.

Zabreyko, P. P., Koshelev, A. I., Krasnosel'skii, M. A., Mikhlin, S. G. et al. (1975). *Integral Equations—A Reference Text.* Noordhoff International Publishing, Leiden.

Zucker, D. M. and Karr, A. F. (1990). Nonparametric survival analysis with time-dependent covariate effects: a penalized partial likelihood approach. *The Annals of Statistics* **18**, 329–353.

C. R. Rao and R. Chakraborty, eds., *Handbook of Statistics, Vol. 8*
© Elsevier Science Publishers B.V. (1991) 515–538

16

Sample Size Determination in Clinical Research

Jürgen Bock and H. Toutenburg

1. Introduction

The statistical tools used in medical research are mainly tests for the comparison of means and response or failure rates, confidence intervals, survival analysis methods and correlation and regression techniques. Multidimensional methods like discriminant analysis, factor analysis, multidimensional analysis of variance, cluster analysis, multidimensional scaling are infrequently applied in clinical research and there exist only few results dealing with power. Consequently these methods will not be considered here.

Although economic and logistic requirements and limitations play an important role in the design and planning of clinical trials, the primary reasons for using appropriate sample sizes are ethical. It is ethical to expose only a small number of patients or volunteers to a new drug in the early stage of development (phase I: first experiments in man) and to increase that number with growing knowledge of the safety and efficacy of the drug.

On the other hand false negative results can lead to a no-go decision in the drug development process and thus deprive patients of effective treatment. Freimann et al. (1978) reviewed the power of 71 published randomized clinical trials which had failed to detect a significant difference between groups and found that 50 could have missed a 50% therapeutic improvement because of insufficient sample size.

Nowadays many regulatory authority guidelines require power calculations for clinical trials. This coincides with the philosophy of successful pharmaceutical companies, because long-term marketing success can only be achieved with an effective and safe drug. Sample size calculations support effective drug development and research planning.

Sample size formulas have been derived for a broad area of applications (see e.g. Mace, 1964; Cohen, 1969; Rasch et al., 1978, 1980). Therefore one might ask whether it is necessary to reconsider sample size problems in the framework of medical applications. What are the special difficulties peculiar to clinical research?

There are a number of issues in clinical research which critically influence the determination of sample size. Important among these are the facts that patients are not recruited simultaneously, are not all given the same duration of therapy and both the recruitment and therapy can be of considerable duration. In some studies a subsequent follow-up period is added to the observation period with or without continuation of treatment for safety reasons. Observations may be incomplete at the time of termination of the study.

Patients have the right to decide to discontinue therapy or may withdraw for other reasons [withdrawal] (e.g. patients must be withdrawn in the case of severe adverse events). They may change treatments [drop out (from one treatment), drop in (into another treatment)] without discontinuing to participate in the study or may not comply with the trial protocol (drug administration, diet, excercises, follow-up assessments) [non-complier, lost to follow up]. Death may occur. Patients may not be evaluable (e.g. because a specific pathogen could not be found, an assessment has not been conducted inside the foreseen time window, baseline values are missing etc.).

In clinical trials with the time to a critical event (survival time) as endpoint, the observed times may be censored. This may be because the critical event did not occur until the end of the observational or follow-up period or because of other reasons (hopefully not related to the treatment or stage of disease).

Needless to say, special attention has to be paid to these problems in the development of sample size formulas for long term clinical trials.

Another point is that simple sample size formulas must be extended to multi-center study designs. This can easily be done for continuous response variables, but one is confronted with some non-realistic assumptions (e.g. about equal odds ratios) in the case of binary response.

Unlike other application fields such as agriculture, economics or natural sciences, most clinical trials have either binary outcomes or time to a critical event as response variable. Special models (logistic regression, proportional hazards) have been developed to take into account the influence of concomitant variables. The response may depend on the cumulative dose over time or be related in another way to time (e.g. after a surgery).

The aims of this paper are to provide a survey of the most frequently applied sample size formulas and to give an insight how to tackle the more complex problems. For details the reader is referred to the original papers.

Excellent surveys have been published by Lachin (1981), Donner (1984), Dupont and Plummer (1990) and Moussa (1988, 1990) with emphasis on trials with categorical outcomes or time to a critical event as outcome.

We restrict our consideration to randomized clinical studies, because otherwise the underlying statistical assumptions are not justified.

2. Continuous response variables

2.1. Difference testing

In general, the difference in the efficacy of two drugs or therapies is described by the difference between the distributions of some efficacy criterion, such as change to baseline in systolic blood pressure. Most frequently this is expressed as the difference of location parameters of the distributions, but sometimes even the shape of the distribution changes. Let us assume that the test statistic t used to decide between two hypotheses (null and alternative hypothesis) is distributed with the cumulative distribution function $F(t)$ and density $f(t)$, if the null hypothesis H_0 is true, and with $G(t)$ and $g(t)$ under the alternative hypothesis H_A. The fundamental relationship of the parameters used to determine sample size is given in Figure 1.

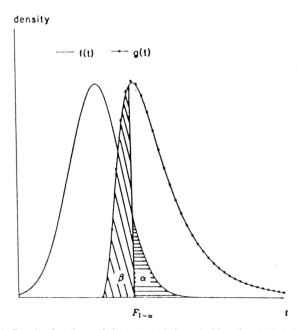

Fig. 1. Density functions of the test statistic and risks of a statistical test.

At the level α, the null hypothesis is rejected if the observed value of t exceeds the $(1 - \alpha)$-quantile $F_{1 - \alpha} = F^{-1}(1 - \alpha)$. The power is calculated as $1 - \beta = 1 - G(F_{1 - \alpha})$. In general, both distributions depend on the sample size or group sample sizes, and one can choose these to achieve a prefixed power, say $1 - \beta_0$, i.e. one has to solve the equation

$$F^{-1}(1 - \alpha) = G^{-1}(\beta_0),\tag{2.1}$$

or equivalently

$$G(F_{1 - \alpha}) = \beta_0,\tag{2.2}$$

with respect to the sample size or sample sizes. Here and in the further sections we regard the smallest integer larger or equal to the real-valued solution as integer solution of the corresponding sample size equation.

The quantiles $G^{-1}(\beta_0)$ of the non-central distribution depend for the commonly applied parametric distribution models on a non-centrality parameter nc which is a function of the true parameters and the sample sizes. Then (2.1) or (2.2) are first solved for nc_0 and $\text{nc} = \text{nc}_0$ for the sample sizes with fixed parameters or parameter differences.

Often the distributions of continuous response variables (e.g. supine diastolic blood pressure, serum level of an oral dose of a drug, haemoglobin) can be well fitted by Gaussian distributions $N(\mu, \sigma^2)$. The efficacy of the administered dose of a drug is then measured by the difference between the mean response μ_2 to the drug and the mean response μ_1 to placebo or a standard drug. In the simple case of two independent patient samples of size n_1, n_2 with response means $\bar{y}_1$, $\bar{y}_2$ and known variances σ_1^2, σ_2^2 the test statistic

$$z = (\bar{y}_2 - \bar{y}_1)/\sqrt{\sigma_1^2/n_1 + \sigma_2^2/n_2},$$

is used to decide between H_0: $\mu_2 \leq \mu_1$ and H_A: $\mu_2 > \mu_1$. The level α of the test is achieved for $\mu_1 = \mu_2$, then $z \sim N(0, 1)$. The non-central distribution of z (under H_A) is $N(\text{nc}, 1)$ with the non-centrality parameter

$$\text{nc} = (\mu_2 - \mu_1)/\sqrt{\sigma_1^2/n_1 + \sigma_2^2/n_2}.$$

Then (2.1) reads

$$z_{1-\alpha} = \text{nc} + z_{\beta_0}, \tag{2.3}$$

where $z_{1-\alpha}$, z_{β_0} denote the quantiles of the standard normal distribution.

Mainly for ethical reasons the ratio k of group sizes, i.e. $n_2 = kn_1$, cannot be chosen optimally, but has to be stipulated (e.g. to treat fewer patients with placebo). From (2.3) follows

$$n_1 = \frac{\sigma_1^2 + \sigma_2^2/k}{\delta^2}(z_{1-\alpha} + z_{1-\beta_0})^2, \tag{2.4}$$

with $\delta = \mu_2 - \mu_1 > 0$.

In the more realistic situation of equal but unknown variances $\sigma_1^2 = \sigma_2^2 = \sigma^2$ Student's t-test is the appropriate test. Then (2.1) has no explicit solution, but approximately holds (Rasch et al., 1978)

$$n_1 = [\sigma^2/\delta^2](1 + 1/k)(t_{1-\alpha} + t_{1-\beta_0})^2, \tag{2.5}$$

where $t_{1-\alpha} = t(1-\alpha, \text{df})$, $t_{1-\beta_0} = t(1-\beta_0, \text{df})$ denote the quantiles of Student's t distribution with $\text{df} = n_1 + n_2 - 2$ degrees of freedom.

Because nowadays in several software packages the cumulative distribution functions and quantile functions of the common central and non-central distributions are available, sample sizes can be determined exactly by solving (2.1) or (2.2) iteratively. The approximate formulas provide the starting values. For example, SAS gives the standard normal quantile $z_p = \text{PROBIT}(p)$, the cumulative distribution functions $\text{PROBT}(t, \text{df}, \text{nc})$ and $\text{PROBF}(F, \text{df}_1, \text{nc})$ of the non-central t- and F-distribution respectively and the corresponding quantile functions $\text{TINV}(p, \text{df}, \text{nc})$, $\text{FINV}(p, \text{df}_1, \text{df}_2, \text{nc})$. If nc is not specified or zero, the central distribution is used.

Due to (2.2) the inequality

$$\text{PROBT}[\text{TINV}(1 - \alpha, \text{df}), \text{df}, \text{nc}] \leqslant \beta_0, \tag{2.6}$$

with $\text{df} = n_1 + n_2 - 2$, $n_2 = k n_1$ and the non-centrality parameter

$$\text{nc} = \frac{\delta}{\sigma \sqrt{1 + 1/k}} \sqrt{n_1}, \tag{2.7}$$

has to be solved for the smallest integer by a search procedure.

The starting value corresponding to (2.4) is

$$n_{10} = \text{CEIL}\,[\sigma^2(1 + 1/k)/\delta^2[\text{PROBIT}(1 - \alpha) + \text{PROBIT}(1 - \beta_0)]^2]. \tag{2.8}$$

(CEIL (x) provides the smallest integer larger or equal to x.) This is a lower bound for the solution, therefore we only have to proceed for n_{10}, $n_{10} + 1$, $n_{10} + 2$, $\ldots$. In practical situations, n_1 usually lies in the range $n_{10} \leqslant n_1 \leqslant n_{10} + 2$.

In the case of a two-sided alternative, i.e.

$$H_0: \mu_1 = \mu_2, \qquad H_A: \mu_1 \neq \mu_2,$$

one rejects the null hypothesis if $F = t^2 > F(1 - \alpha, 1, \text{df})$ i.e. if the square of the t-statistic is larger than the $(1-\alpha)$-quantile of the central F-distribution. The same procedure can be applied to solve

$$\text{PROBF}[\text{FINV}(1 - \alpha, \text{df}_1, \text{df}_2), \text{df}_1, \text{df}_2, \text{nc}] = \beta_0 \tag{2.9}$$

with $\text{df}_1 = 1$, $\text{df}_2 = \text{df}$ and the non-centrality parameter

$$\text{nc} = \delta^2 n_1 / [\sigma^2(1 + 1/k)]. \tag{2.10}$$

Again the search starts from (2.8) with $\frac{1}{2}\alpha$ instead of α. A generalization to more than two groups is straightforward (see next section).

Sample sizes for non-parametric tests have only been derived using the normal approximation for large sample sizes. Noether (1987) proposed the corresponding formulas for the sign-test, Wilcoxon's one- and two-sample test (for one-sided alternatives) and for Kendall's test of independence.

Sign test:

$$n = \frac{(z_{1-\alpha} + z_{1-\beta_0})^2}{4(p - \frac{1}{2})^2} \qquad (2.11)$$

(*p* denotes the probability that an observation exceeds the hypothetical median). *Wilcoxon's one-sample test* of symmetry (centered at zero) of a distribution:

$$n = \frac{(z_{1-\alpha} + z_{1-\beta_0})^2}{3(p' - \frac{1}{2})^2} \qquad (2.12)$$

(*p'* denotes the probability $p' = \mathrm{Pr}(y + y' > 0)$, *y* and *y'* being two independent observations).

Wilcoxon's two-sample test for the comparison of the distributions of two independent random variables *x*, *y*:

$$N = \frac{(z_{1-\alpha} + z_{1-\beta_0})^2}{12c(1-c)(p'' - \frac{1}{2})^2} \qquad (2.13)$$

($N = n_1 + n_2$, $n_1 = cN$, $p'' = \mathrm{Pr}(y > x)$).
Kendall's test of independence of two random variables:

$$n = \frac{(z_{1-\alpha} + z_{1-\beta_0})^2}{9(p_c - \frac{1}{2})^2} \qquad (2.14)$$

(p_c denotes the probability of concordance, where two sample observations (x, y) and (x', y') are said to be concordant or disconcordant, depending on whether $(x - x')(y - y')$ is positive or negative).

2.2. Linear models

The procedure described above can easily be generalized to all *F*-tests in the scope of general linear models. One only has to specify the degrees of freedom df_1, df_2 and the non-centrality parameter nc of (2.9). At a first sight this seems to be the complete solution to many problems. Indeed it is, but in practice the problem arises to transform the practical requirements into requirements about the non-centrality parameter. In the regression analysis and in the case of more complex designs the structural formula of the non-centrality parameter depends on the structure of the design. The total sample size has to be partioned into partial sample sizes. This will be demonstrated by the following examples.

In a one-way analysis of variance with a factor levels (e.g. dose levels), equal group sizes n, expectations μ_i, $i = 1, \ldots, a$ and error variance σ^2 we obtain $df_1 = a - 1$, $df_2 = a(n - 1)$ and

$$nc = \frac{n}{\sigma^2} \sum_{i=1}^{a} (\mu_i - \bar{\mu})^2 , \tag{2.15}$$

with $\bar{\mu} = \sum \mu_i / a$. To determine the sample size one replaces the μ_i and σ^2 with estimates from previous studies.

Frequently these are not available and one wants to detect a difference δ if at least two of the expectations differ by δ. The worst case, which leads to the minimum of the non-centrality parameter, occurs if two of the expectations differ by δ and the other equal their arithmetic mean. Then $nc = n\delta^2/(2\sigma^2)$ which is the same as (2.10) for $k = 1$. The worst case group sample size decreases as a function of the number of groups only.

As an example for the regression analysis (model I: non-random influence variables) we regard the problem of comparing several dose-response curves, assuming a linear relationship between x, the logarithm of the dose level, and a continuous response variable y, say the blood serum concentration at a fixed time point after administration of an oral dose. To compare the biovailability of a drugs entails to compare the slopes γ_i, $i = 1, \ldots, a$ of the regression lines $\alpha_i + \gamma_i x$,

$$H_0: \gamma_1 = \gamma_2 = \cdots = \gamma_a .$$

The alternative hypothesis states that at least two of the slopes differ. The usual F-test of parallelity of a regression lines has $df_1 = a - 1$, $df_2 = N - 2a$ degrees of freedom ($N = \sum n_i$, n_i = group sizes for drug i, and x_{ij} = logarithm of dose level j of drug i) and the non-centrality parameter

$$nc = \frac{1}{\sigma^2} \sum_{i=1}^{a} w_i (\gamma_i - \bar{\gamma})^2 , \tag{2.16}$$

with the weighted mean $\bar{\gamma}$ and the weights $w_i = \sum (x_{ij} - \bar{x}_{i.})^2$, i.e. the non-centrality parameter depends on the design (see Bock, 1984). The optimal design uses the intended minimum and maximum (logarithm of) dose levels $x_{i, \min}$, $x_{i, \max}$ only, with equal frequencies. These may vary from drug to drug. For the optimal design $w_i = \frac{1}{4} n_i \Delta_i^2$, $\Delta_i = x_{i, \max} - x_{i, \min}$ holds. The precision of the estimates of the slopes depends on the ranges Δ_i and the error variance σ^2, which is assumed to be the same for all regression lines.

Bock (1984) investigated the optimal partition of the total sample size $N = \sum n_i$ and the worst case if at least two slopes differ by d. The following results were obtained:

If the ranges do not differ so much that one of the slopes is much more precisely estimated than the others, i.e. if the $\theta_{\max} \leqslant \frac{1}{2}$, where the weights θ_i

$(i = 1, 2, \ldots, a)$ are defined by $\theta_i = \Delta_i^{-2}/\sum \Delta_j^{-2}$, then the optimal partition is

$$n_1 \Delta_1^2 = n_2 \Delta_2^2 = \cdots = n_a \Delta_a^2 = \text{const} = w, \qquad (2.17)$$

and the total sample size can be determinated from

$$\text{PROBF}(\text{FINV}(1 - \alpha, a - 1, N - 2a), a - 1, N - 2a, \text{nc}) \approx \beta_0 \qquad (2.18)$$

according to (2.2) with

$$\text{nc} = \frac{Nd^2}{4\sigma^2} \frac{1}{2 \sum \Delta_i^{-2}}. \qquad (2.19)$$

The formulas are somewhat more complicated if $\theta_{\max} > \frac{1}{2}$ and the reader is referred to Bock's original paper except for the case of only two regression lines $(a = 2)$, where we have $\theta_{\max} > \frac{1}{2}$ since $\theta_1 + \theta_2 = 1$. The optimal partition is then

$$n_1 \Delta_1 = n_2 \Delta_2, \qquad (2.20)$$

i.e. the group sizes are now proportional to the reciprocal of the length of the logarithmic dose ranges, but not to the squares as above. The non-centrality parameter to be used in (2.17) is here

$$\text{nc} = \frac{Nd^2}{4\sigma^2} \frac{1}{(\Delta_1^{-1} + \Delta_2^{-1})^2}. \qquad (2.21)$$

In the limit case $\Delta_1 = \Delta_2$, the two partitions and non-centrality parameters coincide, of course.

2.3. Equivalence testing

The equivalence problem has received considerable attention recently in the statistical literature. To prove equivalence (e.g. bioequivalence of two formulations of a drug or equivalent efficacy of two therapies) one should not apply the difference tests. By testing the null hypothesis of exact equality, the F-test is testing the wrong hypothesis. What we want to explore is, whether the mean value difference of the bioavailability of the drugs or the mean difference in the efficacy exceeds prefixed limits $\pm \delta$ or not, i.e. we intend to test interval-hypotheses.

Let us start with the much more simple problem of at-least-equivalence assuming approximate normality of the random observations. If we want to study whether an experimental drug E with mean response μ_E is at least equivalent to a control drug C with mean response μ_C, we have to test the following hypotheses at level α:

H_0: The mean response to the experimental [drug, treatment, ...] is smaller than that to the control by more than δ ($\mu_E - \mu_C < -\delta$), versus

H_A: The experimental [drug, treatment, ...] is at least equivalent to the control i.e. the mean response to experimental [drug, treatment, ...] is not smaller than that to the control by more than δ ($\mu_E - \mu_C \geqslant -\delta$).

Note that compared to significance testing, H_0 and H_1 are interchanged. The alternative hypothesis H_A will be accepted if the lower $(1 - \alpha)$-confidence limit of $\mu_E - \mu_C$ exceeds $-\delta$, or equivalently, if $t = (\bar{d} + \delta)/s_d > t(1 - \alpha, \mathrm{df}) = t_{1-\alpha}$, where $\bar{d} = \bar{y}_E - \bar{y}_C$ denotes the difference of the arithmetic means of the observations. The estimate s_d of the standard deviation σ_d of $\bar{d}$ and the degrees of freedom have to be choosen according to the design (two independent groups, paired observations or cross over design). If we stipulate the power, i.e. the probability of accepting at-least-equivalence, for equal means $\mu_E = \mu_C$ to be $1 - \beta_0$, then the same formula (2.5) applies as for the one-sided difference test in the case of two independent groups of observations, and the same procedure as in Section 2.1 is used to compute the exact sample size.

Equivalence in the strict sense means that the mean difference $\mu_E - \mu_C$ lies within the interval $[-\delta_1, \delta_2]$. The equivalence limits $-\delta_1 < 0$ and δ_2 are stipulated based on practical considerations (e.g. clinical relevant changes). Most frequently $\delta_1 = \delta_2 = \delta$ is assumed.

Then the following hypotheses have been tested at level α:

H_0: The [drugs, treatments, ...] are non-equivalent i.e. the mean responses to treatments differ by more than δ ($|\mu_E - \mu_C| > \delta$), versus

H_1: The [drugs, treatments, ...] are equivalent i.e. the mean responses to treatments do not differ by more than δ ($|\mu_E - \mu_C| \leqslant \delta$).

The 2α-Westlake-procedure (Westlake, 1979, 1981) requires that the $(1-2\alpha)$-confidence interval of the mean difference be computed and equivalence be accepted if this is fully included in $[-\delta, +\delta]$. Equivalently one could carry out the two one-sided α-level t-tests of the hypotheses H_{01}: $\mu_d \geqslant -\delta$ versus H_{11}: $\mu_d < -\delta$ and H_{02}: $\mu_d > \delta$ versus H_{12}: $\mu_d \leqslant \delta$ and accept equivalence only if both lead to significance (Schuirmann, 1981). This can be performed more easily by computing the P-value

$$P = \mathrm{PROBT}([|\bar{d}| - \delta]/s_d, \mathrm{df}),\tag{2.22}$$

and accepting equivalence if $P < \alpha$ (PROBT denotes as above the cumulative distribution function of Student's t). Anderson and Hauck (1983) proposed a new procedure. They compute the P-value from

$$P = \mathrm{PROBT}([\delta + |\bar{d}|]/s_d, \mathrm{df}) - \mathrm{PROBT}([\delta - |\bar{d}|]/s_d, \mathrm{df}),\tag{2.23}$$

and accept equivalence (H_1) if $P < \alpha$.

The 2α-Westlake procedure is sometimes preferred because of its straightforward interpretation, but it can be very conservative (i.e. the true α is smaller than nominal α). For a more detailed discussion, see Schuirmann (1987) and Müller-Cohrs (1990).

The power function has been calculated by Frick (1987), using numerical integration. A comparison of the power of both procedures has been given by Müller-Cohrs (1990). Because arithmetic means and sample variances of normally distributed observations are independent, one computes the power by conditioning first on $x = s_d^2/\sigma_d^2$, which has the density function

$$g(x) = \kappa^\kappa e^{-\kappa x} x^{\kappa - 1}/\Gamma(\kappa), \tag{2.24}$$

with $\kappa = \frac{1}{2}\mathrm{df}$. To derive the unconditioned power one takes the expectation with respect to the distribution of x. The inequality $P < \alpha$ is equivalent to $|\bar{d}| < c(x)$ with

$$c(x) = \delta - t_{1-\alpha}\sigma_d\sqrt{x} \tag{2.25}$$

for the 2α-Westlake procedure, and

$$c(x) = h\left(\frac{\delta}{\sigma_d\sqrt{x}}\right)\sigma_d\sqrt{x} \tag{2.26}$$

for the Anderson–Hauck procedure, where $h = h(w)$ denotes the solution of

$$\mathrm{PROBT}(w + h, \mathrm{df}) - \mathrm{PROBT}(w - h, \mathrm{df}) = \alpha. \tag{2.27}$$

Therefore the power is given by

$$\Pr(|\bar{d}| < c(x)) = \int_0^\xi \left[\Phi\left(\frac{\mu_E - \mu_C + c(x)}{\sigma_d}\right) \right.$$
$$\left. - \Phi\left(\frac{\mu_E - \mu_C + c(x)}{\sigma_d}\right) \right] g(x)\,\mathrm{d}x \tag{2.28}$$

with the upper limit $\xi = \delta^2/(\sigma_d^2 t^2(1 - \alpha, \mathrm{df}))$ for the Westlake-procedure, $\xi = \infty$ for the Anderson–Hauck procedure and the distribution function $\Phi(u) = \mathrm{PROBT}(u)$ of the standard normal distribution. The latter integral has to be calculated by numerical integration.

Anderson and Hauck (1983) recommended calculating the sample size by considering the maximum power only, i.e. the power at $\mu_E - \mu_C = 0$. They described an iterative procedure to determine the sample size approximately for a given value, say $1 - \beta_0$, which is equivalent to the procedure described in Section 2.1 for the two-sided t-test. One has simply to interchange α and β and to replace the difference to be detected by the half length δ of the equivalence region, i.e.

$$n = \frac{2\sigma^2}{\delta^2} (t_{1 - \beta_0/2} + t_{1 - \alpha})^2 , \qquad (2.29)$$

holds approximately for a parallel group design ($n_1 = n_2 = n$).

Stipulating the maximum power only is a very weak requirement. The power at the border of the equivalence region is approximately given by the nominal level α. Although differences near to $\pm \delta$ belong to the defined equivalence region, there is only a small chance of deciding for equivalence. Therefore the power should be calculated at a $\delta^* < \delta$. Using the power calculations, mentioned above, one can iteratively determine the sample size for which the power at $\pm \delta^*$ equals $1 - \beta_0$. Then the power is larger than $1 - \beta_0$ for all differences within the region $[-\delta^*, +\delta^*]$.

A similar approach has been published by Patel and Gupta (1984) using the statistic $F = \bar{d}^2/s_d^2$ in a parallel group design. They propose to reject H_0^*: $|\mu_E - \mu_C| = \delta$ and thereby to accept H_A^*: $|\mu_E - \mu_C| < \delta$ if

$$F < F_\alpha^*(\sigma, \delta) = \text{FINV}[\alpha, 1, 2(n - 1), \text{nc}] \qquad (2.30)$$

with the non-centrality parameter nc $= n\delta^2/(2\sigma^2)$. The sample size n is obtained by solving

$$1 - \beta_0 = \text{PROBF}[F_\alpha^*(\sigma, \delta^*), 1, 2(n - 1), n\delta^{*2}/(2\sigma^2)] , \qquad (2.31)$$

iteratively.

However, $F_\alpha^*(\sigma, \delta)$ cannot be calculated in practice since σ is unknown. Replacing σ by the error standard deviation s_R introduces the random bound $F_\alpha^*(s_R, \delta)$ whereby the level and the power of the test are changed. Therefore (2.31) holds only approximately.

3. Binary responses

3.1. Difference testing of proportions

The majority of clinical trials conducted to explore the efficacy of drugs or therapies use rates (success rate, failure rate, improvement rate, clinical global impression rates, cure rate) as primary criteria. In epidemiological research exposure rates, incidence rates, prevalence rates play an important role. Therefore there is a huge number of papers dealing with this topic.

Two response rates from independent samples are usually compared using Fisher's exact test or alternatively the approximate normal distribution test and its squared version, the χ^2-test, with and without Yate's correction of continuity.

Lachin (1981) considered the general family of statistics, say z, that are normally distributed under a null hypothesis (H_0) as $N(\mu_1, v_1^2)$ and under an alternative hypothesis (H_A) as $N(\mu_2, v_2^2)$ and where v_1^2, v_2^2 are some functions of the variances of the single observations and sample sizes. From (2.1) follows the basic equation

$$|\delta| = z_{1-\alpha} v_1 + z_{1-\beta_0} v_2, \tag{3.1}$$

with $\delta = \mu_2 - \mu_1$.

In the case of two independent samples of sizes n_1, n_2 of binary outcomes from two populations with probabilities π_1 and π_2 respectively the relative frequencies p_1, p_2 are approximately normally distributed. As above we choose $n_2 = kn_1$ $(k \geqslant 1)$. To test the hypotheses $H_0: \pi_2 \leqslant \pi_1$ and $H_A: \pi_2 > \pi_1$ we could use the difference $z = p_2 - p_1$, which is approximately normally distributed as $N(0, v_1^2)$ under H_0 for $\pi_1 = \pi_2 = \pi$ and as $N(\delta, v_2^2)$ under H_A with $\delta = \pi_2 - \pi_1$, $v_1^2 = \pi(1-\pi)(1+k)/n_2$ and $v_2 = [k\pi_1(1-\pi_1) + \pi_2(1-\pi_2)]/n_2$. Substituting into (3.1) and replacing the standard deviations v_1, v_2 through estimates with p_1, p_2 replacing π_1, π_2 and $\bar{p} = (p_1 + kp_2)/(1+k)$ as estimator of π yields the well known formula of Schneidermann (1964), Fleiss (1973) $[k = 1]$ and Fleiss et al. (1980):

$$n_1 = \frac{1}{k\delta^2} [z_{1-\alpha}\sqrt{\bar{p}(1-\bar{p})(1+k)} + z_{1-\beta_0}\sqrt{kp_1(1-p_1) + p_2(1-p_2)}]^2. \tag{3.2}$$

Various other sample size formulas, found in the literature, can be derived from the basic equation by replacing the variance estimators, e.g. (for $k = 1$, $n_1 = n_2 = n$)

$$n = \frac{1}{\delta^2} [z_{1-\alpha} - z_{1-\beta_0}]^2 [p_1(1-p_1) + p_2(1-p_2)], \tag{3.3}$$

with equal variance estimators or accounting for the worst case

$$n = \frac{1}{2\delta^2} [z_{1-\alpha} + z_{1-\beta_0}]^2. \tag{3.4}$$

The very rough estimate $n = 5/\delta^2$ was recommended by Walther (1980).

For small π_i and $np_2 > 10$, Gail (1974) obtained from a combined normal and Poisson distribution approximation the formula

$$n = [(p_1 + p_2)/\delta^2][z_{1-\alpha} + z_{1-\beta_0}]^2. \tag{3.5}$$

Silitto (1949) exploited the variance stabilizing arcsin transformation:

$$n = \frac{[z_{1-\alpha} + z_{1-\beta_0}]^2}{2[\arcsin\sqrt{p_2} - \arcsin\sqrt{p_1}]^2}. \tag{3.6}$$

For the two-sided test problem α has to be replaced by $\frac{1}{2}\alpha$. These formulas will be applied for tests without continuity correction. In the case of continuity correction Casagrande et al. (1978) recommend to adjust the sample size by

$$n_1^* = \tfrac{1}{4} n_1 \left[1 + \sqrt{1 + 2(1 + k)/(k n_1 \delta)} \right]^2 , \tag{3.7}$$

whereas Fleiss et al. (1980) state that the adjustment

$$n^* = n + 2/\delta , \tag{3.8}$$

has a remarkable accuracy ($k = 1$, $n = n_1 = n_2$, $n\delta \geqslant 4$).

Power calculations for Fisher's exact test are to be found in the papers of Bennet and Hsu (1960), tables have been computed by Gail and Gart (1973), which were revised by Hasemann (1978). Failing and Victor (1981) compared the sample sizes from several formulas including (3.6) with that of Hasemann. They concluded that the combination of formulas (3.2) and (3.7) provides nearly accurate sample sizes for $0.05 \leqslant \pi_1 < \pi_2 \leqslant 0.95$, whereas considerable deviations were produced by most other formulas. Of course (3.4) behaves badly, whereas (3.6) always provides too small numbers, but it could be adjusted also corresponding to (3.7). Thus various other formulas are merely of historical interest. In the case of small probabilities no generally well performing approximation formula exists. Therefore the exact distributions should be used.

Large sample size tables including tables for equivalence testing, means and survival curves have been proposed by Machin and Campbell (1987), based on the approximate formulas. They can serve to provide a quick overview over the range of reasonable sample sizes, while discussing the possible strategies in a development process.

Sample sizes to detect a deviation of the relative risk π_1/π_2 from 1 are derived by an algebraic transformation of (3.2) (Fleiss, 1981).

A very simple method of adjusting sample size requirements for an anticipated drop-out rate d has been proposed by Lachin (1981). He argues that in a comparison of a treatment (T) with placebo or control treatment (C) the effective event rate is $\pi_T^* = d\pi_C + (1 - d)\pi_T$, i.e. the effective difference $\delta^* = (1 - d)\delta$ has to be put into (3.2). Therefore the sample size has to be inflated by dividing by $(1 - d)^2$. This is true for patients who refrain from treatment, but do not withdraw from the study. If w denotes the rate of withdrawals one would adjust by dividing by $(1 - w)$, if it can be assured that the rate is the same for both groups. Similarly the sample size could be inflated by dividing by $(1 - e)$, if e denotes the anticipated rate of non-evaluable patients.

A more sophisticated approach is described by Schork and Remington (1967). They deal with shifts from the treatment to the placebo group (and also in the reverse direction) in fixed time units, say years. Suppose that the anticipated yearly event rates in study with the duration of L years are constantly p_{T_y} and p_{C_y} and the early drop-out rates from the treatment group are d_i, $i = 1, 2, \ldots, L$.

Then the effective L-years event rate in the treatment group is

$$\pi_T^* = \sum_{i=1}^{L} d_i [1 - (1 - p_{T_y})^{i - 1/2}(1 - p_{C_y})^{L - i + 1/2}] + c[1 - (1 - p_{T_y})^2], \tag{3.9}$$

where $c = 1 - \sum d_i$ is the anticipated proportion of patients in the treatment group who complete the study (see Donner, 1984). $\delta^* = \pi_T^* - \pi_C$, $p_1 = \pi_C$, $p_2 = \pi_T^*$ should be substituted into (3.2).

Halperin et al. (1968) derived a theoretical model to adjust the event rate in long term clinical trials. They assume that the time-to-event among control group subjects and the time-to-drop out among treatment group subjects each follow an exponential distribution.

In multicenter studies one has to account for the stratification of subjects by centers, but there are many other possibilities for stratification e.g. sex, age groups, indication groups, risk groups etc.). Gail (1973) addresses this problem. Let p_{T_j}, p_{C_j} $(i = 1, 2, \ldots, a)$ denote the events rates among treatment and control group subjects respectively and n_j the size of the j-th stratum. The Mantel–Haenszel χ^2-test presumes a common odds ratio

$$OR = [p_{T_j}(1 - p_{C_j})]/[p_{C_j}(1 - p_{T_j})],$$

to test the null hypothesis H_0: $OR = 1$ versus H_A: $OR > 1$. The total size $N = \sum n_j = N \sum f_j$, where f_j denotes the fraction of observations contained in the j-th table, can be determined from

$$N = (z_{1 - \alpha} + z_{1 - \beta_0})^2 / [\sum_{j=1}^{a} g_j f_j], \tag{3.10}$$

with

$$g_j = \frac{\ln^2(OR)}{[p_{C_j}(1 - p_{C_j})]^{-1} + [p_{T_j}(1 - p_{T_j})]^{-1}}, \quad j = 1, 2, \ldots, a. \tag{3.11}$$

Donner (1984) states that this formula depends on the assumption that the marginal totals of the (2×2)tables are random variables and thus it most strictly corresponds to the unconditional analysis by Cox (cf. Fleiss, 1981, p. 81). However, Gail suggests that the formula yields an appropriate sample size, provided $n_j > 15$ and $0.1 < p_{C_j}$, $p_{T_j} < 0.9$.

The sample size corresponding to the McNemar test was investigated by Miettinen (1968), Bennett and Underwood (1970), Schlesselmann (1982), Duffy (1984), Connett et al. (1987), Connor (1987) and Feuer and Kessler (1989).

A generalization of the formulas (3.2) and (3.7) for the Cochran–Armitage test of trends in dose-response relationships has been derived by Jun-mo Nam (1987). Let x_i, $i = 0, \ldots, a - 1$, be mutually independent binomial variates based on sample sizes of n_i at dose level d_i. Assume that the probability of response follows a linear trend on the logistic scale

$$\pi_i = e^{\gamma + \lambda d_i}/(1 + e^{\gamma + \lambda d_i}). \tag{3.12}$$

An approximate test of $H_0: \lambda = 0$ versus $H_A: \lambda > 0$ is based on the asymptotically normal deviate

$$z = (U' - \tfrac{1}{2}\Delta)/\sqrt{\mathrm{Var}(U'\,|\,X)_0}, \tag{3.13}$$

where $X = \Sigma x_i$, $U = \Sigma x_i d_i$, $N = \Sigma n_i$, $\bar{p} = X/N$, $\bar{q} = 1 - \bar{p}$, $\bar{d} = \Sigma n_i d_i/N$, $U' = \Sigma x_i (d_i - \bar{d})$ and $\mathrm{Var}(U'\,|\,X)_0 = \bar{p}\bar{q} \Sigma n_i (d_i - \bar{d})^2$. The continuity correction $\tfrac{1}{2}\Delta = \tfrac{1}{2}(d_{i+1} - d_i)$ applies only in the case of equally spaced doses. z^2 is the familiar Cochran–Armitage statistic.

For stipulated ratios $r_i = n_i/n_0$ of sample size n_0 of the control group can be determined from

$$n_0 = \tfrac{1}{4}n_0^*[1 + \sqrt{1 + 2\Delta/(An_0^*)}]^2, \tag{3.14}$$

where

$$n_0^* = [z_{1-\alpha}\sqrt{\bar{\pi}(1 - \bar{\pi})\,\Sigma\,r_i(d_i - \bar{d})^2}$$
$$+ z_{1-\beta}\sqrt{\Sigma\,\pi_i(1 - \pi_i)r_i(d_i - \bar{d})^2}]^2/A^2 \tag{3.15}$$

is the sample size for the test without continuity correction and $\bar{\pi} = \Sigma n_i \pi_i/N$, $A = \Sigma r_i \pi_i(d_i - \bar{d})$.

3.2. Equivalence testing

The problem of 'proving the null hypothesis' has been considered by several authors. The hypotheses to be tested can be stated in the same way as described in Section 2.3 replacing the means μ_E, μ_C with the response rates π_E and π_C respectively. Makuch and Simon (1978), Blackwelder (1982, 1984) recommended the following formula for the one-sided case (*at-least-equivalence*):

$$n = \frac{[z_{1-\alpha} + z_{1-\beta_0}]^2[\pi_E(1 - \pi_E) + \pi_C(1 - \pi_C)]}{[\delta - (\pi_C - \pi_E)]^2}. \tag{3.16}$$

Tables based on this formula are proposed by Machin and Campbell (1987).

4. Survival time studies

Any time to a critical event (e.g. death, reinfection, infarct, recovery, improvement, relapse, drop out) from a fixed time point (e.g. first administration of a drug, surgery) or the time between two critical events (e.g. time interval between the first and second infarct) is regarded as survival time in clinical studies. In general the survival times are modelled by parametric failure time distributions like exponential and Weibull distribution or proportional-hazards regression model of Cox (1972), which accounts for confounding variables. The most frequently used tests are Cox F-test (cf. Lee, 1980) for the comparison of exponential distributions and the logrank-test (Tarone–Ware, 1977) for the comparison of distribution-free survival function estimates.

One of the milestones with regard to sample size considerations in this area was the paper of Bernstein and Lagakos (1978) on sample size and power for stratified clinical trials. Each patient is initially classified into one of a strata (e.g. risk groups). It is assumed that N patients are accrued into the trial uniformly over a period of T time units and followed for additional τ time units. Within each stratum the patients are randomly assigned to the control (C) or experimental (E) treatment in proportions θ and $1 - \theta$ respectively. Therefore the total expected numbers of control and experimental patients in stratum j are $n_{C_j} = NTp_j\theta$ and $n_{E_j} = NTp_j(1 - \theta)$, where p_j denotes the expected proportion of patients in stratum j. Bernstein and Lagakos assume exponential distributions with rate parameters λ_{C_j} and λ_{E_j} for the control and experimental group respectively in stratum j, which may vary over strata, but have a constant ratio $\Delta = \lambda_{C_j}/\lambda_{E_j}$ over strata. To allow for censoring, let π_{ij} $(i = C, E; j = 1, 2, \ldots, a)$ denote the probability that a patient in stratum j on treatment i will fail before the trial is finished. This can be calculated from the assumed exponential failure time distributions:

$$\pi_{ij} = \pi(\lambda_{ij}) = 1 - (\lambda_{ij}T)^{-1} e^{-\lambda_{ij}\tau}[1 - e^{-\lambda_{ij}T}], \quad i = C, E. \tag{4.1}$$

Since the logarithm of a sum of independent, identically distributed exponential random variables is approximately normally distributed, the sample size for a stipulated level α and power $1 - \beta_0$ is approximately determined by

$$N = \frac{[z_{1-\alpha}\gamma^{-1/2}(1) + z_{1-\beta_0}\gamma^{-1/2}(\Delta)]^2}{T\theta(1 - \theta)\ln^2(\Delta)}, \tag{4.2}$$

where

$$\gamma(\Delta) = \sum_{j=1}^{a} \frac{p_j\pi_{C_j}\pi_{E_j}}{\theta\pi_{C_j} + (1 - \theta)\pi_{E_j}}. \tag{4.3}$$

Moussa (1988) compared this sample size with that of George and Desu (1974),

$$n = 2(z_{1-\alpha} + z_{1-\beta_0})^2/\ln^2(\Delta), \tag{4.4}$$

for a Poisson process entry of patients, non-censoring ($\pi_E = \pi_C = 1$), $\tau = 0$, $a = 1$ and equal allocation ($\theta = \frac{1}{2}$, $N = 2n$, $T = 1$), that of Makuch and Simon (1982) for the comparison of m treatment groups with a χ^2-test with $m - 1$ degrees of freedom

$$n = 2\delta(m - 1, 1 - \alpha, 1 - \beta_0)/\ln^2(\Delta), \tag{4.5}$$

where δ denotes the non-centrality parameter to ensure the stipulated power, that of Pasternack and Gilbert (1971) for uniform patients accrual and exponential survival times

$$n = \frac{[z_{1-\alpha}\sqrt{(\Delta + 1)^2/2} + z_{1-\beta_0}\sqrt{\Delta^2 + 1}]^2}{(\Delta - 1)^2}, \tag{4.6}$$

that of Freedmann (1982) for the distribution-free case (logrank-test)

$$N = 2n = \frac{d(1 + \zeta)}{\zeta(1 - P_C) + (1 - P_E)} \tag{4.7}$$

with the T-years survival rates P_C, P_E and the total number of events required in the two groups

$$d = \frac{(z_{1-\alpha} + z_{1-\beta_0})^2(1 + \Delta\zeta)^2}{\zeta(1 - \Delta)^2}, \tag{4.8}$$

where $\zeta = \theta/(1 - \theta)$, and that of Schoenfeld (1981, 1983) for the proportional hazard regression model

$$n = \frac{(z_{1-\alpha} + z_{1-\beta_0})^2}{\theta(1 - \theta)D\ln^2(\Delta)}, \tag{4.9}$$

with the proportion D of deaths. He discussed the similarities of the formulas and the influence of the parameters.

Schoenfeld showed that the formula of Bernstein and Lagakos (4.2) is also valid for the logrank-test when comparing treatments with proportional hazards assuming equal censoring distributions, and extended this to heterogeneous patient populations. The corresponding nomograms are presented by Schoenfeld and Richter (1982).

Schuhmacher developed a distribution-free procedure based on the NcNemar test and compared the power with that of other methods.

Assuming exponentially distributed failure times and single strata Lachin (1981) has shown that

$$N = \frac{[z_{1-\alpha}\sqrt{\Psi(\bar{\lambda})(\theta^{-1} + (1 - \theta)^{-1})} + z_{1-\beta}\sqrt{\Psi(\lambda_C)\theta^{-1} + \Psi(\lambda_E)(1 - \theta)^{-1}}]^2}{(\lambda_E - \lambda_C)^2}, \tag{4.10}$$

where $\Psi(\lambda) = \lambda^2/\pi(\lambda)$ (see (4.1)) and $\bar{\lambda} = \theta\lambda_{\mathrm{E}} + (1 - \theta)\lambda_{\mathrm{C}}$ is the total sample size $N = n_{\mathrm{E}} + n_{\mathrm{C}}$ needed to test H_0: $\lambda_{\mathrm{E}} - \lambda_{\mathrm{C}} = 0$ versus a one-sided alternative using the asymptotic Wald-statistic. Lachin and Foulkes (1986) presented a general method for the evaluation of sample size and power of survival studies allowing for non-uniform patient entry, losses to follow up, noncompliance and stratification modifying the function $\Psi(\lambda)$.

Wu et al. (1980) generalized the above method of Halperin et al. (1968) to allow for different rates in different time intervals. The computations involve numerical integration. Lakatos (1986) developed a general non-stationary Markov model for complex clinical trials, which permits the experimental and control group event probabilities P_{E} and P_{C} to be modelled depending on non-compliance, lag time, staggered entry etc. These are then substituted into (3.2) to determine the sample size. In a later paper (Lakatos, 1988) the Markov model has been applied to estimate sample sizes for the logrank test. While the effort required to numerically solve the continuous time model may be formidable, a discrete time formulation leads to simple numerical computation and equivalent results. The computer program is given by Lakatos (1986).

5. Correlations

In clinical studies questions of the following kind frequently arise:

(a) Are the investigated variables (e.g. blood pressure, cholesterol, age and body weight) correlated at all?

(b) Are there variables in a set of variables which are related to the target variable?

That means, correlations have to be compared with zero. Cohen (1969) proposed tables of the necessary sample size to achieve with level α the stipulated power $1 - \beta_0$ using the statistic $t = r\sqrt{n - 2}/\sqrt{1 - r^2}$, which depends on the sample correlation coefficient r. The tables are based on the approximate formula

$$n - 2 = (z_{1-\alpha} + z_{1-\beta_0})^2/\psi^2(\rho_d), \tag{5.1}$$

where $\psi(r) = 0.5[\ln(1 + r) - \ln(1 - r)]$ denotes the transformation introduced by R. A. Fisher and ρ_d the correlation coefficient to be detected. Bock (1977) has proved that the same formula can be applied for the t-test to compare the slope β_1 of the regression line with a stipulated constant β_{10}, assuming a two-dimensional normal distribution $\mathrm{N}(\mu_x, \mu_y, \sigma_x^2, \sigma_y^2, \rho)$. In that case ρ_d has to be replaced by

$$\rho_d = \frac{\Delta_d}{\sqrt{1 + \Delta_d^2}}, \qquad \Delta_d = \frac{\beta_1 - \beta_{10}}{\sigma}\sigma_x,$$

where $\sigma^2 = \sigma_y^2(1 - \rho^2)$ denotes the error variance.

The generalization of this result to the F-test of the hypothesis that the multiple

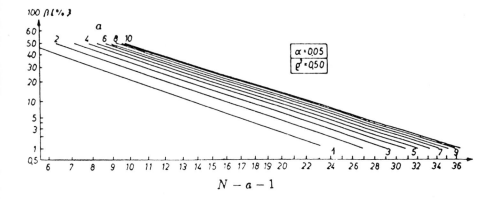

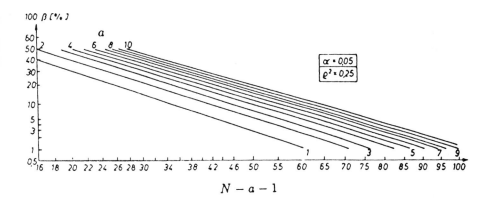

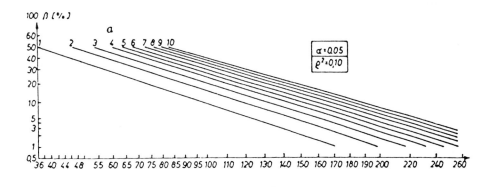

Fig. 2. Probability β to accept H_0: $\rho = 0$, $df_1 = a$-number of influence variables, $df_2 = N - a - 1$, N-sample size (normal probability plot).

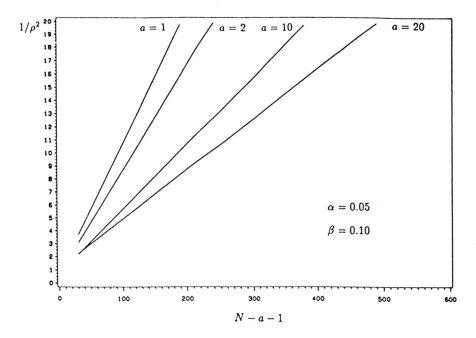

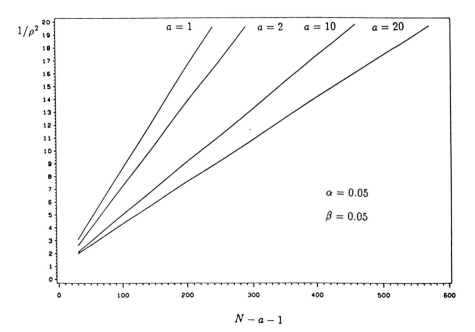

Fig. 3. Denominator degrees of freedom $N - a - 1$ (test correlation), N-sample size, $df_1 = a$-number of influence variables.

regression coefficients vanish simultaneously or equivalently that the multiple correlation coefficient ρ equals zero has been presented by Bock (1984). The conditional power of this test at level α with df_1 and df_2 degrees of freedom for fixed values of the influence variables can be obtained from the non-central F-distribution. The expectation with respect to the distribution of the conditioning variables provides the unconditioned power. Using the series expansion of the cumulative distribution function of the non-central F-distribution by Tiku (1966) one obtains a series expansion for the unconditional non-central distribution. This rapidly converging series allows for recursive calculations. Thus the power can be computed from

$$1 - \beta = \alpha + \sum_{j=1}^{\infty} w_j(A_1 + A_2 + \cdots + A_j), \qquad (5.2)$$

with

$$w_0 = (1 - \rho^2)^{p+q}, \qquad w_j = \frac{p+q+j-1}{j} \rho^2 w_{j-1}, \quad j = 1, 2, \ldots,$$

$$A_1 = \frac{z^p(1-z)^q}{pB(p,q)}, \qquad A_j = z\frac{p+q+j-2}{p+j-1} A_{j-1}, \quad j = 2, 3, \ldots,$$

$$p = \tfrac{1}{2}df_1, \qquad q = \tfrac{1}{2}df_2, \qquad z = \frac{df_1 F(df_1, df_2, 1 - \alpha)}{df_2 + df_1 F(df_1, df_2, 1 - \alpha)},$$

where $F(df_1, df_2, 1 - \alpha)$ denotes the $(1 - \alpha)$ quantile of the central F-distribution and $B(p,q)$ the Beta-function

$$B(p,q) = \int_0^1 t^{p-1}(1-t)^{q-1} \, dt.$$

The degrees of freedom are determined by the number a of influence variables $df_1 = a$ and $df_2 = N - a - 1$, where N denotes the sample size.

From (5.1) follows that the power should nearly produce straight lines in a normal-probability plot versus $\sqrt{df_2}$ for $a = 1$. As Figure 2 shows this is also true for $a > 1$. A PASCAL-program has been used to calculate df_2 or N respectively for given power. Plotting $1/\rho^2$ versus df_2 again straight lines are obtained (Figure 3).

It has been shown that the results can be used also for partial correlation coefficients by replacing ρ with the partial correlation coefficient and using the corresponding degrees of freedom.

References

Anderson, S. and Hauck, W. W. (1983). A new procedure for testing equivalence in comparative bioavailability and other clinical trials. *Communications in Statistics – Theory and Methods* **12**, 2663–2692.

Bennet, B. M. and Hsu, P. (1960). On the power function of the exact test for the 2×2 contingency table. *Biometrika* **47**, 393–398.

Bennett, B. M. and Underwood, R. E. (1970). On McNemar's test for 2×2 table and its power function. *Biometrics* **26**, 339–343.

Bernstein, D. and Lagakos, S. W. (1978). Sample size and power determination for stratified clinical trials. *Journal of Statistical Computing and Simulation* **8**, 65–73.

Blackwelder, W. C. (1982). 'Proving the null hypothesis' in clinical trials. *Controlled Clinical Trials* **3**, 345–353.

Blackwelder, W. C. (1984). Sample size graphs for 'Proving the null hypothesis'. *Controlled Clinical Trials* **5**, 97–105.

Bock, J. (1977). Sample size determination for the comparison of the regression coefficient with a constant in the case of simple linear regression modell II. *Biometrical Journal* **19**, 23–29.

Bock, J. (1984). Die Bestimmung des Stichprobenumfangs in der Regressionsanalyse. *Nova Acta Leopoldina* **254** (55).

Casagrande, J. T., Pike, M. C. and Smith, P. G. (1978). An improved approximate formula for calculating sample sizes for comparing two binomial distributions. *Biometrics* **34**, 483–486.

Cohen, J. (1969). *Statistical Power Analysis for the Behavioural Sciences.* Academic Press, New York.

Connett, J. E., Smith, J. A. and McHugh, R. B. (1987). Sample size and power for pair-matched case-controll studies. *Statistics in Medicine* **6**, 53–59.

Connor, R. J. (1987). Sample size for testing differences in proportions for the paired-sample design. *Biometrics* **43**, 207–211.

Cox, D. R. (1972). Regression models and life tables (with discussion). *J. R. Stat. Soc. B* **34**, 187–220.

Donner, A. (1984). Approaches to sample size estimation in the design of clinical trials—a review. *Statistics in Medicine* **3**, 199–214.

Duffy, S. W. (1984). Asymptotic and exact power for the McNemar test and its analogue with R controls per case. *Biometrics* **40**, 1005–1015.

Dupont, W. D. and Plummer Jr., W. D. (1990). Power and sample size calculations. *Controlled Clinical Trials* **11**, 116–128.

Failing, K. and Victor, N. (1981). Die Schätzung des benötigten Stichprobenumfangs für Therapie-Studien, wenn Erfolgsraten verglichen werden. Vortrag auf der GMDS-Tagung 1981.

Feuer, E. J. and Kessler, L. G. (1989). Test statistic and sample size for a two-sample McNemar test. *Biometrics* **45**, 629–636.

Fleiss, J. L. (1973). Determining sample sizes needed to detect a difference between 2 proportions. In: *Statistical Methods for Rates and Proportions, Vol. 3*, Wiley, New York.

Fleiss, J. L. (1981). *Statistical Methods for Rates and Proportions.* Wiley, New York.

Fleiss, J. L., Tytun, A. and Ury (1980). A simple approximation for calculating sample sizes for comparing independent proportions. *Biometrics* **36**, 343–346.

Freedman, L. S. (1982). Tables of the number of patients required in clinical trials using the logrank test. *Statistics in Medicine* **1**, 121–130.

Freimann, J. A., Chalmers, T. C., Smith, H. Jr., et al (1978). The importance of beta, the type II error and sample size in the design and interpretation of the randomized control trial: Survey 'negative trials'. *New England Journal of Medicine* **299**, 690–694.

Frick, H. (1987). On level and power of Anderson and Hauck's procedure for testing equivalence in comparative bioavailability. *Communications in Statistics – Theory and Methods* **16**, 2771–2778.

Gail, M. (1973). The determination of sample sizes for trials involving several independent 2×2 tables. *Journal of Chronic Diseases* **26**, 669–673.

Gail, M. (1974). Power computations for design comparative Poisson trials. *Biometrics* **30**, 231–237.

Gail, M. and Gart, J. J. (1973). The determination of sample sizes for use with the exact conditional test in 2×2 comparative trials. *Biometrics* **29**, 441–448.

George, S. L. and Desu, M. M. (1974). Planning the size and duration of clinical trials. *Journal of Chronic Diseases* **27**, 15–24.

Halperin, M., Rogot, E., Gurian, J. and Ederer, F. (1968). Sample size for medical trials with special reference to long-term therapy. *Journal of Chronic Diseases* **21**, 13–24.

Hasemann, J. K. (1978). Exact sample sizes for use with Fisher–Irwin test of 2×2 tables. *Biometrics* **34**, 106–109.

Hauck, W. W. and Anderson, S. (1983). A new statistical procedure for testing equivalence in two-group comparative bioavailability trials. *Journal of Pharmacokinetics and Biopharmaceutics* **12**, 83–91.

Jun-mo Nam (1987). A simple approximation for calculating sample sizes for detecting linear trend in proportions. *Biometrics* **43**, 701–705.

Lachin, J. M. (1981). Introduction to sample size determination and power analysis for clinical trials. *Controlled Clinical Trials* **2**, 93–113.

Lachin, J. M. and Foulkes, M. A. (1986). Evaluation of sample size and power for analyses of survival with allowance for nonuniform patient entry' losses to follow-up, noncompliance and stratification. *Biometrics* **42**, 507–519.

Lakatos, E. (1986). Sample sizes for clinical trials with time-dependent rates of losses and non-compliance. *Controlled Clinical Trials* **7**, 189–199.

Lakatos, E. (1988). Sample sizes based on the log-rank statistic in complex clinical trials. *Biometrics* **44**, 229–241.

Lee, E. T. (1980). *Statistical Methods for Survival Data Analysis*. Lifetime Learning, California, pp. 133–135.

Mace, A. (1964). *Sample Size Determination*. Robert E. Krieger Publishing Company, Huntington, New York.

Machin, D. and Campbell, M. J. (1987). *Statistical Tables for the Design of Clinical Trials*. Blackwell Scientific Publications, Oxford.

Makuch, R. W. and Simon, R. M. (1978). Sample size requirements for evaluating a conservative therapy. *Cancer Treatment Reports* **62**, 1037–1040.

Makuch, R. W. and Simon, R. M. (1982). Sample size requirements for comparing time-to-failure among K treatment groups. *Journal of Chronic Diseases* **35**, 861–867.

Miettinen, O. S. (1968). The matched pairs design in the case of all-or-none responses. *Biometrics* **24**, 339–352.

Moussa, M. A. A. (1988). Planning the size of survival time clinical trials with allowance for stratification. *Statistics in Medicine* **7**, 559–569.

Moussa, M. A. A. (1990). Planning the size of clinical trials with allowance for patients non-compliance. *Methods of Information in Medicine* **29**, 243–246.

Müller-Cohrs, J. (1990). The power of the Anderson–Hauck test and the double *t*-test. *Biometrical Journal* **32**, 259–266.

Noether, G. E. (1987). Sample size determination for some common nonparametric test. *Journal of the American Statistical Association* **82**, 645–647.

Pasternack, B. S. and Gilbert, H. S. (1971). Planning the duration of long-term survival time studies designed for accrual by cohorts. *Journal of Chronic Diseases* **24**, 681–700.

Patel, H. I. and Gupta, G. D. (1984). Problem of equivalence in clinical trials. *Biometrical Journal* **26**, 471–474.

Rasch, D., Herrendörfer, G., Bock, J. and Busch, K. (1978, 1980). *Verfahrensbibliothek Versuchsplanung und -auswertung*. VEB Deutscher Landwirtschaftsverlag Berlin (Bd. I, II 1978, Bd. III 1980).

Schlesselmann, J. J. (1982). *Case-Control Studies: Design, Conduct, Analysis*. Oxford University Press, New York.

Schneidermann, M. A. (1964). The proper size of a clinical trial: 'Grandma's strudel' method. *The Journal of New Drugs* **4**, 3–11.

Schoenfeld, D. A. (1981). The asymptotic properties of nonparametric tests for comparing survival distributions. *Biometrika* **68**, 316–319.

Schoenfeld, D. A. (1983). A sample size formula for the proportional hazards – regression model. *Biometrics* **39**, 499–503.

Schoenfeld, D. A. and Richter, J. R. (1982). Nomograms for calculating the number of patients needed for a clinical trial with survival as endpoint. *Biometrics* **38**, 163–170.

Schork, M. A. and Remington, R. D. (1967). The determination of sample size in treatment-control comparisons for chronic disease studies in which dropout or non-adherence is a problem. *Journal of Chronic Diseases* **20**, 233–239.

Schuirmann, D. J. (1981). On hypothesis testing to determine if the mean of a normal distribution is contained in a known interval (Abstract). *Biometrics* **37**, 617.

Schuirmann, D. J. (1987). A comparison of the one-sided test procedure and the power approach for assessing the equivalence of average bioavailability. *Journal of Pharmacokinetics and Biopharmaceutics* **15**, 657–680.

Taron, R. E. and Ware, J. (1977). On distribution-free tests for equality of survival distributions. *Biometrika* **64**, 156–160.

Tiku, M. L. (1966). A note on approximating the non-central *F*-distribution. *Biometrika* **53**, 606–610.

Wu, M., Fisher, M. and DeMets, D. (1980). Sample sizes for long-term medical trials with time-dependent dropout and event rates. *Controlled Clinical Trials* **1**, 111–123.

Walther, E. (1980). *Not published minutes, Mathematisches Forschungsinstitut Oberwolfach.*

Westlake, W. J. (1979). Statistical aspects of comparative bioavailability. *Biometrics* **35**, 273–289.

Westlake, W. J. (1981). Bioequivalence testing—A need to rethink (Reader Reaction Response). *Biometrics* **37**, 580–594.

Subject Index

Handbook of Statistics
Contents of Previous Volumes

Volume 1. Analysis of Variance
Edited by P. R. Krishnaiah
1980 xviii + 1002 pp.

Volume 2. Classification, Pattern Recognition and Reduction of Dimensionality
Edited by P. R. Krishnaiah and L. N. Kanal
1982 xxii + 903 pp.

Volume 3. Time Series in the Frequency Domain
Edited by D. R. Brillinger and P. R. Krishnaiah
1983 xiv + 485 pp.

Volume 4. Nonparametric Methods
Edited by P. R. Krishnaiah and P. K. Sen
1984 xx + 968 pp.

Volume 5. Time Series in the Time Domain
Edited by E. J. Hannan, P. R. Krishnaiah and M. M. Rao
1985 xiv + 490 pp.

Volume 6. Sampling
Edited by P. R. Krishnaiah and C. R. Rao
1988 xvi + 594 pp.

Volume 7. Quality Control and Reliability
Edited by P. R. Krishnaiah and C. R. Rao
1988 xiv + 503 pp.